TEACHER'S EDITION

ADDISON-WESLEY GENERAL MATHEMATICS

MERVIN L. KEEDY

MARVIN L. BITTINGER

STANLEY A. SMITH

PAUL A. ANDERSON

ADDISON-WESLEY PUBLISHING COMPANY

Menlo Park, California
Reading, Massachusetts
London
Amsterdam
Don Mills, Ontario
Sydney

CONTENTS

ISBN 0-201-10301-X

BCDEFGHIJK-VH-887654

HIGHLIGHTS

1. FOUR-STEP SKILL DEVELOPMENT PRESENTED IN TWO-PAGE LESSONS

- Flow-chart presentations
- Specific examples
- TRY THIS exercises—with answers—for immediate reinforcement
- Practice for each lesson

2. VARIED EXERCISE FORMATS

- After each lesson
- More Practice keyed to lessons at ends of chapters
- Cumulative Skills Reviews
- Periodic calculator corners
- Supplemental drill masters available

3. FREQUENT WORD PROBLEMS

- Problem corners concluding most lessons
- Problem-Solving lessons to develop problem-solving skills
- Math applications twice in each chapter
- Careers using math in each chapter

4. COMPLETE TESTING PROGRAM

- Pre-Tests, Chapter Reviews, and Chapter Tests with every chapter
- Test items keyed to student pages
- Year-end basic skills and applications exams
- Supplemental quiz and test masters available

5. CONTROLLED READING

- Carefully controlled readability level
- Graphic explanations to facilitate easy understanding

6. MOTIVATING ART AND DESIGN

- Colorful student pages
- Clear, open, and easy-to-understand format

BEFORE THE LESSON

PRE-TEST

The PRE-TEST may be given before each chapter to determine those skills a student may have already mastered. All test items are keyed to student pages for easy management.

GETTING STARTED

Each chapter begins with a GETTING STARTED activity. GETTING STARTED deals with an important chapter skill and establishes the focus for the mathematics that will follow.

CHAPTER 9 PRE-TEST

Convert to a decimal.

1. 37% **2.** 136% **3.** 9% **4.** 4.2%

Convert to a percent.

5. 0.17 **6.** 2.47 **7.** 0.03 **8.** 0.004

Convert to a fraction, mixed number, or whole number.

9. 37% **10.** 45% **11.** 120% **12.** $12\frac{1}{2}\%$

Convert to a percent.

13. $\frac{4}{5}$ **14.** $\frac{19}{20}$ **15.** $3\frac{1}{2}$ **16.** 6

17. $\frac{2}{3}$ **18.** $\frac{1}{8}$ **19.** $2\frac{1}{3}$ **20.** $1\frac{3}{7}$

Translate. Do not solve.

21. 18% of 38 is what number?

22. 42 is 75% of what number?

23. 5 is what percent of 90?

24. What percent of 10 is 3?

Solve.

25. What is 25% of 40?

26. 20% of 32 is what number?

27. 8 is what percent of 12?

28. What percent of 9 is 3?

29. 4 is 20% of what number?

30. 36 is $16\frac{2}{3}\%$ of what number?

Find the percent of increase or decrease.

31. From 10 to 12

32. From 12 to 10

Application

33. Matt earns $ 300 per month plus 9% of sales over $ 1000. How much does he earn if his total sales are $ 3300 in a month?

34. A tennis racquet regularly sells for $ 35. It is on sale for 20% off. What is the sale price?

GETTING STARTED

MAXIMIZE

WHAT'S NEEDED The 30 squares of paper with digits marked on them as in GOTCHA, Chapter 1.

THE RULES

1. On a sheet of paper each player writes the following.
   ___% of ____ is what number?
2. A leader draws a digit from the box. Each player writes the digit on any one of the blanks. Once it is written, it cannot be changed or moved.
3. The leader draws four more digits, one at a time. Write each digit on a blank as it is drawn.
4. Each player is now ready to solve the problem.

SAMPLE 3 8 % of 9 4 0 is what number?

5. Who has the largest number? Find out by multiplying.

SAMPLE

$$\begin{array}{r} 940 \\ \times\ 0.38 \\ \hline 7520 \\ 28200 \\ \hline 357.20 \end{array}$$

The player with the largest answer is the winner.

THE LESSON PAGES

Each lesson begins with a flowchart presentation and clearly worked out examples. Next comes TRY THIS, an opportunity to try exercises similar to the examples. Answers are available in the student text for immediate reinforcement.

Each lesson contains frequent and varied practice. Exercises are divided into sections that correspond to skills taught in preceding TRY THIS problems and worked out examples. A PROBLEM CORNER follows the practice with real-life problems using the skills covered in the lesson.

ADDING DECIMALS

Add. 21.668 + 5.491

Line up the decimal points. → Add as with whole numbers → Place the decimal point in the sum.

```
 21.668      21.668      21.668
+ 5.491     + 5.491     + 5.491
             27159       27159
```

EXAMPLE 1. Add. 18.9 + 9.376

```
 18.900    Line up the decimal points.
+ 9.376    Write 0's to fill the places.
 28.276    Add.
```

TRY THIS

Add.

1. 18.76 + 19.9
2. 9.786 + 0.895
3. 165.97 + 36.075
4. 0.765 + 0.89

EXAMPLE 2. Add. 9.7 + 0.89 + 16.756 + 8.6

```
  9.700    Write 0's to fill the places.
  0.890
 16.756    Add.
+ 8.600
 35.946    Place the decimal point in the sum.
```

TRY THIS

Add.

5. 4.3 + 0.876 + 21.42
6. 0.96 + 0.096 + 9.6
7. 4.95 + 0.879 + 19.8 + 0.007
8. 0.15 + 0.19 + 0.875 + 0.919

PRACTICE

Add.

1. 19.6 + 7.9
2. 27.9 + 16.1
3. 37.3 + 19.8
4. 0.9 + 8.8
5. 47.98 + 16.84
6. 2.87 + 9.03
7. 6.87 + 9.9
8. 3.76 + 0.9
9. 7.398 + 0.475
10. 37.485 + 18.08
11. 2.786 + 0.9
12. 63.284 + 19.986
13. 8.736 + 9.49
14. 36.892 + 18.08
15. 0.7 + 0.6
16. 297.8 + 96.88
17. 13.767 + 4.382 + 7.689
18. 3.4 + 9.6 + 7.3
19. 7.86 + 4.08 + 0.9
20. 13.621 + 9.78 + 0.5
21. 0.78 + 6.94 + 8.72 + 4.81
22. 16.284 + 9.786 + 4.395 + 5.007
23. 9.8 + 19.4 + 0.98 + 5.9
24. 874.1 + 87.41 + 8.741 + 0.874
25. 7.9 + 6.84 + 14.3 + 8.9
26. 16.9 + 3.18 + 0.8 + 0.9
27. 0.68 + 6.8 + 3.9 + 0.395
28. 4.875 + 0.129 + 97.6 + 0.976

PROBLEM CORNER

Use the chart.

1. Which city usually gets more rain in January? June?
2. How much rain does Los Angeles usually get in the winter? summer?
3. Does Albany get more rain in December or July?

Average Expected Rainfall in Centimeters

	Winter			Summer		
	Dec	Jan	Feb	Jun	Jul	Aug
Los Angeles, Cal.	6.07	6.4	5.89	0.08	0.03	0.05
Albany, N.Y.	7.44	5.59	5.36	7.62	7.92	7.29

More Practice, page 35

AFTER THE LESSON

CHAPTER REVIEW

Success with the material in the CHAPTER REVIEW indicates that the student understands the chapter and is ready for the CHAPTER TEST.

CHAPTER TEST

A CHAPTER TEST follows the CHAPTER REVIEW and can be used for evaluation. Items on the test correspond exactly with those on the PRE-TEST and the CHAPTER REVIEW, and they are keyed to the student pages for easy management.

SKILLS TEST

A comprehensive test of math skills appears at the end of the text. An APPLICATIONS TEST is also provided to test students' competencies in applying math to everyday situations.

SKILLS REVIEW

These periodic cumulative reviews cover the material of several chapters. Items are keyed to the student pages for easy management.

CHAPTER 9 REVIEW

Convert to a decimal. p. 248

1. 43% **2.** 120% **3.** 8% **4.** 6.3%

Convert to a percent. p. 248

5. 0.19 **6.** 1.78 **7.** 0.05 **8.** 0.004

Convert to a fraction, mixed number, or whole number. p. 250

9. 49% **10.** 24% **11.** 145% **12.** $33\frac{1}{3}\%$

Convert to a percent. p. 252

13. $\frac{3}{5}$ **14.** $\frac{4}{25}$ **15.** $2\frac{1}{4}$ **16.** 5

17. $\frac{1}{3}$ **18.** $\frac{3}{8}$ **19.** $2\frac{2}{3}$ **20.** $1\frac{2}{7}$

Translate. Do not solve. p. 254

21. 17% of 35 is what number?

22. What percent of 20 is 4?

23. 54 is what percent of 96?

24. 7 is $33\frac{1}{3}\%$ of what number?

Solve. pp. 256-262

25. 36% of 50 is what number?

26. What is 90% of 120?

27. What percent of 12 is 6?

28. 7 is what percent of 21?

29. 40 is $33\frac{1}{3}\%$ of what number?

30. 9 is 75% of what number?

Find the percent of increase or decrease. p. 264

31. From 15 to 10

32. From 10 to 15

Application pp. 260, 266

33. Sandra earns 8% commission on her monthly sales. How much commission does she earn on total monthly sales of $ 8450?

34. Skateboards usually cost $ 24. On sale they cost $ 18. Find the percent of decrease in price.

270 CHAPTER 9

CHAPTER 9 TEST

Convert to a decimal.

1. 97% **2.** 326% **3.** 7% **4.** 2.7%

Convert to a percent.

5. 0.37 **6.** 1.62 **7.** 0.03 **8.** 0.007

Convert to a fraction, mixed number, or whole number.

9. 67% **10.** 15% **11.** 125% **12.** $66\frac{2}{3}\%$

Convert to a percent.

13. $\frac{4}{5}$ **14.** $\frac{7}{20}$ **15.** $1\frac{3}{4}$ **16.** 4

17. $\frac{3}{8}$ **18.** $\frac{3}{7}$ **19.** $1\frac{1}{6}$ **20.** $2\frac{5}{9}$

Translate. Do not solve.

21. 15 is what percent of 45?

22. 76% of 19 is what number?

23. 9 is 50% of what number?

24. What percent of 9 is 5?

Solve.

25. What is 25% of 32?

26. $16\frac{2}{3}\%$ of 42 is what number?

27. 9 is what percent of 45?

28. What percent of 18 is 12?

29. 50 is 25% of what number?

30. 7 is 75% of what number?

Find the percent of increase or decrease.

31. From 20 to 25

32. From 25 to 20

Application

33. Russ earns $ 3.25 an hour in salary. He averages $ 4.75 an hour in tips. How much does he earn in an 8-hour day?

34. Ski jackets are on sale for $\frac{1}{4}$ off. The regular price is $ 30. Find the sale price.

PERCENT 271

MORE PRACTICE

Each chapter ends with MORE PRACTICE. These exercises are grouped by the skills covered in the chapter. A student who has difficulty with any lesson should complete the exercises on the indicated page.

MORE PRACTICE

Write standard notation. Use after page 7.

1. six hundred forty-three
2. five hundred ninety-eight
3. one thousand, six hundred ten
4. eight thousand, eighty
5. three thousand, five hundred
6. nine thousand, nine
7. twelve thousand, ninety-nine
8. ten thousand, two hundred eight
9. eighty thousand, one hundred eleven
10. twenty-six thousand, four hundred nineteen

Write money notation. Use after page 9.

11. eight dollars and ninety-five cents
12. seventeen dollars
13. eighty-one dollars and fifty cents
14. thirty dollars and six cents
15. one hundred twenty dollars and twenty cents
16. seventy-seven dollars and seven cents
17. five dollars and eighty-nine cents
18. sixty-five dollars
19. seventeen dollars and ninety-eight cents
20. two hundred dollars and sixty cents

Write standard notation. Use after page 13.

21. 112 thousand, 600
22. 800 thousand, 4
23. 26 million, 455 thousand
24. 11 million

Add. Use after page 15.

25. 74 + 96
26. 49 + 5
27. 746 + 298
28. 348 + 67
29. 348, 267, + 98
30. 948, 76, 18, + 7
31. $ 4.78, 2.97, + 6.49
32. $ 8.47, 0.49, 6.40, + 0.86

Add. Use after page 17.

1. 7284 + 3796
2. 14,781 + 39,499
3. 347,821 + 981,294
4. 349,278 + 97,146
5. 7269, 3784, 6985, + 3784
6. 47,264, 69,209, + 15,875
7. 349,287, 190,180, 762,498, + 875,287
8. $ 746.84, 198.78, + 58.74

Write decimal notation. Use after page 21.

9. 1 and 6 tenths
10. 12 and 24 hundredths
11. 6 and 7 hundredths
12. 10 and 326 thousandths
13. 4 and 28 thousandths
14. 3 and 4248 ten-thousandths
15. 86 thousandths
16. 9 tenths
17. 9 hundredths
18. 9 ten-thousandths

Add. Use after page 23.

19. 8.7 + 9.2
20. 6.4 + 8.7
21. 3.78 + 9.65
22. 0.918 + 6.942
23. 4.98, 3.7, + 5.876
24. 9.7, 6.82, + 9.487
25. 7.328, 0.09, 1.4268, + 0.6
26. 4.298, 5.29, 6.2, + 7.0

27. 5.62 + 0.039 + 37.8
28. 0.2 + 29.6 + 7.12 + 0.594

Estimate the sums. Use after page 29.

29. 86, 94, + 38
30. 376, 298, + 815
31. 324, 58, + 63
32. 4286, 750, + 310
33. $ 8.76 + 9.48
34. $ 3.98, 2.47, 6.50, + 1.25
35. 1.78 + 3.25
36. 4.298, 7.9, 6.48, + 7.007

APPLICATION LESSONS

APPLICATIONS

Each chapter contains two APPLICATION lessons that cover skills used in real-life situations. Students discover the reasons for mastering skills and learn how to apply them to everyday problems.

CAREER

Each chapter also contains a CAREER that presents some of the mathematics used in that field. While learning about the CAREER, students see how math skills are applied, and then they practice those skills.

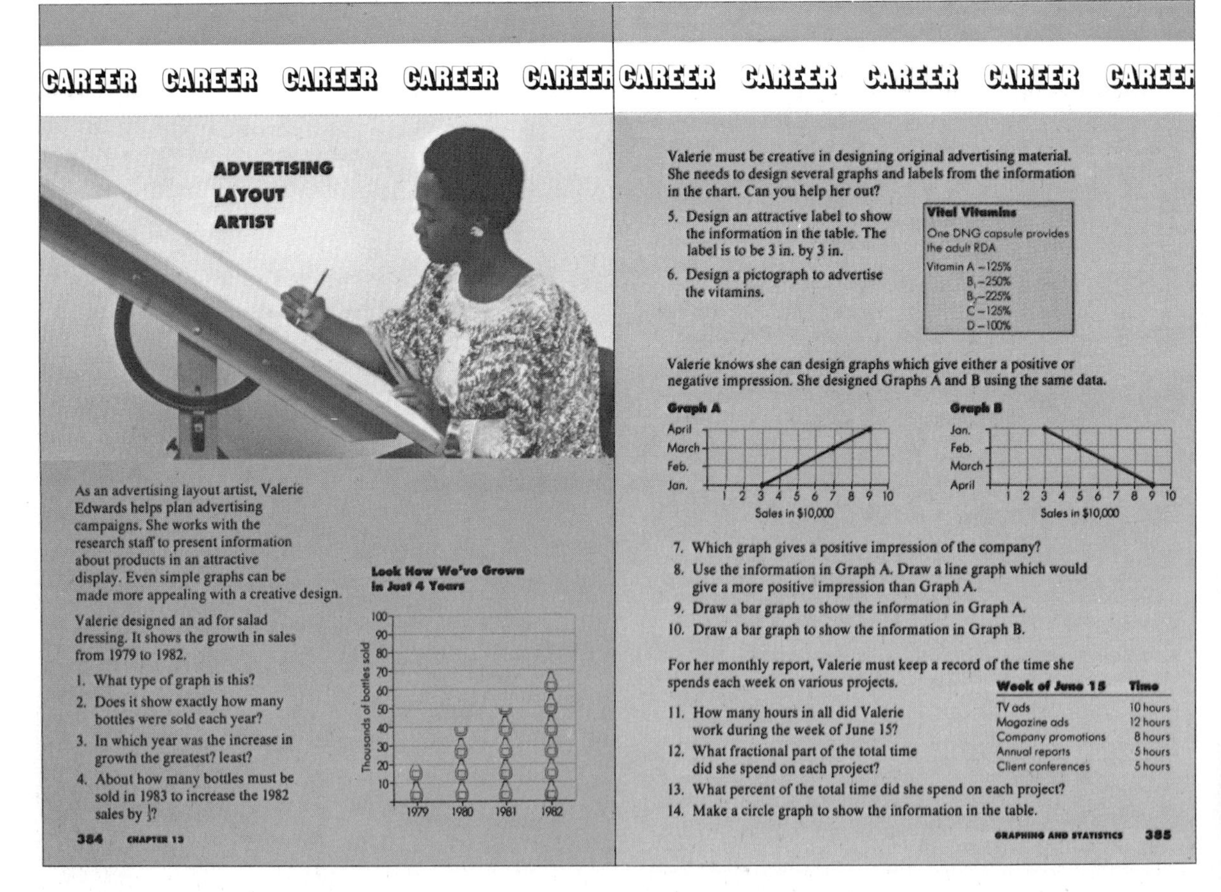

CAREER CAREER CAREER CAREER CAREER

ADVERTISING LAYOUT ARTIST

As an advertising layout artist, Valerie Edwards helps plan advertising campaigns. She works with the research staff to present information about products in an attractive display. Even simple graphs can be made more appealing with a creative design.

Valerie designed an ad for salad dressing. It shows the growth in sales from 1979 to 1982.

1. What type of graph is this?
2. Does it show exactly how many bottles were sold each year?
3. In which year was the increase in growth the greatest? least?
4. About how many bottles must be sold in 1983 to increase the 1982 sales by $\frac{1}{3}$?

384 CHAPTER 13

CAREER CAREER CAREER CAREER CAREER

Valerie must be creative in designing original advertising material. She needs to design several graphs and labels from the information in the chart. Can you help her out?

5. Design an attractive label to show the information in the table. The label is to be 3 in. by 3 in.
6. Design a pictograph to advertise the vitamins.

Vital Vitamins
One DNG capsule provides the adult RDA
Vitamin A – 125%
B_1 – 250%
B_2 – 225%
C – 125%
D – 100%

Valerie knows she can design graphs which give either a positive or negative impression. She designed Graphs A and B using the same data.

7. Which graph gives a positive impression of the company?
8. Use the information in Graph A. Draw a line graph which would give a more positive impression than Graph A.
9. Draw a bar graph to show the information in Graph A.
10. Draw a bar graph to show the information in Graph B.

For her monthly report, Valerie must keep a record of the time she spends each week on various projects.

Week of June 15	Time
TV ads	10 hours
Magazine ads	12 hours
Company promotions	8 hours
Annual reports	5 hours
Client conferences	5 hours

11. How many hours in all did Valerie work during the week of June 15?
12. What fractional part of the total time did she spend on each project?
13. What percent of the total time did she spend on each project?
14. Make a circle graph to show the information in the table.

GRAPHING AND STATISTICS 385

SOLVING PROBLEMS

Each SOLVING PROBLEMS lesson focuses on a problem-solving skill. These seven lessons highlight and reinforce the major areas of problem solving. Students can master the basic steps to solve everyday problems.

CALCULATOR CORNER

Each CALCULATOR CORNER is an activity for practicing and sharpening skills. Many interesting number patterns and relationships can be discovered by using a calculator. Either the whole class or a small group can perform some of the activities.

SOLVING PROBLEMS

When solving problems, keep these steps in mind.

1. **UNDERSTAND** What must I find?
   What information is given?
   Do I have enough information?

2. **PLAN** Have I solved a problem like this before?
   Should I restate the problem?
   What operations are needed?
   Is an approximate answer sufficient?
   Should I use a calculator?

3. **CARRY OUT** Did I calculate correctly?
   Did I check my calculations?

4. **LOOK BACK** Is my answer reasonable?
   Does the answer fit the problem?

EXAMPLE. Mr. and Mrs. Capan plan to rent an apartment. The basic rent is $310 with a $24 maintenance fee. A $55 security deposit is required. What is the total payment for the first month?

UNDERSTAND I must find the total.
I have enough information.

PLAN I must add to find the total.

CARRY OUT

```
  $310
    24   It checks.
 +  55
  $389
```

LOOK BACK The answer is reasonable.
It makes sense in the problem.

CALCULATOR CORNER

THREE OF A KIND

Players one or more
Materials one calculator, paper and pencil

Suppose a calculator has 345.6 on the display.
How many ways can you get a number □□□. 8 on the display where the three boxes contain the same digit?

Try these with your calculator.

1. Add 321.2 to 345.6
   Result: 666.8
2. Subtract 122.8 from 345.6
   Result: 222.8

Find 4 more ways to get such a number on the display. Write down your steps.

EXERCISES Write down your steps for each of the following.

1. Start with 236.5 on the display. Find 5 ways to get a number □□□.7 on the display, where the three boxes contain the same digit.
2. Start with 342.16 on the display. Find 5 ways to get a number □□□.48 on the display, where the three boxes contain the same digit.
3. Start with 423.152 on the display. Find 5 ways to get number 6□□.□66 on the display, where the three boxes contain the same digit.

SUPPLEMENTS

PRACTICE AND PROJECTS

MAKING PRACTICE FUN consists of 87 reproducible games, puzzles, and other self-correcting activities. Each deals with a specific skill and is keyed to the student pages.

PROJECTS IN GENERAL MATHEMATICS carries the high interest of the APPLICATIONS and CAREERS into 60 pages of related projects.

QUIZZES AND TESTS

A separate testing package is available on blackline masters. These quizzes and tests parallel the PRE-TESTS, CHAPTER REVIEWS, and CHAPTER TESTS found in the text. Test items are keyed to objectives and student pages for easy management.

Making Practice Fun 1 Name__________

Whole Numbers—Adding, Subtracting, Multiplying, Dividing

CROSS NUMBER PUZZLE

1. Work each exercise.
2. Write the answer in the correct squares.

Across

1. 1307 × 6 = 7842
3. 906 − 477
5. 344 + 195 + 349
6. 4419 ÷ 3
7. 100,000 − 43,861
10. 9964 ÷ 212
11. 1009 + 992 + 1245
13. 5000 − 1329
15. 12 × 72
17. 16 + 48 + 29
20. 223 × 8
22. 5145 ÷ 35
23. 1111 − 995
24. 69 + 169 + 269
25. 11 × 232
26. 600 − 14
29. 48,210 ÷ 3214
31. 45,575 + 106,259
34. 917 − 333
36. 622 × 12
38. 8000 − 921
40. 103,200 ÷ 30
41. 28,800 + 346 + 27,519

Down

1. 149 × 5
2. 520 − 49
3. 95 + 88 + 291
4. 8484 − 6111
5. 2807 × 3
6. 4047 ÷ 213
8. 12 × 54
9. 1000 − 606
11. 26 + 55 + 292
12. 10,000 − 1290
14. 1157 × 6
16. 12,930 ÷ 2
18. 507 + 161 + 174
19. 11 × 101
21. 24,332 ÷ 316
22. 63 × 25
24. 8121 − 2467
27. 272 × 30
28. 555 + 255 + 120
30. 909 − 335
32. 293 × 3
33. 95 × 5
35. 25 × 19
37. 15,444 ÷ 351
39. 1016 − 920

7 8 4 2

General Mathematics Name__________

QUIZ 7 Class__________ Grade______

Use the chart to find the answers.

	Bills		Coins		
Denomination	$5	$1	25¢	10¢	5¢
Number of Coins or Bills	6	8	9	3	7

1. How much of the money is in bills?
2. How much of the money is in coins?
3. How much of the money is there in all?

Estimate.

4. 2645 × 781
5. $3.67 × 5
6. 49.75 × 7.93

7. List the change that should be given. Use words and money notation.
   Charge: $2.75 Paid with: $5.00

Clerk gives	Clerk says

Multiply.

8. 9.5 × 0.8
9. 51.7 × 6.3
10. 90.1 × 0.27
11. 100 × 0.72
12. 0.01 × 83.6
13. 0.1 × 0.52

TEACHER'S EDITION

The Teacher's Edition includes lesson objectives, a guide to the text, and additional answers. Authors' notes appear in the margins.

Student pages are printed full size in full color. Answers to all exercises appear in color next to the exercises.

REFERENCES

Objective numbers are printed near the lesson titles. Appropriate use of the quizzes and MAKING PRACTICE FUN is noted in color. Page references to MORE PRACTICE appear on the practice pages.

ESTIMATING SUMS

Estimate the sum. 436 + 297 + 548

Line up vertically. → Round each number. → Add.

Line up vertically	Round each number	Add
436	400	400
297	300	300
+ 548	+ 500	+ 500
		1200

We usually round to the highest place in the smallest number.

EXAMPLES. Estimate the sums.

1. 426 + 31 + 85 → 430 + 30 + 90 = 550
2. 4296 + 3500 + 7301 → 4000 + 4000 + 7000 = 15,000
3. 17.8 + 3.14 + 6.359 → 18 + 3 + 6 = 27 *Round each number to the nearest whole number. Add.*
4. $29.15 + 32.75 + 5.98 → $29 + 33 + 6 = $68 *Round each amount to the nearest dollar. Add.*

TRY THIS

Estimate the sums.

1. 78 + 25 + 34
2. 4568 + 379 + 615
3. 36.781 + 9.31 + 6.8
4. $36.49 + 5.50 + 7.11

PRACTICE

Estimate the sums.

1. 27 + 34
2. 39 + 28
3. 748 + 258
4. 698 + 307
5. 9286 + 6148
6. 4736 + 8367
7. 3,748,976 + 3,487,296
8. 4,286,987 + 8,628,146
9. 74 + 28 + 78
10. 348 + 296 + 755 + 508
11. 2748 + 3476 + 4985 + 6297
12. 3,287,496 + 5,349,891 + 6,548,196
13. 871 + 364 + 58
14. 725 + 75 + 86
15. 4297 + 385 + 115
16. 7284 + 6813 + 743
17. 8.7 + 6.5
18. 8.743 + 3.289
19. 15.87 + 7.3
20. 36.784 + 45.39
21. 3.9 + 4.8 + 8.9
22. 3.76 + 5.98 + 7.42
23. 1.742 + 3.9 + 6.1
24. 15.8 + 9.87 + 3.147
25. $7.75 + 2.98
26. $8.37 + 9.65
27. $16.19 + 24.78
28. $38.79 + 73.86
29. $9.15 + 6.25 + 8.60
30. $3.49 + 2.78 + 3.85
31. $24.73 + 9.68 + 5.45
32. $28.79 + 38.44 + 8.50

PROBLEM CORNER

1. Estimate the cost of a skirt and two blouses.
2. You have $10. Can you buy a shirt and two ties? Estimate to find the answer.

blouse $9.51 · tie $1.98 · skirt $19.85 · shirt $5.95

More Practice, page 35

ANSWERS AND REFERENCES

Answers are found in several places.

1. In the Student Text, TRY THIS answers begin on page 461. By providing answers to these exercises, students have immediate reinforcement. They can then determine if they understand the skill and are ready to go on to the Practice exercises. The TRY THIS answers are also included in the Teacher's Edition so the teacher will know which answers are available to students.
2. The Teacher's Edition contains overprinted answers for all problems, including TRY THIS exercises, APPLICATIONS, and CAREERS. When answers are too long to be included on the page, they appear in the Additional Answers section which begins on page T26.

References to the lesson pages and supplements are found in several places.

1. The PRE-TESTS, CHAPTER REVIEWS, MORE PRACTICE, SKILLS REVIEWS, and SKILLS TEST are correlated to the lesson by page numbers. These references appear in both the Teacher's Edition and the Student Edition.
2. Objective numbers appear only in the Teacher's Edition. They are printed near the lesson titles.
3. Page references to MORE PRACTICE appear on the practice pages in both the Teacher's Edition and the Student Edition.
4. Appropriate use of the quizzes and MAKING PRACTICE FUN are indicated on the lesson and practice pages only in the Teacher's Edition.

GUIDE TO THE OBJECTIVES

Objectives are included for the computational skill lessons as well as the applications lessons. You can use the right-hand column to record dates when an objective is introduced and when it is mastered. These pages may be reproduced for your record keeping.

Objective Number	Objective	Page Number	Exposure/	Mastery
1	To write dates 3 ways	4		
2	To tell the likely use of groups of numbers	4		
3	To write time notation	4		
4	To find the place and value of digits to thousands	6		
5	To write word names for numbers to thousands	6		
6	To write standard notation for numbers to thousands	6		
7	To write word names for money	8		
8	To write money notation	8		
9	To write money notation in terms of dollars	8		
10	To find the place and value of digits to billions	12		
11	To write short word names for numbers to billions	12		
12	To write standard notation for numbers to billions	12		
13	To add up to four 3-digit whole numbers	14		
14	To add using money notation	14, 16		
15	To add up to four 6-digit whole numbers	16		
16	To find the place and value of a digit to ten-thousandths	20		
17	To write word names for decimals	20		
18	To write decimal notation	20		
19	To add decimals	22		
20	To round numbers to a given place	26		
21	To round to the nearest dollar	26		
22	To estimate sums	28		
23	To estimate sums using money notation	28		
24	To subtract whole numbers to hundreds	40		
25	To subtract using money notation	40, 42		

Objective Number	Objective	Page Number	Exposure/ Mastery	
26	To subtract whole numbers to hundred-thousands	42		
27	To subtract with special zero cases	44		
28	To compare decimals using $=$, $>$, and $<$	48		
29	To subtract decimals	50		
30	To estimate differences	54		
31	To estimate differences using money notation	54		
32	To multiply by a 1-digit number	68		
33	To multiply using money notation	68, 70		
34	To multiply by a 2-digit number	70		
35	To multiply by a 3-digit number	72		
36	To multiply with special zero cases	76		
37	To estimate products of whole numbers	76		
38	To estimate products using money notation	78		
39	To estimate products of decimals	78		
40	To multiply decimals	82		
41	To multiply decimals by 10, 100 and 1000	84		
42	To multiply decimals by 0.1, 0.01, and 0.001	84		
43	To divide by a 1-digit divisor with a 1- or 2-digit quotient	98		
44	To divide by a 1-digit divisor with a 3- or 4-digit quotient	100		
45	To divide using money notation	100, 102		
46	To divide by a 2-digit divisor	102		
47	To divide by a 3-digit divisor	104		
48	To divide a decimal by a whole number	108		
49	To divide a whole number by a whole number with a decimal quotient	108		
50	To divide by 10, 100, or 1000	110		
51	To estimate quotients	112		
52	To divide a decimal by a decimal	116		
53	To divide a whole number by a decimal	116		
54	To find quotients to the nearest tenth, hundredth, or thousandth	118		
55	To identify what fractional part of a figure is shaded	136		
56	To simplify fractions such as $\frac{8}{8}$, $\frac{0}{8}$, and $\frac{8}{1}$	138		

Objective Number	Objective	Page Number	Exposure/ Mastery	
57	To determine whether two fractions are equal	140		
58	To compare fractions using $<$ and $>$	142		
59	To convert decimals to fractions	144		
60	To convert fractions to decimals	146		
61	To convert mixed numbers to fractions	150		
62	To convert fractions to mixed numbers	152		
63	To convert decimals to mixed numbers	154		
64	To convert mixed numbers to decimals	154		
65	To multiply fractions	168		
66	To find equivalent fractions	170		
67	To find an equivalent fraction with a given denominator	170		
68	To determine which fraction in a set is simplest	172		
69	To simplify fractions	172		
70	To multiply two fractions and simplify the result	174		
71	To multiply with mixed numbers	176		
72	To find the reciprocal of a number	180		
73	To divide fractions	180		
74	To divide with mixed numbers	182		
75	To add fractions with the same denominators	196		
76	To find the least common denominator of fractions	198		
77	To add fractions with different denominators	200		
78	To add with mixed numbers	202		
79	To subtract fractions	206		
80	To subtract with mixed numbers	208		
81	To subtract fractions with renaming	210		
82	To write ratios using fractions	224		
83	To simplify ratios	224		
84	To find unit rates	226		
85	To determine whether proportions are true	230		
86	To solve proportions	230		
87	To solve problems using proportions	232		
88	To find a measure given the corresponding measures of a similar figure	234		

Objective Number	Objective	Page Number	Exposure/	Mastery
89	To convert percents to decimals	248		
90	To convert decimals to percents	248		
91	To convert percents to fractions, mixed numbers, or whole numbers	250		
92	To convert fractions and mixed numbers to percents	252		
93	To translate to number sentences	254		
94	To replace percent in a translation with its decimal or fractional equivalent	254		
95	To find a percent of a given number	256		
96	To find the percent one number is of another	258		
97	To find a number when a percent is known	262		
98	To find the percent of increase	264		
99	To find the percent of decrease	264, 266		
100	To identify the likely metric measure of length	284		
101	To measure length using millimeters and centimeters	284		
102	To change metric measures of length	286		
103	To identify the likely metric measure of capacity	290		
104	To change metric measures of capacity	290		
105	To identify the likely metric measure of mass	292		
106	To change metric measures of mass	292		
107	To identify the likely temperature in Celsius or Fahrenheit	296		
108	To change customary measures	298		
109	To identify right angles, angles greater or less than a right angle	300		
110	To measure angles with a protractor	300		
111	To find the perimeters of polygons	312		
112	To find the circumferences of circles	314		
113	To find the areas of rectangles	318		
114	To find the areas of squares	318		
115	To find the areas of parallelograms	320		
116	To find the areas of triangles	320		
117	To find the areas of trapezoids	322		
118	To find the areas of circles	326		
119	To find the volumes of rectangular prisms	340		

Objective Number	Objective	Page Number	Exposure/ Mastery	
120	To find the volumes of cubes	340		
121	To find the volumes of cylinders	344		
122	To find the volumes of pyramids	346		
123	To find the volumes of cones	346		
124	To find the volumes of spheres	348		
125	To graph an ordered pair on a grid	368		
126	To find an ordered pair for a point on a grid	368		
127	To construct a line graph	370		
128	To interpret a line graph	370		
129	To construct a bar graph	372		
130	To interpret a bar graph	372		
131	To construct a circle graph	374		
132	To interpret a circle graph	374		
133	To find the mean of a set of numbers	378		
134	To find the median of a set of numbers	380		
135	To find the mode of a set of numbers	380		
136	To give a positive or negative number or a given situation	394		
137	To compare positive and negative numbers $<$ or $>$	394		
138	To add using a number line	396		
139	To add positive and negative numbers	398		
140	To subtract using a number line	402		
141	To subtract positive and negative numbers	404		
142	To multiply positive and negative numbers	406		
143	To divide positive and negative numbers	408		
144	To evaluate expressions	422		
145	Be able to solve equations of the type $x + a = b$ and $x - a = b$ using the addition principle	426		
146	Be able to solve equations of the type $ax = b$ and $\frac{x}{a} = b$ using the multiplication principle	428		
147	To solve equations using both the addition and multiplication principles	430		
148	To solve problems using equations	432		
149	To graph equations	436		
150	To write checks	10		

Objective Number	Objective	Page Number	Exposure/	Mastery
151	To fill out deposit slips	18		
152	To complete check stubs and check registers	46		
153	To reconcile monthly bank statements	52		
154	To sort and count money	74		
155	To determine whether change was counted correctly	80		
156	To make change using words	80		
157	To compute rates and wages	106		
158	To read a pay stub	106		
159	To compute items on a pay stub	106		
160	To compute job and flat rates	114		
161	To convert measures of time	148		
162	To add time	148		
163	To subtract time	148		
164	To convert a given time to the other time zones	156		
165	To find how much time has passed	156		
166	To interpret a budget	178		
167	To find Recommended Daily Allowances	184		
168	To compute calories and carbohydrates in foods	184		
169	To compare nutrients in food	204		
170	To interpret information from drawings to solve problems	212		
171	To use unit prices to find the most economical buy	228		
172	To use a shipping chart	236		
173	To find the total cost of catalog items	236		
174	To compute commissions and total earnings	260		
175	To compute tips and total salaries	260		
176	To find the sale price	266		
177	To find the unit sale price	266		
178	To find the percent off	266		
179	To find the original price	266		
180	To read and interpret time schedules	288		
181	To read and interpret rate schedules	294		
182	To compute sales tax	294		

Objective Number	Objective	Page Number	Exposure/ Mastery	
183	To estimate rent budgets	316		
184	To compute the costs to rent an apartment	316		
185	To compute construction costs	324		
186	To compute unit rates of electricity	342		
187	To compute operation costs	342		
188	To calculate the savings in money and energy resources	350		
189	To compute travel costs	376		
190	To compute sports statistics	382		
191	To read and interpret charts	400		
192	To read and interpret maps	410		
193	To compute simple interest	424		
194	To compute compound interest	424		
195	To compute credit costs	434		
196	To use problem solving skills in consumer-related word problems	24, 56, 86, 120, 280, 328, 364		

GUIDE TO THE TEXT

This guide shows the sequence of the lessons in relation to More Practice, Skills Reviews, and the supplemental materials. Page numbers are listed to help you find the appropriate section. The charts may also be used to record student assignments or progress in mastering skills. They may be reproduced for your record keeping.

Lesson	√	**More Practice**	√	**Skills Review**	√	**Making Practice Fun (MPF)**	√
Using Numbers						MPF 1	
Place Value to Thousands		34					
Writing Money Notation		34				MPF 2	
Application: Writing Checks							
Place Value to Billions		34					
Adding Whole Numbers		34		128, 274		MPF 3	
Adding Larger Numbers		35		128, 274		MPF 4	
Application: Making Deposit Slips							
Decimal Place Value		35					
Adding Decimals		35		128, 274		MPF 5	
Solving Problems						MPF 6	
Rounding Numbers							
Estimating Sums		35		128		MPF 7	
Career: Agricultural Pilot							
Chapter 1 Review						MPF 8	
Chapter 1 Test							
Subtracting Whole Numbers		62		129, 274		MPF 9	
Subtracting Larger Numbers		62		129, 274			
Zeros in Subtracting		62		129, 274		MPF 10	
Application: Keeping a Running Balance							
Comparing Decimals		63				MPF 11	
Subtracting Decimals		63		129, 274		MPF 12	
Application: Reconciling Monthly Statements							
Estimating Differences		63		129			
Solving Problems						MPF 13	
Career: Worm Farming							
Chapter 2 Review						MPF 14	

Lesson	√	More Practice	√	Skills Review	√	Making Practice Fun (MPF)	√
Chapter 2 Test							
Multiplying by 1-digit Numbers		92		130, 275		MPF 15	
Multiplying by 2-digit Numbers		92		130, 275		MPF 16	
Multiplying by 3-digit Numbers		92		130, 275		MPF 17	
Application: Counting Money							
Special Products and Estimating		92, 93		130, 131			
More on Estimating Products		93		131		MPF 18	
Application: Receiving and Giving Change							
Multiplying Decimals		93		130, 275		MPF 19	
Other Special Products		93		131			
Solving Problems						MPF 20	
Career: Fast Food Restaurant Manager							
Chapter 3 Review						MPF 21	
Chapter 3 Test							
1-digit Divisors		126		131, 275		MPF 22	
Dividing Larger Numbers		126		131, 275			
2-digit Divisors		126		131, 275		MPF 23	
3-digit Divisors		126		132, 275		MPF 24	
Application: Rates and Wages							
Dividing a Decimal by a Whole Number		127		132, 275		MPF 25	
Special Quotients		127		132, 275			
Estimating Quotients		127		132, 275		MPF 26	
Application: Job and Flat Rates							
Dividing a Decimal by a Decimal		127		132, 275		MPF 27	
Rounding Quotients		127		132, 275		MPF 28	
Solving Problems						MPF 29	
Career: Maintenance Service							
Chapter 4 Review						MPF 30	
Chapter 4 Test							
Using Fractions		162		276		MPF 31	
Fractions Mean Division		162				MPF 32	
Equality		162					

Lesson	√	More Practice	√	Skills Review	√	Making Practice Fun (MPF)	√
Comparing Fractions		162		276		MPF 33	
Writing Decimals as Fractions		162					
Writing Fractions as Decimals		162, 163				MPF 34	
Application: Measuring Time							
Writing Mixed Numbers as Fractions		163		276			
Writing Fractions as Mixed Numbers		163		276		MPF 35	
Mixed Numbers and Decimals		163		276			
Application: Time Zones							
Career: Operations Manager							
Chapter 5 Review						MPF 36	
Chapter 5 Test							
Multiplying		190		358		MPF 37	
Equivalent Fractions		190				MPF 38	
Simplifying		190		358			
Multiplying and Simplifying		190		276, 358		MPF 39	
Multiplying with Mixed Numbers		191		277, 358		MPF 40	
Application: Budgeting							
Reciprocals and Dividing		191		277, 358			
Dividing with Mixed Numbers		191		277, 358		MPF 41	
Application: Food and Nutrition							
Career: Home Economist							
Chapter 6 Review						MPF 42	
Chapter 6 Test							
Adding with the Same Denominators		218		277, 358		MPF 43	
Least Common Denominator		218		277, 358		MPF 44	
Adding with Different Denominators		218		277, 358		MPF 45	
Adding with Mixed Numbers		218		277, 358		MPF 46	
Application: Food Values							
Subtracting		219		277, 358			
Subtracting with Mixed Numbers		219		278, 358		MPF 47	
Subtracting with Renaming		219		278, 358		MPF 48	
Application: Household Problems							

Lesson	√	More Practice	√	Skills Review	√	Making Practice Fun (MPF)	√
Career: Van Customizing							
Chapter 7 Review						MPF 49	
Chapter 7 Test							
Ratios		242		278			
Unit Rates		242		278		MPF 50	
Application: Comparison Shopping							
Proportions		242		278, 359		MPF 51	
Solving Problems Using Proportions		243		278, 359		MPF 52	
Similar Figures and Proportions		243					
Application: Catalog Shopping							
Career: Merchandising							
Chapter 8 Review						MPF 53	
Chapter 8 Test							
Converting Percents and Decimals		272		279		MPF 54	
Converting Percents to Fractions		272		279		MPF 55	
Converting Fractions to Percents		272		279		MPF 56	
Translating to Number Sentences		272					
Finding a Percent of a Number		273		279, 359		MPF 57	
What Percent A Number is of Another		273		279, 359		MPG 58	
Application: Working for Commissions and Tips							
Finding a Number Given a Percent		273		279, 359		MPF 59	
Finding Percent of Increase or Decrease		273		279			
Application: Buying on Sale							
Career: Beautician							
Chapter 9 Review						MPF 60	
Chapter 9 Test							
Solving Problems						MPF 61	
Metric Length		306		360		MPF 62	
Changing Metric Units of Length		306		444			
Application: Time Schedules							
Metric Capacity		306		360, 444		MPF 63	

Lesson	√	More Practice	√	Skills Review	√	Making Practice Fun (MPF)	√
Metric Mass		306		360, 444		MPF 64	
Application: Rate Schedules							
Temperature							
Customary Measures		307		360, 444		MPF 65	
Measuring Angles		307					
Career: Civil Service							
Chapter 10 Review						MPF 66	
Chapter 10 Test							
Perimeter		334		361, 444		MPF 67	
Circumference		334		361, 444		MPF 68	
Application: Renting an Apartment							
Area of Rectangles and Squares		334		362, 445		MPF 69	
Area of Parallelograms and Triangles		335		362, 445		MPF 70	
Area of Trapezoids		335		362, 445			
Application: Buying a House							
Area of Circles		335		362, 445			
Solving Problems						MP 71	
Career: Architect							
Chapter 11 Review						MPF 72	
Chapter 11 Test							
Rectangular Prisms and Cubes		356		363, 445		MPF 73	
Application: Energy Bills							
Cylinders		356		363, 445			
Pyramids and Cones		357		363, 445		MPF 74	
Spheres		357		363, 445			
Application: Energy Conservation							
Career: Oceanographer							
Chapter 12 Review						MPF 75	
Chapter 12 Test							
Solving Problems						MPF 76	
Graphing Ordered Pairs		388					
Making Line Graphs		388		446		MPF 77	

Lesson	√	More Practice	√	Skills Review	√	Making Practice Fun (MPF)	√
Making Bar Graphs		388, 389		446			
Making Circle Graphs		389		446		MPF 78	
Application: Travel Costs							
Finding the Mean		389		446		MPF 79	
Finding the Median and Mode		389		446, 447		MPF 80	
Application: Sports Statistics							
Career: Advertising Layout Technician							
Chapter 13 Review						MPF 81	
Chapter 13 Test							
Positive and Negative Numbers		416		447			
Adding on a Number Line							
Adding		416		447		MPF 82	
Application: Reading Charts							
Subtracting on a Number Line							
Subtracting		417		447		MPF 83	
Multiplying		417		447			
Dividing		417		448		MPF 84	
Application: Reading Maps							
Career: Well Driller							
Chapter 14 Review						MPF 85	
Chapter 14 Test							
Evaluating Expressions		442		448		MPG 86	
Application: Interest							
The Addition Principle		442		448			
The Multiplication Principle		442		448			
Using the Principles Together		443		449		MPF 87	
Solving Problems with Equations		443		449			
Application: Credit Costs							
Graphing Equations		443		449			
Career: Optometrist							
Chapter 15 Review						MPF 88	
Chapter 15 Test							

ADDITIONAL ANSWERS

CHAPTER 1

Chapter 1 Pre-Test, page 1

21.

October 12, 19 81

Pay to the Order of Acme Fuel Company $ 276.79

Two hundred seventy-six and 79/100 DOLLARS

For fuel Jason Sanders

Try This, page 6

1. tens' place, 40 **2.** thousands' place, 5000 **3.** ones' place, 5 **4.** ninety-eight **5.** three hundred seventy-six **6.** five thousand, two hundred eighty-seven **7.** six thousand, six

Practice, page 7

1. thousands' place, 4000 **2.** hundreds' place, 800 **3.** ones' place, 5 **4.** hundreds' place, 700 **5.** hundreds' place, 200 **6.** thousands' place, 4000 **7.** thousands' place, 9000 **8.** tens' place, 70 **9.** sixty-eight **10.** eighty-six **11.** forty-one **12.** ninety **13.** three hundred seventy-four **14.** two-hundred seventy-six **15.** four hundred seventy **16.** four hundred seven **17.** three thousand, nine hundred forty-eight **18.** two thousand, seven hundred eighty-nine **19.** five thousand, seven hundred eighty **20.** six thousand, seventy-four

Try This, pages 8-9

1. one cent **2.** seventeen dollars and forty-six cents **3.** two hundred thirty-six dollars **7.** nine and 89/100 dollars **8.** fifty-six and 04/100 dollars **9.** one hundred seventy-four and 00/100 dollars

Practice, page 9

1. seventy-nine cents **2.** eighty-seven cents **3.** seven cents **4.** eight cents **5.** nineteen dollars **6.** forty-two dollars **7.** seven hundred thirty-eight dollars **8.** four hundred four dollars **9.** twenty-six dollars and thirty-seven cents **10.** nine dollars and eighty-four cents **11.** thirty-eight dollars and ninety-eight cents **12.** forty-seven dollars and fifty cents **13.** five hundred six dollars and five cents **14.** eight hundred seventy dollars and seventy cents **15.** one hundred dollars and one cent **16.** three hundred ninety dollars and nine cents **31.** eight and 78/100 dollars **32.** four and 25/100 dollars **33.** thirty-nine and 56/100 dollars **34.** nineteen and 75/100 dollars **35.** three hundred twenty-six and 48/100 dollars **36.** four hundred twenty-seven and 86/100 dollars **37.** one hundred twenty-five and 87/100 dollars **38.** nine hundred ninety-nine and 99/100 dollars **39.** seven and 04/100 dollars **40.** sixteen and 09/100 dollars **41.** four hundred seven and 09/100 dollars **42.** three hundred twenty-eight and 05/100 dollars **43.** five and 00/100 dollars **44.** 95/100 dollars **45.** ten and 00/100 dollars **46.** 08/100 dollars

Application, page 11

1.

September 9, 19 84

Pay to the Order of Western Mortgage Company $ 258.54

Two hundred fifty-eight and 54/100 DOLLARS

For house payment Lucy Lee

2.

October 25, 19 81

Pay to the Order of Anderson Food Services $ 38.24

Thirty-eight and 24/100 DOLLARS

For groceries Kazuo Sata

3.

December 18, 19 80

Pay to the Order of Ace Catering Service $ 89.00

Eighty-nine and 00/100 DOLLARS

For party Alice Washington

4.

February 19, 19 82

Pay to the Order of Joe's Cycle Shop $ 158.45

One hundred fifty-eight and 45/100 DOLLARS

For bike Paulo Lopez

Try This, page 12

1. ten-thousands' place, 70,000 **2.** hundred-millions' place, 700,000,000 **3.** hundred-billions place, 800, 000,000,000 **4.** 496 million, 287 thousand, 104 **5.** 380 billion, 87 million, 6 thousand, 100

Practice, page 13

1. ten-thousands' place, 30,000 **2.** ten-thousands' place, 20,000 **3.** thousands' place, 9000 **4.** millions' place, 6,000,000 **5.** millions' place, 7,000,000 **6.** hundred-millions' place, 300,000,000 **7.** billions' place, 9,000,000,000 **8.** ten-billions' place, 80,000, 000,000 **9.** billions' place, 9,000,000,000 **10.** hundred-thousands' place, 700,000 **11.** ten-millions'place, 80,000,000 **12.** ten-billions' place, 80,000,000,000

13. 37 thousand, 826 **14.** 428 thousand, 874 **15.** 5 thousand, 687 **16.** 80 thousand **17.** 984 million, 126 thousand, 481 **18.** 82 million, 476 thousand, 125 **19.** 9 million, 9 thousand, 9 **20.** 80 million, 800 thousand, 8 **21.** 9 billion, 8 million, 7 thousand, 6 **22.** 73 billion, 816 million, 479 thousand, 100 **23.** 728 million **24.** 7 billion, 7

Application, page 19

7.

		Dollars	Cents
Cash	Currency	23	00
	Coin	2	00
Checks	94-18	19	50
	25-16	75	00
Total		119	50
Less Cash Received			-0-
Net Deposit		119	50

8.

		Dollars	Cents
Cash	Currency	77	00
	Coin	22	00
Checks	46-3	17	85
Total		116	85
Less Cash Received			-0-
Net Deposit		116	85

9.

		Dollars	Cents
Cash	Currency		
	Coin	5	40
Checks	21-20	127	05
	24-20	84	46
Total		216	91
Less Cash Received			-0-
Net Deposit		216	91

10.

		Dollars	Cents
Cash	Currency	400	00
	Coin		
Checks	92-18	375	00
	52-12	795	95
Total		1570	95
Less Cash Received			-0-
Net Deposit		1570	95

Try This, page 20

1. tenths' place, 0.4 **2.** hundredths' place, 0.09 **3.** thousandths' place, 0.006

Practice, page 21

1. tenths' place, 0.4 **2.** tenths' place, 0.4 **3.** tenths' place, 0.5 **4.** tenths' place, 0.5 **5.** hundredths' place, 0.07 **6.** thousandths' place, 0.006 **7.** ten-thousandths' place, 0.0005 **8.** tenths' place, 0.4 **9.** ten-thousandths' place, 0.0009 **10.** hundredths' place, 0.01 **11.** thousandths' place, 0.006 **12.** ten-thousandths' place, 0.0004 **13.** 3 and 7 tenths **14.** 49 and 8 tenths **15.** 73 and 28 hundredths **16.** 9 and 326 thousandths **17.** 17 and 4295 ten-thousandths **18.** 95 and 875 ten-thousandths **19.** 38 hundredths **20.** 1 and 141 thousandths **21.** 295 and 6 tenths **22.** 36 and 9853 ten-thousandths **23.** 8 and 7 tenths **24.** 2 and 74 hundredths **25.** 3 and 95 thousandths **26.** 785 thousandths **27.** 7 and 7 ten-thousandths **28.** 7 thousand and 7 tenths

Career, page 31

13.

August 7, 19 81

Pay to the Order of Chemical Supply Corp. $ 303.16

Three hundred three and 16/100 DOLLARS

For chemicals Jason Sanders

Chapter 1 Review, page 32

21.

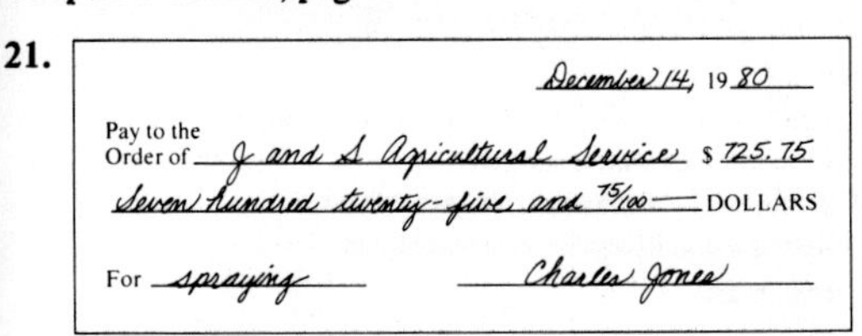

December 14, 19 80

Pay to the Order of J and S Agricultural Service $ 725.75

Seven hundred twenty-five and 75/100 DOLLARS

For spraying Charles Jones

Chapter 1 Test, page 33

21.

March 22, 19 83

Pay to the Order of Ace Chemical Company $ 78.50

Seventy-eight and 50/100 DOLLARS

For chemicals Sherie Sanders

CHAPTER 3

Application, page 81

7.
1 quarter $8.50
1 half-dollar 9.00
1 dollar 10.00

8.
1 quarter $14.00
1 dollar 15.00
1 $5 bill 20.00

9.
1 penny $3.69
1 penny 3.70
1 nickel 3.75
1 quarter 4.00
1 dollar 5.00

10.
1 penny $ 2.65
1 dime 2.75
1 quarter 3.00
1 dollar 4.00
1 dollar 5.00

11.
1 dollar $1.00
1 half-dollar 1.50
1 quarter 1.75

12.
1 $5 bill $ 5.00
1 dollar 6.00
1 dollar 7.00
1 half-dollar 7.50
1 quarter 7.75
1 dime 7.85

13.
1 dollar $1.00
1 dollar 2.00
1 dime 2.10
1 dime 2.20
1 penny 2.21
1 penny 2.22
1 penny 2.23
1 penny 2.24

14.
1 dollar $ 1.00
1 half-dollar 1.50
1 quarter 1.75
1 nickel 1.80
1 penny 1.81
1 penny 1.82
1 penny 1.83

15.
1 $5 bill $5.00
1 dollar 6.00
1 half-dollar 6.50
1 quarter 6.75
1 nickel 6.80
1 penny 6.81

16.
1 dollar $ 1.00
1 dollar 2.00
1 dollar 3.00
1 half-dollar 3.50
1 dime 3.60
1 nickel 3.65
1 penny 3.66
1 penny 3.67
1 penny 3.68

CHAPTER 7

Application, page 205

16. Whole Milk, 0.001; Skimmed Milk, 0.002; Whole Wheat Bread, 0.009; Peanut Butter, 0.003; Swiss Cheese, 0.003; Spinach, 0.080; Vanilla Ice Cream, 0.001 **17.** Whole Milk, 0.056; Skimmed Milk, 0.100; Whole Wheat Bread, 0.036; Peanut Butter, 0.044; Swiss Cheese, 0.076; Spinach, 0.133; Vanilla Ice Cream, 0.018 **18.** Whole Milk, 1.800; Skimmed Milk, 3.367; Whole Wheat Bread, 0.400; Peanut Butter, 0.133; Swiss Cheese, 2.495; Spinach, 4.956; Vanilla Ice Cream, 0.606 **19.** Whole Milk, 0.019; Skimmed Milk, 0.033; Whole Wheat Bread, 0; Peanut Butter, 0; Swiss Cheese, 0; Spinach, 1.2; Vanilla Ice Cream, 0.006

CHAPTER 13

Pre-Test 13, page 365

1.

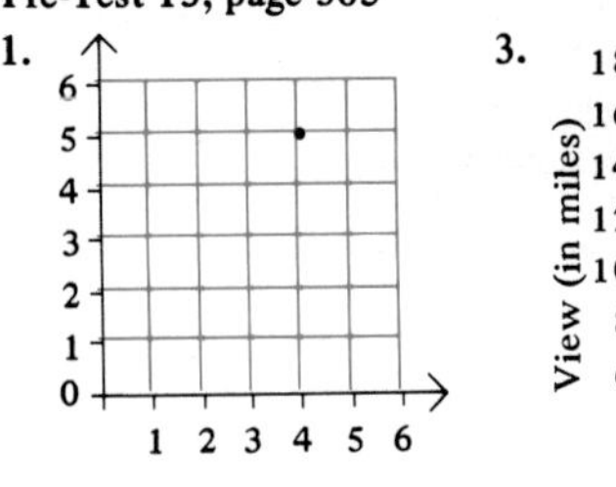

3.

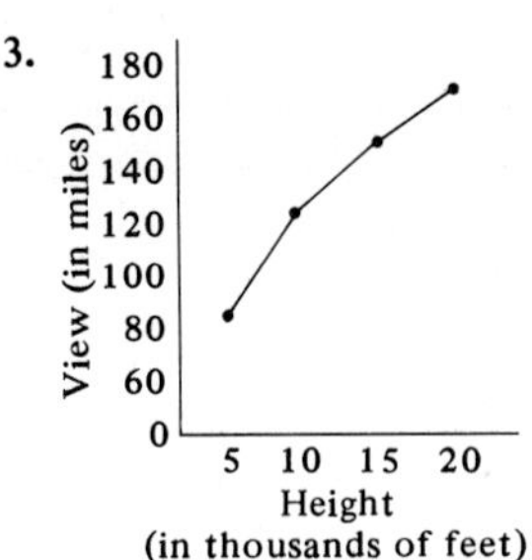

4. Cause of Death

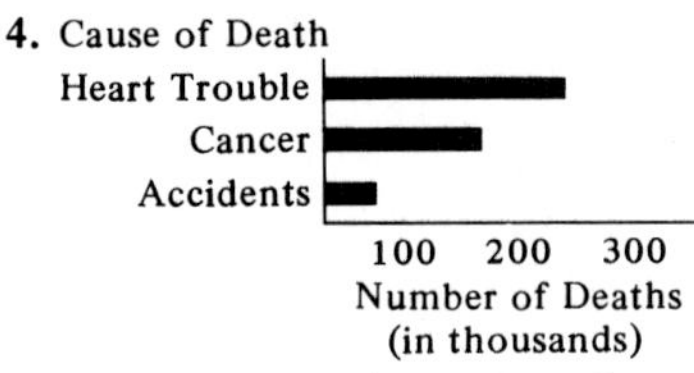

5. Angle: 48%, 173°; 34%, 122°; 16%, 58°; 2%, 7°

House Color

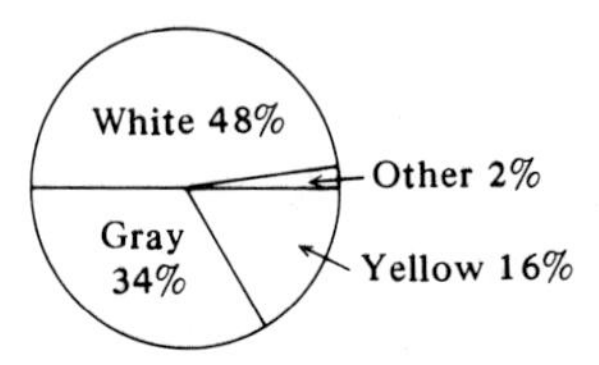

Try This, page 368
1-4.

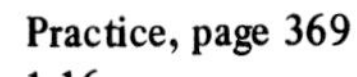

Practice, page 369
1-16.

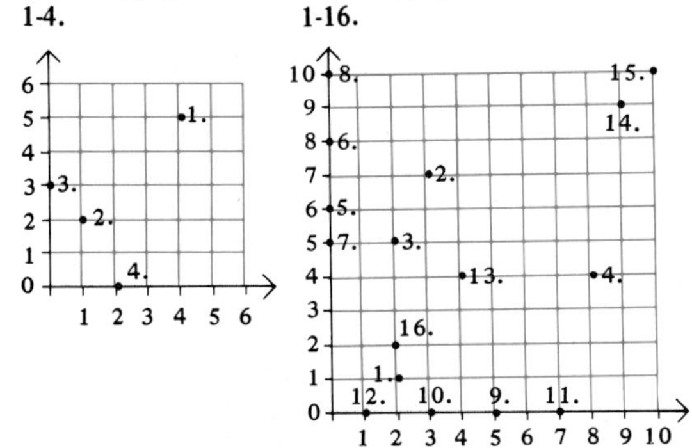

Problem Corner, page 369

The completed graph should show only four dots. A is 2 units to the right and 3 up, B is 7 units to the right and 9 up, C is 10 units to the right and 5 up, and D is 5 units to the right exactly.

Try This, page 370

The years should be listed along the horizontal axis. Dollars should be shown on the vertical axis by $1000 groups. $900 is above 1940, $1500 is above 1950, $2500 is above 1960, and $6300 is above 1970.

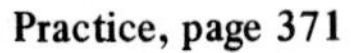

Practice, page 371

1.

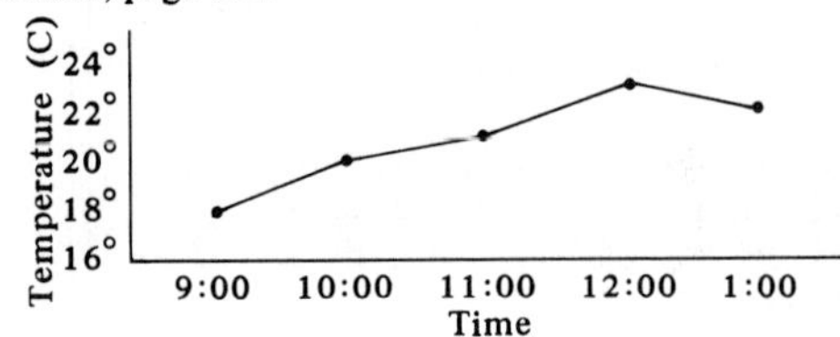

2.

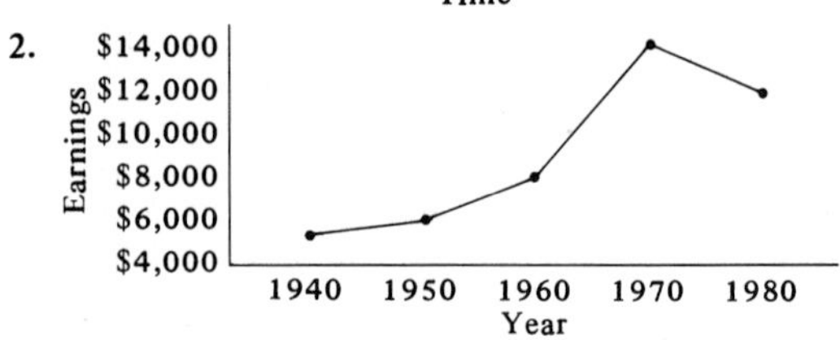

3.

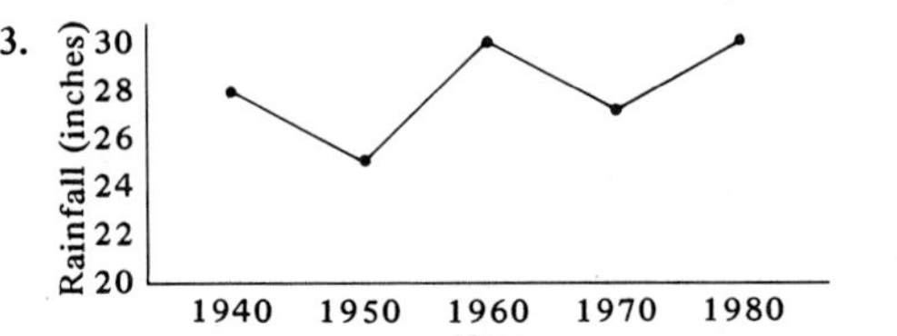

4.

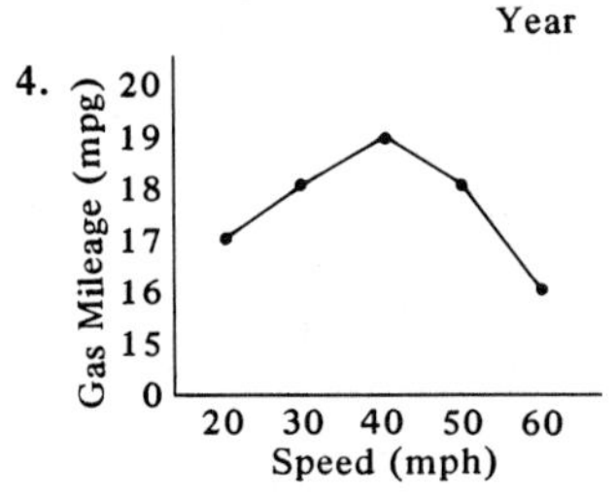

Try This, page 372

1.

Ice Cream Flavor	Calories in a One-Scoop Cone
Chocolate Fudge	230
French Vanilla	220
Strawberry	170
Peach	170
Butter Pecan	200
Chocolate Mint	190

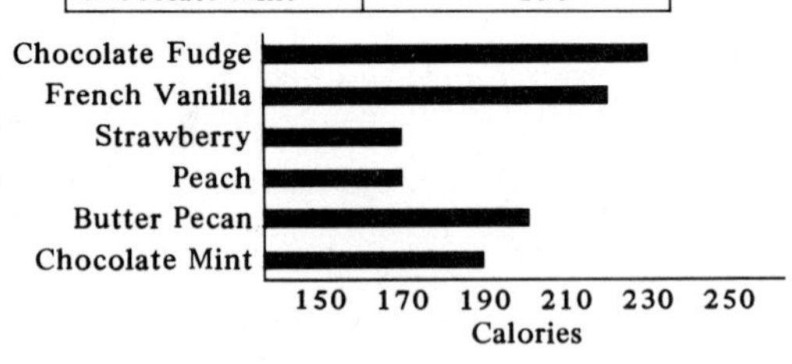

Practice, page 373

1.

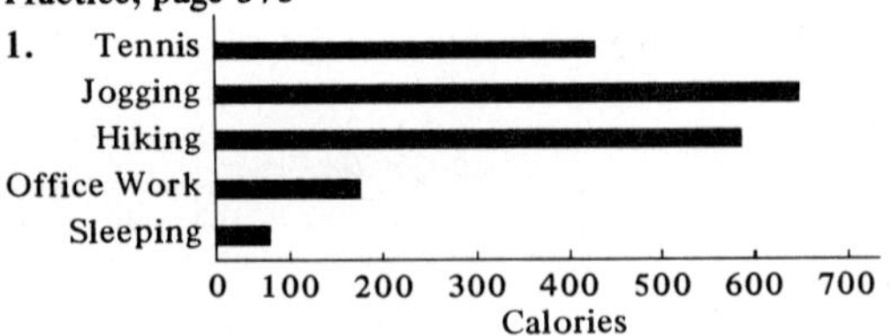

2. **3.**

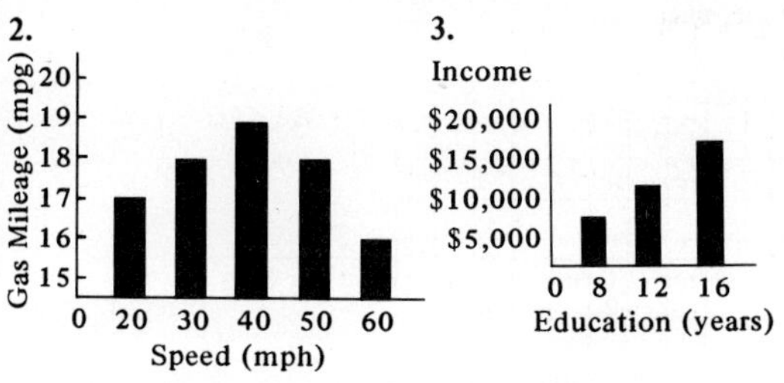

4.

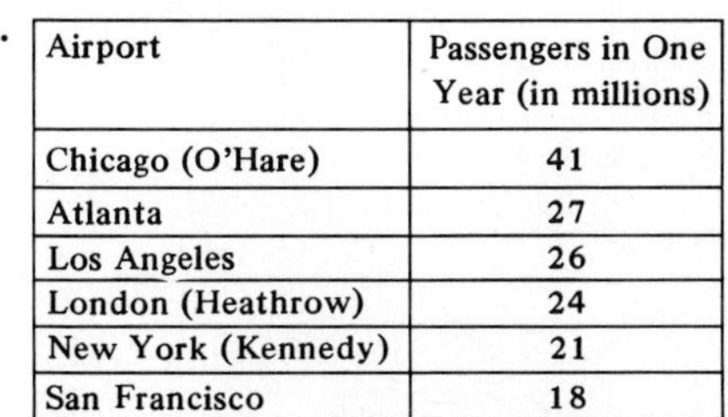

Airport	Passengers in One Year (in millions)
Chicago (O'Hare)	41
Atlanta	27
Los Angeles	26
London (Heathrow)	24
New York (Kennedy)	21
San Francisco	18

Chicago (O'Hare)
Atlanta
Los Angeles
London (Heathrow)
New York (Kennedy)
San Francisco

10 15 20 25 30 35 40
Passengers (in millions)

Try This, page 374

1. Angle: 20%, 72°; 14%, 50°; 10%, 36°; 26%, 94°; 30%, 108°

Auto 14%
Clothing 10%
Food 20%
Housing 26%
Other 30%

2. Angle: 10%, 36°; 50%, 180°; 15%, 54°; 5%, 18°; 20%, 72°

Practice, page 375

1. Angle: 24%, 86°; 38%, 137°; 20%, 72°; 18%, 65°

Study Time

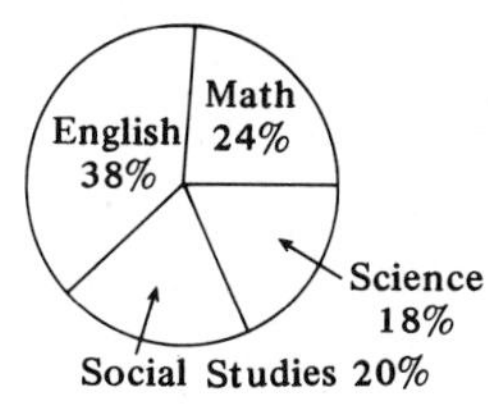

3. Angle: 12%, 43°; 20%, 72°; 4%, 14°; 16%, 58°; 20%, 72°; 28%, 101°

Sales — Favorite TV Shows

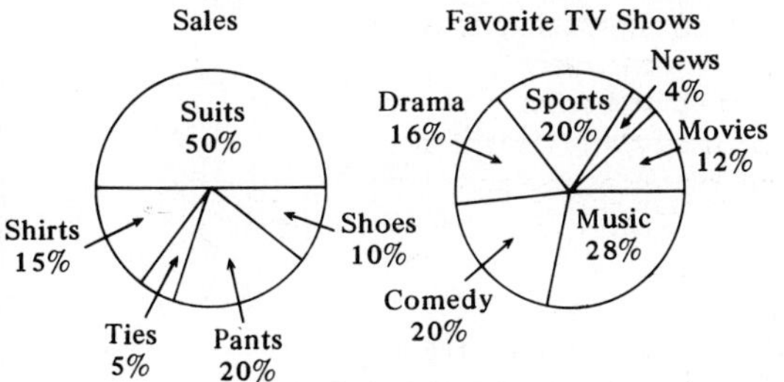

4. Angle: 13%, 47°; 33%, 119°; 23%, 83°; 12%, 43°; 19%, 68°

Personal Income in 1980

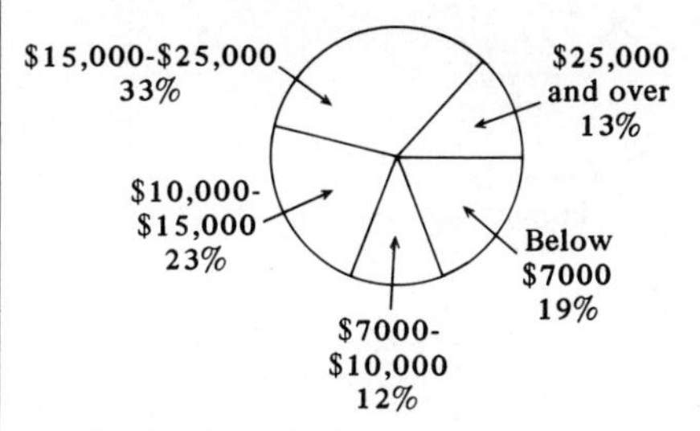

Application, page 376

8.

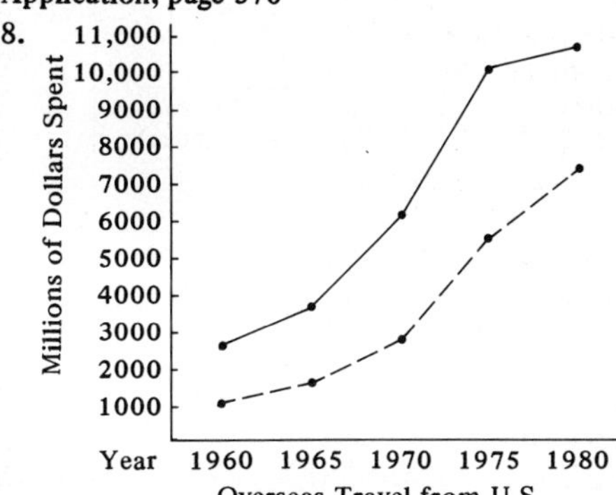

Application, page 377

18.

	Transportation	Food	Total
Jerry	$83.00	$35.00	$118.00
Jill	$107.90	$28.00	$136.90

Career, page 385

8. 9.

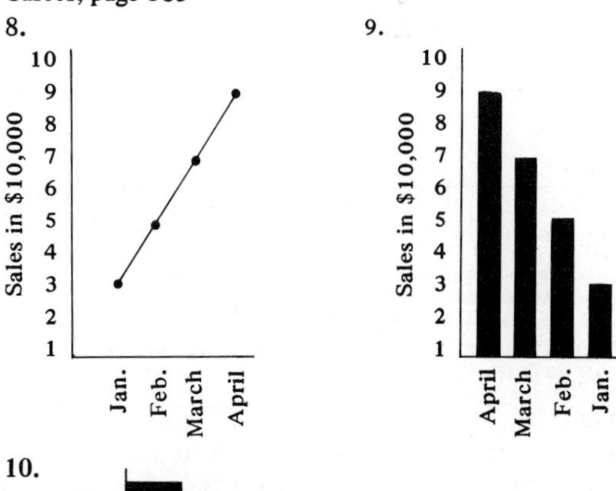

10.

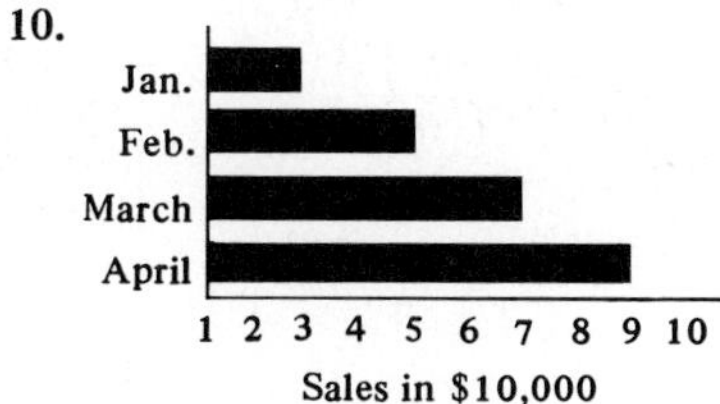

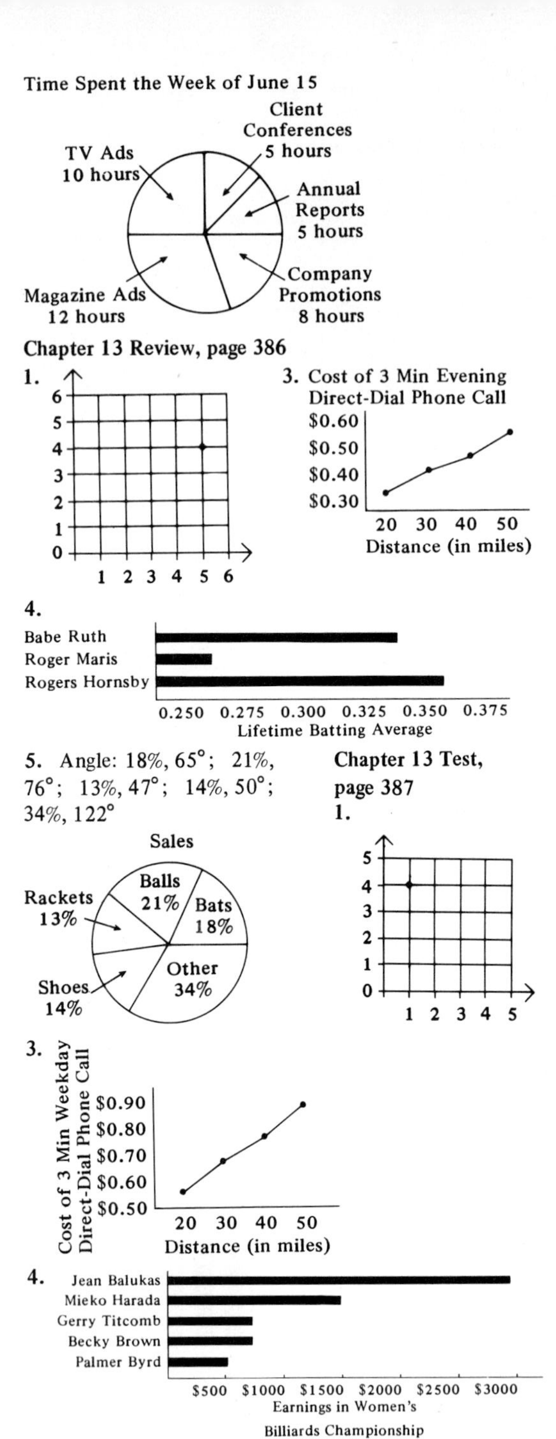

5. Angle: 78%, 281°; 20%, 72°; 2%, 7°

Batters

Right-Handed 78%

Both 2%

Left-Handed 20%

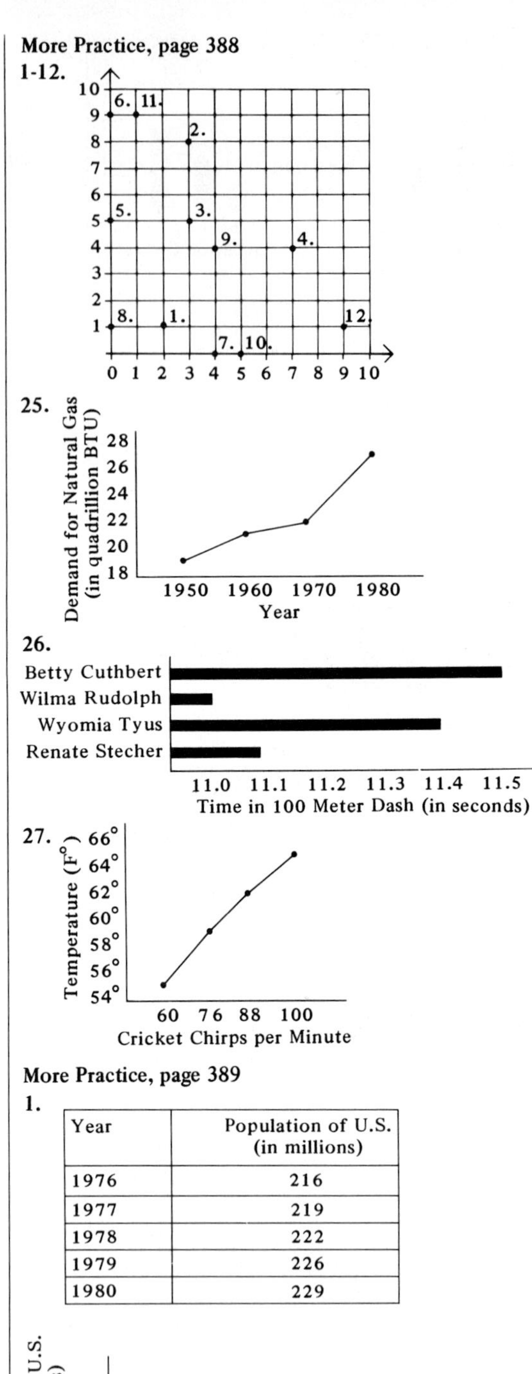

Year	Population of U.S. (in millions)
1976	216
1977	219
1978	222
1979	226
1980	229

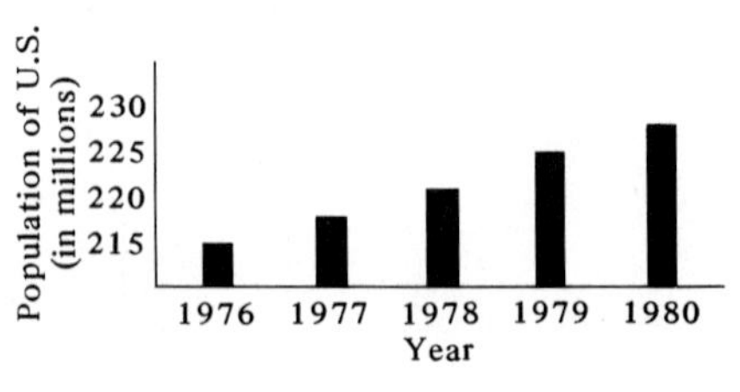

2. Angle: 22%, 79°; 18%, 65°; 11%, 40°; 25%, 90°; 10%, 36°; 14%, 50°

3. Angle: 48%, 173°; 12%, 43°; 16%, 58°; 24%, 86°

Family Expenses

Auto 18%
Food 22%
Clothing 11%
Housing 25%
Recreation 10%
Other 14%

Investment

Savings Account 48%
Stock 12%
Mutual Funds 16%
Retirement Fund 24%

CHAPTER 15

Pre-Test 15, page 419

17.

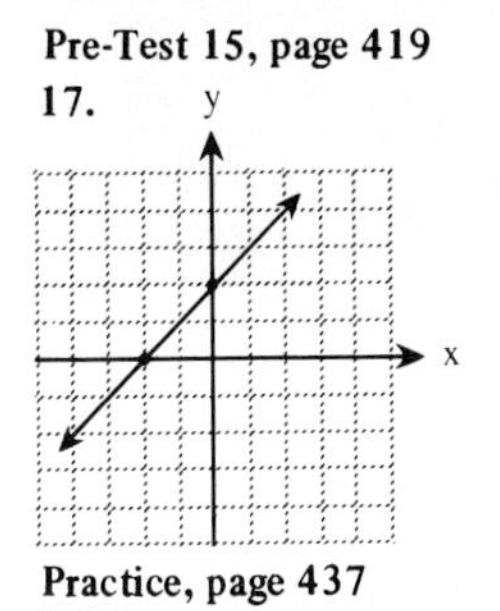

Try This, page 437

1-2.

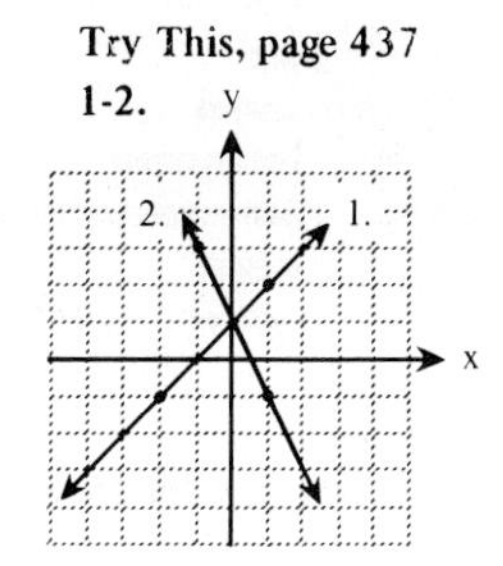

Practice, page 437

1-6.

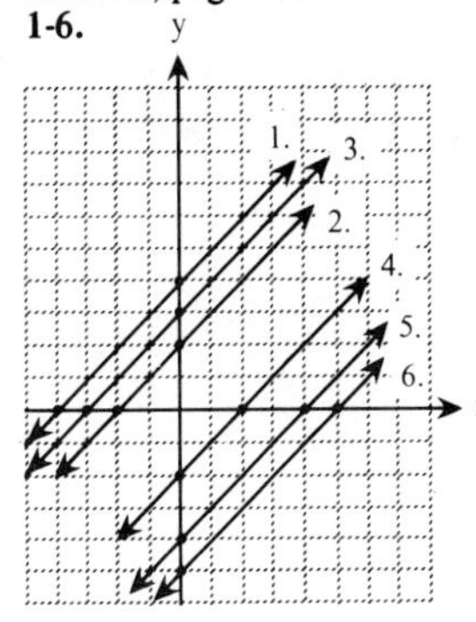

7-9.

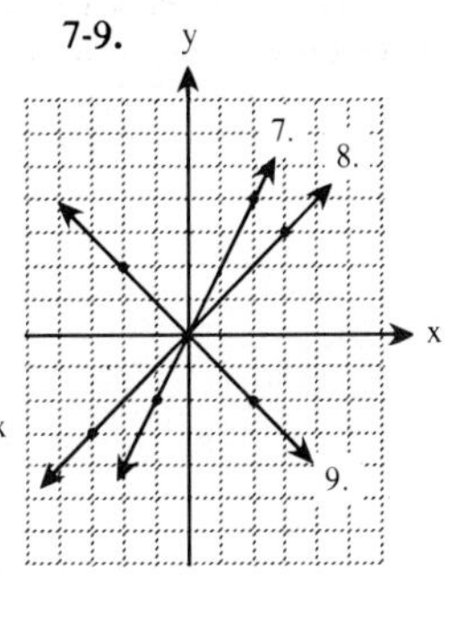

10-11.

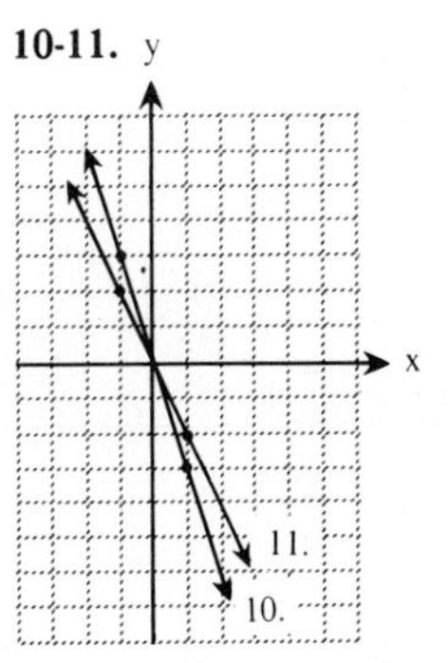

12-14.

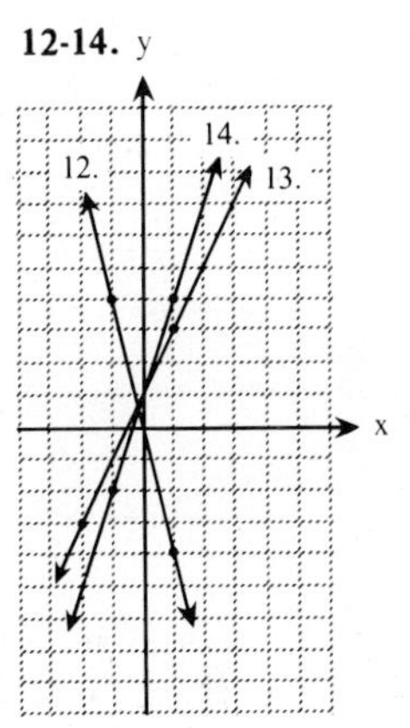

15-18.

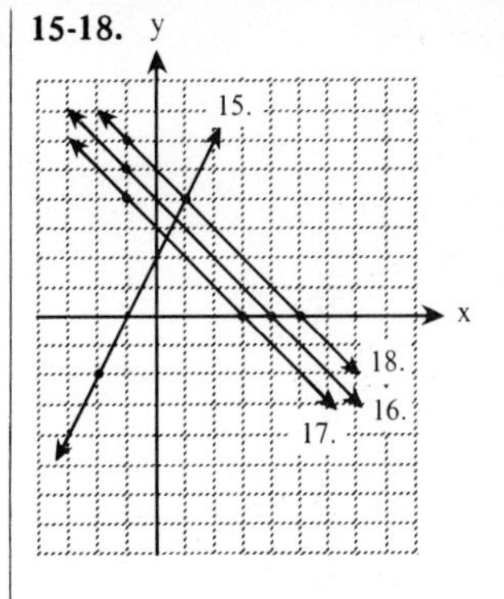

Practice, page 437

19-21.

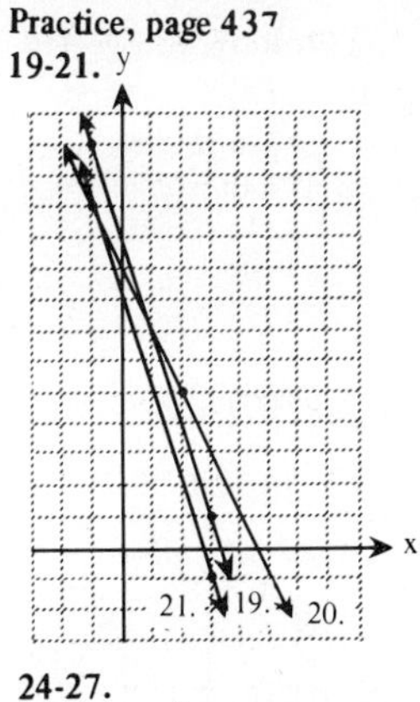

22-23.

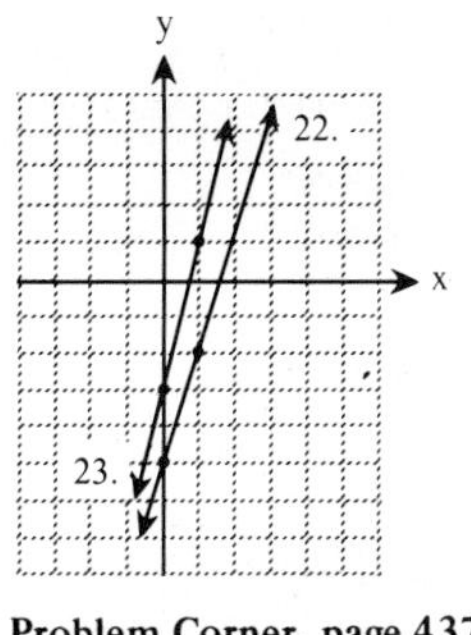

24-27.

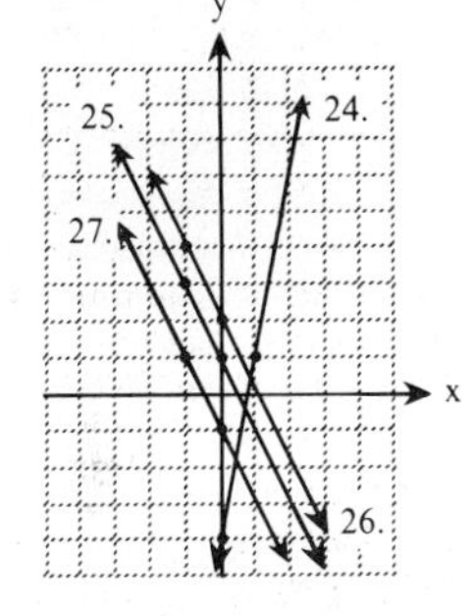

Problem Corner, page 437

1.

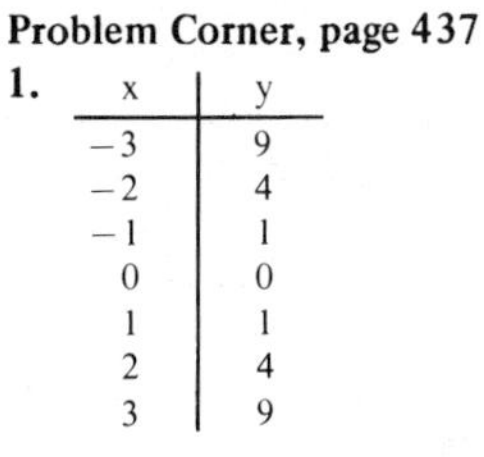

x	y
−3	9
−2	4
−1	1
0	0
1	1
2	4
3	9

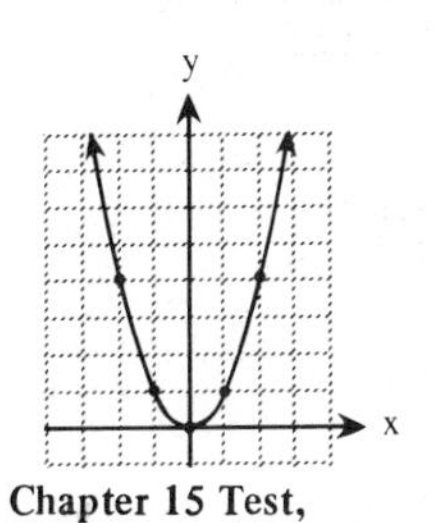

Chapter 15 Review, page 440

17.

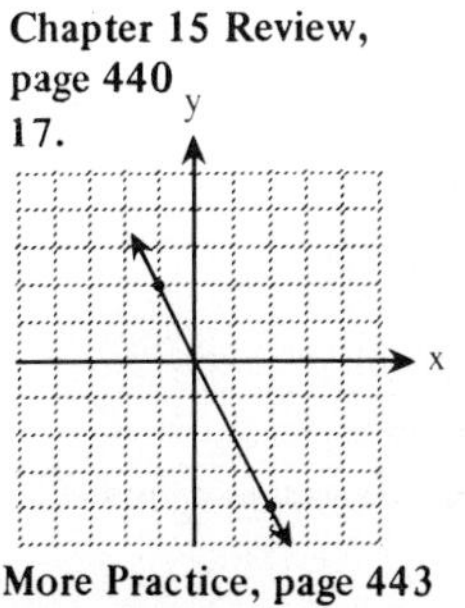

Chapter 15 Test, page 441

17.

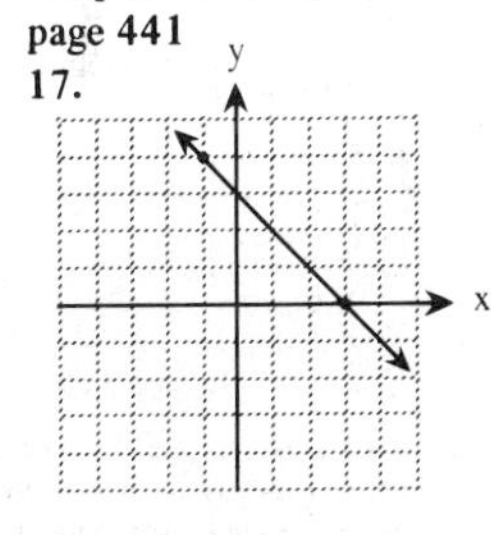

More Practice, page 443

31-34.

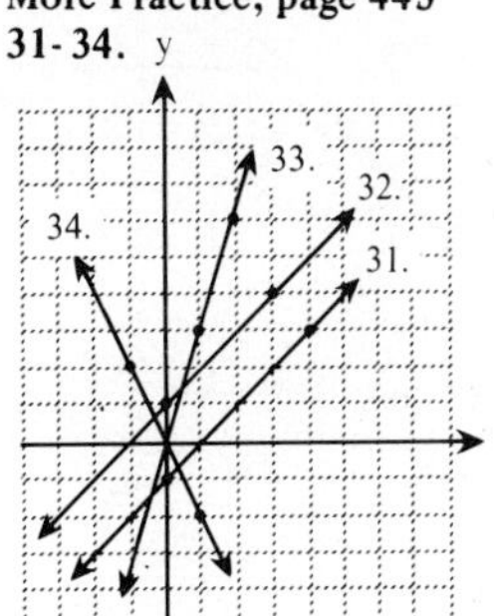

35-38.

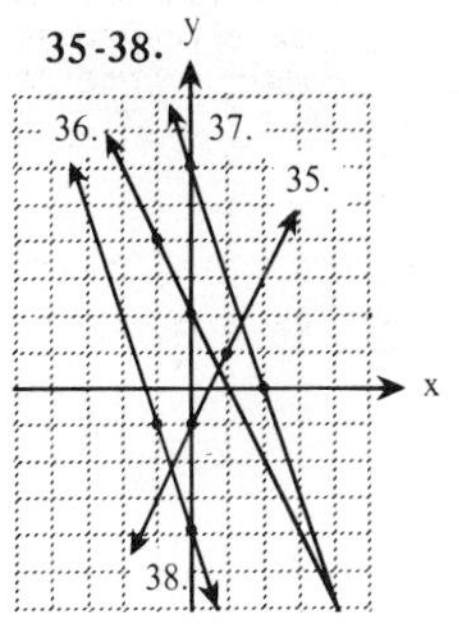

Skills Review, page 446

1.

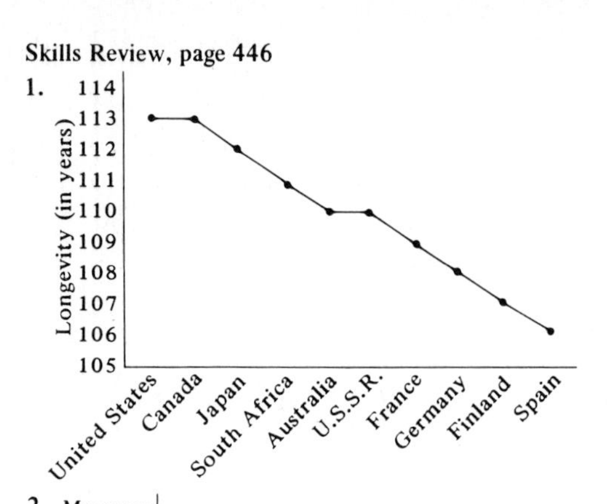

2.

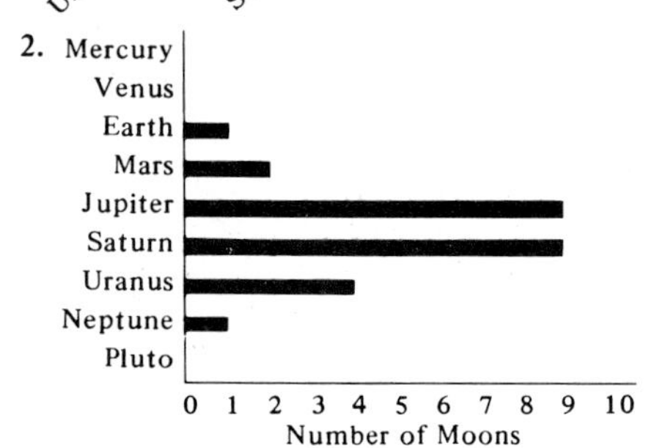

3. Angle: 40%, 144°; 25%, 90°; 25%, 90°; 4%, 14°; 6%, 22°

Time Spent in One Week

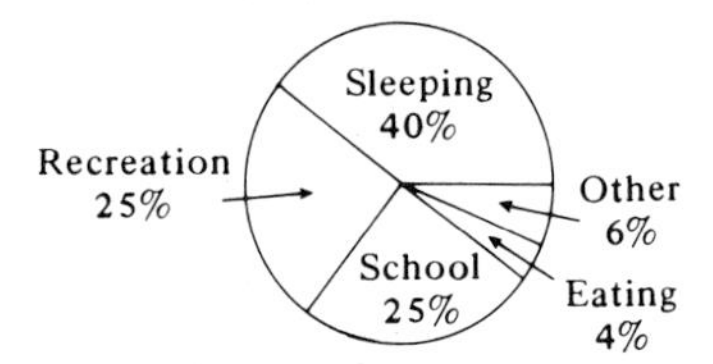

Skills Review, page 449

36-41.

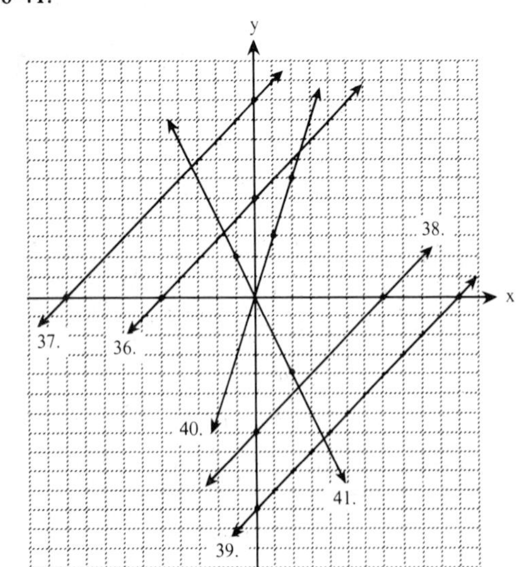

42-43.

44.

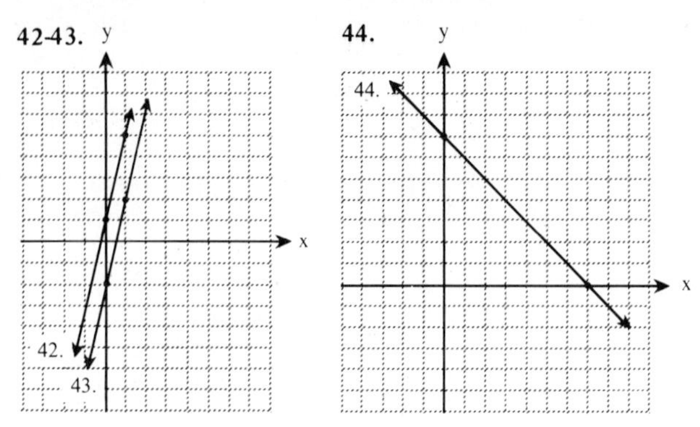

Skills Test, page 455

100.

101.

127-128.

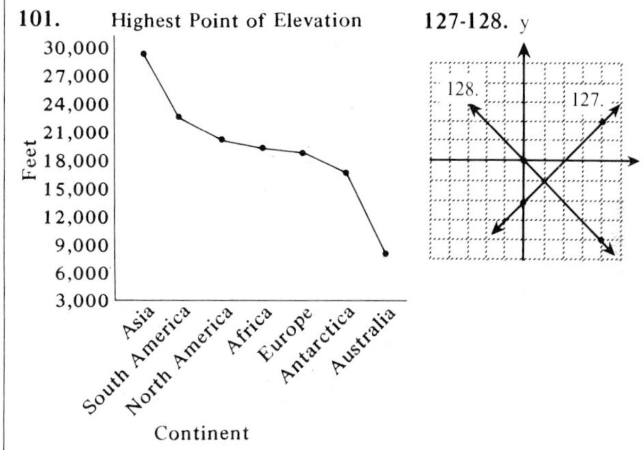

Applications Skills Test, page 456

1.

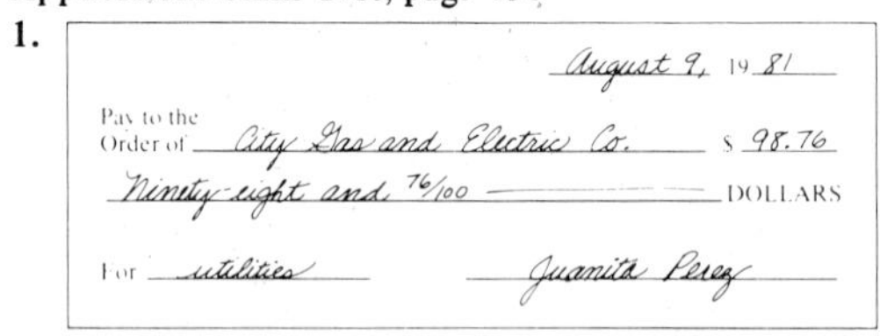

August 9, 19 81

Pay to the Order of City Gas and Electric Co. $ 98.76

Ninety-eight and 76/100 DOLLARS

For utilities Juanita Perez

2.

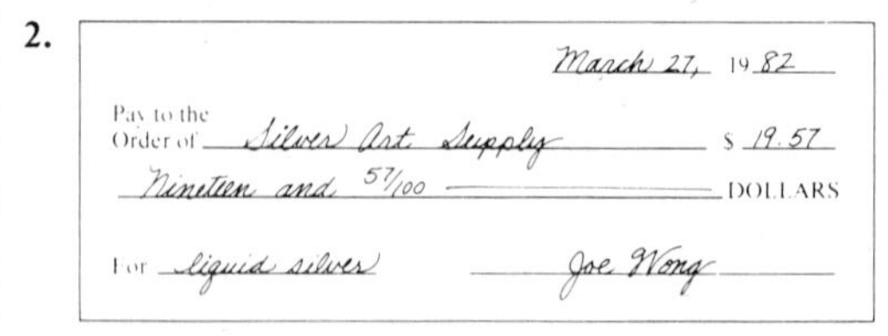

March 27, 19 82

Pay to the Order of Silver Art Supply $ 19.57

Nineteen and 57/100 DOLLARS

For liquid silver Joe Wong

ADDISON-WESLEY GENERAL MATHEMATICS

MERVIN L. KEEDY

MARVIN L. BITTINGER

STANLEY A. SMITH

PAUL A. ANDERSON

ADDISON-WESLEY PUBLISHING COMPANY

Menlo Park, California
Reading, Massachusetts
London
Amsterdam
Don Mills, Ontario
Sydney

ILLUSTRATIONS

Teresa Camozzi: 17, 25, 29, 45, 49, 55, 57, 74, 77, 83, 85, 99, 103, 105, 109, 111, 113, 117, 137, 141, 145, 153, 155, 169, 171, 175, 177, 181, 183, 184, 185, 197, 201, 203, 209, 224, 228, 229, 236, 253, 257, 265, 287, 291, 293, 295, 297, 299, 351, 369, 381, 397, 403, 405, 407, 409, 423, 427, 429, 437

Bill Shields: Cover and chapter openers.

All other illustrations by Addison-Wesley staff.

PHOTOGRAPHS

Elihu Blotnick*: 280, 324

George B. Fry III*: 30, 58, 80, 81, 86, 87, 88, 114, 122, 123, 158, 179, 186, 204, 212, 214, 228, 236, 238, 260, 268, 269, 288, 302, 316, 330, 342, 352, 376, 383, 384, 424, 434, 438

Historical Pictures Service, Inc.: 121

© Leo Touchet/Van Cleve Photography: 350, 412

*Photographs provided expressly for the publisher.

ACKNOWLEDGMENTS

p. 236 left: from Recreational Equipment, Inc. The Co-op Spring 1979 Catalogue, © 1979 by Recreational Equipment, Inc. Reprinted by permission of R.E.I. Co-op, Seattle, Washington.

ISBN 0-201-10300-1

CDEFGHIJK-VH-887654

AUTHORS

MERVIN L. KEEDY is Professor of Mathematics at Purdue University. He received his Ph.D. degree at the University of Nebraska, and formerly taught at the University of Maryland. He has also taught mathematics and science in junior and senior high schools. He is the author of many books on mathematics, and he is co-author with Marvin L. Bittinger and Stanley A. Smith of *Algebra One, Algebra Two,* and *Algebra Two and Trigonometry* (Addison-Wesley, 1982).

MARVIN L. BITTINGER is Professor of Mathematics Education at Indiana University-Purdue University at Indianapolis. He earned his Ph.D. degree at Purdue University. He is the author of several books and is co-author with Mervin L. Keedy and Stanley A. Smith of *Algebra One, Algebra Two,* and *Algebra Two and Trigonometry* (Addison-Wesley, 1982).

STANLEY A. SMITH is Coordinator, Office of Mathematics (K–12), for Baltimore County Public Schools, Maryland. He has taught junior high school mathematics and science and senior high school mathematics. He earned his M.A. degree at the University of Maryland. He is co-author with Mervin L. Keedy and Marvin L. Bittinger of *Algebra One, Algebra Two,* and *Algebra Two and Trigonometry* (Addison-Wesley, 1982).

PAUL A. ANDERSON is a teacher in the Clark County School District, Las Vegas, Nevada. He earned his M.Ed. degree at the University of Nevada. He is a former principal and instructor of mathematics education at the University of Nevada, Las Vegas. He is a co-author of several K–8 mathematics series.

SPECIAL THANKS

To Marilyn Ohriner, Las Vegas, Nevada, for her research and preparation of the Application and Career lessons.

To Ray Chayo, Norte Del Rio High School, Sacramento, California, for his work on the *Making Practice Fun* supplement.

To all the administrators and teachers who aided in reviewing this text:

Wendell L. Dain
Mathematics Teacher
Earl Wooster High School
Reno, Nevada

Dyanne B. Dandridge
Instructional Services Coordinator
Chicago Public Schools
Chicago, Illinois

Bill R. Fisher
Mathematics Teacher
Arlington High School
Indianapolis, Indiana

Philip P. Halloran
Supervisor of Mathematics
Springfield Public Schools
Springfield, Massachusetts

Peggy Pase
Mathematics Teacher
Trevor G. Browne High School
Phoenix, Arizona

Georgia Strecker
Mathematics Program Supervisor
Berkeley School District
Berkeley, California

Geraldine L. Thompson
Mathematics Teacher
William Penn High School
New Castle, Delaware

Curtis B. Wolf
Mathematics Department Chairman
Golden Ring Junior High School
Baltimore, Maryland

AN INTRODUCTION TO THE TEXT

This text has been designed to help you learn and use mathematics skills.

Here are some of the special features that should make learning easy and effective for you.

1. Full-Color Art and Photography
2. Flowchart Presentation
3. Examples and Try This Exercises
4. Lots of Practice
5. Frequent Problems and Applications
6. Complete Testing Program
7. Career and Calculator Strands

BEFORE THE LESSONS

Each chapter begins with a Pre-Test and a Getting Started activity.

First take the Pre-Test to see which skills you have already mastered.

Then enjoy Getting Started practicing mathematics skills. The whole class or a small group can work together in the activity.

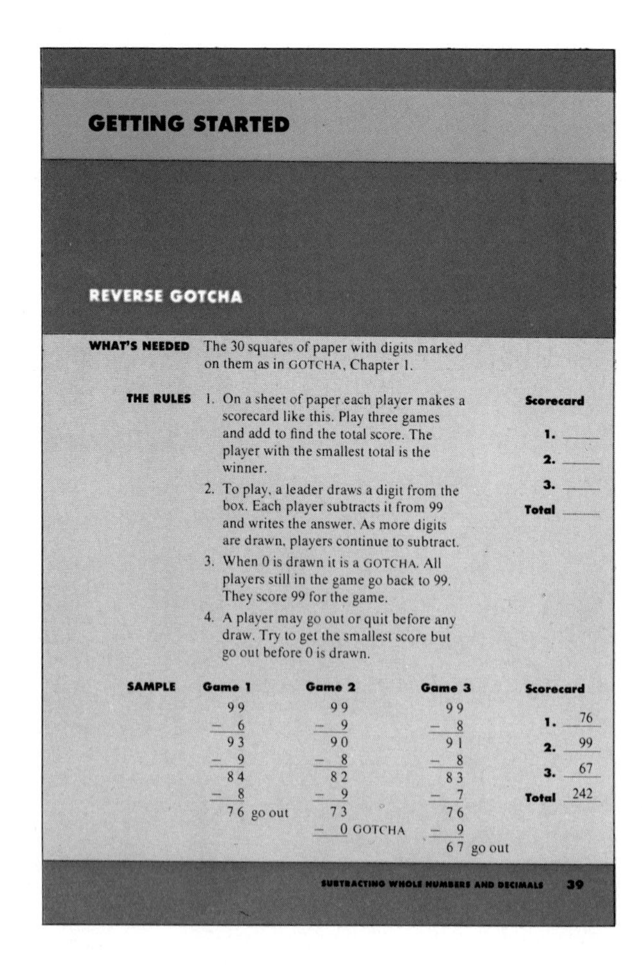

GETTING STARTED

REVERSE GOTCHA

WHAT'S NEEDED The 30 squares of paper with digits marked on them as in GOTCHA, Chapter 1.

THE RULES

1. On a sheet of paper each player makes a scorecard like this. Play three games and add to find the total score. The player with the smallest total is the winner.
2. To play, a leader draws a digit from the box. Each player subtracts it from 99 and writes the answer. As more digits are drawn, players continue to subtract.
3. When 0 is drawn it is a GOTCHA. All players still in the game go back to 99. They score 99 for the game.
4. A player may go out or quit before any draw. Try to get the smallest score but go out before 0 is drawn.

Scorecard

1. ____
2. ____
3. ____

Total ____

SAMPLE

Game 1	Game 2	Game 3
99	99	99
− 6	− 9	− 8
93	90	91
− 9	− 8	− 8
84	82	83
− 8	− 9	− 7
76 go out	73	76
	− 0 GOTCHA	− 9
		67 go out

Scorecard

1. 76
2. 99
3. 67

Total 242

SUBTRACTING WHOLE NUMBERS AND DECIMALS 39

LESSONS

The lessons are presented in an easy-to-read style. First, follow the Flowchart steps. Then work through some Examples. Finally, do the TRY THIS exercises to see if you have learned the skill.

Now you are ready to Practice the skills by completing the exercises and Problem Corner applications which follow the lessons.

If you need More Practice, do the exercises on the indicated pages.

You can Review and Test your skills with exercises at the end of each chapter.

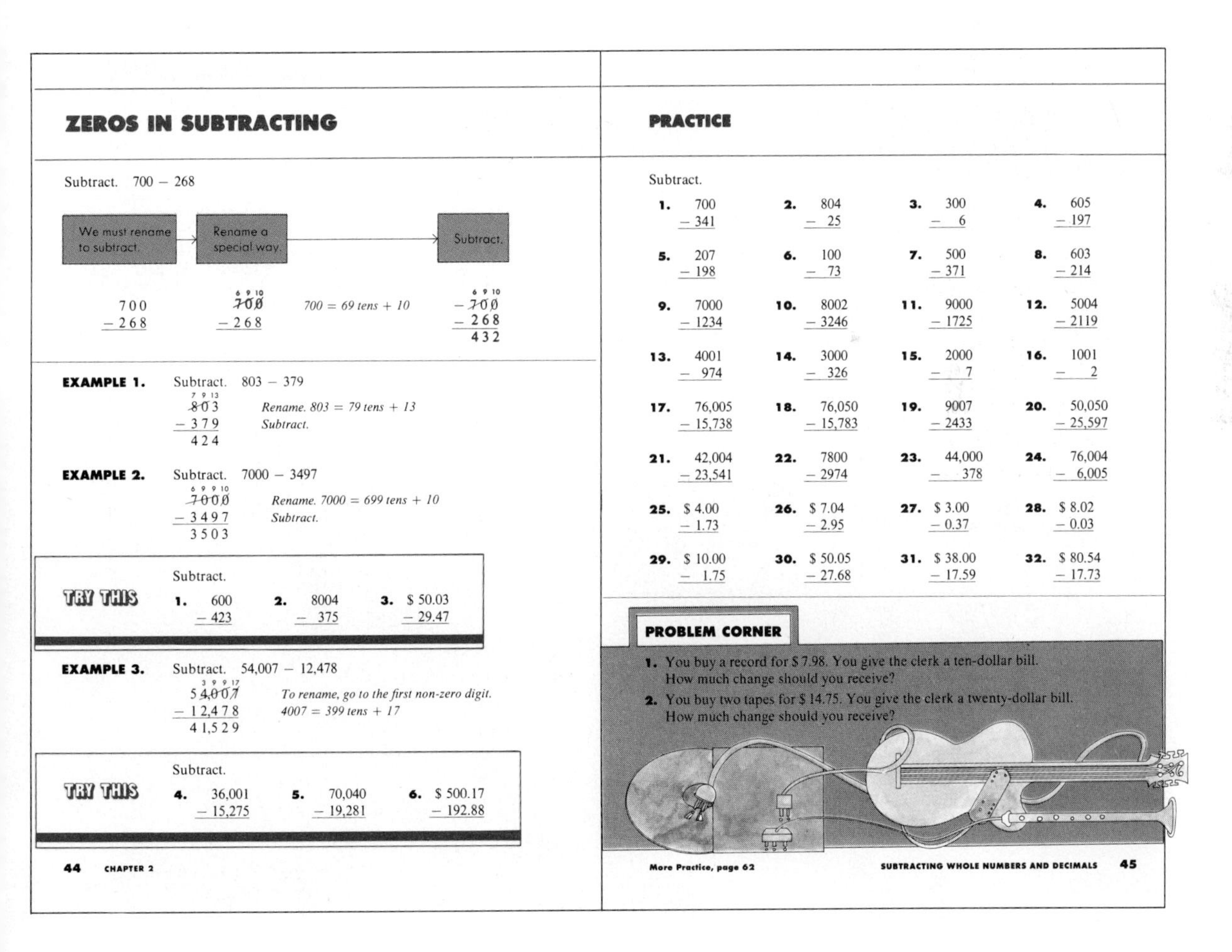

ZEROS IN SUBTRACTING

Subtract. 700 − 268

$\begin{array}{r} 700 \\ -268 \end{array}$ $\begin{array}{r} \overset{6\,9\,10}{\not7\not0\not0} \\ -268 \end{array}$ *700 = 69 tens + 10* $\begin{array}{r} \overset{6\,9\,10}{\not7\not0\not0} \\ -268 \\ \hline 432 \end{array}$

EXAMPLE 1. Subtract. 803 − 379

$\begin{array}{r} \overset{7\,9\,13}{\not8\not0\not3} \\ -379 \\ \hline 424 \end{array}$ *Rename. 803 = 79 tens + 13*
Subtract.

EXAMPLE 2. Subtract. 7000 − 3497

$\begin{array}{r} \overset{6\,9\,9\,10}{\not7\not0\not0\not0} \\ -3497 \\ \hline 3503 \end{array}$ *Rename. 7000 = 699 tens + 10*
Subtract.

TRY THIS Subtract.

1. 600 − 423
2. 8004 − 375
3. $ 50.03 − 29.47

EXAMPLE 3. Subtract. 54,007 − 12,478

$\begin{array}{r} 5\overset{3\,9\,9\,17}{\not4,\not0\not0\not7} \\ -12,478 \\ \hline 41,529 \end{array}$ *To rename, go to the first non-zero digit.*
4007 = 399 tens + 17

TRY THIS Subtract.

4. 36,001 − 15,275
5. 70,040 − 19,281
6. $ 500.17 − 192.88

44 CHAPTER 2

PRACTICE

Subtract.

1. 700 − 341
2. 804 − 25
3. 300 − 6
4. 605 − 197
5. 207 − 198
6. 100 − 73
7. 500 − 371
8. 603 − 214
9. 7000 − 1234
10. 8002 − 3246
11. 9000 − 1725
12. 5004 − 2119
13. 4001 − 974
14. 3000 − 326
15. 2000 − 7
16. 1001 − 2
17. 76,005 − 15,738
18. 76,050 − 15,783
19. 9007 − 2433
20. 50,050 − 25,597
21. 42,004 − 23,541
22. 7800 − 2974
23. 44,000 − 378
24. 76,004 − 6,005
25. $ 4.00 − 1.73
26. $ 7.04 − 2.95
27. $ 3.00 − 0.37
28. $ 8.02 − 0.03
29. $ 10.00 − 1.75
30. $ 50.05 − 27.68
31. $ 38.00 − 17.59
32. $ 80.54 − 17.73

PROBLEM CORNER

1. You buy a record for $ 7.98. You give the clerk a ten-dollar bill. How much change should you receive?
2. You buy two tapes for $ 14.75. You give the clerk a twenty-dollar bill. How much change should you receive?

More Practice, page 62   SUBTRACTING WHOLE NUMBERS AND DECIMALS 45

SOLVING PROBLEMS AND CALCULATOR ACTIVITIES

Solving Problems is an important skill you use every day. There are several lessons to help you develop and practice this ability. Master the steps to become a good problem solver.

The Calculator activities will help you practice and sharpen your skills. Discover interesting number patterns and relationships by using a calculator. Some of the activities can be performed by the whole class or a small group.

SOLVING PROBLEMS

When solving a problem, we sometimes need to decide whether to do a mental estimate, a paper and pencil calculation, or use a calculator. For each of the following problem situations, tell whether you would use a calculator, do a paper and pencil calculation, or make a mental estimate.

SITUATION 1 You are at the ballgame. Do you have enough money to buy two hot dogs and a bag of peanuts?

SITUATION 2 You are completing your income tax return.

SITUATION 3 You are deciding about how much it will cost to buy furniture for a bedroom.

SITUATION 4 You are planning a vacation. From a map you want to estimate the approximate number of kilometers you will drive.

SITUATION 5 You are balancing your checkbook.

SITUATION 6 You are in a pastry shop with $ 6. You are to purchase a dozen pastries for a party. Pastries sell for $ 1.20/dozen, but there is a sign telling that there is a 10% discount for the purchase of 5 dozen or more. Should you take advantage of the discount and buy 5 dozen?

CALCULATOR CORNER

WIPE OUT

Players two
Materials two calculators, paper and pencils

SAMPLE Chin enters a five-digit whole number, such as 59,762, on the display. Kristen enters a five-digit whole number, such as 48,829, on the display.

They exchange calculators. Each player must now attempt to "wipe out" the number and get 0 on the display.

To do so they may add, subtract, multiply, or divide by any one or two-digit whole number or decimal except 0.

	Kristen		Chin
Play 1	59,762 −62 = 59,700	**Play 1**	48,829 −99 = 48,730
Play 2	×0.01 = 597	**Play 2**	−30 = 48,700
Play 3	−97 = 500	**Play 3**	×0.01 = 487
Play 4	×0.1 = 50	**Play 4**	−87 = 400
Play 5	−50 = 0	**Play 5**	×0.01 = 4
		Play 6	−4 = 0

Kristen's score is 5. Chin's score is 6.

Kristen wins the game, because she reaches 0 in fewer plays.

APPLICATIONS AND CAREERS

Learn how skills are used in real-life situations in the Applications. There are two lessons in each chapter in which you can apply mathematics to problems you encounter every day. Discover why you need to master mathematics skills.

You will also see how skills are applied in various Careers. Each chapter contains a Career which focuses on some of the mathematics used in that job. You will learn about the Career while practicing many important skills.

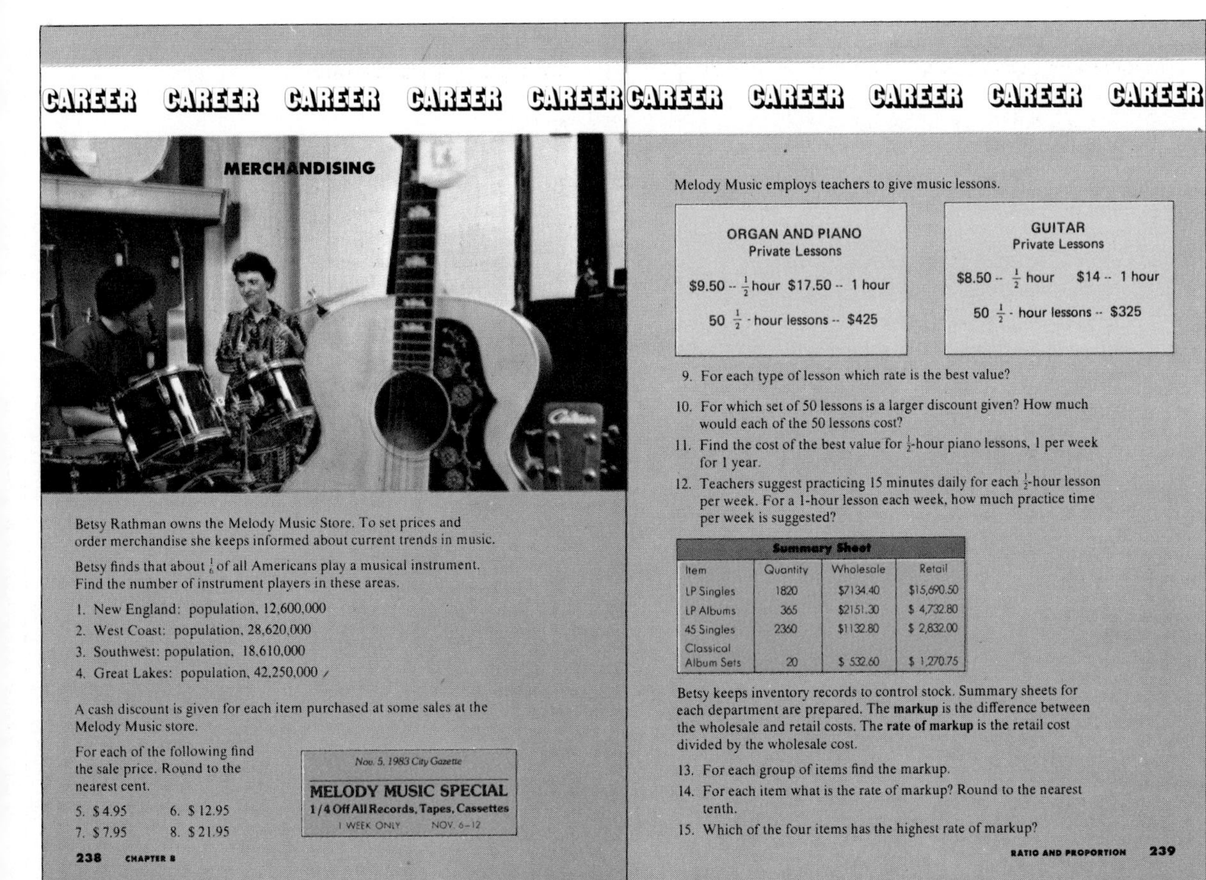

CAREER CAREER CAREER CAREER CAREER CAREER CAREER CAREER CAREER CAREER

MERCHANDISING

Betsy Rathman owns the Melody Music Store. To set prices and order merchandise she keeps informed about current trends in music.

Betsy finds that about $\frac{1}{6}$ of all Americans play a musical instrument. Find the number of instrument players in these areas.

1. New England: population, 12,600,000
2. West Coast: population, 28,620,000
3. Southwest: population, 18,610,000
4. Great Lakes: population, 42,250,000

A cash discount is given for each item purchased at some sales at the Melody Music store.

For each of the following find the sale price. Round to the nearest cent.

5. $4.95
6. $12.95
7. $7.95
8. $21.95

Nov. 5, 1983 City Gazette

MELODY MUSIC SPECIAL
1/4 Off All Records, Tapes, Cassettes
1 WEEK ONLY NOV. 6–12

238 CHAPTER 8

Melody Music employs teachers to give music lessons.

ORGAN AND PIANO
Private Lessons
$9.50 -- $\frac{1}{2}$ hour $17.50 -- 1 hour
50 $\frac{1}{2}$ - hour lessons -- $425

GUITAR
Private Lessons
$8.50 -- $\frac{1}{2}$ hour $14 -- 1 hour
50 $\frac{1}{2}$ - hour lessons -- $325

9. For each type of lesson which rate is the best value?
10. For which set of 50 lessons is a larger discount given? How much would each of the 50 lessons cost?
11. Find the cost of the best value for $\frac{1}{2}$-hour piano lessons, 1 per week for 1 year.
12. Teachers suggest practicing 15 minutes daily for each $\frac{1}{2}$-hour lesson per week. For a 1-hour lesson each week, how much practice time per week is suggested?

Summary Sheet

Item	Quantity	Wholesale	Retail
LP Singles	1820	$7134.40	$15,690.50
LP Albums	365	$2151.30	$ 4,732.80
45 Singles	2360	$1132.80	$ 2,832.00
Classical Album Sets	20	$ 532.60	$ 1,270.75

Betsy keeps inventory records to control stock. Summary sheets for each department are prepared. The **markup** is the difference between the wholesale and retail costs. The **rate of markup** is the retail cost divided by the wholesale cost.

13. For each group of items find the markup.
14. For each item what is the rate of markup? Round to the nearest tenth.
15. Which of the four items has the highest rate of markup?

RATIO AND PROPORTION 239

TABLE OF CONTENTS

CHAPTER 1

Adding Whole Numbers and Decimals

CHAPTER 2

Subtracting Whole Numbers and Decimals

CHAPTER 3

Multiplying Whole Numbers and Decimals

CHAPTER 4

Dividing Whole Numbers and Decimals

CHAPTER 5

Fractions

CHAPTER 6

Multiplying and Dividing Fractions

CHAPTER 7

Adding and Subtracting Fractions

CHAPTER 8

Ratios and Proportions

CHAPTER 9

Percent

CHAPTER 10

Measurement

CHAPTER 11

Perimeter and Area

CHAPTER 12

CHAPTER 13

CHAPTER 14

CHAPTER 15

Expressions and Equations

APPENDIX

APPLICATIONS AND CAREERS CONTENTS

APPLICATIONS

CAREERS

CHAPTER 1 PRE-TEST

Write standard notation. p. 6

1. six thousand, three hundred forty-five 6345
2. five thousand, seventy-one 5071
3. 724 thousand, 607 724,607
4. 41 million, 600 thousand 41,600,000

Write money notation. p. 8

5. fifty-five dollars and thirty-four cents $55.34
6. seventy dollars and seven cents $70.07

Write decimal notation. p. 20

7. 13 and 424 thousandths 13.424
8. 5 and 24 ten-thousandths 5.0024

Add. pp. 14–22

9. 849 + 276 = 1125

10. 8749 + 9828 = 18,577

11. 498 + 756 + 49 + 8 = 1311

12. $7.49 + 6.24 + 0.29 = $14.02

13. 6.9 + 7.8 = 14.7

14. 14.248 + 9.876 = 24.124

15. 3.98 + 1.8 + 7.349 = 13.129

16. 19.748 + 3.98 + 7.8 + 0.72 = 32.248

Estimate the sums. p. 28

17. 948 + 658 + 850 — 2500

18. 3486 + 2785 + 320 — 6600

19. $7.98 + 6.78 + 3.64 — $19.00

20. 3.176 + 4.92 + 3.8 — 12

Application pp. 10, 18

21. Write a check from Jason Sanders to Acme Fuel Company in the amount of $ 276.79. Date it October 12, 1981. See answer section.

22. Find the net deposit. $261.67

		Dollars	Cents
Cash	Currency	86	00
	Coin	18	75
Checks	12 - 24	8	76
	1 - 107	148	16
Total			
Less Cash Received			-0-
Net Deposit			

1

ADDING WHOLE NUMBERS AND DECIMALS

GETTING STARTED

GOTCHA

WHAT'S NEEDED 30 squares of paper or cardboard about this size

Write 0 on three of them, 1 on three of them, 2 on three of them, and so on up to 9. Put them in a box and mix well.

THE RULES

1. On a sheet of paper each player makes a scorecard like this. Play three games and add to find the total score. The player with the largest total wins.

Scorecard

1. ______

2. ______

3. ______

Total ______

2. To play, a leader draws digits from the box, one at a time. All players write them down. They are to be added as they are drawn to get the total for the game.
3. When 0 is drawn, it is a GOTCHA. Each player still in the game gets a 0 for that game.
4. A player can quit or go out at any time. Try to get a big score but go out before 0 is drawn.

The probability of drawing 0 is 1 out of 10.

SAMPLE

Game 1	Game 2	Game 3
4	5	9
+ 6	+ 1	+ 8
10	6	17
+ 8	+ 8	+ 9
18	14	26
+ 9	+ 0 GOTCHA	+ 4
27 go out		30
		+ 8
		38 go out

Scorecard

1. 27

2. 0

3. 38

Total 65

Objectives 1, 2, and 3

USING NUMBERS

We use numbers in many ways.

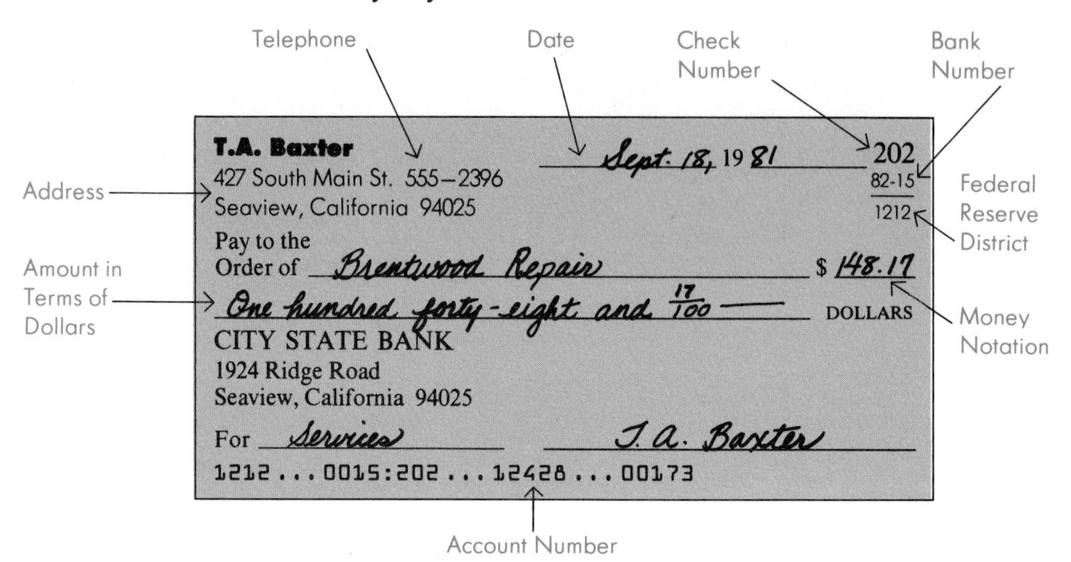

EXAMPLE 1. Write August 14, 1979, in three ways.

We may also write *14 August 79.*

8–14–79	month-day-year	
79–8–14	year-month-day	*(used on some forms)*
14–8–79	day-month-year	*(used by government agencies)*

TRY THIS Write these dates in three ways.

1. October 21, 1936 **10-21-36, 36-10-21, 21-10-36.**

2. December 25, 1981 **12-25-81, 81-12-25, 25-12-81.**

EXAMPLES. Tell a likely use.

2. 452–367–2184 Telephone Number *(area code–number)*
3. 312–44–0786 Social Security Number
4. 89102 Zip Code

TRY THIS Tell a likely use.

3. 512–62–1407 **Social Security Number**

4. 46035 **Zip Code**

5. 801–457–0826 **Telephone Number**

To write time notation we use colons to separate hours, minutes, and seconds.

EXAMPLES. Write time notation.

Electronic timers at sports events record elapsed times in this way.

5. 10 hours and 7 minutes 10:07
6. 9 minutes and 29 seconds 9:29
7. 15 hours, 8 minutes, and 47 seconds 15:08:47
8. 12 minutes after 6 o'clock 6:12

TRY THIS

Write time notation.

6. 28 minutes and 9 seconds 28:09
7. 16 minutes after 7 o'clock 7:16
8. 11 hours and 15 minutes 11:15
9. 9 hours, 9 minutes and 9 seconds 9:09:09

PRACTICE

For extra practice, use Making Practice Fun 1 with this lesson.

Write these dates in three ways.

1. January 15, 1980 1-15-80, 80-1-15, 15-1-80
2. February 7, 1979 2-7-79, 79-2-7, 7-2-79
3. March 21, 1982 3-21-82, 82-3-21, 21-3-82
4. May 2, 1981 5-2-81, 81-5-2, 2-5-81
5. December 7, 1941 12-7-41, 41-12-7, 7-12-41
6. July 4, 1963 7-4-63, 63-7-4, 4-7-63

Tell a likely use.

7. 317–463–6129 Telephone Number
8. 46032 Zip Code
9. 427–86–5104 Soc. Sec. Number
10. 736–81–4104 Soc. Sec. Number
11. 21107 Zip Code
12. 415–854–3000 Telephone Number

Write time notation.

13. 7 hours and 45 minutes 7:45
14. 11 minutes and 30 seconds 11:30
15. 30 minutes after 3 o'clock 3:30
16. 5 hours, 18 minutes, and 4 seconds 5:18:04
17. 9 hours and 9 minutes 9:09
18. 15 minutes after 1 o'clock 1:15
19. 10 hours, 50 minutes and 10 seconds 10:50:10
20. 40 minutes and 40 seconds 40:40

Objectives 4, 5, and 6

PLACE VALUE TO THOUSANDS

In this book four-digit numbers are written without a comma. Individual teachers may choose to have their classes use the comma if they wish.

To name numbers we use place value.

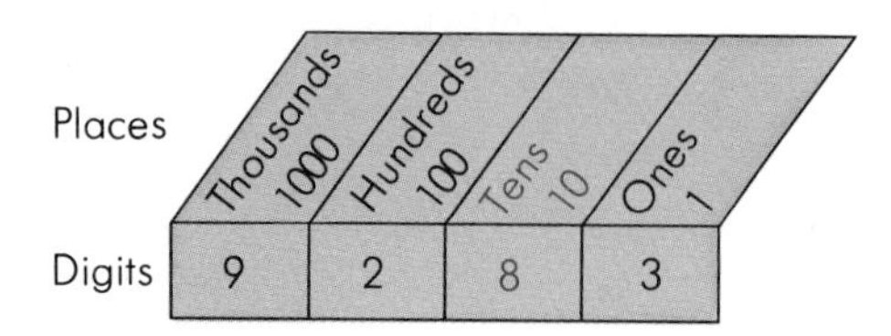

The 8 is in the tens' place. It means 8 × 10 (digit times place), so its value is 80.

EXAMPLE 1. Find the place and value of the underlined digit.

8<u>2</u>79 The 2 is in the hundreds' place.
Its value is 200.

TRY THIS Find the place and value of the underlined digit.

1. 37<u>4</u>1 **2.** <u>5</u>286 **3.** 401<u>5</u> See answer section.

To write word names we place a comma after thousands and a hyphen between tens and ones.

EXAMPLES. Write word names.

2. 9283 nine thousand, two hundred eighty-three
3. 580 five hundred eighty
4. 508 five hundred eight

TRY THIS Write word names. See answer section.

4. 98 **5.** 376 **6.** 5287 **7.** 6006

EXAMPLES. Write standard notation.

5. eight thousand, three hundred forty-nine 8349
6. five hundred five 505
7. seven thousand, seventy 7070

TRY THIS

Write standard notation.

8. three thousand, six hundred fifty-one 3651
9. five thousand, eighty-eight 5088
10. three thousand, three 3003

PRACTICE

Find the place and value of the underlined digit. See answer section.

1. $\underline{4}286$	**2.** $9\underline{8}74$	**3.** $398\underline{5}$	**4.** $2\underline{7}05$
5. $6\underline{2}94$	**6.** $\underline{4}876$	**7.** $\underline{9}000$	**8.** $40\underline{7}8$

Write word names. See answer section.

9. 68	**10.** 86	**11.** 41	**12.** 90
13. 374	**14.** 276	**15.** 470	**16.** 407
17. 3948	**18.** 2789	**19.** 5780	**20.** 6074

Write standard notation.

21. nine hundred sixty-three 963

22. seven hundred forty-two 742

23. one hundred twelve 112

24. six hundred six 606

25. four thousand, four hundred forty-four 4444

26. eight thousand, one hundred ninety-three 8193

27. one thousand, five hundred seventy 1570

28. nine thousand, six hundred five 9605

29. eight thousand, nine hundred 8900

30. seven thousand, ninety 7090

31. six thousand, nine 6009

32. five thousand 5000

33. eight hundred thirty-seven 837

34. three hundred sixteen 316

35. two thousand, eleven 2011

36. one thousand, sixty-two 1062

More Practice, page 34

Objectives 7, 8, and 9

WRITING MONEY NOTATION

Money Notation	Word Name
73¢ or $ 0.73	seventy-three cents
$ 18 or $ 18.00	eighteen dollars
$ 18.73	eighteen dollars and seventy-three cents
$ 307.90	three hundred seven dollars and ninety cents

We write *and* for the decimal point to separate dollars and cents.

EXAMPLES. Write word names.

1. $ 0.04 four cents **2.** $ 6.00 six dollars
3. $ 9.05 nine dollars and five cents

TRY THIS Write word names. See answer section.

1. $ 0.01 **2.** $ 17.46 **3.** $ 236.00

EXAMPLES. Write money notation.

4. twenty-five cents 25¢ or $ 0.25
5. twenty-five dollars $ 25 or $ 25.00
6. five dollars and eighteen cents $ 5.18
7. ninety dollars and nine cents $ 90.09

TRY THIS Write money notation.

4. thirty cents 30¢ or $0.30
5. thirty dollars $30 or $30.00
6. one dollar and seven cents $1.07

For checks we write money notation in terms of dollars. Cents are written as hundredths of a dollar.

EXAMPLES. Write in terms of dollars.

8. $ 73.56 seventy-three and $\frac{56}{100}$ dollars
9. $ 901.07 nine hundred one and $\frac{07}{100}$ dollars
10. $ 7.00 seven and $\frac{00}{100}$ dollars

Sometimes this is written *seven and $\frac{no}{100}$ dollars.*

TRY THIS Write in terms of dollars. See answer section.

7. $ 9.89 **8.** $ 56.04 **9.** $ 174.00

PRACTICE

For extra practice, use Making Practice Fun 2 with this lesson.

Write word names. See answer section.

1. 79¢ **2.** $ 0.87 **3.** $ 0.07 **4.** 8¢

5. $ 19.00 **6.** $ 42 **7.** $ 738 **8.** $ 404.00

9. $ 26.37 **10.** $ 9.84 **11.** $ 38.98 **12.** $ 47.50

13. $ 506.05 **14.** $ 870.70 **15.** $ 100.01 **16.** $ 390.09

Write money notation.

17. four cents
4¢ or $0.04

18. thirteen cents
13¢ or $0.13

19. sixty-five cents
65¢ or $0.65

20. ninety cents
90¢ or $0.90

21. three dollars
$3 or $3.00

22. twenty-one dollars
$21 or $21.00

23. five dollars and forty-two cents
$5.42

24. two dollars and thirteen cents
$2.13

25. fifty-three dollars and fifteen cents
$53.15

26. seventy-five dollars and eleven cents
$75.11

27. fifty dollars and fifty cents
$50.50

28. five dollars and five cents
$5.05

29. sixty dollars and six cents
$60.06

30. one hundred dollars and one cent
$100.01

Write in terms of dollars. See answer section.

31. $ 8.78 **32.** $ 4.25 **33.** $ 39.56 **34.** $ 19.75

35. $ 326.48 **36.** $ 427.86 **37.** $ 125.87 **38.** $ 999.99

39. $ 7.04 **40.** $ 16.09 **41.** $ 407.09 **42.** $ 328.05

43. $ 5.00 **44.** $ 0.95 **45.** $ 10.00 **46.** $ 0.08

More Practice, page 34

Objective 150

APPLICATION Many local banks have literature available concerning checking accounts.

WRITING CHECKS

Most payments are made by checks. Checks are used for safety, time saving and record keeping.

2. Write the name of payee.

1. Write today's date.

The date may be written in other forms.

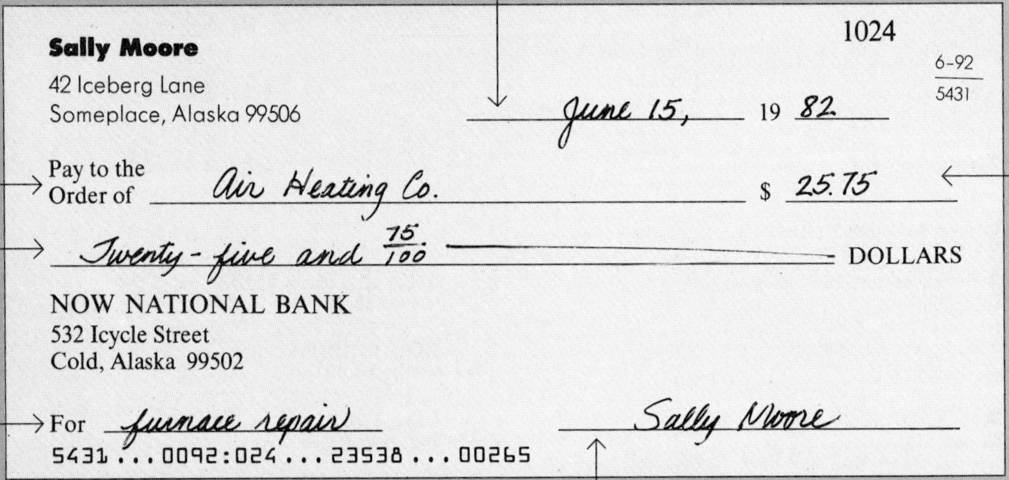

Sally Moore
42 Iceberg Lane
Someplace, Alaska 99506

1024

6-92
5431

June 15, 19 82

Pay to the Order of Air Heating Co. $ 25.75

Twenty-five and 75/100 DOLLARS

NOW NATIONAL BANK
532 Icycle Street
Cold, Alaska 99502

For furnace repair Sally Moore

5431...0092:024...23538...00265

This is used for record keeping and budgeting.

6. Write what check is for.

5. Signature

4. Write amount in terms of dollars.

3. Write money notation.

CONSUMER HINTS

1. Use the correct date. Do not date ahead.
2. The payee is the person to whom the money is paid.
3. How much in money notation? Start at the left.
4. How much in terms of dollars? Fill up the space.
5. Sign the same way each time.
6. This is a reminder of what the check was for.

These hints refer to the numbered items above.

Make copies of checks like this. You may wish to make a duplicating master of blank checks. Sometimes local banks will donate old checkbooks to schools.

_______________ 19 ________

Pay to the
Order of ____ $ ________

____ DOLLARS

For ______________________ ______________________________

Write these checks. Use the given information. See answer section.

1. Lucy Lee made a house payment of $ 258.54 to Western Mortgage Company on September 9, 1984.
2. On October 25, 1981 Kazuo Sata wrote a $ 38.24 check to Anderson Food Services for groceries.
3. Ace Catering Service sent Alice Washington a bill for $ 89.00 for a party. Alice paid the bill on December 18, 1980.
4. Paulo Lopez bought a bike from Joe's Cycle Shop on February 19, 1982. It cost $ 158.45. He paid by check.

Write these checks. Use your own name and today's date. Choose the payee and the amount. Answers will vary.

5. A restaurant for dinner for two.
6. A friend for payment of a loan.
7. A radio shop for stereo equipment.
8. A sporting goods store for camping gear.

Objectives 10, 11, and 12

PLACE VALUE TO BILLIONS

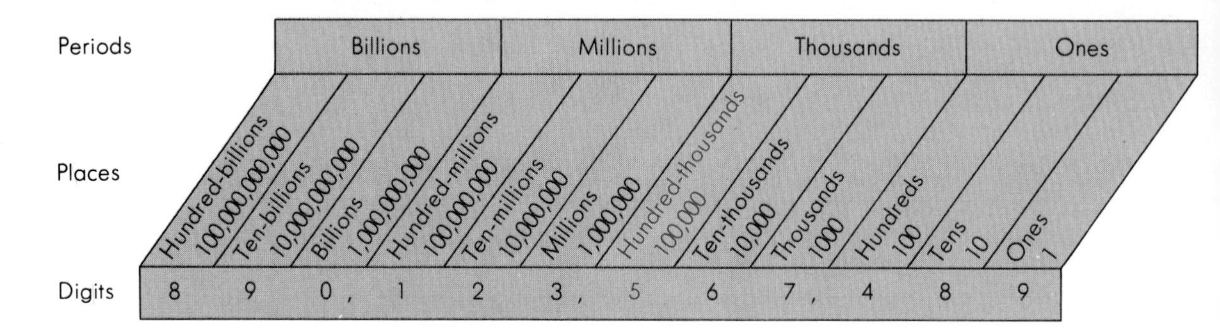

The 5 is in the hundred-thousands' place. It means 5 × 100,000 (digit times place), so its value is 500,000.

EXAMPLE 1. Find the place and value of the underlined digit.

276,4<u>8</u>9,124,019 The 8 is in the ten-millions' place. Its value is 80,000,000.

TRY THIS Find the place and value of the underlined digit.

1. 1,4<u>7</u>8,269 **2.** <u>7</u>65,078,947 **3.** <u>8</u>76,491,024,000

See answer section.

For large numbers we often use short word names.

Period names are used except for ones. Notice that 3,000,405 is written as 3 million, 405. The thousands' period is not used.

EXAMPLES. Write short word names.

2. 123,456,789 123 million, 456 thousand, 789
3. 40,350,027,080 40 billion, 350 million, 27 thousand, 80

TRY THIS Write short word names. See answer section.

4. 496,287,104 **5.** 380,087,006,100

EXAMPLES. Write standard notation.

4. 76 million, 280 thousand, 400 76,280,400
5. 900 billion 900,000,000,000
6. 9 million, 8 thousand 9,008,000

There are no more than 3 places in each period.

0's must be used here.

TRY THIS

Write standard notation.

6. 986 million, 280 thousand, 186 986,280,186

7. 4 billion 4,000,000,000

8. 20 million, 20 20,000,020

PRACTICE

Use Quiz 1 after this Practice.

Find the place and value of the underlined digit. See answer section.

1. 37,486 **2.** 425,807 **3.** 19,875

4. 36,498,248 **5.** 7,864,297 **6.** 368,270,027

7. 9,846,701,425 **8.** 89,678,130,031 **9.** 429,878,000,000

10. 784,965 **11.** 86,000,000 **12.** 86,000,000,000

Write short word names. See answer section.

13. 37,826 **14.** 428,874 **15.** 5687

16. 80,000 **17.** 984,126,481 **18.** 82,476,125

19. 9,009,009 **20.** 80,800,008 **21.** 9,008,007,006

22. 73,816,479,100 **23.** 728,000,000 **24.** 7,000,000,007

Write standard notation.

25. 28 thousand, 514 28,514

26. 348 thousand, 704 348,704

27. 36 thousand, 15 36,015

28. 999 thousand 999,000

29. 7 million 7,000,000

30. 28 million 28,000,000

31. 18 million, 238 thousand, 375 18,238,375

32. 326 million, 400 thousand, 18 326,400,018

33. 5 million, 5 thousand, 5 5,005,005

34. 18 million, 18 thousand 18,018,000

35. 1 billion 1,000,000,000

36. 38 billion 38,000,000,000

More Practice, page 34

Objectives 13 and 14

ADDING WHOLE NUMBERS

Add. 378 + 519

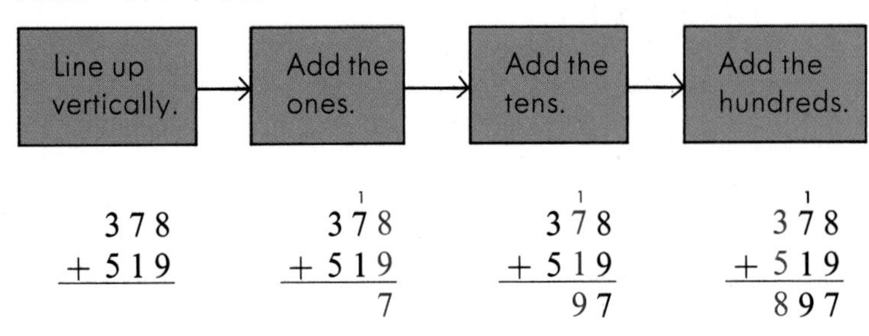

EXAMPLES. Add.

1.
```
  316
+  43
  359
```

2.
```
  1 1
  489
+ 637
 1126
```

TRY THIS Add.

1. 476 + 13 489
2. 878 + 239 + 18 1135
3. 708 + 897 1605
4. 348 + 16 + 7 371

EXAMPLE 3. Add. $6.58 + $0.97 + $0.06. A dollar sign is usually placed in both the top addend and the sum.

```
  1 2
$ 6.58     Add as with whole numbers.
  0.97
+ 0.06     Write money notation in the answer.
 $7.61
```

TRY THIS Add.

5. $4.96 + $0.86 + $0.05 $5.87
6. $0.76 + $0.09 + $0.87 $1.72
7. $4.93 + 0.29 + 6.34 $11.56
8. $9.48 + 7.04 + 0.09 $16.61

PRACTICE

For extra practice, use Making Practice Fun 3 with this lesson.

Add.

1. 14 + 72 86

2. 81 + 18 99

3. 36 + 33 69

4. 14 + 42 56

5. 238 + 611 849

6. 976 + 22 998

7. 80 + 914 994

8. 7 + 720 727

9. 98 + 76 = 174

10. 76 + 8 = 84

11. 47 + 9 = 56

12. 66 + 77 = 143

13. 347 + 286 = 633

14. 975 + 87 = 1062

15. 498 + 74 = 572

16. 787 + 9 = 796

17. 731 + 979 + 486 = 2196

18. 829 + 331 + 97 = 1257

19. 648 + 926 + 9 = 1583

20. 999 + 99 + 9 = 1107

21. 498 + 786 + 594 + 687 = 2565

22. 79 + 98 + 98 + 42 = 317

23. 876 + 429 + 15 + 4 = 1324

24. 987 + 36 + 8 + 7 = 1038

25. \$8.76 + 4.29 = \$13.05

26. \$9.47 + 3.86 = \$13.33

27. \$7.49 + 0.87 = \$8.36

28. \$7.93 + 0.08 = \$8.01

29. \$0.76 + 0.94 + 0.07 = \$1.77

30. \$8.94 + 0.75 + 9.27 = \$18.96

31. \$1.23 + 4.56 + 0.10 = \$5.89

32. \$9.87 + 6.54 + 0.46 = \$16.87

PROBLEM CORNER

Use the bar graph.

1. How many cars were parked on the third floor? 88 cars

2. On which floor were the most cars parked? 1st

3. The top three floors were reserved for all-day parking.
How many cars parked all day? 166 cars

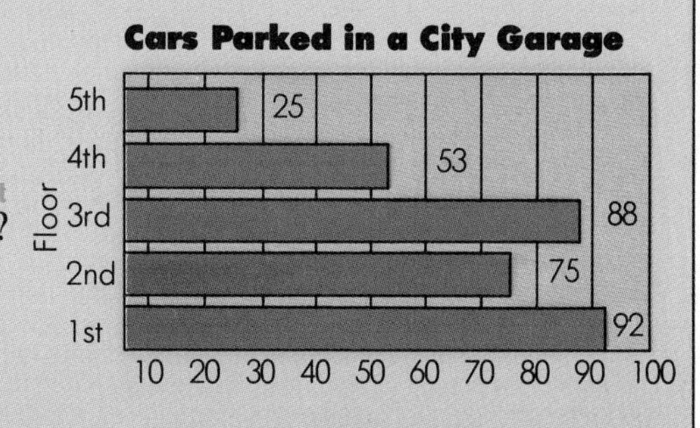

Objectives 14 and 15

ADDING LARGER NUMBERS

Add. 6297 + 8059

To be successful with column addition, we must be able to add mentally a two-digit number plus a one-digit number. Use oral drill on sums like 13 + 8, 12 + 9, 26 + 7, 38 + 5, etc.

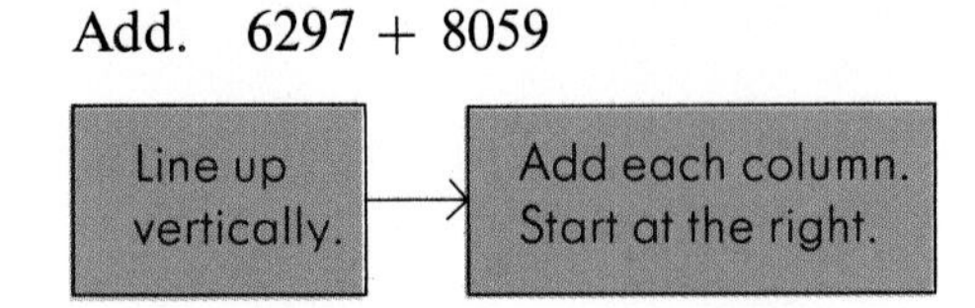

$$\begin{array}{r} 6297 \\ +\,8059 \\ \hline \end{array} \qquad \begin{array}{r} {}^{1\,1} \\ 6297 \\ +\,8059 \\ \hline 14{,}356 \end{array}$$

EXAMPLE 1. Add. 897,348 + 37,496

$$\begin{array}{r} {}^{1\,1\quad 1\,1} \\ 897{,}348 \\ +\quad 37{,}496 \\ \hline 934{,}844 \end{array}$$

Line up.
Add.

TRY THIS

Add.

1. 13,247 + 9129 22,376

2. 427,986 + 794,129 = 1,222,115

3. \$ 764.97 + 274.86 = \$1,039.83

EXAMPLE 2. Add. 734,198 + 98,374 + 9,479

$$\begin{array}{r} {}^{1\,2\,1\,2\,2} \\ 734{,}198 \\ 98{,}374 \\ +\quad 9{,}479 \\ \hline 842{,}051 \end{array}$$

Line up.
Add.

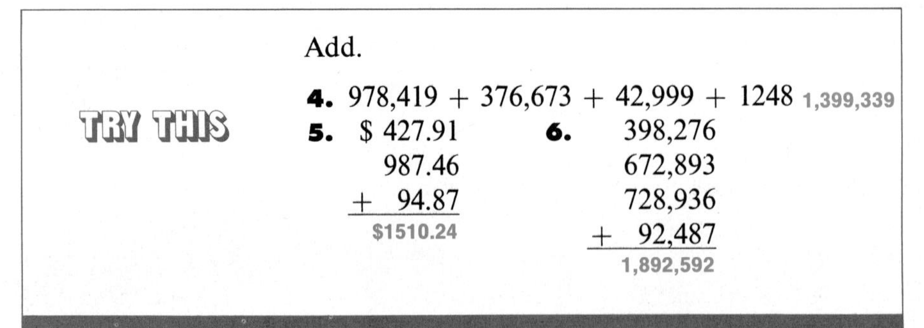

TRY THIS

Add.

4. 978,419 + 376,673 + 42,999 + 1248 1,399,339

5. \$ 427.91 + 987.46 + 94.87 = \$1510.24

6. 398,276 + 672,893 + 728,936 + 92,487 = 1,892,592

PRACTICE

For extra practice, use Making Practice Fun 4 with this lesson.

Add.

1. 7486 + 9281 16,767	**2.** 4236 + 6143 10,379	**3.** 7281 + 3409 10,690	**4.** 9876 + 1234 11,110
5. 46,287 + 94,789 141,076	**6.** 72,287 + 63,943 136,230	**7.** 74,198 + 82,798 156,996	**8.** 63,078 + 92,937 156,015
9. 498,276 + 987,409 1,485,685	**10.** 763,297 + 429,846 1,193,143	**11.** 943,217 + 629,805 1,573,022	**12.** 348,296 + 296,348 644,644
13. 73,846 + 9,017 82,863	**14.** 498,766 + 9,487 508,253	**15.** 368,499 + 386 368,885	**16.** 369,421 + 86,493 455,914
17. 14,798 29,287 + 98,075 142,160	**18.** 424,688 327,941 + 327,198 1,079,827	**19.** 8294 3789 + 6081 18,164	**20.** 728,495 62,481 + 9,987 800,963
21. 9824 7698 5247 + 4098 26,867	**22.** 18,476 29,487 38,265 + 47,154 133,382	**23.** 268,497 862,794 321,543 + 629,980 2,082,814	**24.** 798,476 420,895 15,746 + 9,285 1,244,402
25. \$ 85.98 + 37.68 \$123.66	**26.** \$ 246.97 + 579.23 \$826.20	**27.** \$ 17.47 + 9.88 \$27.35	**28.** \$ 258.17 + 9.49 \$267.66

29. \$ 16.49 + \$ 87.68 + \$ 15.45 **\$119.62**

30. \$ 37.48 + \$ 26.49 + \$ 5.97 + \$ 0.74 **\$70.68**

PROBLEM CORNER

Watch for extra information.

1. Mrs. James bought a dress for \$ 19.95, marked down from \$ 30.00. She also bought a scarf for \$ 3.29. How much did she spend in all? **\$23.24**

2. There were 29,348 adults and 15,478 children at the game. There were seats for 54,345 people. How many people were at the game? **44,826 people**

Objective 151

APPLICATION

MAKING DEPOSIT SLIPS

It is often possible to obtain sample deposit slips from local banks.

When you deposit money in your checking account you must fill out a deposit slip.

1. Total paper money

2. Total coins

3. List checks separately.

For Deposit to the Checking Account of
Sally Moore
42 Iceberg Lane
Someplace, Alaska 99506

Date Oct. 20, 1981

Sign Here
for Less Cash ____________

NOW NATIONAL BANK
532 Icycle Street
Cold, Alaska 99502

5431 . . . 0092:114 . . . 23538 . . . 00265

Account Number

		Dollars	Cents
Cash	Currency	18	00
	Coin	7	25
Checks	26-28	320	87
	94-15	19	75
Total		365	87
Less Cash Received			-0-
Net Deposit		365	87

6. Amount of deposit

5. Subtract cash received.

4. Add money and checks.

These hints refer to the numbered items above.

CONSUMER HINTS

1. Arrange bills from greatest on top to least.
2. Put coins in paper rolls, if you have a large number of coins.
3. Endorse each check in ink. List them separately by bank number and amount. See page 4 for the location of the bank number.
4. Add currency, coins and checks.
5. Subtract the amount you want back. Sign your name to indicate you received the money.
6. Record the amount actually deposited.

Find the net deposits. Provide duplicated copies of deposit slips or sample bank deposit slips.

1.

		Dollars	Cents
Cash	Currency	12	00
	Coin	7	45
Checks	21-17	18	24
	16-12	7	98
Total			
Less Cash Received			
Net Deposit		45	67

2.

		Dollars	Cents
Cash	Currency	5	00
	Coin		50
Checks	21-87	8	75
Total			
Less Cash Received			
Net Deposit		14	25

3.

		Dollars	Cents
Cash	Currency		
	Coin		
Checks	19-29	7	49
	18-26	14	98
Total			
Less Cash Received			
Net Deposit		22	47

4.

		Dollars	Cents
Cash	Currency	8	00
	Coin		
Checks	21-86	195	00
Total			
Less Cash Received			
Net Deposit		203	00

5.

		Dollars	Cents
Cash	Currency	28	00
	Coin	18	95
Checks			
Total			
Less Cash Received			
Net Deposit		46	95

6.

		Dollars	Cents
Cash	Currency	10	00
	Coin	7	86
Checks	59-18	185	45
	62-52	30	00
	28-21	156	00
Total			
Less Cash Received			
Net Deposit		389	31

Make copies of deposit slips like the ones above. Write these deposits. See answer section.

7. 1 ten-dollar bill, 2 five-dollar bills, 3 one-dollar bills, 3 half-dollar coins, 1 quarter, 2 dimes, 1 nickel, check 94-18 for $ 19.50, check 25-16 for $ 75.00.

8. 2 twenty-dollar bills, 3 ten-dollar bills, 7 one-dollar bills, 1 two-dollar roll of nickels, 2 ten-dollar rolls of dimes, check 46-3 for $ 17.85.

9. 10 half-dollars, 8 nickels, check 21-20 for $ 127.05, check 24-20 for $ 84.46.

10. 2 one-hundred-dollar bills, 3 fifty-dollar bills, 5 ten-dollar bills, check 92-18 for $ 375.00, check 52-12 for $ 795.95.

Objectives 16, 17, and 18

DECIMAL PLACE VALUE

The decimal point separates the whole number from the part less than 1.

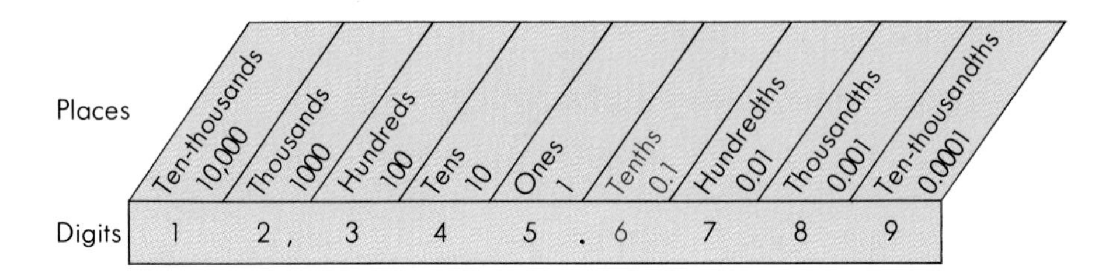

The 6 is in the tenths' place. It means 6 × 0.1 (digit times place), so its value is 0.6.

EXAMPLE 1. Find the place and value of the underlined digit.

7.04<u>8</u>2 The 8 is in the thousandths' place. Its value is 0.008.

TRY THIS Find the place and value of the underlined digit. See answer section.

1. 73.<u>4</u>294 **2.** 6.7<u>9</u>6 **3.** 927.48<u>6</u>3

To write short word names we write *and* for the decimal point. We use the name of the last decimal place. In some applications we say *point* instead of *and* for the decimal point. Example: 6 point 5

EXAMPLES. Write short word names.

2. 6.5 6 and 5 tenths
3. 6.54 6 and 54 hundredths
4. 6.05 6 and 5 hundredths
5. 6.543 6 and 543 thousandths
6. 6.1002 6 and 1002 ten-thousandths
7. 0.0007 7 ten-thousandths

TRY THIS Write short word names.

4. 7.3 7 and 3 tenths
5. 27.386 27 and 386 thousandths
6. 7.0492 7 and 492 ten-thousandths

EXAMPLES. Write decimal notation.

8. 9 and 6 tenths 9.6
9. 7 and 42 hundredths 7.42
10. 87 thousandths 0.087 *Write 0's to fill the places.*
11. 5 and 126 ten-thousandths 5.0126

TRY THIS Write decimal notation.

7. 6 hundredths 0.06 **8.** 47 and 58 thousandths 47.058

PRACTICE

Find the place and value of the underlined digit. See answer section.

1. $9.\underline{4}872$ **2.** $47.\underline{4}6$ **3.** $0.\underline{5}98$ **4.** $376.\underline{5}$

5. $46.9\underline{7}24$ **6.** $5.07\underline{6}$ **7.** $31.907\underline{5}$ **8.** $286.\underline{4}98$

9. $94.704\underline{9}$ **10.** $8.0\underline{1}24$ **11.** $0.00\underline{6}$ **12.** $3.123\underline{4}$

Write short word names. See answer section.

13. 3.7 **14.** 49.8 **15.** 73.28 **16.** 9.326

17. 17.4295 **18.** 95.0875 **19.** 0.38 **20.** 1.141

21. 295.6 **22.** 36.9853 **23.** 8.7 **24.** 2.74

25. 3.095 **26.** 0.785 **27.** 7.0007 **28.** 7000.7

Write decimal notation.

29. 7 and 15 hundredths 7.15 **30.** 95 and 7 tenths 95.7

31. 16 and 7 hundredths 16.07 **32.** 3 and 3 thousandths 3.003

33. 26 and 275 thousandths 26.275 **34.** 1 ten-thousandth 0.0001

35. 8 and 27 hundredths 8.27 **36.** 5 tenths 0.5

37. 784 thousandths 0.784 **38.** 19 and 48 thousandths 19.048

More Practice, page 35

Objective 19

ADDING DECIMALS

In decimals between 1 and −1, we write 0 before the decimal point, so that the decimal point will not be overlooked.

Add. 21.668 + 5.491

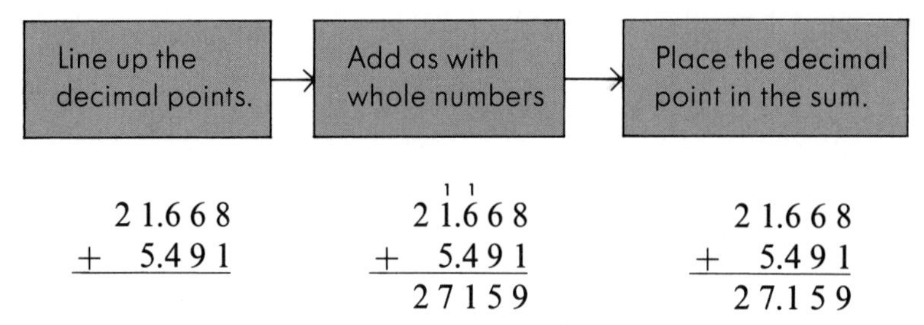

EXAMPLE 1. Add. 18.9 + 9.376

$$\begin{array}{r} \overset{1}{1}\overset{1}{8}.900 \\ +\ \ 9.376 \\ \hline 28.276 \end{array}$$

Line up the decimal points.
Write 0's to fill the places.
Add.

TRY THIS

Add.

1. 18.76 + 19.9 38.66
2. 9.786 + 0.895 10.681

3.
$$\begin{array}{r} 165.97 \\ +\ \ 36.075 \\ \hline 202.045 \end{array}$$

4.
$$\begin{array}{r} 0.765 \\ +\ 0.89 \\ \hline 1.655 \end{array}$$

EXAMPLE 2. Add. 9.7 + 0.89 + 16.756 + 8.6

$$\begin{array}{r} ^{2}\ \overset{2}{9}.\overset{1}{7}00 \\ 0.890 \\ 16.756 \\ +\ \ 8.600 \\ \hline 35.946 \end{array}$$

Write 0's to fill the places.

Add.

Place the decimal point in the sum.

TRY THIS

Add.

5. 4.3 + 0.876 + 21.42 26.596 **6.** 0.96 + 0.096 + 9.6 10.656

7.
$$\begin{array}{r} 4.95 \\ 0.879 \\ 19.8 \\ +\ \ 0.007 \\ \hline 25.636 \end{array}$$

8.
$$\begin{array}{r} 0.15 \\ 0.19 \\ 0.875 \\ +\ 0.919 \\ \hline 2.134 \end{array}$$

PRACTICE

For extra practice, use Making Practice Fun 5 with this lesson.
Use Quiz 2 after this Practice.

Add.

1. 19.6 + 7.9 27.5
2. 27.9 + 16.1 44.0
3. 37.3 + 19.8 57.1
4. 0.9 + 8.8 9.7

5. 47.98 + 16.84 64.82
6. 2.87 + 9.03 11.90
7. 6.87 + 9.9 16.77
8. 3.76 + 0.9 4.66

9. 7.398 + 0.475 = 7.873
10. 37.485 + 18.08 = 55.565
11. 2.786 + 0.9 = 3.686
12. 63.284 + 19.986 = 83.270

13. 8.736 + 9.49 = 18.226
14. 36.892 + 18.08 = 54.972
15. 0.7 + 0.6 = 1.3
16. 297.8 + 96.88 = 394.68

17. 13.767 + 4.382 + 7.689 = 25.838
18. 3.4 + 9.6 + 7.3 = 20.3
19. 7.86 + 4.08 + 0.9 = 12.84
20. 13.621 + 9.78 + 0.5 = 23.901

21. 0.78 + 6.94 + 8.72 + 4.81 = 21.25
22. 16.284 + 9.786 + 4.395 + 5.007 = 35.472
23. 9.8 + 19.4 + 0.98 + 5.9 = 36.08
24. 874.1 + 87.41 + 8.741 + 0.874 = 971.125

25. 7.9 + 6.84 + 14.3 + 8.9 = 37.94
26. 16.9 + 3.18 + 0.8 + 0.9 = 21.78
27. 0.68 + 6.8 + 3.9 + 0.395 = 11.775
28. 4.875 + 0.129 + 97.6 + 0.976 = 103.580

PROBLEM CORNER

Use the chart.

1. Which city usually gets more rain in January? June? Los Angeles; Albany

2. How much rain does Los Angeles usually get in the winter? summer? 18.36 cm; 0.16 cm

3. Does Albany get more rain in December or July? July

Average Expected Rainfall in Centimeters

	Winter			Summer		
	Dec	Jan	Feb	Jun	Jul	Aug
Los Angeles, Cal.	6.07	6.4	5.89	0.08	0.03	0.05
Albany, N.Y.	7.44	5.59	5.36	7.62	7.92	7.29

Objective 196

SOLVING PROBLEMS

These problem solving steps could be used as the basis of a bulletin board display. Refer constantly to these steps in each problem solving situation.

When solving problems, keep these steps in mind.

1. UNDERSTAND What must I find?
What information is given?
Do I have enough information?

2. PLAN Have I solved a problem like this before?
Should I restate the problem?
What operations are needed?
Is an approximate answer sufficient?
Should I use a calculator?

3. CARRY OUT Did I calculate correctly?
Did I check my calculations?

4. LOOK BACK Is my answer reasonable?
Does the answer fit the problem?

EXAMPLE. Mr. and Mrs. Capan plan to rent an apartment. The basic rent is \$ 310 with a \$ 24 maintenance fee. A \$ 55 security deposit is required. What is the total payment for the first month?

UNDERSTAND I must find the total.
I have enough information.

PLAN I must add to find the total.

CARRY OUT

$$\begin{array}{r} \$\ 310 \\ 24 \\ +\ \ 55 \\ \hline \$\ 389 \end{array}$$

It checks.

LOOK BACK The answer is reasonable.
It makes sense in the problem.

TRY THIS

Terry, a college student, rents an efficiency apartment for $ 145 a month. There is a charge of $ 23 for electricity and gas. Find the total monthly payment. $168

PRACTICE

Before assigning the practice exercises, have your students read the problems silently and then orally.
For extra practice, use Making Practice Fun 6 with this lesson.

1. Makato and Yin Lee bought a new house. The closing costs were:

Title Fee	$ 143.90
State Stamps	54.50
Transfer Tax	304.00

You may need to discuss the idea of closing costs with your students.

Find the total closing costs. $502.40

2. Phyllis paid auto insurance costs of $ 115 for bodily injury liability, $ 72 for property damage, and $ 118 for collision. What was her total annual premium? $305

3. The Emperor Jahangir of India died in 1627. He owned 931,500 carats of emeralds, 376,600 carats of rubies, 279,450 carats of diamonds, and 186,300 carats of jade. How many carats of gems did he own? 1,773,850

4. Recent records show there are 4,600,000 men in the ABC (American Bowling Congress). There are 4,034,000 women in the WIBC (Women's International Bowling Congress). How many bowlers are in these two organizations? 8,634,000

5. Bob Hart bought an old car for $ 325.85. The materials to restore and repair it cost $ 527.47. What was his cash investment? $853.32

6. Cecilia Ortiz bought a pair of shoes for $ 36.50, a dress for $ 47.17, and a purse for $ 21.83. How much did she spend in all? $105.50

Objectives 20 and 21

ROUNDING NUMBERS

EXAMPLE 1. Round to the nearest thousand. 4 3 6,8 2 7

Find the place to be rounded. The digit to the right is 5 or more.

4 3 6,8 2 7

Round up. Change to 0's.

4 3 7,0 0 0

EXAMPLE 2. Round to the nearest hundred-thousand. 4 3 6,8 2 7

Find the place to be rounded. The digit to the right is less than 5.

4 3 6,8 2 7

Keep the digit the same. Change to 0's.

4 0 0,0 0 0

TRY THIS

Round to the nearest hundred.

1. 486 500 **2.** 6713 6700 **3.** 14,354 14,400

EXAMPLE 3. Round to the nearest tenth. 7.3 7 2

Find the place to be rounded. The digit to the right is 5 or more.

7.3 7 2

Round up. Drop the digits to the right.

7.4

EXAMPLE 4. Round to the nearest whole number. 2 8.3 9 1

Find the ones' place. The digit to the right is less than 5.

2 8.3 9 1

Keep the digit the same. Drop the digits to the right.

2 8

EXAMPLE 5. Round to the nearest dollar. $ 1 2 9.5 0

When the number to be rounded up is 9, the next place is also rounded up. So, to the nearest dollar, $1999.80 is $2000.

Dollar. The place to the right is 5 or more.

$ 1 2 9.5 0

Round up. Drop the digits to the right.

$ 1 3 0

TRY THIS

Round to the nearest hundredth.

4. 7.372 7.37 **5.** 9.4891 9.49

Round to the nearest whole number.

6. 8.7 9 **7.** 342.3259 342

Round to the nearest dollar.

8. $3.75 $4 **9.** $41.28 $41

PRACTICE

Round to the nearest ten.

1. 48 50 **2.** 859 860 **3.** 596 600 **4.** 1412 1410

Round to the nearest hundred.

5. 563 600 **6.** 215 200 **7.** 32,850 32,900 **8.** 43,119 43,100

Round to the nearest thousand.

9. 5482 5000 **10.** 5582 6000 **11.** 428,750 429,000 **12.** 870,991 871,000

Round to the nearest tenth.

13. 8.71 8.7 **14.** 0.85 0.9 **15.** 14.98 15.0 **16.** 3.7099 3.7

Round to the nearest hundredth.

17. 8.765 8.77 **18.** 0.423 0.42 **19.** 17.396 17.40 **20.** 0.0075 0.01

Round to the nearest whole number.

21. 7.3 7 **22.** 8.9 9 **23.** 91.476 91 **24.** 15.72 16

Round to the nearest dollar.

25. $8.88 $9 **26.** $17.45 $17 **27.** $29.51 $30 **28.** 225.50 $226

29. $53.29 $53 **30.** $124.32 $124 **31.** $37.67 $38 **32.** $1.15 $1

Objectives 22 and 23

ESTIMATING SUMS

There are no commonly agreed-on rules for estimating. These rules help students round to a given place. Thus they avoid estimates like 32,657 + 91 = 30,000 + 90 = 30,090.

Estimate the sum. 436 + 297 + 548

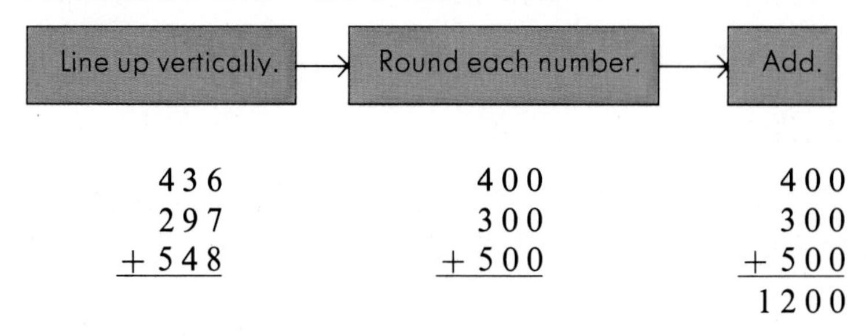


```
  436        400        400
  297        300        300
+ 548      + 500      + 500
                       1200
```

We usually round to the highest place in the smallest number.

EXAMPLES. Estimate the sums.

The highest place in 31 is tens, so round to tens.

```
1.   426     430     2.   4296     4000
      31      30          3500     4000
   +  85   +  90        + 7301   + 7000
             550                 15,000
```

```
3.   17.8     18
      3.14     3
   +  6.359 +  6
              27
```

Round each number to the nearest whole number. Add.

```
4. $ 29.15   $ 29
     32.75     33
   +  5.98   +  6
             $ 68
```

Round each amount to the nearest dollar. Add.

TRY THIS

Estimate the sums.

```
1.   78         2.   4568
     25               379
   + 34             + 615
    140              5600

3.  36.781      4.  $ 36.49
     9.31              5.50
   + 6.8             + 7.11
      53               $49
```

PRACTICE

For extra practice, use Making Practice Fun 7 with this lesson.
Use Quiz 3 after this Practice.

Estimate the sums.

1. 27 + 34 = 60	**2.** 39 + 28 = 70	**3.** 748 + 258 = 1000	**4.** 698 + 307 = 1000
5. 9286 + 6148 = 15,000	**6.** 4736 + 8367 = 13,000	**7.** 3,748,976 + 3,487,296 = 7,000,000	**8.** 4,286,987 + 8,628,146 = 13,000,000
9. 74, 28, + 78 = 180	**10.** 348, 296, 755, + 508 = 1900	**11.** 2748, 3476, 4985, + 6297 = 17,000	**12.** 3,287,496, 5,349,891, + 6,548,196 = 15,000,000
13. 871, 364, + 58 = 1290	**14.** 725, 75, + 86 = 900	**15.** 4297, 385, + 115 = 4800	**16.** 7284, 6813, + 743 = 14,800
17. 8.7 + 6.5 = 16	**18.** 8.743 + 3.289 = 12	**19.** 15.87 + 7.3 = 23	**20.** 36.784 + 45.39 = 82
21. 3.9, 4.8, + 8.9 = 18	**22.** 3.76, 5.98, + 7.42 = 17	**23.** 1.742, 3.9, + 6.1 = 12	**24.** 15.8, 9.87, + 3.147 = 29
25. $ 7.75 + 2.98 = $11	**26.** $ 8.37 + 9.65 = $18	**27.** $ 16.19 + 24.78 = $41	**28.** $ 38.79 + 73.86 = $113
29. $ 9.15, 6.25, + 8.60 = $24	**30.** $ 3.49, 2.78, + 3.85 = $10	**31.** $ 24.73, 9.68, + 5.45 = $40	**32.** $ 28.79, 38.44, + 8.50 = $76

PROBLEM CORNER

1. Estimate the cost of a skirt and two blouses. $40

2. You have $ 10. Can you buy a shirt and two ties? no
Estimate to find the answer.

CAREER CAREER CAREER CAREER CAREER

AGRICULTURAL PILOT

Agricultural pilots are sometimes called cropdusters. They do dust crops to kill insects. They also spray to kill weeds, apply fertilizer, and so on. The job is only for skillful pilots. Most of the flying is close to the ground.

Jason and Sherry Sanders own "J & S Agricultural Flying Service." One of their big expenses is airplane fuel. Their three gas storage tanks are refilled each month. They use a chart to record monthly fuel deliveries.

1. How much gas was delivered to tank #2 in May? 2876 L
2. How much gas was put in the three tanks in March? 4593 L
3. How much gas was put in the three tanks in April? 5499 L
4. In which month was more gas used, May or June? June

Monthly Fuel Deliveries – 1983 (in liters)

	Mar 30	Apr 30	May 30	Jun 30
Tank #1	1138	2210	3344	3461
Tank #2	2161	2131	2876	2827
Tank #3	1294	1158	3851	3798
Total	4593			

CAREER CAREER CAREER CAREER CAREER

J & S Service can be hired to check irrigation ditches. They fly low to spot any problems with the ditches. This is faster than checking the ditches by truck. This map shows some check routes. The distances are given in miles.

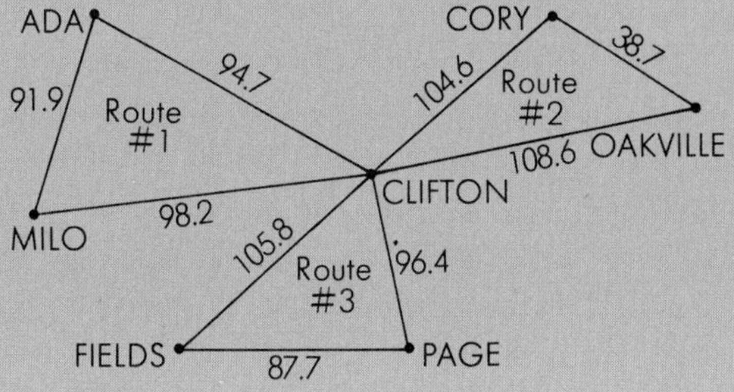

5. How long is Route #2? 251.9 miles
6. Which is longer, Route #1 or Route #3? Route #3
7. One plane flew all three routes. Estimate the distance flown.
   825 miles

Ed Bluespruce is a farmer. He wants to have his fields sprayed to prevent insects. He compares estimates for two spraying methods.

J & S Estimate	
Time	3 hours
Chemicals	$165
Pilot and Plane	180
Assistant	30
Total	

Estimate of Ground Spraying	
Time	15 hours
Sprayer Rental	$ 75
Truck Costs	35
Labor	75
Chemicals	140
Total	

8. Which method takes less time? J & S
9. Which method costs less? ground spraying

On Aug. 7, 1981, J & S Service had $ 352.48 in a checking account. Jason wanted to buy the following items from Chemical Supply Corp.

1 barrel	P2-7	$ 92.53
1 barrel	TB804	$ 118.18
100 liters	filler fluid	$ 51.95
50 kg	pest powder	$ 40.50

10. Estimate Jason's cost. $300
11. Can Jason pay by check before making an account deposit? yes
12. Calculate Jason's exact cost. $303.16
13. Using the information above, copy and complete a check like the one on page 11.
    See answer section

CHAPTER 1 REVIEW

For extra practice, use Making Practice Fun 8 with this lesson.

Write standard notation. p. 6

1. nine thousand, four hundred seventy-one 9471
2. eight thousand, six 8006
3. 320 thousand, 720 320,720
4. 800 million, 8 thousand 800,008,000

Write money notation. p. 8

5. eighty-six dollars and fifteen cents $86.15
6. forty dollars and five cents $40.05

Write decimal notation. p. 20

7. 18 and 14 hundredths 18.14
8. 7 and 6 thousandths 7.006

Add. pp. 14–22

9. 376 + 984 = 1360
10. 9824 + 6958 = 16,782
11. 732 + 599 + 57 + 6 = 1394
12. $8.95 + 6.87 + 0.78 = $16.60
13. 5.4 + 8.7 = 14.1
14. 17.291 + 8.789 = 26.080
15. 4.72 + 3.9 + 1.426 = 10.046
16. 18.984 + 7.3 + 8.47 + 0.987 = 35.741

Estimate the sums. p. 28

17. 2725 + 3895 + 4350 = 11,000
18. 428 + 67 + 41 = 540
19. $8.75 + 2.30 + 1.98 = $13.00
20. 4.2 + 8.75 + 3.199 = 16

Application pp. 10, 18

21. Write a check from Charles Jones to J and S Agricultural Service in the amount of $725.75. Date it December 14, 1980. See answer section.

22. Find the net deposit. $202.03

		Dollars	Cents
Cash	Currency	25	00
	Coin	6	10
Checks	16-28	37	48
	11-32	128	45
	17-94	5	00
Total			
Less Cash Received			-0-
Net Deposit			

CHAPTER 1 TEST

Write standard notation.

1. two thousand, six hundred fifty-eight 2658
2. nine thousand, ninety 9090
3. 386 thousand, 317 386,317
4. 40 million, 400 thousand 40,400,000

Write money notation.

5. seventy-four dollars and fifty cents \$74.50
6. sixteen dollars and seven cents \$16.07

Write decimal notation.

7. 6 and 8 tenths 6.8
8. 29 and 123 ten-thousandths 29.0123

Add.

9. $429 + 738$ 1167

10. $4298 + 8764$ 13,062

11. $862 + 291 + 87 + 9$ 1249

12. $\$7.49 + 8.50 + 0.73$ \$16.72

13. $9.72 + 8.69$ 18.41

14. $3.429 + 0.984$ 4.413

15. $5.762 + 8.7 + 0.94$ 15.402

16. $37.382 + 7.48 + 9.9 + 8.739$ 63.501

Estimate the sums.

17. $498 + 878 + 212$ 1600

18. $7384 + 921 + 650$ 9000

19. $\$9.34 + 6.50 + 7.75$ \$24.00

20. $7.6 + 8.34 + 6.149$ 22

Application

21. Write a check to Ace Chemical Company from Sherie Sanders in the amount of \$ 78.50. Date it March 22, 1983. See answer section.

22. Find the net deposit. \$177.43

		Dollars	Cents
Cash	Currency	5	00
	Coin	8	75
Checks	19-25	18	00
	8-406	10	00
	4-15	135	68
Total			
Less Cash Received			-0-
Net Deposit			

MORE PRACTICE

Write standard notation. Use after page 7.

1. six hundred forty-three 643

2. five hundred ninety-eight 598

3. one thousand, six hundred ten 1610

4. eight thousand, eighty 8080

5. three thousand, five hundred 3500

6. nine thousand, nine 9009

7. twelve thousand, ninety-nine 12,099

8. ten thousand, two hundred eight 10,208

9. eighty thousand, one hundred eleven 80,111

10. twenty-six thousand, four hundred nineteen 26,419

Write money notation. Use after page 9.

11. eight dollars and ninety-five cents $8.95

12. seventeen dollars $17.00

13. eighty-one dollars and fifty cents $81.50

14. thirty dollars and six cents $30.06

15. one hundred twenty dollars and twenty cents $120.20

16. seventy-seven dollars and seven cents $77.07

17. five dollars and eighty-nine cents $5.89

18. sixty-five dollars $65.00

19. seventeen dollars and ninety-eight cents $17.98

20. two hundred dollars and sixty cents $200.60

Write standard notation. Use after page 13.

21. 112 thousand, 600 112,600

22. 800 thousand, 4 800,004

23. 26 million, 455 thousand 26,455,000

24. 11 million 11,000,000

Add. Use after page 15.

25. $\begin{array}{r} 74 \\ +\ 96 \\ \hline \end{array}$ 170

26. $\begin{array}{r} 49 \\ +\ 5 \\ \hline \end{array}$ 54

27. $\begin{array}{r} 746 \\ +\ 298 \\ \hline \end{array}$ 1044

28. $\begin{array}{r} 348 \\ +\ 67 \\ \hline \end{array}$ 415

29. $\begin{array}{r} 348 \\ 267 \\ +\ 98 \\ \hline \end{array}$ 713

30. $\begin{array}{r} 948 \\ 76 \\ 18 \\ +\ 7 \\ \hline \end{array}$ 1049

31. $\begin{array}{r} \$\ 4.78 \\ 2.97 \\ +\ 6.49 \\ \hline \end{array}$ $14.24

32. $\begin{array}{r} \$\ 8.47 \\ 0.49 \\ 6.40 \\ +\ 0.86 \\ \hline \end{array}$ $16.22

Add. Use after page 17.

1. 7284 + 3796 = 11,080

2. 14,781 + 39,499 = 54,280

3. 347,821 + 981,294 = 1,329,115

4. 349,278 + 97,146 = 446,424

5. 7269 + 3784 + 6985 + 3784 = 21,822

6. 47,264 + 69,209 + 15,875 = 132,348

7. 349,287 + 190,180 + 762,498 + 875,287 = 2,177,252

8. $ 746.84 + 198.78 + 58.74 = $1004.36

Write decimal notation. Use after page 21.

9. 1 and 6 tenths 1.6

10. 12 and 24 hundredths 12.24

11. 6 and 7 hundredths 6.07

12. 10 and 326 thousandths 10.326

13. 4 and 28 thousandths 4.028

14. 3 and 4248 ten-thousandths 3.4248

15. 86 thousandths 0.086

16. 9 tenths 0.9

17. 9 hundredths 0.09

18. 9 ten-thousandths 0.0009

Add. Use after page 23.

19. 8.7 + 9.2 = 17.9

20. 6.4 + 8.7 = 15.1

21. 3.78 + 9.65 = 13.43

22. 0.918 + 6.942 = 7.860

23. 4.98 + 3.7 + 5.876 = 14.556

24. 9.7 + 6.82 + 9.487 = 26.007

25. 7.328 + 0.09 + 1.4268 + 0.6 = 9.4448

26. 4.298 + 5.29 + 6.2 + 7.0 = 22.788

27. 5.62 + 0.039 + 37.8 43.459

28. 0.2 + 29.6 + 7.12 + 0.594 37.514

Estimate the sums. Use after page 29.

29. 86 + 94 + 38 = 220

30. 376 + 298 + 815 = 1500

31. 324 + 58 + 63 = 440

32. 4286 + 750 + 310 = 5400

33. $ 8.76 + 9.48 = $18.00

34. $ 3.98 + 2.47 + 6.50 + 1.25 = $14.00

35. 1.78 + 3.25 = 5

36. 4.298 + 7.9 + 6.48 + 7.007 = 25

CALCULATOR CORNER

ADDITION ESTIMATION

Players two
Materials two calculators, paper and pencils

SAMPLE

1. Nick asks Devon to estimate a sum, such as 422 + 576 + 182. Devon asks Nick to estimate a sum, such as 386 + 94 + 214.
2. Devon and Nick write down their problems. They estimate and add.
3. Devon and Nick calculate the exact answers to their problems.
4. They then calculate the differences between their estimates and the exact answers.

Devon

422	400
576	600
+ 182	200
	1200
estimate	1200
exact answer	− 1180
	20

Nick

386	390
94	90
+ 214	210
	690
exact answer	694
estimate	− 690
	4

Since Nick's estimate was closer to the exact answer, he gets 1 point.

Play continues until a player gets 5 points.

It is important to have your students actually do the estimating. Students should realize that estimates are not right or wrong. They are only close to or far from the exact solution.

The choice of the level of sums to estimate will depend on the class.

If the supply of calculators is limited, you could adapt the activity by having the class determine estimates of a sum. Use one calculator to calculate the exact sum. Students could then compare their estimates with this answer.

CHAPTER 2 PRE-TEST

Subtract. pp. 40–50

1. 85 − 23 62	**2.** 72 − 39 33	**3.** 421 − 368 53	**4.** \$ 6.48 − 2.89 \$3.59
5. 6287 − 1938 4349	**6.** 62,486 − 19,197 43,289	**7.** 427,685 − 388,747 38,938	**8.** \$ 238.46 − 95.88 \$142.58
9. 400 − 146 254	**10.** 5001 − 3247 1754	**11.** 75,003 − 19,437 55,566	**12.** \$ 10.00 − 2.45 \$7.55
13. 7.43 − 4.68 2.75	**14.** 3.786 − 1.9 1.886	**15.** 6.3 − 1.428 4.872	**16.** 8 − 7.34 0.66

Compare. Use =, >, or <. p. 48

17. 0.2 and 0.24 < **18.** 0.7 and 0.07 > **19.** 0.87 and 0.870 =

20. 0.2 and 0.200 = **21.** 0.3 and 0.29 > **22.** 0.367 and 0.37 <

23. 0.62 and 0.620 = **24.** 0.03 and 0.003 > **25.** 0.465 and 0.473 <

Estimate the differences. p. 54

26. 471 − 182 300	**27.** 6349 − 2481 4000	**28.** 758 − 86 670	**29.** 9.1 − 3.986 5

Application p. 46

Find the balance carried forward.

30. Date *July 5,* 19*82* *\$72.46* 291
To *Foreign Auto Repair*
For *tune-up*

	Dollars	Cents
Balance Forward	348	26
Amount of Deposit		-0-
Total		
Amount of Check		
Balance Carried Forward	275	80

31. Date *Aug. 16,* 19*82* *\$28.50* 325
To *Long's Market*
For *groceries*

	Dollars	Cents
Balance Forward	268	.74
Amount of Deposit	198	70
Total		
Amount of Check		
Balance Carried Forward	438	94

2
SUBTRACTING WHOLE NUMBERS AND DECIMALS

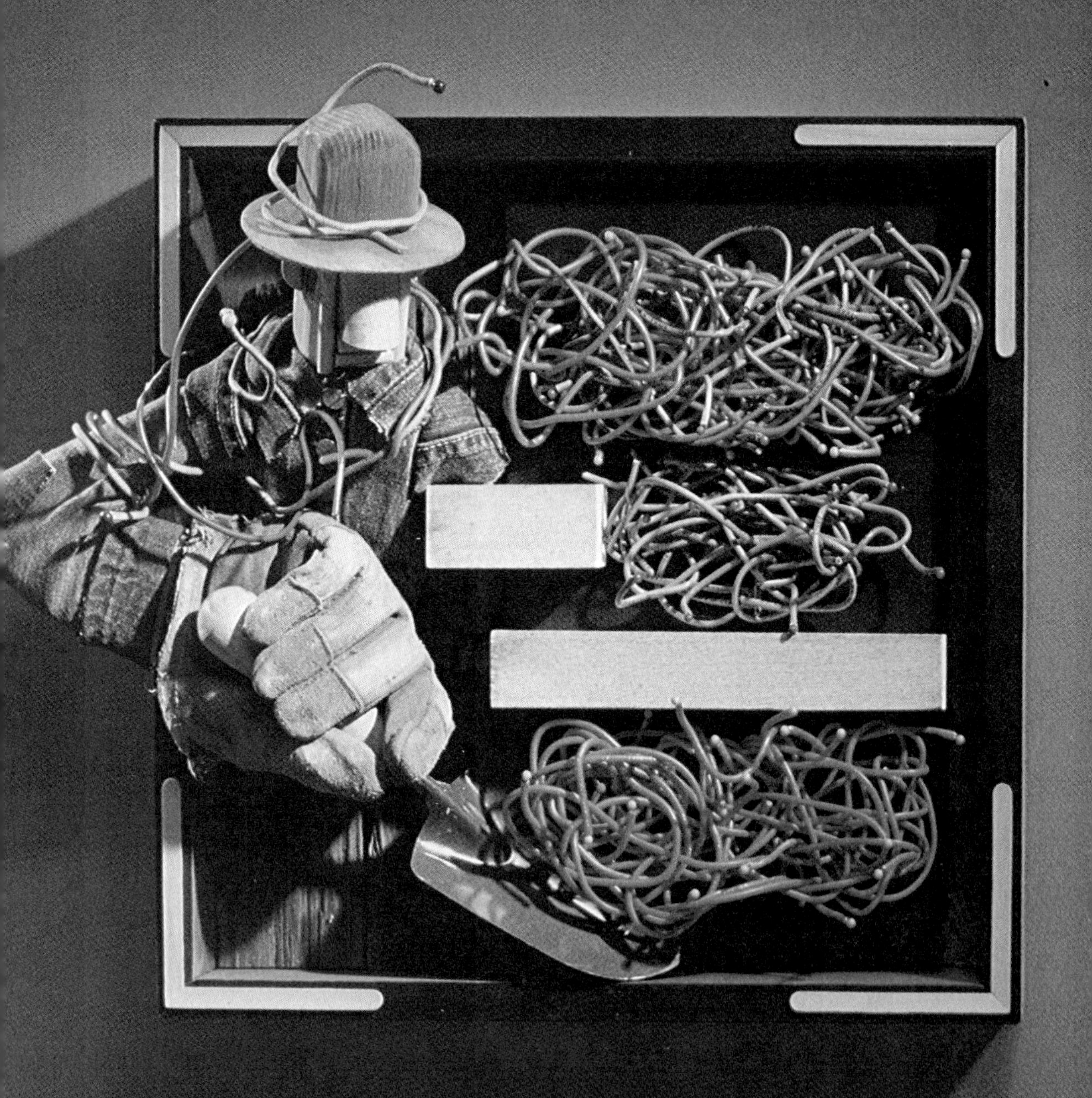

GETTING STARTED

REVERSE GOTCHA

WHAT'S NEEDED The 30 squares of paper with digits marked on them as in GOTCHA, Chapter 1.

THE RULES

1. On a sheet of paper each player makes a scorecard like this. Play three games and add to find the total score. The player with the smallest total is the winner.
2. To play, a leader draws a digit from the box. Each player subtracts it from 99 and writes the answer. As more digits are drawn, players continue to subtract.
3. When 0 is drawn it is a GOTCHA. All players still in the game go back to 99. They score 99 for the game.
4. A player may go out or quit before any draw. Try to get the smallest score but go out before 0 is drawn.

Scorecard

1. ______
2. ______
3. ______

Total ______

SAMPLE

Game 1	Game 2	Game 3
99	99	99
− 6	− 9	− 8
93	90	91
− 9	− 8	− 8
84	82	83
− 8	− 9	− 7
76 go out	73	76
	− 0 GOTCHA	− 9
		67 go out

Scorecard

1. 76
2. 99
3. 67

Total 242

Objectives 24 and 25

SUBTRACTING WHOLE NUMBERS

Notice that problems involving 0's in the upper numbers are presented in the third lesson of this chapter.

Subtract. 486 − 157

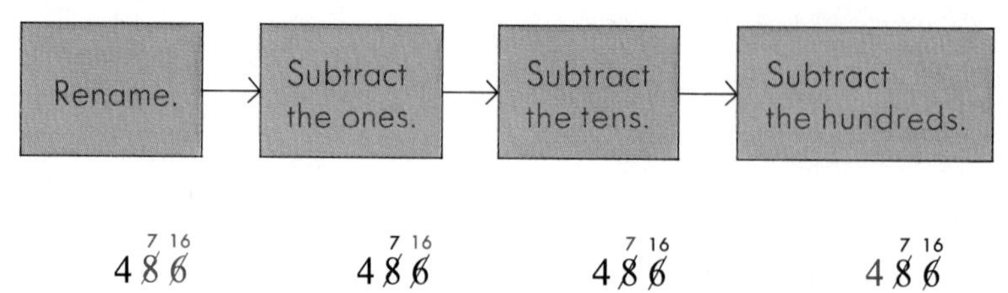

$$\begin{array}{r} \overset{7\ 16}{4\,\cancel{8}\,\cancel{6}} \\ -\,1\,5\,7 \\ \hline \end{array} \qquad \begin{array}{r} \overset{7\ 16}{4\,\cancel{8}\,\cancel{6}} \\ -\,1\,5\,7 \\ \hline 9 \end{array} \qquad \begin{array}{r} \overset{7\ 16}{4\,\cancel{8}\,\cancel{6}} \\ -\,1\,5\,7 \\ \hline 2\,9 \end{array} \qquad \begin{array}{r} \overset{7\ 16}{4\,\cancel{8}\,\cancel{6}} \\ -\,1\,5\,7 \\ \hline 3\,2\,9 \end{array}$$

EXAMPLE 1. Subtract. 759 − 327

$$\begin{array}{r} 7\,5\,9 \\ -\,3\,2\,7 \\ \hline 4\,3\,2 \end{array}$$

Subtract the ones.
Subtract the tens.
Subtract the hundreds.

EXAMPLE 2. Subtract. 942 − 148

$$\begin{array}{r} \overset{8\ 13\ 12}{\cancel{9}\,\cancel{4}\,\cancel{2}} \\ -\,1\,4\,8 \\ \hline 7\,9\,4 \end{array}$$

Rename. Subtract the ones.
Rename. Subtract the tens.
Subtract the hundreds.

TRY THIS Subtract.

1. 78 − 24 = 54   **2.** 876 − 45 = 831   **3.** 78 − 39 = 39   **4.** 673 − 385 = 288

EXAMPLE 3. Subtract. \$ 8.48 − \$ 3.29

$$\begin{array}{r} \$\ 8.\overset{3}{\cancel{4}}\overset{18}{\cancel{8}} \\ -\ 3.2\,9 \\ \hline \$\ 5.1\,9 \end{array}$$

Subtract as with whole numbers.
Write money notation in the answer.

TRY THIS Subtract.

5. \$ 7.67 − 1.31 = \$6.36   **6.** \$ 1.19 − 0.98 = \$0.21   **7.** \$ 7.64 − 1.89 = \$5.75

PRACTICE

For extra practice, use Making Practice Fun 9 with this lesson.

Subtract.

1. 95 − 31 = 64	**2.** 76 − 25 = 51	**3.** 86 − 40 = 46	**4.** 78 − 45 = 33
5. 426 − 123 = 303	**6.** 837 − 501 = 336	**7.** 298 − 236 = 62	**8.** 425 − 113 = 312
9. 75 − 38 = 37	**10.** 86 − 37 = 49	**11.** 42 − 17 = 25	**12.** 72 − 43 = 29
13. 725 − 118 = 607	**14.** 477 − 269 = 208	**15.** 914 − 382 = 532	**16.** 328 − 148 = 180
17. 721 − 389 = 332	**18.** 424 − 398 = 26	**19.** 724 − 377 = 347	**20.** 711 − 222 = 489
21. 83 − 8 = 75	**22.** 424 − 19 = 405	**23.** 613 − 38 = 575	**24.** 386 − 8 = 378
25. \$0.86 − 0.19 = \$0.67	**26.** \$7.62 − 3.47 = \$4.15	**27.** \$8.71 − 1.90 = \$6.81	**28.** \$9.47 − 9.38 = \$0.09
29. \$4.27 − 1.88 = \$2.39	**30.** \$3.76 − 0.87 = \$2.89	**31.** \$6.92 − 5.93 = \$0.99	**32.** \$6.50 − 3.75 = \$2.75

PROBLEM CORNER

Use the chart.

1. What is the yearly electricity cost of a tube type color TV? $35.59

2. Which is the least expensive TV to watch per year? B & W Solid State

3. How much more does it cost per year to watch a tube type color TV than a solid state color TV? $17.52

Yearly Electricity Cost For TV	
Black and White (tube type)	$19.71
Black and White (solid state)	2.74
Color (tube type)	35.59
Color (solid state)	18.07

These costs are based on 6 hours of use per day and a rate of 5¢ per kilowatt-hour.

More Practice, page 62

Objectives 25 and 26

SUBTRACTING LARGER NUMBERS

Subtract. 8148 − 2529

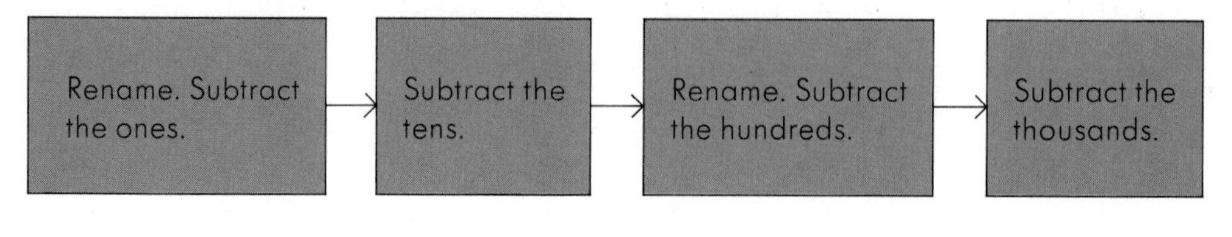

$$\begin{array}{r} {\scriptstyle 3\ 18} \\ 8148 \\ -\ 2529 \\ \hline 9 \end{array} \qquad \begin{array}{r} {\scriptstyle 3\ 18} \\ 8148 \\ -\ 2529 \\ \hline 19 \end{array} \qquad \begin{array}{r} {\scriptstyle 7\ 11\ 3\ 18} \\ 8148 \\ -\ 2529 \\ \hline 619 \end{array} \qquad \begin{array}{r} {\scriptstyle 7\ 11\ 3\ 18} \\ 8148 \\ -\ 2529 \\ \hline 5619 \end{array}$$

EXAMPLE 1. Subtract. 73,206 − 38,543

$$\begin{array}{r} {\scriptstyle 6\ 12\ 11\ 10} \\ 73{,}206 \\ -\ 38{,}543 \\ \hline 34{,}663 \end{array}$$

Subtract the ones.
Rename. Subtract the tens.
Rename. Subtract the hundreds.
Rename. Subtract the thousands.
Subtract the ten-thousands.

TRY THIS

Subtract.

1. 4873 − 2379 = 2494

2. 28,497 − 9,598 = 18,899

3. 346,291 − 253,834 = 92,457

EXAMPLE 2. Subtract. $ 794.61 − $ 345.87

$$\begin{array}{r} {\scriptstyle 8\ 13\ 15\ 11} \\ \$\ 794.61 \\ -\ 345.87 \\ \hline \$\ 448.74 \end{array}$$

Subtract as with whole numbers.
Write money notation in the answer.

TRY THIS

Subtract.

4. $ 47.28 − 26.42 = $20.86

5. $ 4276.94 − 3148.75 = $1128.19

6. $ 7385.65 − 948.79 = $6436.86

PRACTICE

Subtract.

1. 8472 − 5380 = 3092

2. 5298 − 3876 = 1422

3. 9412 − 7308 = 2104

4. 1476 − 925 = 551

5. 1525 − 776 = 749

6. 3641 − 1278 = 2363

7. 4371 − 1719 = 2652

8. 7526 − 1958 = 5568

9. 36,297 − 18,429 = 17,868

10. 73,248 − 28,076 = 45,172

11. 42,573 − 19,861 = 22,712

12. 44,444 − 27,705 = 16,739

13. 72,486 − 29,549 = 42,937

14. 34,937 − 17,679 = 17,258

15. 52,648 − 37,959 = 14,689

16. 68,860 − 9,493 = 59,367

17. 498,276 − 379,818 = 118,458

18. 769,287 − 589,549 = 179,738

19. 315,264 − 276,581 = 38,683

20. 728,651 − 354,974 = 373,677

21. 864,590 − 127,498 = 737,092

22. 726,627 − 398,893 = 327,734

23. 345,215 − 68,758 = 276,457

24. 421,742 − 8,756 = 412,986

25. $ 79.48 − 27.39 = $52.09

26. $ 48.36 − 19.28 = $29.08

27. $ 68.14 − 39.25 = $28.89

28. $ 81.46 − 7.98 = $73.48

29. $ 182.87 − 94.38 = $88.49

30. $ 361.27 − 284.82 = $76.45

31. $ 432.87 − 76.98 = $355.89

32. $ 153.82 − 77.77 = $76.05

PROBLEM CORNER

1. Which is the largest island on Earth? Greenland

2. How much larger is New Guinea than Borneo? 76,664 km²

3. The area of Texas is 692,405 km². How much larger is Borneo than Texas? 51,961 km²

Three Largest Islands on Earth	
Greenland	2,175,600 km²
New Guinea	821,030 km²
Borneo	744,366 km²

Notice that the areas are written in terms of square kilometers (km²).

Objective 27

ZEROS IN SUBTRACTING

To rename, we go to the first non-zero digit. It becomes one less in value, and the passed 0's become 9's.

Subtract. 700 − 268

We must rename to subtract. → Rename a special way. → Subtract.

```
  700          6 9 10
- 268          7 0 0      700 = 69 tens + 10       6 9 10
             - 2 6 8                               7 0 0
                                                 - 2 6 8
                                                   4 3 2
```

EXAMPLE 1. Subtract. 803 − 379

```
  7 9 13
  8 0 3     Rename. 803 = 79 tens + 13
- 3 7 9     Subtract.
  4 2 4
```

EXAMPLE 2. Subtract. 7000 − 3497

```
  6 9 9 10
  7 0 0 0   Rename. 7000 = 699 tens + 10
- 3 4 9 7   Subtract.
  3 5 0 3
```

TRY THIS

Subtract.

1. 600 − 423 = 177

2. 8004 − 375 = 7629

3. \$ 50.03 − 29.47 = \$20.56

EXAMPLE 3. Subtract. 54,007 − 12,478

```
     3 9 9 17
  5  4,0 0 7    To rename, go to the first non-zero digit.
- 1  2,4 7 8    4007 = 399 tens + 17
  4  1,5 2 9
```

TRY THIS

Subtract.

4. 36,001 − 15,275 = 20,726

5. 70,040 − 19,281 = 50,759

6. \$ 500.17 − 192.88 = \$307.29

PRACTICE

For extra practice, use Making Practice Fun 10 with this lesson.
Use Quiz 4 after this Practice.

Subtract.

1. 700 − 341 = 359	**2.** 804 − 25 = 779	**3.** 300 − 6 = 294	**4.** 605 − 197 = 408
5. 207 − 198 = 9	**6.** 100 − 73 = 27	**7.** 500 − 371 = 129	**8.** 603 − 214 = 389
9. 7000 − 1234 = 5766	**10.** 8002 − 3246 = 4756	**11.** 9000 − 1725 = 7275	**12.** 5004 − 2119 = 2885
13. 4001 − 974 = 3027	**14.** 3000 − 326 = 2674	**15.** 2000 − 7 = 1993	**16.** 1001 − 2 = 999
17. 76,005 − 15,738 = 60,267	**18.** 76,050 − 15,783 = 60,267	**19.** 9007 − 2433 = 6574	**20.** 50,050 − 25,597 = 24,453
21. 42,004 − 23,541 = 18,463	**22.** 7800 − 2974 = 4826	**23.** 44,000 − 378 = 43,622	**24.** 76,004 − 6,005 = 69,999
25. $ 4.00 − 1.73 = $2.27	**26.** $ 7.04 − 2.95 = $4.09	**27.** $ 3.00 − 0.37 = $2.63	**28.** $ 8.02 − 0.03 = $7.99
29. $ 10.00 − 1.75 = $8.25	**30.** $ 50.05 − 27.68 = $22.37	**31.** $ 38.00 − 17.59 = $20.41	**32.** $ 80.54 − 17.73 = $62.81

PROBLEM CORNER

1. You buy a record for $ 7.98. You give the clerk a ten-dollar bill. How much change should you receive? $2.02

2. You buy two tapes for $ 14.75. You give the clerk a twenty-dollar bill. How much change should you receive? $5.25

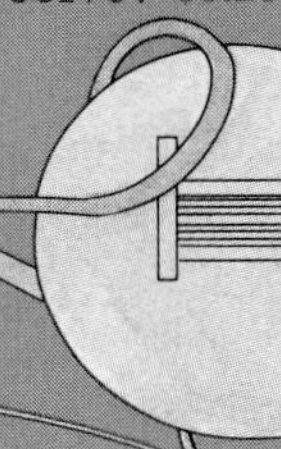

More Practice, page 62

Objective 152

It would be helpful to have samples of check stubs and registers from some local banks.

A transparency of each might help with the presentation of the idea.

KEEPING A RUNNING BALANCE

One method of recording checks is the check stub attached to the check.

CHECK STUB

Date Sept. 15, 1981 $43.20 101
To Paper Box, Inc.
For gross worm boxes

	Dollars	Cents
Balance Forward	648	96
Amount of Deposit	215	50
Total		
Amount of Check	43	20
Balance Carried Forward		

Money in account → Balance Forward
Add → Amount of Deposit
Money after deposit → Total
Subtract → Amount of Check
Money after transaction → Balance Carried Forward

FIRST NATIONAL CITY BANK
of
Crete, Illinois 60417

Pay to the Order of
Paper Box, Inc.
Forty-three and 20/100

Carl Sanchez
1843 Vista Road
Crete, Illinois, 60417

Use the check stub above.

1. Find the account total by adding. $864.46
2. Find the balance forward by subtracting. $821.26

Checks can also be recorded in a check register. It is in a booklet separate from your checks.

Void is written if an error is made so that the check cannot be used.

A voided check should be torn up.

CHECK REGISTER

This part shows a running balance.

Check Number	Date	Checks issued to or description of deposit	Amount of check		✓	Amount of deposit		Balance Forward		
									$648	96
101	9-15	To Paper Box, Inc. For gross worm boxes	43	20				Check or deposit	43	20
								Balance		
		To Deposit For receipts 9-14-80				215	50	Check or deposit	215	50
								Balance		
103	9-16	To City Electric Co. For electricity through 9-12	72	95				Check or deposit	72	95
								Balance		

Use the register.

3. Find the balance after subtracting check 101. $605.76
4. Find the balance after adding the deposit on 9-15-80. $821.26
5. Find the balance after subtracting check 103. $748.31

Provide duplicated copies of the stubs and register, if necessary.

Copy and complete the right-hand portion of each check stub.

6.

Date Oct. 1, 19 81	$24.95	135
To Sun Lighting		
For Lamp		
	Dollars	Cents
Balance Forward	$ 487	56
Amount of Deposit		0
Total	487	56
Amount of Check	24	95
Balance Carried Forward	462	61

Balance Carried Forward to Next Stub

7.

Date Oct. 2, 19 81	$16.85	136
To Joe's Market		
For Food		
	Dollars	Cents
Balance Forward	462	61
Amount of Deposit		0
Total	462	61
Amount of Check	$ 16	85
Balance Carried Forward	445	76

8.

Date Jan. 12, 19 82	$18.75	258
To John Burke		
For Birthday gift		
	Dollars	Cents
Balance Forward	$ 392	15
Amount of Deposit	65	00
Total	457	15
Amount of Check	18	75
Balance Carried Forward	438	40

Balance Carried Forward to Next Stub

9.

Date Jan. 14, 19 82	$106.11	259
To IRS		
For Tax payment		
	Dollars	Cents
Balance Forward	438	40
Amount of Deposit	$10	95
Total	449	35
Amount of Check	106	11
Balance Carried Forward	343	24

10. Copy and complete the right-hand portion of this check register.

Check Number	Date	Checks issued to or description of deposit	Amount of check		✓	Amount of deposit		Balance Forward		
									$398	59
273	2-15	To Sun Fashions	16	77				Check or deposit	16.77	
		For white skirt						Balance	381.82	
	2-15	To Deposit				175	80	Check or deposit	175.80	
		For						Balance	557.62	
274	2-17	To Ready Start	68	25				Check or deposit	68.25	
		For battery						Balance	489.37	
275	2-18	To Altos Apts.	295	00				Check or deposit	295.00	
		For March rent payment						Balance	194.37	
	2-22	To Deposit				175	80	Check or deposit	175.80	
		For						Balance	370.17	
276	2-23	To Northern Gas Co.	31	18				Check or deposit	31.18	
		For gas bill 2-19						Balance	338.99	

Objective 28

COMPARING DECIMALS

In some decimals writing or dropping zeros shows that the decimals are equal.

EXAMPLE 1. Write a decimal in hundredths equal to 0.5.

$0.5 = 0.50$ *Write one 0.*

EXAMPLE 2. Write a decimal in tenths equal to 0.300.

$0.300 = 0.3$ *Drop two 0's.*

TRY THIS Write each decimal in hundredths.

1. 0.2 0.20 **2.** 0.700 0.70 **3.** 0.90000 0.90

We can compare decimals using $=$ (is equal to), $>$ (is greater than), or $<$ (is less than).

Compare. 0.8 and 0.73

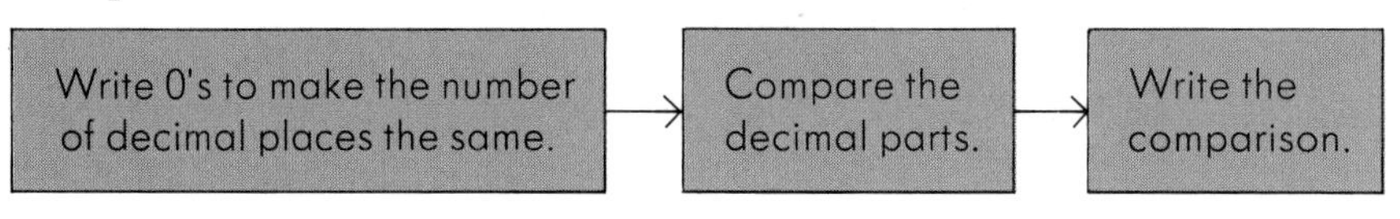

$0.8 = 0.80$
$0.73 = 0.73$

$80 > 73$

$0.8 > 0.73$

EXAMPLE 3. Compare. 0.54 and 0.6

$0.54 = 0.54$ *Write 0's to make the number of decimal*
$0.6 = 0.60$ *places the same.*
$54 < 60$ *Compare the decimal parts.*
$0.54 < 0.6$ *Write the decimals.*

TRY THIS Compare. Use $=$, $>$ or $<$.

4. 0.86 and 0.91 $<$
5. 0.6 and 0.529 $>$
6. 0.7 and 0.700 $=$

PRACTICE

For extra practice, use Making Practice Fun 11 with this lesson.

Write each decimal in tenths.

1. 0.80 0.8
2. 0.800 0.8
3. 0.8000 0.8
4. 0.70 0.7
5. 0.7000 0.7
6. 0.300 0.3
7. 0.60 0.6
8. 0.10 0.1

Write each decimal in hundredths.

9. 0.1 0.10
10. 0.2 0.20
11. 0.8 0.80
12. 0.9 0.90
13. 0.300 0.30
14. 0.4000 0.40
15. 0.150 0.15
16. 0.3800 0.38

Write each decimal in thousandths.

17. 0.1 0.100
18. 0.20 0.200
19. 0.3 0.300
20. 0.40 0.400
21. 0.32 0.320
22. 0.48 0.480
23. 0.5000 0.500
24. 0.9000 0.900

Compare. Use =, >, or <.

25. 0.3 and 0.300 =
26. 0.4 and 0.3 >
27. 0.72 and 0.7 >
28. 0.319 and 0.391 <
29. 0.319 and 0.32 <
30. 0.45 and 0.450 =
31. 0.4 and 0.04 >
32. 0.007 and 0.07 <
33. 0.310 and 0.31 =
34. 0.8 and 0.8000 =
35. 0.9 and 0.09 >
36. 0.08 and 0.008 >
37. 0.07 and 0.7 <
38. 0.914 and 0.92 <
39. 0.5 and 0.500 =
40. 0.421 and 0.42 >
41. 0.628 and 0.625 >
42. 0.24 and 0.27 <

PROBLEM CORNER

1. Ana's fish weighed 0.73 kilograms. Luis's fish weighed 0.712 kilograms. Whose fish weighed more? Ana's

2. Cory ran 6.51 kilometers. Devon ran 6.53 kilometers. Who ran further? Devon

SUBTRACTING DECIMALS

Subtract. 28.35 − 16.57

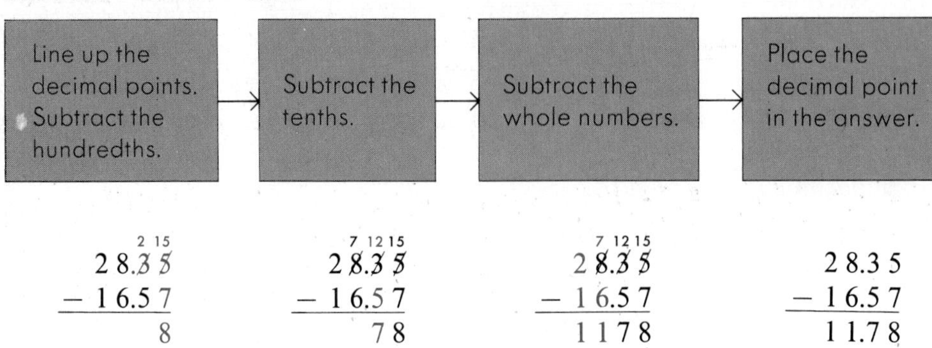


```
    2 15        7 12 15       7 12 15
  2 8.3 5      2 8.3 5      2 8.3 5      2 8.3 5
− 1 6.5 7    − 1 6.5 7    − 1 6.5 7    − 1 6.5 7
        8          7 8      1 1 7 8      1 1.7 8
```

Sometimes we write 0's to fill the places.

EXAMPLES. Subtract.

1. 7.48 − 3.7 **2.** 9.3 − 3.754

```
  6 14          8 12 9 10
  7.4 8         9.3 0 0
− 3.7 0       − 3.7 5 4
  3.7 8         5.5 4 6
```

Line up the decimal points.
Subtract.

TRY THIS

Subtract.

1. 3.71 − 2.16 1.55

2. 15.287 − 7.6 7.687

3. 8.7 − 4.24 4.46

When one number is a whole number, we write a decimal point and 0's.

EXAMPLES. Subtract.

3. 6.743 − 2 **4.** 8 − 3.62

```
                 7 9 10
  6.7 4 3        8.0 0
− 2.0 0 0      − 3.6 2
  4.7 4 3        4.3 8
```

Line up.
Subtract.
Place the decimal point in the answer.

TRY THIS

Subtract.

4. 7.48 − 3 4.48

5. 9 − 3.7 5.3

6. 14 − 6.21 7.79

PRACTICE

For extra practice, use Making Practice Fun 12 with this lesson.

Subtract.

1. 7.3 − 3.9 3.4

2. 13.71 − 4.52 9.19

3. 9.386 − 3.739 5.647

4. 6.74 − 3.81 2.93

5. 9.36 − 4.7 4.66

6. 12.84 − 6.9 5.94

7. 9.04 − 3.2 5.84

8. 17.61 − 7.8 9.81

9. 8.76 − 3.8 4.96

10. 3.743 − 1.92 1.823

11. 4.765 − 3.9 0.865

12. 9.38 − 2.5 6.88

13. 4.26 − 3.9 0.36

14. 4.27 − 1.835 2.435

15. 6.7 − 4.381 2.319

16. 7.2 − 3.84 3.36

17. 9.3 − 7 2.3

18. 6.75 − 3 3.75

19. 8.764 − 5 3.764

20. 6.98 − 5 1.98

21. 6 − 1.8 4.2

22. 8 − 3.27 4.73

23. 7 − 3.843 3.157

24. 5 − 1.25 3.75

25. 7.9 − 3 4.9

26. 15.1 − 8 7.1

27. 9 − 4.6 4.4

28. 11 − 6.7 4.3

PROBLEM CORNER

1. Which is the heaviest book; the lightest book? Dictionary Notebook

2. How much heavier is the math book than the English book? 0.207 kg

3. How much lighter is the math book than the dictionary? 0.454 kg

4. How much heavier is the English book than the notebook? 0.415 kg

Weighty Knowledge	
Math Book	0.72 kg
English Book	0.513 kg
Note Book	0.098 kg
Dictionary	1.174 kg

Objective 153

Develop this lesson carefully. Since there is so much information on a monthly statement, there are many opportunities for confusion to occur. Some actual samples of monthly statements would be helpful in presenting this lesson.

RECONCILING MONTHLY STATEMENTS

Each month you receive a bank statement. Seeing if it agrees with your records is called *reconciling*.

FIRST NATIONAL CITY BANK
of
Crete, Illinois 60417

Account #13-05762-2173-94

page 1 of 1 Enclosures-4

Stella Davis
1843 Vista Road
Crete, Illinois 60417

Account Summary and Detail

For period ending 02-03-81

513.81	Beginning balance this period
428.39	Plus deposits/credits
394.26	Less checks/debits
3.25	Less service charges/fees
544.69	Ending balance this period

Non-check transactions

Amount	Date	Description
275.19	01-01-81	deposit
153.20	10-14-81	deposit

Check transactions

Amount	Check	Date
97.18	100	01-05-81
35.70	101	01-15-81
172.15	103	01-19-81
89.23	105	01-26-81

Daily Balance

Amount	Date	+/−
789.00	01-01-81	+
691.82	01-05-81	-
845.02	01-14-81	+
809.32	01-15-81	-
637.17	01-19-81	-
547.94	01-26-81	-
544.69	02-03-81	-

1. What was the balance in Stella Davis's account on 02-03-81? $544.69
2. What was the service charge? $3.25
3. Which checks have not been deducted? 102, 104

CONSUMER HINTS
1. Reconcile the account immediately upon receiving the statement.
2. Sort and match the checks to your stubs or check register.

The bank provides a form for reconciling your account. Stella Davis completes this form after receiving her statement.

Checks not shown on statement		
Number	Dollars	Cents
117	48	19
120	56	00
Total		

Bank balance shown on statement	$ 631.42
Add deposits not shown on statement.	$ 89.16
TOTAL	$
Subtract checks not shown on statement.	$
BALANCE (Should agree with check stub or register)	$

4. What is the total after deposits are added? $720.58
5. What is the total of checks not shown on statement? $104.19
6. Subtract checks not shown on statement. $616.39
7. There was no service charge. Is the account reconciled? yes

Stella's next check stub shows:

Date ______ 19__		121
To		
For		
Balance Forward	Dollars 616	Cents 39
Amount of Deposit		
Total		
Amount of Check		
Balance Carried Forward		

The following is information from bank statements. Do the two records agree? Complete copies of the bank form above to help you decide.

	Bank Balance	Deposits Not Shown By Bank	Total Checks Not Shown On Statement	Service Charge	Your Balance	
8.	$ 173.95	$187.15	$ 58.15	0	$ 302.95	Yes
9.	$ 315.00	0	$128.25	$2.25	$ 184.50	Yes
10.	$ 463.19	$ 92.00	$106.22	0	$ 458.97	No
11.	$ 36.21	$126.14	0	$1.75	$ 160.60	Yes
12.	$1627.58	$743.29	$615.98	0	$1745.98	No
13.	$ 598.74	$315.85	$130.67	$2.25	$ 781.67	Yes

Objectives 30 and 31

ESTIMATING DIFFERENCES

To estimate, we usually round to the highest place in the smallest number. Compare with the lesson on page 28.

Estimate the difference. 874 − 386

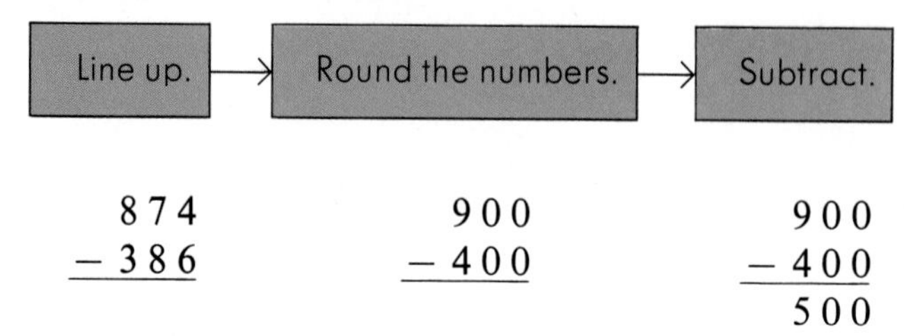

EXAMPLES. Estimate the differences.

1. 7286 − 3500

$$\begin{array}{r} 7286 \\ -\,3500 \\ \hline \end{array} \qquad \begin{array}{r} 7000 \\ -\,4000 \\ \hline 3000 \end{array}$$

2. 876 − 57

$$\begin{array}{r} 876 \\ -\ \ 57 \\ \hline \end{array} \qquad \begin{array}{r} 880 \\ -\ \ 60 \\ \hline 820 \end{array}$$

3. 25.3 − 9.476

$$\begin{array}{r} 25.3 \\ -\ \ 9.476 \\ \hline \end{array} \qquad \begin{array}{r} 25 \\ -\ \ 9 \\ \hline 16 \end{array}$$

Round to the nearest whole number.
Subtract.

4. \$78.98 − \$6.89

$$\begin{array}{r} \$\,78.98 \\ -\ \ \ 6.89 \\ \hline \end{array} \qquad \begin{array}{r} \$\,79 \\ -\ \ 7 \\ \hline \$\,72 \end{array}$$

Round to the nearest dollar.
Subtract.

TRY THIS

Estimate the differences.

1. $\begin{array}{r} 87 \\ -\,38 \\ \hline \end{array}$ 50

2. $\begin{array}{r} 4428 \\ -\ \ 735 \\ \hline \end{array}$ 3700

3. $\begin{array}{r} 627 \\ -\ \ 43 \\ \hline \end{array}$ 590

4. $\begin{array}{r} 8.14 \\ -\,3.2 \\ \hline \end{array}$ 5

5. $\begin{array}{r} \$\,12.85 \\ -\ \ \ 3.90 \\ \hline \end{array}$ \$9

6. $\begin{array}{r} 12.6 \\ -\ \ 8.1 \\ \hline \end{array}$ 5

PRACTICE

Estimate the differences.

1. 83 − 24 (60)	**2.** 78 − 39 (40)	**3.** 69 − 31 (40)	**4.** 74 − 15 (50)
5. 428 − 139 (300)	**6.** 672 − 386 (300)	**7.** 811 − 298 (500)	**8.** 689 − 203 (500)
9. 4276 − 1398 (3000)	**10.** 5984 − 1190 (5000)	**11.** 8743 − 6819 (2000)	**12.** 8612 − 2500 (6000)
13. 378 − 49 (330)	**14.** 265 − 36 (230)	**15.** 731 − 88 (640)	**16.** 349 − 85 (260)
17. 2744 − 349 (2400)	**18.** 6273 − 580 (5700)	**19.** 3811 − 299 (3500)	**20.** 7313 − 827 (6500)
21. 9.382 − 6.45 (3)	**22.** 8.194 − 3.9 (4)	**23.** 3.156 − 2.2 (1)	**24.** 7.1 − 3.985 (3)
25. 12.75 − 3.864 (9)	**26.** 14.382 − 8.4 (6)	**27.** 16.482 − 9.39 (7)	**28.** 17.3 − 8.498 (9)
29. $ 7.98 − 2.99 ($5)	**30.** $ 9.15 − 3.20 ($6)	**31.** $ 7.53 − 2.65 ($5)	**32.** $ 9.07 − 1.98 ($7)
33. $ 36.47 − 8.49 ($28)	**34.** $ 74.36 − 38.40 ($36)	**35.** $ 35.15 − 8.89 ($26)	**36.** $ 63.95 − 40.16 ($24)

PROBLEM CORNER

1. You pay for a record with a $ 10 bill. Estimate how much change you should receive. 5
2. You pay for a tape with a $ 20 bill. Estimate how much change you should receive. 13

Objective 196

SOLVING PROBLEMS

This type of problem is extremely important since most problems encountered in life do not have all information at hand. Discuss this idea with your class and have students provide other examples.

Many real life problems do not have enough information for an immediate solution.

EXAMPLE 1. What is the cost for two new snow tires?

UNDERSTAND I must find the total cost.
I do not know the cost of one tire or the tax.

PLAN Find an ad showing prices and tax.
I will add when the price is known.

CARRY OUT The ad shows tires for $ 32 each, plus $ 2.18 tax.

$$\begin{array}{r} \$\,32.00 \\ 32.00 \\ 2.18 \\ \underline{2.18} \\ \$\,68.36 \end{array}$$

LOOK BACK The answer is reasonable.

Sometimes there is too much information. We need to decide what to use.

An important skill is being able to decide what information is needed to solve the problem.

EXAMPLE 2. Mr. Weimer bought a chain saw with a 17 in. blade and a 3.7 cubic inch motor chain saw for $ 199.99. He had to pay $ 10 sales tax. What was the total bill?

UNDERSTAND I must find the total cost.
The blade length and engine size are extra information.
I know the cost and the tax.

PLAN I will add.
I do not need a calculator.

CARRY OUT

$$\begin{array}{r} \$\,199.99 \\ \underline{+\quad 10.00} \\ \$\,209.99 \end{array}$$

LOOK BACK The answer is reasonable.

PRACTICE

For extra practice, use Making Practice Fun 13 with this lesson.
Use Quiz 5 after this Practice.

Make up or find missing information. Solve each problem.

1. Jeri ordered a pair of nylon jogging shoes and two pairs of striped knee highs. What will the items cost? Answers will vary.

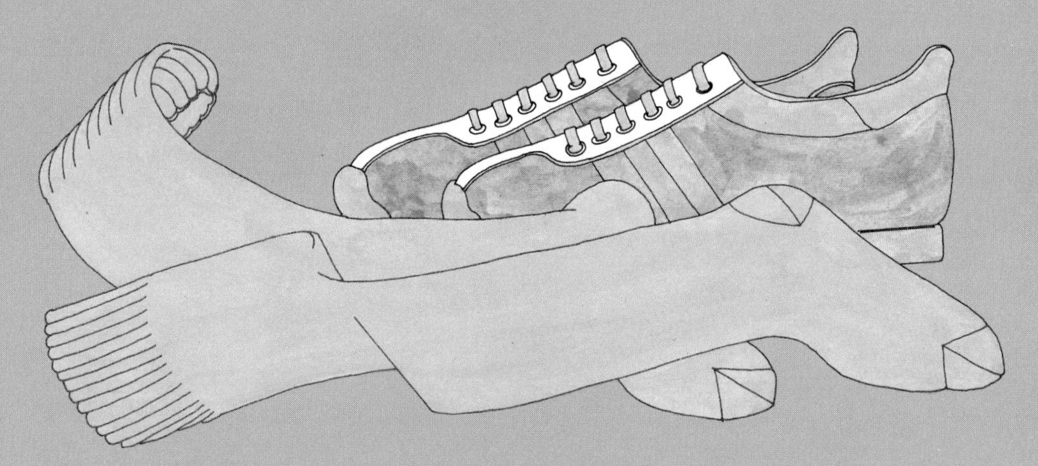

2. Bill Engelmeyer bought 100 shares of utility stock. What was the total cost? Answers will vary.

3. Rhode Island is the smallest state; it has an area of 1214 square miles. Alaska is the largest state. How much larger is Alaska than Rhode Island? 585,198 sq mi

4. Mrs. Bruder's phone bill was $ 18.50, and her gas bill was $ 21.20. Her electric bill was the highest. How much did she spend for phone and utilities? Answers will vary.

Decide what information is extra. Solve each problem.

5. Kris King bought a 32-gallon trash container with a 4-year warranty. The trash container was on sale for $ 13.99. The regular price was $ 19.85. How much was saved? extra: container size and warranty; $5.86

6. In 1965 there were about 127 million dollars in two-dollar bills in circulation in the 48 contiguous states. By 1975 there were $ 135,000,000 in two-dollar bills in circulation. How much more was in circulation in 1975 than in 1965? extra: number of states; $8,000,000

7. Recently, Yale University library spent $ 2,298,882 for new books and $ 6,676,271 for salaries. Stanford University spent $ 2,427,272 for books and $ 5,799,732 for salaries. Which university spent more for books? How much more? extra: amount for salaries; Stanford—$128,390

CAREER CAREER CAREER CAREER CAREER

WORM FARMING

Carl Sanchez owns the Blue Box Worm Farm. He began his business while he was in high school.

Worms are used for bait and in gardening. People say the African Night Crawler is the best worm for fishing. The Red Worm is used by gardeners to improve the soil.

Worm Sales—June

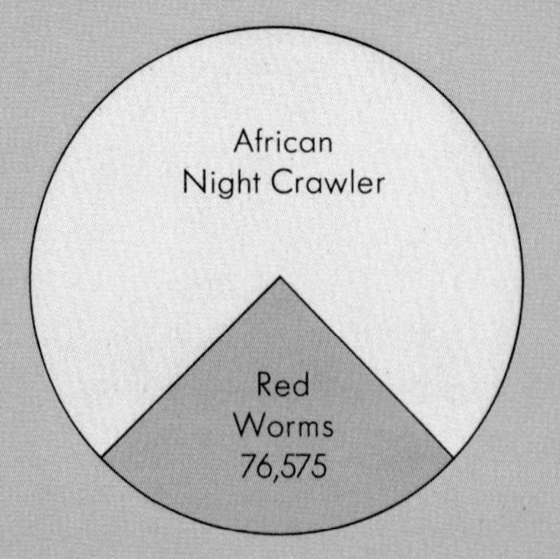

1. Last June the Blue Box Worm Farm sold a total of 225,000 worms. How many African Night Crawlers were sold? 148,425
2. A large bait company bought 75,500 worms. How many worms were sold to other customers? 149,500

CAREER CAREER CAREER CAREER CAREER

The best worms are handpicked and sold individually. The Blue Box Worm Farm hires pickers to handpick worms.

3. How many worms did Rhea pick? 3973
4. Who picked the least? Sandy
5. How many more worms did Mike pick than Sal? 2534
6. How many worms were picked in all? 15,706

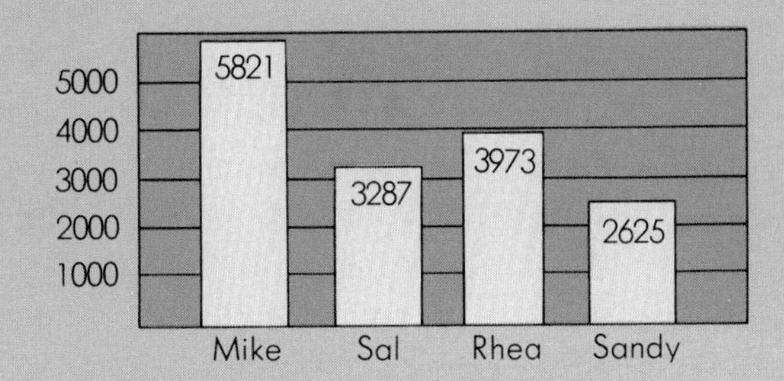

Worms must be fed properly to be fat, healthy, and saleable. The Blue Box Farm uses this recipe.

7. How many kilograms of food do they mix in one batch? 254.75 kg
8. How much more almond dust do they use than walnut meal? 11.5 kg
9. How much more rolled oats do they use than brown sugar? 0.75 kg

Worm Food Recipe: 1 batch

100 kg dry manure
74 kg almond dust
62.5 kg walnut meal
9.5 kg rolled oats
8.75 kg brown sugar

Mix well. Sprinkle over worm bed.

Worms eat less on cool days than on warm days.

10. How much was eaten on Sunday? 6 kg
11. Which was the coolest day of the week? Friday
12. Which was the warmest day of the week? Thursday
13. It rained 1.5 centimeters on Friday. How much less was eaten on Friday than Saturday? 6 kg
14. How much more was eaten on Thursday than on Monday? 9 kg

Problem 13 contains extra information.

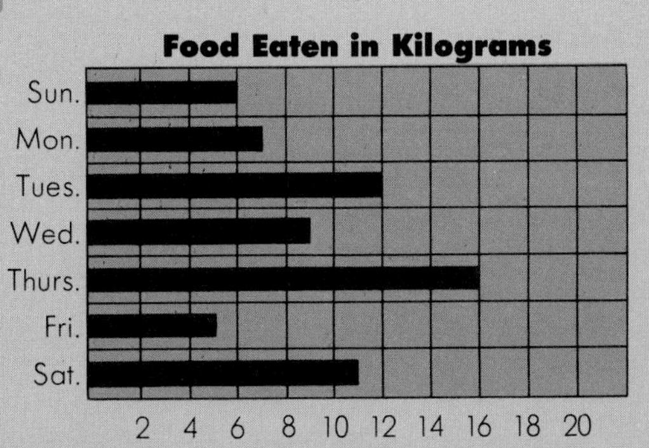

CHAPTER 2 REVIEW

For extra practice, use Making Practice Fun 14 with this lesson.

Subtract. pp. 40–50

1. 746 − 125 = 621

2. 86 − 37 = 49

3. 713 − 385 = 328

4. \$4.15 − 3.86 = \$0.29

5. 8246 − 3948 = 4298

6. 71,387 − 24,198 = 47,189

7. 726,414 − 345,547 = 380,867

8. \$486.79 − 87.80 = \$398.99

9. 500 − 148 = 352

10. 8004 − 5628 = 2376

11. 37,000 − 19,756 = 17,244

12. \$30.05 − 27.86 = \$2.19

13. 9.7 − 3.8 = 5.9

14. 4.86 − 1.9 = 2.96

15. 7.2 − 3.65 = 3.55

16. 5. − 1.75 = 3.25

Compare. Use =, >, or <. p. 48

17. 0.8 and 0.08 >

18. 0.004 and 0.4 <

19. 0.5 and 0.50 =

20. 0.700 and 0.7 =

21. 0.16 and 0.163 <

22. 0.27 and 0.268 >

23. 0.03 and 0.030 =

24. 0.07 and 0.16 <

25. 0.259 and 0.248 >

Estimate the differences. p. 54

26. 85 − 36 → 50

27. 4500 − 1684 → 3000

28. 5482 − 675 → 4800

29. 6.7 − 3.84 → 3

Application

Find the balance carried forward. p. 46.

30.

Date *Nov. 18,* 19*82* *\$25.00* 473
To *Sports, Inc.*
For *magazine subscription*

	Dollars	Cents
Balance Forward	219	15
Amount of Deposit		0
Total	219	15
Amount of Check	25	00
Balance Carried Forward	194	15

31.

Date *Dec. 24,* 19*81* *\$36.75* 519
To *Eaton's*
For *shoes*

	Dollars	Cents
Balance Forward	736	58
Amount of Deposit	75	00
Total	811	58
Amount of Check	36	75
Balance Carried Forward	774	83

CHAPTER 2 TEST

Subtract.

1. 84 − 33
 51

2. 92 − 28
 64

3. 821 − 657
 164

4. $ 5.34 − 3.76
 $1.58

5. 9487 − 3891
 5596

6. 82,491 − 28,735
 53,756

7. 487,241 − 398,655
 88,586

8. $ 728.41 − 649.58
 $78.83

9. 900 − 261
 639

10. 4006 − 1288
 2718

11. 63,005 − 18,247
 44,758

12. $ 90.00 − 16.43
 $73.57

13. 3.82 − 1.94
 1.88

14. 7.21 − 3.4
 3.81

15. 9.21 − 3.786
 5.424

16. 4 − 3.281
 0.719

Compare. Use =, >, or <.

17. 0.6 and 0.06 >

18. 0.27 and 0.271 <

19. 0.08 and 0.080 =

20. 0.12 and 0.21 <

21. 0.504 and 0.54 <

22. 0.600 and 0.6 =

23. 0.03 and 0.030 =

24. 0.07 and 0.7 <

25. 0.847 and 0.816 >

Estimate the differences.

26. 432 − 141
 300

27. 7876 − 2904
 5000

28. 756 − 87
 670

29. 9.68 − 3.7
 6

Application

Find the balance carried forward.

30.

Date *Jan. 25,* 19 *81* *$76.25* 265
To *Maxwell's*
For *tires*

	Dollars	Cents
Balance Forward	*287*	*00*
Amount of Deposit		*-0-*
Total		
Amount of Check		
Balance Carried Forward	210	75

31.

Date *Oct. 21,* 19 *84* *$16.84* 787
To *Sound, Inc.*
For *records*

	Dollars	Cents
Balance Forward	*325*	*16*
Amount of Deposit	*276*	*15*
Total		
Amount of Check		
Balance Carried Forward	584	47

MORE PRACTICE

Subtract. Use after page 41.

1. 87 − 69 18	**2.** 47 − 38 9	**3.** 60 − 47 13	**4.** 91 − 36 55
5. 346 − 278 68	**6.** 831 − 129 702	**7.** 243 − 69 174	**8.** 715 − 47 668
9. $ 0.76 − 0.47 $0.29	**10.** $ 6.48 − 1.94 $4.54	**11.** $ 8.75 − 2.98 $5.77	**12.** $ 4.53 − 0.68 $3.85

Subtract. Use after page 43.

13. 4287 − 3196 1091	**14.** 8421 − 6789 1632	**15.** 6426 − 1238 5188	**16.** 7134 − 2929 4205
17. 73,197 − 28,088 45,109	**18.** 61,217 − 19,073 42,144	**19.** 46,810 − 38,134 8676	**20.** 93,241 − 8,645 84,596
21. 691,286 − 424,878 266,408	**22.** 876,215 − 198,378 677,837	**23.** 921,486 − 67,498 853,988	**24.** 175,034 − 68,318 106,716
25. $ 63.40 − 19.65 $43.75	**26.** $ 476.15 − 138.20 $337.95	**27.** $ 5286.15 − 3874.66 $1411.49	**28.** $ 2410.25 − 976.43 $1433.82

Subtract. Use after page 45.

29. 600 − 143 457	**30.** 801 − 645 156	**31.** 4000 − 1286 2714	**32.** 5003 − 647 4356
33. 42,000 − 12,876 29,124	**34.** 43,008 − 9,429 33,579	**35.** 60,080 − 12,345 47,735	**36.** 7006 − 1809 5197
37. 80,000 − 34,261 45,739	**38.** 900,008 − 6,789 893,219	**39.** 86,040 − 19,428 66,612	**40.** 31,000 − 2,499 28,501
41. $ 20.00 − 16.57 $3.43	**42.** $ 30.00 − 21.83 $8.17	**43.** $ 400.50 − 399.86 $0.64	**44.** $ 100.00 − 62.86 $37.14

Compare. Use $=$, $>$, or $<$. Use after page 49.

1. 0.7 and 0.70 $=$ **2.** 0.3 and 0.03 $>$ **3.** 0.04 and 0.4 $<$

4. 0.12 and 0.120 $=$ **5.** 0.6 and 0.59 $>$ **6.** 0.10 and 0.1 $=$

7. 0.08 and 0.079 $>$ **8.** 0.9 and 0.90 $=$ **9.** 0.012 and 0.021 $<$

10. 0.460 and 0.46 $=$ **11.** 0.104 and 0.12 $<$ **12.** 0.3 and 0.34 $<$

13. 0.68 and 0.612 $>$ **14.** 0.740 and 0.704 $>$ **15.** 0.4 and 0.400 $=$

16. 0.20 and 0.200 $=$ **17.** 0.539 and 0.568 $<$ **18.** 0.27 and 0.264 $>$

Subtract. Use after page 51.

19. $8.6 - 1.9$ 6.7 **20.** $7.38 - 1.29$ 6.09 **21.** $6.071 - 4.268$ 1.803 **22.** $4.37 - 0.19$ 4.18

23. $9.87 - 1.9$ 7.97 **24.** $6.742 - 3.87$ 2.872 **25.** $6.123 - 3.9$ 2.223 **26.** $9.26 - 2.7$ 6.56

27. $8.7 - 1.93$ 6.77 **28.** $6.74 - 0.986$ 5.754 **29.** $4.3 - 2.87$ 1.43 **30.** $10.6 - 7.93$ 2.67

31. $5.78 - 3.9$ 1.88 **32.** $14.3 - 6.58$ 7.72 **33.** $5.5 - 1.79$ 3.71 **34.** $16.471 - 8.62$ 7.851

35. $18.427 - 9$ 9.427 **36.** $26.32 - 7$ 19.32 **37.** $4 - 1.123$ 2.877 **38.** $6 - 0.84$ 5.16

Estimate the differences. Use after page 55.

39. $93 - 34$ 60 **40.** $685 - 290$ 400 **41.** $448 - 149$ 300 **42.** $6285 - 3396$ 3000

43. $643 - 78$ 560 **44.** $8764 - 385$ 8400 **45.** $4785 - 67$ 4720 **46.** $7684 - 498$ 7200

47. $4.7 - 3.87$ 1 **48.** $6.482 - 1.39$ 5 **49.** $15.3 - 6.1$ 9 **50.** $27.6 - 19.4$ 9

51. \$8.78 − 1.98 \$7 **52.** \$6.14 − 3.15 \$3 **53.** \$40.43 − 28.89 \$11 **54.** \$17.51 − 9.49 \$9

CALCULATOR CORNER

This activity focuses on estimation and approximation skills.

ADDITION – SUBTRACTION "IN BETWEEN"

Players two

Materials two calculators, paper and pencils

The interval involved can be altered depending on the abilities of the students.

SAMPLE Juan gives Carlos a calculator with a number, such as 174, on the display. Juan tells Carlos that by adding a number to 174, he must get a number on the display which is between 285 and 295.

Carlos gives Juan a calculator with a number, such as 475, on the display. Carlos tells Juan that by subtracting a number from 475, he must get a number on the display which is between 345 and 355.

ROUND 1 **Carlos** Carlos writes down 174 and the interval 285–295.

Play 1 He adds 90 to 174. He records 90.
Result: 264—too small

Play 2 He adds 100 to 174. He records 100.
Result: 274—too small

Play 3 He adds 110 to 174. He records 110.
Result: 284—too small

Play 4 He adds 115 to 174. He records 115.
Result: 289–OK

It took Carlos 4 plays, so his score is 4.

Juan Juan writes down 475 and the interval 345–355.

Play 1 Juan subtracts 100 from 475. He records 100.
Result: 375—too large

Play 2 Juan subtracts 120 from 475. He records 120.
Result: 355—too large

Play 3 Juan subtracts 125 from 475. He records 125.
Result: 350—OK

It took Juan 3 plays, so his score is 3.

In the next round Carlos subtracts, and Juan adds.

Play continues for 6 rounds. The player with the lowest score wins.

CHAPTER 3 PRE-TEST

Multiply. pp. 68–84

1. 41 × 6 = 246

2. 87 × 4 = 348

3. 8394 × 5 = 41,970

4. \$ 9.78 × 7 = \$68.46

5. 85 × 30 = 2550

6. 73 × 26 = 1898

7. 2745 × 89 = 244,305

8. \$ 5.38 × 64 = \$344.32

9. 827 × 500 = 413,500

10. 3827 × 567 = 2,169,909

11. 498 × 703 = 350,094

12. \$ 9.42 × 358 = \$3372.36

13. 80 × 70 = 5600

14. 900 × 30 = 27,000

15. 4000 × 60 = 240,000

16. 8000 × 700 = 5,600,000

17. 9.6 × 8 = 76.8

18. 3.42 × 0.7 = 2.394

19. 0.6 × 0.3 = 0.18

20. 0.04 × 0.02 = 0.0008

21. 10 × 8.7 87

22. 100 × 9.4 940

23. 0.1 × 36 3.6

24. 0.001 × 1.2 0.0012

Estimate. pp. 76–78

25. 82 × 9 = 720

26. 73 × 68 = 4900

27. 694 × 72 = 49,000

28. 4215 × 876 = 3,600,000

29. \$ 7.14 × 39 = \$280

30. \$ 27.98 × 41 = \$1200

31. 8.94 × 6 = 54

32. 3.1427 × 8.81 = 27

Application p. 80

33. Is the change counted correctly? yes
Charge: \$ 2.85. Paid with: \$ 5.00.

Clerk gives	Clerk says
Item	\$ 2.85
1 nickel	2.90
1 dime	3.00
1 dollar	4.00
1 dollar	5.00

34. Complete the list to give change.
Charge: \$ 7.75. Paid with: \$ 10.00.
Machine computes \$ 2.25 change.

Clerk gives	Clerk says
1 dollar	\$ 1.00
1 dollar	2.00
1 quarter	2.25

3

MULTIPLYING WHOLE NUMBERS AND DECIMALS

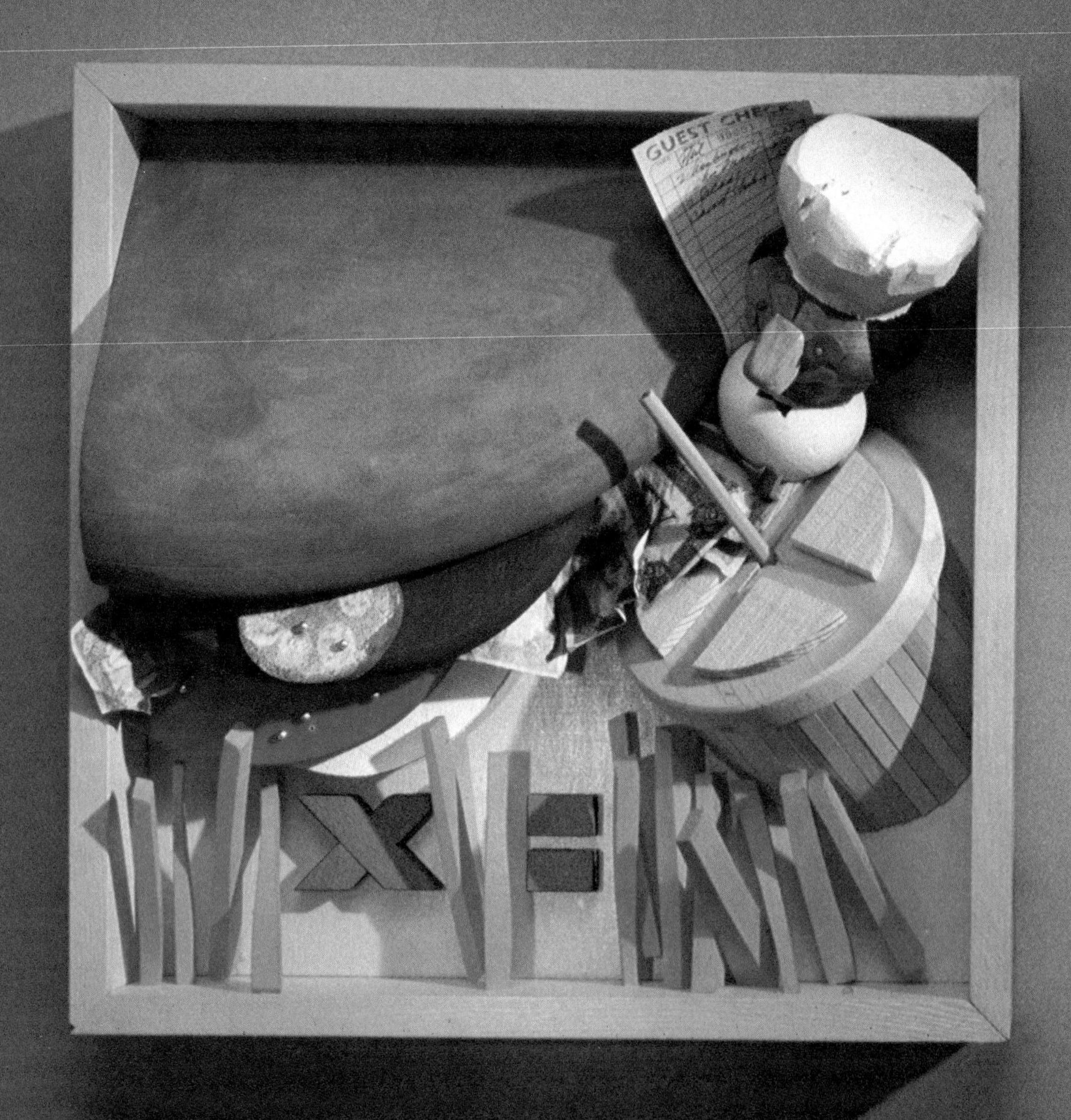

GETTING STARTED

MULTIPLICATION GOTCHA

WHAT'S NEEDED The squares of paper with digits marked on them as used in GOTCHA, Chapter 1. Use only 10 squares, one of each digit.

THE RULES

1. On a sheet of paper each player makes a scorecard like this. Play three games and add to find the total score. The player with the highest total score wins.
2. To play, a leader draws digits from the box, one at a time. All players write them down. They are to be multiplied as they are drawn to get the total for the game.
3. When 0 is drawn, it is a GOTCHA. Each player still in the game gets a 0 for that game.
4. A player may go out or quit before any draw. Try to get the biggest score but go out before 0 is drawn.

Scorecard

1. ______

2. ______

3. ______

Total ______

SAMPLE

Game 1	Game 2	Game 3	Scoreboard
7	9	7	**1.** 189
× 3	× 8	× 9	**2.** 0
21	72	63	**3.** 2520
× 9	× 6	× 8	**Total** 2709
189 go out	432	504	
	× 0	× 5	
	0 GOTCHA	2520 go out	

Objectives 32 and 33

MULTIPLYING BY 1-DIGIT NUMBERS

Multiply. 714×6

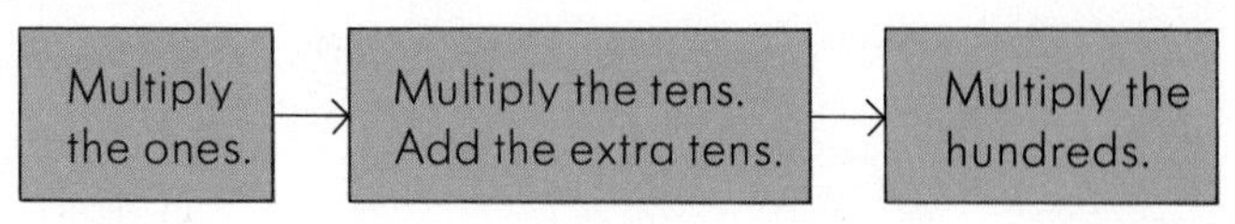

$$\begin{array}{r} \overset{2}{7\,1}\,4 \\ \times \quad 6 \\ \hline 4 \end{array} \qquad \begin{array}{r} \overset{2}{7\,1}\,4 \\ \times \quad 6 \\ \hline 84 \end{array} \qquad \begin{array}{r} \overset{2}{7\,1}\,4 \\ \times \quad 6 \\ \hline 4284 \end{array}$$

EXAMPLE 1. Multiply. 52×3

$$\begin{array}{r} 52 \\ \times \ 3 \\ \hline 156 \end{array}$$

Multiply the ones.
Multiply the tens.

EXAMPLE 2. Multiply. 784×8

$$\begin{array}{r} {}^{6\,3} \\ 784 \\ \times \ 8 \\ \hline 6272 \end{array}$$

Multiply the ones.
Multiply the tens.
Multiply the hundreds.

TRY THIS

1. $\begin{array}{r} 643 \\ \times \ 2 \\ \hline \end{array}$ 1286

2. $\begin{array}{r} 39 \\ \times \ 4 \\ \hline \end{array}$ 156

3. $\begin{array}{r} 8015 \\ \times \ 9 \\ \hline \end{array}$ 72,135

EXAMPLE 3. Multiply. $\$3.73 \times 5$

$$\begin{array}{r} {}^{3\,1} \\ \$\ 3.73 \\ \times \ 5 \\ \hline \$\ 18.65 \end{array}$$

Multiply as with whole numbers.
Write money notation in the answer.

TRY THIS

Multiply.

4. $\begin{array}{r} \$\,0.71 \\ \times \ 8 \\ \hline \end{array}$ \$5.68

5. $\begin{array}{r} \$\,7.58 \\ \times \ 4 \\ \hline \end{array}$ \$30.32

6. $\begin{array}{r} \$\,235.19 \\ \times \ 7 \\ \hline \end{array}$ \$1646.33

PRACTICE

For extra practice, use Making Practice Fun 15 with this lesson.

Multiply.

1. 74 × 2 = 148	**2.** 32 × 3 = 96	**3.** 71 × 6 = 426	**4.** 90 × 8 = 720
5. 731 × 3 = 2193	**6.** 824 × 2 = 1648	**7.** 121 × 4 = 484	**8.** 811 × 5 = 4055
9. 78 × 6 = 468	**10.** 42 × 8 = 336	**11.** 65 × 7 = 455	**12.** 83 × 9 = 747
13. 374 × 2 = 748	**14.** 876 × 3 = 2628	**15.** 347 × 5 = 1735	**16.** 986 × 4 = 3944
17. 3287 × 2 = 6574	**18.** 9827 × 4 = 39,308	**19.** 3270 × 6 = 19,620	**20.** 8947 × 8 = 71,576
21. 12,497 × 3 = 37,491	**22.** 24,685 × 5 = 123,425	**23.** 73,496 × 7 = 514,472	**24.** 32,083 × 9 = 288,747
25. 425,681 × 8 = 3,405,448	**26.** 542,614 × 6 = 3,255,684	**27.** 320,521 × 4 = 1,282,084	**28.** 681,299 × 7 = 4,769,093
29. $ 0.87 × 2 = $1.74	**30.** $ 3.76 × 4 = $15.04	**31.** $ 18.78 × 8 = $150.24	**32.** $ 316.74 × 5 = $1583.70

PROBLEM CORNER

A $ 50.00 United States Savings Bond costs $ 37.50 to purchase.
It must earn interest for 5 years before its value is $ 50.00.
Joseph Shemanski celebrated becoming a U.S. citizen.
He bought seven $ 50.00 bonds.

1. How much did Joseph pay for the bonds? $262.50

2. How much will they be worth in 5 years? $350.00

3. How much interest will they earn in 5 years? $87.50

More Practice, page 92

MULTIPLYING BY 2-DIGIT NUMBERS

Multiply. 36 × 43

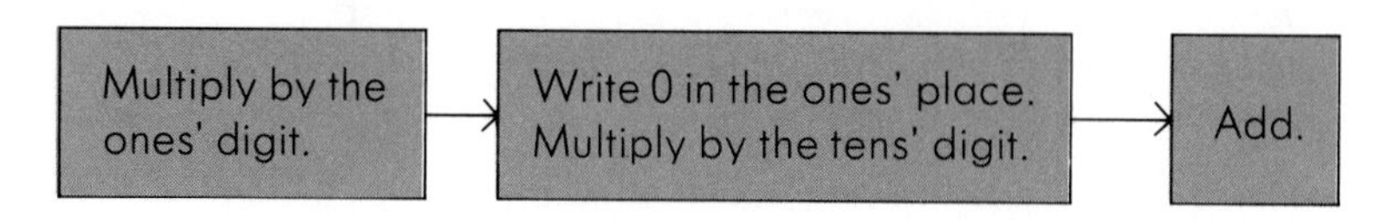

36
× 43
108

36
× 43
108
1440

36
× 43
108
1440
1548

EXAMPLE 1. Multiply. 84 × 60

84
× 60
5040

Write 0 in the ones' place.
Multiply by the tens' digit.

EXAMPLE 2. Multiply. 246 × 38

Here we say that 0 is used as a place holder.

246
× 38
1968 — *Multiply by the ones' digit.*
7380 — *Write 0 in the ones' place. Multiply by the tens' digit.*
9348 — *Add.*

TRY THIS

Multiply.

1. 892 × 70 = 62,440
2. 98 × 27 = 2646
3. 4764 × 12 = 57,168

EXAMPLE 3. Multiply. $7.48 × 23

$7.48
× 23
2244 — *Multiply as with whole numbers.*
14960
$172.04 — *Write money notation in the answer.*

TRY THIS

Multiply.

4. $\$0.73 \times 12$ = $8.76	**5.** $\$6.25 \times 34$ = $212.50	**6.** $\$56.79 \times 56$ = $3180.24

PRACTICE

For extra practice, use Making Practice Fun 16 with this lesson.

Multiply.

1. 73×20 = 1460	**2.** 84×40 = 3360	**3.** 76×60 = 4560	**4.** 73×80 = 5840
5. 249×30 = 7470	**6.** 648×50 = 32,400	**7.** 240×70 = 16,800	**8.** 704×90 = 63,360
9. 3789×10 = 37,890	**10.** 1827×80 = 146,160	**11.** 4286×20 = 85,720	**12.** 3219×40 = 128,760
13. 46×24 = 1104	**14.** 38×42 = 1596	**15.** 75×15 = 1125	**16.** 19×48 = 912
17. 728×31 = 22,568	**18.** 436×13 = 5668	**19.** 806×44 = 35,464	**20.** 980×16 = 15,680
21. 5284×33 = 174,372	**22.** 8047×27 = 217,269	**23.** 7297×64 = 467,008	**24.** 9040×87 = 786,480
25. $\$7.25 \times 30$ = $217.50	**26.** $\$8.76 \times 60$ = $525.60	**27.** $\$4.29 \times 50$ = $214.50	**28.** $\$7.08 \times 40$ = $283.20
29. $\$0.98 \times 21$ = $20.58	**30.** $\$4.27 \times 43$ = $183.61	**31.** $\$17.38 \times 54$ = $938.52	**32.** $\$46.50 \times 75$ = $3487.50

PROBLEM CORNER

This problem contains extra information, namely, the time involved.

In a leap-frog marathon fourteen students covered 400 laps around a track in 23 hours and 11 minutes. They averaged 42 leaps each lap. How many times did the students leap in the marathon? 16,800 leaps

More Practice, page 92

Objective 35

MULTIPLYING BY 3-DIGIT NUMBERS

Multiply. 478 × 293

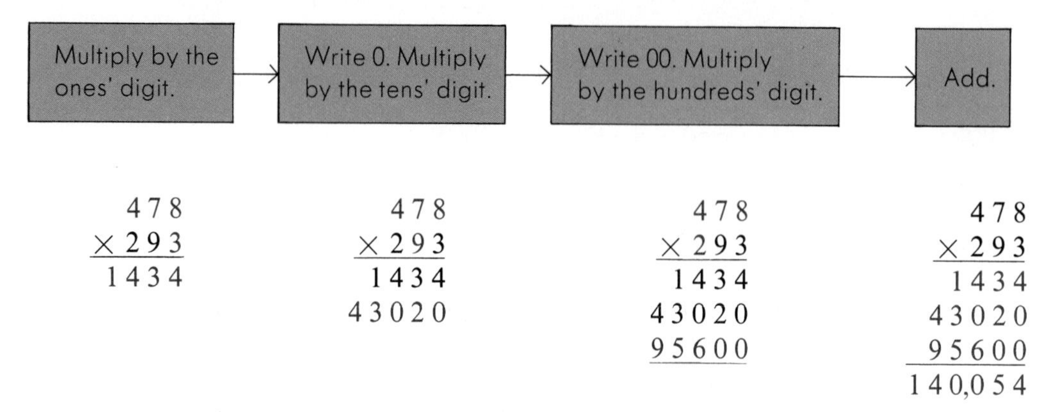

EXAMPLE 1. Multiply. 246 × 300

$$\begin{array}{r} 246 \\ \times\ 300 \\ \hline 73{,}800 \end{array}$$

Write 00.
Multiply by the hundreds' digit.

EXAMPLE 2. Multiply. 724 × 631

$$\begin{array}{r} 631 \\ \times\ 724 \\ \hline 2524 \\ 12620 \\ 441700 \\ \hline 456{,}844 \end{array}$$

Multiply by the ones' digit.
Write 0. Multiply by the tens' digit.
Write 00. Multiply by the hundreds' digit.
Add.

TRY THIS

1. $\begin{array}{r} 437 \\ \times\ 200 \\ \hline \end{array}$ 87,400

2. $\begin{array}{r} 345 \\ \times\ 234 \\ \hline \end{array}$ 80,730

3. $\begin{array}{r} \$\,4.75 \\ \times\ 321 \\ \hline \end{array}$ $1524.75

EXAMPLE 3. Multiply. 302 × 456

$$\begin{array}{r} 456 \\ \times\ 302 \\ \hline 912 \\ 136800 \\ \hline 137{,}712 \end{array}$$

Multiply by the ones' digit.
We can leave out multiplying by 0.
Write 00.
Multiply by the hundreds' digit.
Add.

TRY THIS

Multiply.

4. 647×405 — 262,035

5. 647×450 — 291,150

6. $\$19.62 \times 303$ — $5944.86

PRACTICE

For extra practice, use Making Practice Fun 17 with this lesson.
Use Quiz 6 after this Practice.

Multiply.

1. 948×200 — 189,600

2. 763×900 — 686,700

3. 6334×400 — 2,533,600

4. 3879×500 — 1,939,500

5. $\$7.68 \times 600$ — $4608.00

6. $\$3.97 \times 400$ — $1588.00

7. $\$72.49 \times 200$ — $14,498.00

8. $\$87.61 \times 800$ — $70,088.00

9. 325×523 — 169,975

10. 761×427 — 324,947

11. 7159×741 — 5,304,819

12. 5828×764 — 4,452,592

13. $\$3.91 \times 428$ — $1673.48

14. $\$8.39 \times 182$ — $1526.98

15. $\$24.27 \times 724$ — $17,571.48

16. $\$77.77 \times 777$ — $60,427.29

17. 624×404 — 252,096

18. 624×440 — 274,560

19. 1398×670 — 936,660

20. 4896×409 — 2,002,464

21. $\$7.65 \times 870$ — $6655.50

22. $\$8.24 \times 703$ — $5792.72

23. $\$19.05 \times 490$ — $9334.50

24. $\$77.30 \times 307$ — $23,731.10

PROBLEM CORNER

Jocelyn Sims is the manager of the school cafeteria. She computes the average daily sales for the 180-day school year.

Average Daily Sales	
Whole Milk	185
Chocolate Milk	118
Orange Juice	150

1. How many whole milks are sold? chocolate milks? 33,300; 21,240

2. How many orange juices are sold? 27,000

3. How many students attend the school? Not enough information

Objective 154

APPLICATION

COUNTING MONEY

Evaluating a pile of bills and coins requires careful counting.

THE A, B, C's OF MONEY COUNTING

Always Sort coins and bills into groups of like denominations.

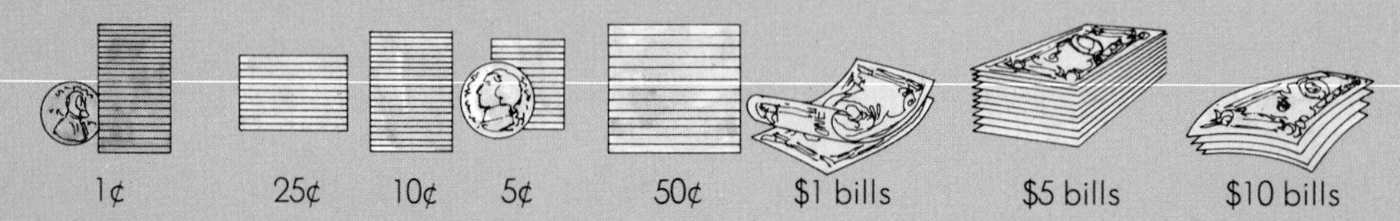

Be Sure to Count the amounts in two ways.

1. Count the number of coins and the number of bills. Then multiply.

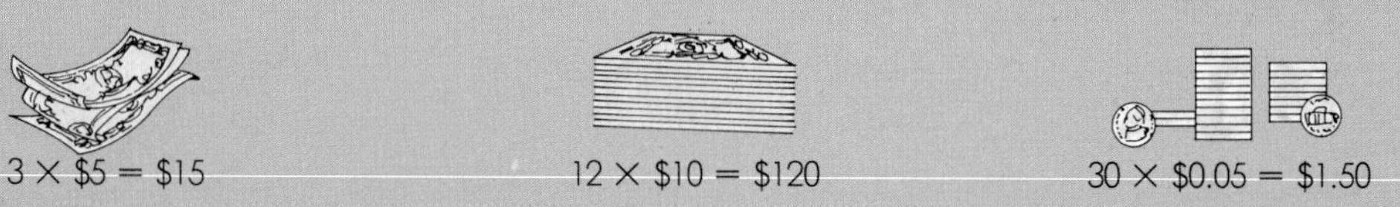

2. Count the total value of the coins and the bills by adding.

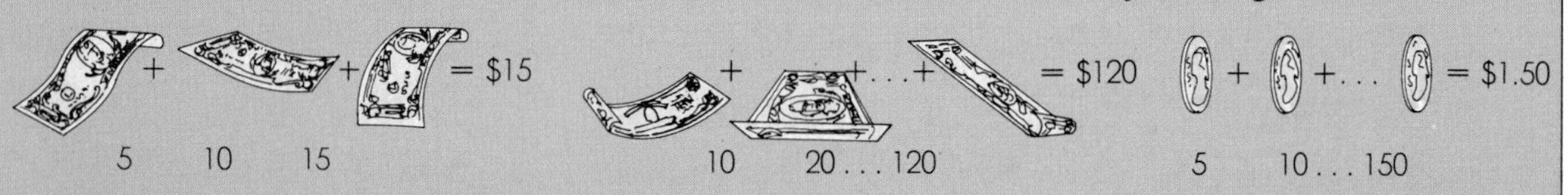

Carefully Add groups together to find the total.

This money has been sorted.

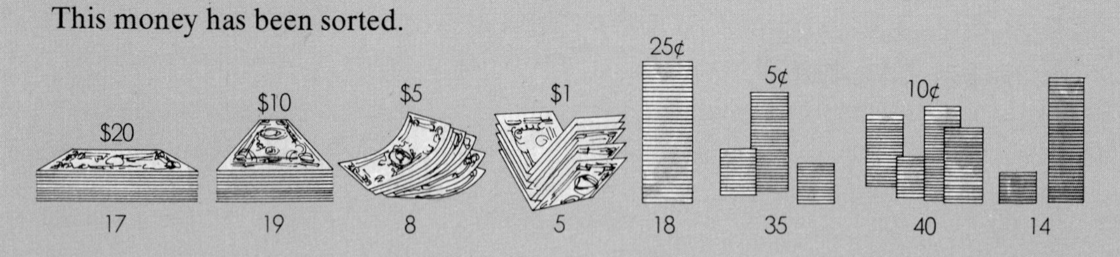

How much money is there:

1. in $ 20 bills? $340 **2.** in $ 10 bills? $190 **3.** in $ 5 bills? $40 **4.** in $ 1 bills? $5

5. in quarters? $4.50 **6.** in nickels? $1.75 **7.** in dimes? $4 **8.** in pennies? $0.14

9. What is the total in bills? $575

10. What is the total in coin? $10.39

11. What is the total amount of money? $585.39

Mr. Stewart manages a restaurant. The small bills and coins he uses to start each day are called the daily bank.

	Bills			**Coins**				
Denomination	$10	$ 5	$ 1	50¢	25¢	10¢	5¢	1¢
Number of Coins or Bills	2	7	17	20	40	50	40	10

12. How much of the daily bank is in bills? $72

13. How much of the daily bank is in coins? $27.10

14. How much is Mr. Stewart's daily bank? $99.10

Students might be asked to interview the manager of some small business to learn how the amount of the daily bank is determined.

Mr. Stewart needs to balance his cash and receipts. He totals the cash, checks and charges. Then he subtracts the daily bank. This tells the total amount of daily sales which should match the cash register sales.

Cash Drawer

Bills	Coins	Checks	Charges
2 – $100	13 – 50¢	$28.44	$25.20
2 – $ 50	23 – 25¢	$14.38	$22.09
16 – $ 20	48 – 10¢	$17.96	
8 – $ 10	15 – 5¢		
11 – $ 5	28 – 1¢		

Cash Register Tape

$796.25 Total Sales

15. How much money is in bills? $755

16. How much money is in coins? $18.08

17. What is the total in checks? $60.78

18. What is the total charges in charges? $47.29

19. What is the total in the cash drawer? $881.15

20. Subtract the amount in the daily bank. $782.05

Ask students to offer explanations for the discrepancy.

21. Does Mr. Stewart's count agree with the cash register tape? No

Objectives 36 and 37

SPECIAL PRODUCTS AND ESTIMATING

Multiply. 800×70

$$\begin{array}{rl} 800 & \textit{two 0's} \\ \times\ \ 70 & \textit{one 0} \\ \hline 56{,}000 & \textit{three 0's} \end{array}$$

1. Count the 0's in the factors.
2. Write the total number of 0's in the answer.
3. Multiply the non-zero digits.

Estimate. 795×68

$$\begin{array}{r} 795 \\ \times\ \ 68 \\ \hline \end{array} \longrightarrow \begin{array}{r} 800 \\ \times\ \ 70 \\ \hline 56{,}000 \end{array}$$

To estimate products:

1. Round the factors.
2. Multiply the rounded factors.

EXAMPLE 1. Multiply. 500×6

$$\begin{array}{r} 500 \\ \times\ \ \ 6 \\ \hline 3000 \end{array}$$

Count two 0's. Write two 0's in the answer.
Multiply the non-zero digits. $6 \times 5 = 30$

EXAMPLE 2. Multiply. 5000×600

$$\begin{array}{r} 5000 \\ \times\ \ 600 \\ \hline 3{,}000{,}000 \end{array}$$

Count five 0's. Write five 0's in the answer.
Multiply the non-zero digits. $6 \times 5 = 30$

TRY THIS

Multiply.

1. $\begin{array}{r} 7000 \\ \times\ \ \ 9 \\ \hline 63{,}000 \end{array}$
2. $\begin{array}{r} 70 \\ \times 10 \\ \hline 700 \end{array}$
3. $\begin{array}{r} 8000 \\ \times\ \ 50 \\ \hline 400{,}000 \end{array}$

EXAMPLES. Estimate.

3. $\begin{array}{r} 486 \\ \times\ \ \ 6 \\ \hline \end{array} \longrightarrow \begin{array}{r} 500 \\ \times\ \ \ 6 \\ \hline 3000 \end{array}$
4. $\begin{array}{r} 3984 \\ \times\ \ 246 \\ \hline \end{array} \longrightarrow \begin{array}{r} 4000 \\ \times\ \ 200 \\ \hline 800{,}000 \end{array}$

TRY THIS

Estimate.

4. $\begin{array}{r} 46 \\ \times\ 8 \\ \hline 400 \end{array}$
5. $\begin{array}{r} 784 \\ \times\ 65 \\ \hline 56{,}000 \end{array}$
6. $\begin{array}{r} 5204 \\ \times\ 395 \\ \hline 2{,}000{,}000 \end{array}$

PRACTICE

Multiply.

1. 80×9 = 720

2. 700×6 = 4200

3. 5000×4 = 20,000

4. 1000×9 = 9000

5. 50×70 = 3500

6. 90×80 = 7200

7. 900×90 = 81,000

8. 300×80 = 24,000

9. 700×100 = 70,000

10. 800×200 = 160,000

11. 7000×100 = 700,000

12. 3000×700 = 2,100,000

Estimate.

13. 76×8 = 640

14. 420×9 = 3600

15. 910×6 = 5400

16. 498×5 = 2500

17. 89×88 = 8100

18. 41×22 = 800

19. 590×38 = 24,000

20. 695×18 = 14,000

21. 888×888 = 810,000

22. 421×395 = 160,000

23. 7284×387 = 2,800,000

24. 6980×785 = 5,600,000

PROBLEM CORNER

These problems require 2 steps to obtain solutions.

1. The average attendance at a basketball game was 8198. The team played 42 games at home and 42 games away. Estimate the total number who attended the basketball games. 640,000

2. Find the exact number who attended the basketball games. 688,632

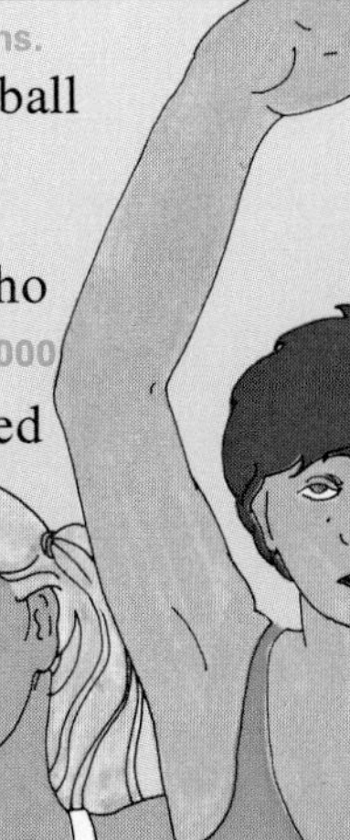

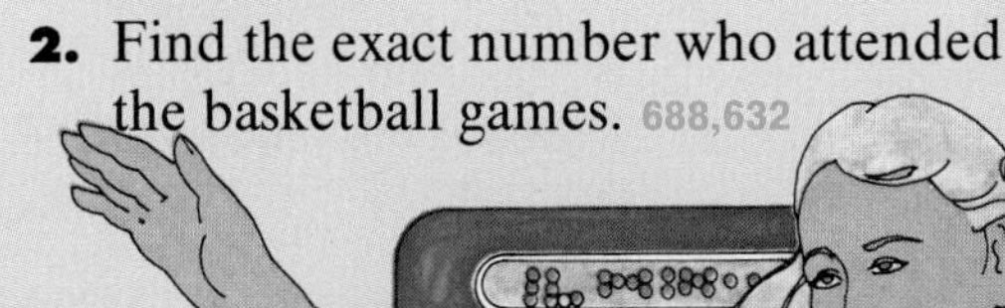

Objectives 38 and 39

MORE ON ESTIMATING PRODUCTS

With Money

We round to the same place, unless this causes unnecessary distortion. To the nearest ten \$8.98 × 79 is \$10 × 80 = \$800, but the method shown here is more accurate.

Estimate. \$8.98 × 79

$$\begin{array}{r} \$8.98 \\ \times\ 79 \\ \hline \end{array} \quad \begin{array}{r} \$\ \ 9 \\ \times\ 80 \\ \hline \$720 \end{array}$$

Round to the nearest dollar.
Round as usual. Multiply.

With Decimals

Estimate. 26.148 × 8.9

$$\begin{array}{r} 26.148 \\ \times\ \ 8.9 \\ \hline \end{array} \quad \begin{array}{r} 26 \\ \times\ 9 \\ \hline \end{array} \quad \begin{array}{r} 30 \\ \times\ 9 \\ \hline 270 \end{array}$$

Round each decimal to the nearest whole number.
Round as usual.
Multiply.

EXAMPLES. Estimate.

1. $\begin{array}{r} \$6.15 \\ \times\ 7 \\ \hline \end{array} \rightarrow \begin{array}{r} \$\ 6 \\ \times\ 7 \\ \hline \$42 \end{array}$

2. $\begin{array}{r} \$2.75 \\ \times 399 \\ \hline \end{array} \rightarrow \begin{array}{r} \$\ \ 3 \\ \times\ 400 \\ \hline \$1200 \end{array}$

3. $\begin{array}{r} \$78.85 \\ \times\ 42 \\ \hline \end{array} \rightarrow \begin{array}{r} \$79 \\ \times 40 \\ \hline \end{array} \rightarrow \begin{array}{r} \$\ \ 80 \\ \times\ 40 \\ \hline \$3200 \end{array}$

Round to the nearest dollar. Round as usual. Multiply.

TRY THIS

Estimate.

1. $\begin{array}{r} \$3.98 \\ \times\ 6 \\ \hline \end{array}$ \$24

2. $\begin{array}{r} \$8.13 \\ \times\ 61 \\ \hline \end{array}$ \$480

3. $\begin{array}{r} \$28.80 \\ \times\ 5 \\ \hline \end{array}$ \$150

4. $\begin{array}{r} \$37.75 \\ \times\ 78 \\ \hline \end{array}$ \$3200

EXAMPLES. Estimate.

4. $\begin{array}{r} 42.93 \\ \times\ 17.8 \\ \hline \end{array} \rightarrow \begin{array}{r} 43 \\ \times 18 \\ \hline \end{array} \rightarrow \begin{array}{r} 40 \\ \times\ 20 \\ \hline 800 \end{array}$

5. $\begin{array}{r} 3.126 \\ \times\ 6.82 \\ \hline \end{array} \rightarrow \begin{array}{r} 3 \\ \times\ 7 \\ \hline 21 \end{array}$

TRY THIS

Estimate.

5.	6.	7.
5.2176 × 68	5.2176 × 3.9	37.245 × 51.17
350	20	2000

PRACTICE

For extra practice, use Making Practice Fun 18 with this lesson.

Estimate.

1. $ 8.95 × 4 $36	**2.** $ 3.15 × 6 $18	**3.** $ 7.75 × 32 $240	**4.** $ 5.87 × 81 $480
5. $ 2.85 × 412 $1200	**6.** $ 7.89 × 878 $7200	**7.** $ 5.12 × 210 $1000	**8.** $ 8.88 × 398 $3600
9. $ 41.80 × 5 $200	**10.** $ 67.50 × 8 $560	**11.** $ 39.95 × 7 $280	**12.** $ 86.10 × 3 $270
13. $ 48.95 × 59 $3000	**14.** $ 73.15 × 43 $2800	**15.** $ 61.12 × 54 $3000	**16.** $ 88.95 × 21 $1800
17. 7.8 × 4 32	**18.** 8.13 × 3 24	**19.** 1.9147 × 31 60	**20.** 4.519 × 44 200
21. 7.9 × 6.1 48	**22.** 3.75 × 5.2 20	**23.** 3.1416 × 8.79 27	**24.** 6.437 × 5.628 36
25. 49.1 × 67.3 3500	**26.** 78.37 × 61.3 4800	**27.** 41.8 × 78.9 3200	**28.** 56.248 × 31.98 1800

PROBLEM CORNER

Problems 2 and 3 require more than one step.

Arrow Construction Company pays unskilled workers $ 4.88 per hour. Estimate what a worker makes during:

1. one 8-hour shift. $40 **2.** an 8-hour shift for a 6-day week. $240

3. an 8-hour shift for a 6-day week for 48 weeks per year. $10,000

Objectives 155 and 156

APPLICATION

It might be worthwhile to have some students act out the giving of change.

RECEIVING AND GIVING CHANGE

Change is usually given in the fewest number of coins and bills. There are two common methods of counting change.

METHOD A The change is counted out from the purchase price, smallest unit to largest.
A charge of $ 0.38 is paid with a $ 1 bill.

Clerk gives	Clerk says
1 penny	$ 0.39
1 penny	0.40
1 dime	0.50
1 half-dollar	1.00

METHOD B The cash register computes the change. Then it is counted out largest unit to smallest.
A charge of $ 0.38 is paid with a $ 1 bill. The register computes $ 0.62 change.

Clerk gives	Clerk says
1 half-dollar	$ 0.50
1 dime	0.60
1 penny	0.61
1 penny	0.62

PRACTICE

Was the change counted correctly?

1. Charge: $ 7.95 Paid with: $ 10

Clerk gives	Clerk says
1 nickel	$ 8.00
1 dollar	9.00
1 dollar	10.00 yes

2. Charge: $ 3.48 Paid with: $ 5

Clerk gives	Clerk says
1 penny	$ 3.49
1 half-dollar	3.50 no

3. Charge: $ 8.15 Paid with: $ 10

Clerk gives	Clerk says
1 quarter	$ 8.25
1 quarter	8.50
1 half-dollar	9.00
1 dollar	10.00 no

4. Charge: $ 4.35 Paid with: $ 20

Clerk gives	Clerk says
1 nickel	$ 4.40
1 dime	4.50
1 half-dollar	5.00
1 $ 5 bill	10.00
1 $ 10 bill	20.00 yes

5. Charge: $ 16.29 Paid with: $ 20
   Machine computes $ 3.71 change.

Clerk gives	Clerk says
1 dollar	$ 1.00
1 dollar	2.00
1 half-dollar	2.50
1 dime	2.60
1 dime	2.70
1 penny	2.71 no

6. Charge: $ 8.38 Paid with: $ 10
   Machine computes $ 1.62 change.

Clerk gives	Clerk says
1 dollar	$ 1.00
1 half-dollar	1.50
1 dime	1.60
1 penny	1.61
1 penny	1.62 yes

List the change that should be given. Use words and money notation. See answer section.

7. Charge: $ 8.25 Paid with: $ 10
8. Charge: $ 13.75 Paid with: $ 20
9. Charge: $ 3.68 Paid with: $ 5
10. Charge: $ 2.64 Paid with: $ 5
11. Charge: $ 8.25 Paid with: $ 10
    Machine computes $ 1.75 change.
12. Charge: $ 12.15 Paid with: $ 20
    Machine computes $ 7.85 change.
13. Charge: $ 17.76 Paid with: $ 20
    Machine computes $ 2.24 change.
14. Charge: $ 3.17 Paid with: $5
    Machine computes $ 1.83 change.
15. Charge: $ 13.19 Paid with: $ 20
    Machine computes $ 6.81 change.
16. Charge: $ 6.32 Paid with: $ 10
    Machine computes $ 3.68 change.

Objective 40

MULTIPLYING DECIMALS

Estimate to check if the products are reasonable.

Multiply. 6.58×7.4

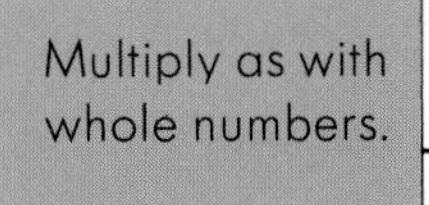

→ Count the number of decimal places in both factors.

→ Place the decimal point in the answer. The answer has the same number of decimal places as the two factors.

```
   6.58            6.58 } 3 decimal          6.58 } 3 places
×   7.4          ×  7.4 } places           ×  7.4 }
  2632             2632                      2632
 46060            46060                     46060
 48692            48692                     48.692   3 places
```

EXAMPLES. Multiply.

```
1.   6.384 } 3 places
   ×     7 }
    44.688   3 places

2.   6.784 } 4 places
   ×   8.1 }
      6784
    542720
    54.9504  4 places

3.     3.9 } 4 places
   × 0.007 }
    0.0273   4 places
```

Multiply.
Count the places.
Write 0's in the answer as needed.

TRY THIS

Multiply.

1. 3.8×6 = 22.8
2. 7.74×0.9 = 6.966
3. 374×0.24 = 89.76
4. 0.008×3 = 0.024
5. 0.07×0.7 = 0.049
6. 0.03×0.003 = 0.00009

PRACTICE

For extra practice, use Making Practice Fun 19 with this lesson.

Multiply.

1. 8.6×7 60.2	**2.** 47×0.9 42.3	**3.** 0.84×7 5.88	**4.** 7.3×0.6 4.38
5. 63×0.04 2.52	**6.** 7.8×0.09 0.702	**7.** 87×0.006 0.522	**8.** 8.7×0.06 0.522
9. 32.6×16 521.6	**10.** 7.28×5.4 39.312	**11.** 0.984×3.3 3.2472	**12.** 7.489×8.2 61.4098
13. 374×0.4 149.6	**14.** 569×0.05 28.45	**15.** 749×0.008 5.992	**16.** 8749×0.003 26.247
17. 43.7×0.8 34.96	**18.** 43.7×0.08 3.496	**19.** 4.37×0.8 3.496	**20.** 4.37×0.08 0.3496
21. 0.7×0.7 0.49	**22.** 0.9×0.4 0.36	**23.** 0.08×8 0.64	**24.** 0.06×7 0.42
25. 0.08×0.8 0.064	**26.** 0.2×0.2 0.04	**27.** 0.3×0.1 0.03	**28.** 0.01×7 0.07
29. 0.06×0.06 0.0036	**30.** 0.04×0.03 0.0012	**31.** 0.004×2 0.008	**32.** 0.007×1 0.007
33. 0.005×0.05 0.00025	**34.** 0.007×0.1 0.0007	**35.** 0.006×1 0.006	**36.** 0.003×0.002 0.000006

PROBLEM CORNER

Myra drinks an average of 5.48 cups of tea per day. How many cups of tea does she drink in a week? in a month (30 days)? 38.36; 164.40

Objectives 41 and 42

OTHER SPECIAL PRODUCTS

When we multiply by 10, 100, or 1000, there is a shorter method.

Multiply. 100×743.684

Count the number of 0's. → Move the decimal point to the right that number of places.

There are two. $100 \times 743.684 = 74,368.4$

EXAMPLES. Multiply.

1. $10 \times 7.98 = 79.8$ **2.** $100 \times 4.37 = 437$
3. $1000 \times 16.500 = 16,500$

TRY THIS

Multiply.

1. 100×6.3 630
2. 1000×9.678 9678
3. 10×8.64 86.4

When we multiply by 0.1, 0.01, or 0.001, there is also a shorter method.

Multiply. 0.01×743.68

Count the number of decimal places. → Move the decimal point to the left that number of places.

There are two. $0.01 \times 743.68 = 7.4368$

EXAMPLES. Multiply.

4. $0.1 \times 52.9 = 5.29$ **5.** $0.01 \times 21 = 0.21$
6. $0.001 \times 004.83 = 0.00483$

TRY THIS

Multiply.

4. 0.01 × 4.3 0.043 **5.** 0.1 × 0.87 0.087 **6.** 0.001 × 7.34 0.00734

PRACTICE

Multiply.

1. 10 × 6.74 67.4	**2.** 10 × 0.76 7.6	**3.** 10 × 9.785 97.85
4. 100 × 7.31 731	**5.** 100 × 7.842 784.2	**6.** 100 × 24.843 2484.3
7. 1000 × 7.6421 7642.1	**8.** 1000 × 97.694 97,694	**9.** 1000 × 82.984 82,984
10. 100 × 7.4 740	**11.** 100 × 19.4 1940	**12.** 100 × 0.7 70
13. 1000 × 9.6 9600	**14.** 1000 × 8.7 8700	**15.** 1000 × 0.3 300
16. 1000 × 9.13 9130	**17.** 1000 × 19.4 19,400	**18.** 1000 × 0.7 700
19. 0.1 × 18.5 1.85	**20.** 0.1 × 36.27 3.627	**21.** 0.1 × 8.7 0.87
22. 0.01 × 17.6 0.176	**23.** 0.01 × 297.4 2.974	**24.** 0.01 × 17.98 0.1798
25. 0.001 × 4748.2 4.7482	**26.** 0.001 × 6284.15 6.28415	**27.** 0.001 × 428.7 0.4287
28. 0.01 × 1.76 0.0176	**29.** 0.01 × 7.4 0.074	**30.** 0.01 × 0.5 0.005
31. 0.001 × 13.4 0.0134	**32.** 0.001 × 28.5 0.0285	**33.** 0.001 × 19.7 0.0197

PROBLEM CORNER

1. There are 100 centimeters in a meter. How many centimeters are there in 8.762 meters? 876.2

2. There are 0.001 kilograms in a gram. How many kilograms are there in 6842 grams? 6.842

More Practice, page 93

Objective 196

SOLVING PROBLEMS

Emphasize the problem solving steps throughout.

Sometimes solutions to problems require more than one step.

EXAMPLE. Finn sold 6 trays of popcorn at the stadium. Each tray contained 15 bags of popcorn. The cost of a bag of popcorn was 75¢. How much did Finn receive for the popcorn?

UNDERSTAND I must find the total for all the popcorn. I have enough information.

PLAN I have solved problems like this before. I must do two multiplications. I do not need a calculator.

CARRY OUT

$$\begin{array}{r} 15 \\ \times\ 6 \\ \hline 90 \end{array} \text{ bags of popcorn} \qquad \begin{array}{r} \$\,0.75 \\ \times\ 90 \\ \hline \$\,67.50 \end{array}$$

Finn received $ 67.50 for the popcorn.

LOOK BACK This answer is reasonable. The answer fits the problem.

When problems require more than one step, we complete each step, one at a time.

TRY THIS Katy bought 3 records at $ 5.95 and 2 records at $ 6.25. What was the total cost for the records? $30.35

PRACTICE

For extra practice, use Making Practice Fun 20 with this lesson. Use Quiz 7 after this Practice.

You may want to have students read these problems orally. Then discuss the steps involved before allowing the class to complete the work.

1. Freda sold 19 push-up packs of flowers at the school fair. Each pack contained 12 flowers. The cost of a flower was $ 0.08. How much did Freda receive for the flowers? $18.24

2. Rico bought 12 cans of soda at $ 0.40 and 6 bags of chips at $ 0.65 for the party. What was the total cost? $8.70

3. Simon earns $ 4.50 per hour for working Monday through Friday. He earns $ 6.75 per hour for weekend work. Last week he worked 8 hours a day on Monday, Wednesday, and Saturday. He worked 6 hours on both Tuesday and Sunday. How much did he earn last week? $193.50

4. Mull Reef, a famous race horse, broke his leg. His fans sent him 30 get well cards a day during the four weeks of his recuperation. How many cards did he receive? 840

5. Humans shed their skin continuously. In fact, there is a new outer layer of skin every 28 days. How many new skins will you have had by your 21st birthday? 273

6. The Booster Club bought 6 cartons of peanuts at $ 3.60 per carton. Each carton contained 72 bags of peanuts. The club sold the peanuts for 20¢ a bag. What was their gross profit after selling all the peanuts? $64.80

7. Kim Su runs 8 km on Monday, Wednesday, and Friday and 16 km on Tuesday, Thursday, and Saturday. How many kilometers does she run per week? 72 km

FAST FOOD RESTAURANT MANAGER

Ana Moreno manages the Apple Cellar Deli. She hires the part-time workers and orders the supplies.

Ana Moreno hires part-time workers for $4.25 per hour. Find the weekly salaries of these part-time workers.

1. Mark works 24 hours per week. $102.00
2. Patty works 8 hours each Saturday and Sunday. $68.00
3. Ben works 3 days per week; 6 hours per day. $76.50
4. Marcia works Monday through Friday for 3 hours each day. $63.75
5. Juan works 4 hours on weekdays and 8 hours on Saturday. $119.00

Ana finds the total cost of supplies ordered for the deli. Find the total cost for each item. Find the total cost of all items.

	Item	Amount	Cost Per Unit	Total Cost
6.	Bread	167 loaves	$0.46	$76.82
7.	Cheese	15 kg	3.88	$58.20
8.	Lettuce	22 heads	0.49	$10.78
9.	Tuna	36 cans	0.89	$32.04
10.	Paper Cups	6 cases	3.07	$18.42
11.	Mayonnaise	12 jars	1.26	$15.12
				Total $211.38

PROFIT = SELLING PRICE − COST

Figuring cost and profit is a major part of Ana Moreno's job.

Cost per Deli Sandwich	
Bread	$0.03
Meat	0.28
Cheese	0.12
Lettuce	0.05
Mayonnaise	0.14

MENU	
Deli Sandwich	$1.49
Tuna Sandwich	0.99
Vegetarian Delight	1.09
Low-Cal Special	0.59
Drinks:	
Small	0.29
Large	0.49

12. What does it cost to make a Deli sandwich? $0.62
13. What is the profit per Deli sandwich? $0.87
14. Last week 1079 Deli sandwiches were sold. What was the profit? $938.73
15. Ana checked this order ticket. What mistakes did she find?
Vegetarian Delight total, large drink cost and total, sub-total, tax, total amount

Apple Cellar Deli Order Form

Quantity	Item	Cost	Total
3	Deli Sandwich	$1.49	$4.47
4	Vegetarian Delight	1.09	5.36
2	Tuna Sandwich	.99	1.98
9	Drinks, large	.29	2.61
		Sub-Total	14.32
		Tax	.59
		Pay This Total Amount	$14.81

CHAPTER 3 REVIEW

For extra practice, use Making Practice Fun 21 with this lesson.

Multiply. pp. 68–84

1. 82 × 4 = 328

2. 96 × 8 = 768

3. 429 × 7 = 3003

4. $48.26 × 5 = $241.30

5. 48 × 60 = 2880

6. 386 × 24 = 9264

7. 7285 × 83 = 604,655

8. $19.48 × 17 = $331.16

9. 849 × 400 = 339,600

10. 859 × 648 = 556,632

11. 3982 × 560 = 2,229,920

12. $74.98 × 603 = $45,212.94

13. 70 × 60 = 4200

14. 800 × 20 = 16,000

15. 5000 × 60 = 300,000

16. 9000 × 400 = 3,600,000

17. 8.74 × 6 = 52.44

18. 4.92 × 6.3 = 30.996

19. 0.5 × 0.6 = 0.3

20. 0.02 × .3 = 0.006

21. 100 × 8.3   830

22. 1000 × 6.76   6760

23. 0.01 × 92   0.92

24. 0.001 × 3.17   0.00317

Estimate. pp. 76–78

25. 89 × 6   540

26. 47 × 61   3000

27. 742 × 891   630,000

28. 6500 × 412   2,800,000

29. $8.74 × 4 $36

30. $37.68 × 59 $2400

31. 9.12 × 32   270

32. 4.291 × 8.76   36

Application p. 80

33. Is the change counted correctly? No
Charge: $13.80. Paid with: $20.00.

Clerk gives	Clerk says
Item	$13.80
1 quarter	14.00
1 dollar	15.00
5 dollars	20.00

34. Complete the list to give change.
Charge: $2.35. Paid with: $5.00
Machine computes $2.65 change.

Clerk gives	Clerk says
1 dollar	$1.00
1 dollar	2.00
1 half-dollar	2.50
1 dime	2.60
1 nickel	2.65

CHAPTER 3 TEST

Multiply.

1. 72×3 216

2. 48×6 288

3. 743×7 5201

4. $\$92.46 \times 5$ \$462.30

5. 72×60 4320

6. 429×81 34,749

7. 7984×68 542,912

8. $\$49.50 \times 43$ \$2128.50

9. 812×900 730,800

10. 629×493 310,097

11. 6284×507 3,185,988

12. $\$42.85 \times 150$ \$6427.50

13. 40×30 1200

14. 700×10 7000

15. 6000×80 480,000

16. 4000×900 3,600,000

17. 9.23×7 64.61

18. 8.76×5.3 46.428

19. 0.7×0.7 0.49

20. 0.02×0.03 0.0006

21. 10×8.3 83

22. 1000×9.67 9670

23. 0.1×8.6 0.86

24. 0.01×36.5 0.365

Estimate.

25. 72×8 560

26. 56×65 4200

27. 525×89 45,000

28. 3994×313 1,200,000

29. $\$6.15 \times 9$ \$54

30. $\$8.79 \times 38$ \$360

31. 6.874×19 140

32. 5.199×6.91 35

Application

33. Is the change counted correctly? Yes
Charge: \$ 4.39. Paid with: \$ 10.00.

Clerk gives	Clerk says
Item	\$ 4.39
1 penny	4.40
1 dime	4.50
1 half-dollar	5.00
5 dollars	10.00

34. Complete the list to give change.
Charge: \$ 8.73. Paid with: \$ 20.00
Machine computes \$ 11.27 change.

Clerk gives	Clerk says
10 dollars	\$10
1 dollar	11
1 quarter	11.25
1 penny	11.26
1 penny	11.27

MORE PRACTICE

Multiply. Use after page 69.

1. 86×3 258 **2.** 97×4 388 **3.** 54×5 270 **4.** 32×6 192

5. 148×9 1332 **6.** 259×8 2072 **7.** 4360×7 30,520 **8.** $\$62.78 \times 4$ \$251.12

Multiply. Use after page 71.

9. 64×30 1920 **10.** 87×50 4350 **11.** 762×40 30,480 **12.** 8246×70 577,220

13. 64×34 2176 **14.** 87×76 6612 **15.** 43×18 774 **16.** 387×49 18,963

17. 425×86 36,550 **18.** 619×57 35,283 **19.** 7324×63 461,412 **20.** 7463×58 432,854

21. $\$5.98 \times 12$ \$71.76 **22.** $\$0.67 \times 24$ \$16.08 **23.** $\$87.96 \times 74$ \$6509.04 **24.** $\$63.87 \times 93$ \$5939.91

Multiply. Use after page 73.

25. 678×600 406,800 **26.** 974×300 292,200 **27.** 891×500 445,500 **28.** 680×800 544,000

29. 968×432 418,176 **30.** 427×628 268,156 **31.** 649×427 277,123 **32.** $\$4.98 \times 123$ \$612.54

33. 874×603 527,022 **34.** 428×640 273,920 **35.** 798×507 404,586 **36.** 640×640 409,600

Multiply. Use after page 77.

37. 30×40 1200 **38.** 60×80 4800 **39.** 400×40 16,000 **40.** 600×90 54,000

41. 300×600 180,000 **42.** 8000×20 160,000 **43.** 4000×600 2,400,000 **44.** 7000×6000 42,000,000

Estimate. Use after page 77.

1. 48×7 350
2. 64×8 480
3. 426×9 3600
4. 7287×8 56,000
5. 65×75 5600
6. 87×23 1800
7. 429×86 36,000
8. 788×55 48,000
9. 428×679 280,000
10. 478×310 150,000
11. 8294×625 4,800,000
12. 6942×7180 49,000,000

Estimate. Use after page 79.

13. $4.98 × 8 $40
14. $38.74 × 5 $200
15. $4.78 × 38 $200
16. $59.48 × 69 $4200
17. 374×6.8 2800
18. 2.9×7 21
19. 11.83×6 72
20. 5.2×37 200
21. 8.74×6.9 63
22. 4.316×6.125 24
23. 87.3×31.9 2700
24. 33.5×18.4 600

Multiply. Use after page 83.

25. 9.7×8 77.6
26. 6.42×7 44.94
27. 3.768×12 45.216
28. 224×8.77 1964.48
29. 6.9×2.7 18.63
30. 5.4×3.8 20.52
31. 1.63×8.2 13.366
32. 4.17×0.09 0.3753
33. 9.8×3.4 33.32
34. 2.76×3.8 10.488
35. 8.14×92.3 751.322
36. 6.483×0.4 2.5932
37. 7.6×5.8 44.08
38. 2.42×1.9 4.598
39. 17.3×0.6 10.38
40. 24.8×0.03 0.744
41. 0.7×0.6 0.42
42. 0.08×0.8 0.064
43. 0.01×0.4 0.004
44. 0.07×0.02 0.0014

Multiply. Use after page 85.

45. 10×7.6 76
46. 10×8.74 87.4
47. 100×9.68 968
48. 1000×3.94 3940
49. 0.1×86 8.6
50. 0.1×9.7 0.97
51. 0.01×94.3 0.943
52. 0.001×87 0.087

CALCULATOR CORNER

The level of difficulty can be varied by placing limits on the factors.

The emphasis should be on the estimate, not the exact answer.

MULTIPLICATION ESTIMATION

Players two
Materials two calculators, paper and pencils

SAMPLE

1. Mark asks Erin to estimate a product, such as 527 × 37. Erin asks Mark to estimate a product, such as 462 × 52.
2. Mark and Erin write down their problems. They estimate and multiply.
3. Mark and Erin calculate the exact answers to their problems.
4. They then calculate the differences between their estimates and the exact answers.

Erin

$$\begin{array}{r} 527 \\ \times\ 37 \\ \hline \end{array} \qquad \begin{array}{r} 500 \\ \times\ 40 \\ \hline 20{,}000 \end{array}$$

estimate	20,000
exact answer	− 19,499
	501

Mark

$$\begin{array}{r} 462 \\ \times\ 52 \\ \hline \end{array} \qquad \begin{array}{r} 460 \\ \times\ 50 \\ \hline 23{,}000 \end{array}$$

exact answer	24,024
estimate	− 23,000
	1,024

Since Erin's estimate was closer to the exact answer, she gets 1 point.

Play continues until a player gets 5 points.

Many times it is difficult to predict that an estimate will be close to the exact answer. Unlike in addition estimation, in multiplication estimation rounding closely to the numbers does not always result in a close estimate. For example, in Erin's problem if you round to the closer estimates of 530 and 40, the answer, 21,200, is less exact than Erin's estimate. In many cases, a close estimate depends on the numbers and how much is lost or gained by rounding.

CHAPTER 4 PRE-TEST

Divide. pp. 98–116

1. 30 ÷ 8 3r6 **2.** 84 ÷ 7 12 **3.** 407 ÷ 6 67r5 **4.** 184 ÷ 9 20r4

5. $5\overline{)920}$ 184 **6.** $7\overline{)45{,}721}$ 6531r4 **7.** $8\overline{)6412}$ 801r4 **8.** $4\overline{)\$\,19.53}$ \$4.88r1

9. $42\overline{)93}$ 2r9 **10.** $33\overline{)1519}$ 46r1 **11.** $28\overline{)14{,}571}$ 520r11 **12.** $56\overline{)\$\,209.64}$ \$3.74r20

13. $276\overline{)2484}$ 9 **14.** $136\overline{)3138}$ 23r10 **15.** $834\overline{)60{,}894}$ 73r12 **16.** $498\overline{)29{,}888}$ 60r8

17. $6\overline{)7.314}$ 1.219 **18.** $42\overline{)365.4}$ 8.7 **19.** $7\overline{)0.049}$ 0.007 **20.** $8\overline{)3}$ 0.375

21. $2.7\overline{)9.18}$ 3.4 **22.** $0.04\overline{)8.76}$ 219 **23.** $0.3\overline{)9}$ 30 **24.** $0.14\overline{)7}$ 50

25. 800 ÷ 10 80 **26.** 732.4 ÷ 100 7.324 **27.** 46,249.4 ÷ 1000 46.2494 **28.** 6 ÷ 10 0.6

Estimate. p. 112

29. $6\overline{)4981}$ 800 **30.** $78\overline{)2359}$ 30 **31.** $7\overline{)\$\,34.72}$ \$5 **32.** $9.3\overline{)46.176}$ 5

Find the quotient to the nearest hundredth. p. 118

33. $7\overline{)9}$ 1.29 **34.** $9\overline{)7}$ 0.78 **35.** $6.4\overline{)9.3}$ 1.45 **36.** $0.03\overline{)2}$ 66.67

Application p. 106

Period Ending	Hours	Rate	Earnings: Amount Earned Regular Rate	Earnings: Overtime and Other	Gross Earnings	Deductions: FICA	Deductions: Federal Withholding Tax	Deductions: State Withholding Tax	Deductions: Others	Net Pay
7-9-81	30	4.50		33.00		10.30	24.50	3.70	40.00	

Statement of Earnings and Deductions for Employee's Record—Detach Before Cashing Check

37. Find the gross earnings. \$168.00

38. Find the net pay. \$89.50

4
DIVIDING WHOLE NUMBERS AND DECIMALS

GETTING STARTED

FISHING FOR FACTORS

WHAT'S NEEDED The 30 squares of paper with digits marked on them as used in GOTCHA, Chapter 1. Take out the 0's.

THE RULES

1. On a sheet of paper each player makes a 5 × 5 grid. Fill in the grid with any 25 numbers.
2. A leader draws two digits from the box, multiplies them and says the product.

   Example: The leader draws 3 and 4 and says 12.
3. Each player crosses out all numbers that are factors of the product just found.

   Example: The factors of 12 are 1, 2, 3, 4, 6 and 12. Each player crosses out a 1, a 2, a 3, a 4, a 6, and a 12.
4. The leader continues to draw digits two at a time. The first player to cross out all 25 numbers is the winner.

1	1	1	1	2
2	2	3	3	4
4	5	5	6	6
7	8	9	10	12
15	25	32	48	56

Objective 43

1-DIGIT DIVISORS

In Example 1, it is not necessary to write 0 in the quotient above 6. But in Example 3 it is necessary to write 0 in the quotient above 7.

Divide and check. $98 \div 5$

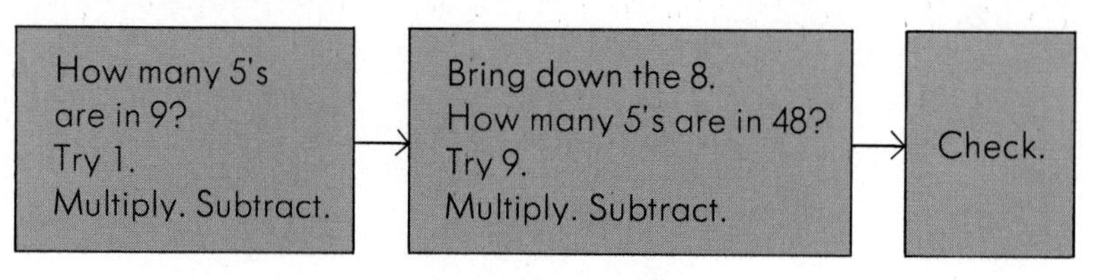


```
    1            19 r3           19 ← quotient
 5)98          5)98            × 5 ← divisor
   5             5↓            ----
   -             --              95
   4             48            + 3 ← remainder
                 45            ----
                 --              98 ← dividend
                  3
```

EXAMPLE 1. Divide and check. $61 \div 8$

```
   7 r5
 8)61      How many 8's are in 6? None.
   56      How many 8's are in 61? Try 7.
   --      Multiply. Subtract.
    5
```

Check.

```
    7
  × 8
  ---
   56
  + 5
  ---
   61
```

EXAMPLE 2. Divide. $87 \div 7$

```
  12 r3
 7)87      How many 7's are in 8? Try 1.
   7       Multiply. Subtract. Bring down the 7.
   -
   17      How many 7's are in 17? Try 2.
   14      Multiply. Subtract.
   --
    3
```

EXAMPLE 3. Divide. $87 \div 8$

```
  10 r7
 8)87      How many 8's are in 8? Try 1.
   8       How many 8's are in 7? Write 0 in the quotient.
   -
   07      Multiply. Subtract.
    0
   --
    7
```

TRY THIS

Divide and check.

1. $5\overline{)31}$ 6r1
2. $7\overline{)42}$ 6
3. $9\overline{)60}$ 6r6
4. $2\overline{)93}$ 46r1
5. $5\overline{)193}$ 38r3
6. $9\overline{)728}$ 80r8

PRACTICE

For extra practice, use Making Practice Fun 22 with this lesson.

Divide.

1. $6\overline{)38}$ 6r2
2. $7\overline{)65}$ 9r2
3. $3\overline{)25}$ 8r1
4. $2\overline{)16}$ 8
5. $4\overline{)31}$ 7r3
6. $5\overline{)42}$ 8r2
7. $8\overline{)42}$ 5r2
8. $6\overline{)42}$ 7
9. $3\overline{)74}$ 24r2
10. $2\overline{)74}$ 37
11. $4\overline{)74}$ 18r2
12. $5\overline{)74}$ 14r4
13. $6\overline{)87}$ 14r3
14. $7\overline{)87}$ 12r3
15. $3\overline{)87}$ 29
16. $4\overline{)87}$ 21r3
17. $2\overline{)196}$ 98
18. $3\overline{)196}$ 65r1
19. $4\overline{)196}$ 49
20. $8\overline{)196}$ 24r4
21. $6\overline{)374}$ 62r2
22. $7\overline{)429}$ 61r2
23. $9\overline{)849}$ 94r3
24. $7\overline{)123}$ 17r4
25. $2\overline{)47}$ 23r1
26. $8\overline{)749}$ 93r5
27. $4\overline{)98}$ 24r2
28. $6\overline{)349}$ 58r1
29. $4\overline{)81}$ 20r1
30. $7\overline{)76}$ 10r6
31. $4\overline{)43}$ 10r3
32. $8\overline{)81}$ 10r1
33. $7\overline{)143}$ 20r3
34. $9\overline{)456}$ 50r6
35. $6\overline{)425}$ 70r5
36. $5\overline{)403}$ 80r3

PROBLEM CORNER

1. There are 90 bottles in cartons. There are 6 bottles in each carton. How many cartons are there? 15 cartons
2. There are 75 people sitting in rows in the class. If there are 5 people in each row, how many rows are there? 15 rows

More Practice, page 126

Objectives 44 and 45

DIVIDING LARGER NUMBERS

Divide. $748 \div 3$

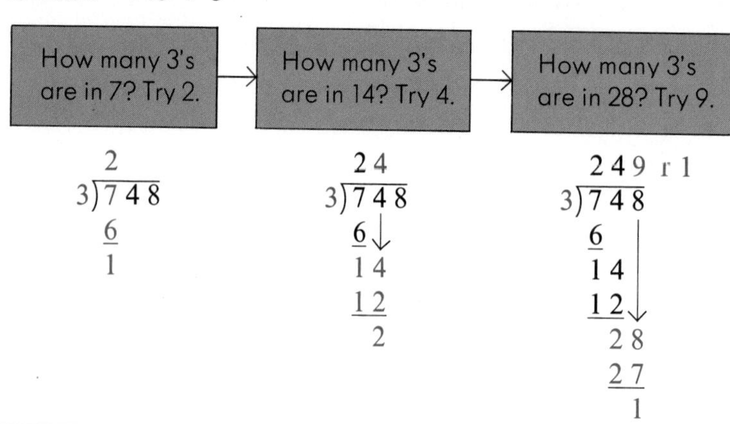

EXAMPLE 1. Divide. $4278 \div 5$

```
   855 r3
5)4278
  40
   27
   25
    28
    25
     3
```

How many 5's are in 4? None.
How many 5's are in 42? Try 8.
How many 5's are in 27? Try 5.
How many 5's are in 28? Try 5.

EXAMPLE 2. Divide. $7258 \div 9$

```
   806 r4
9)7258
  72
   05
    0
    58
    54
     4
```

How many 9's are in 7? None.
How many 9's are in 72? Try 8.
How many 9's are in 5? Write 0.
How many 9's are in 58? Try 6.

TRY THIS

Divide.

1. 6)947 (157r5) **2.** 2)1840 (920) **3.** 6)24,183 (4030r3)

EXAMPLE 3. Divide. $18.45 ÷ 5

```
     $  3.69
5)$ 18.45
    15
     34
     30
      45
      45
```

Divide as with whole numbers. Write money notation for the quotient and remainder.

TRY THIS

Divide.

4. 2)$ 87.46 $43.73 **5.** 8)$ 748.72 $93.59 **6.** 3)$ 18.12 $6.04

PRACTICE

Divide.

1. 798 ÷ 3 266 **2.** 493 ÷ 4 123r1 **3.** 584 ÷ 2 292 **4.** 327 ÷ 5 65r2

5. 824 ÷ 6 137r2 **6.** 930 ÷ 5 186 **7.** 947 ÷ 8 118r3 **8.** 628 ÷ 4 157

9. 2)1587 793r1 **10.** 7)3647 521 **11.** 5)4830 966 **12.** 6)4896 816

13. 3)9486 3162 **14.** 4)9484 2371 **15.** 2)7307 3653r1 **16.** 5)8734 1746r4

17. 8)17,498 2187r2 **18.** 5)26,489 5297r4 **19.** 6)40,294 6715r4 **20.** 9)65,287 7254r1

21. 7)709 101r2 **22.** 6)3665 610r5 **23.** 6)1234 205r4 **24.** 5)3650 730

25. 9)7203 800r3 **26.** 4)16,080 4020 **27.** 6)1805 300r5 **28.** 5)40,198 8039r3

29. 2)$ 5.98 $2.99 **30.** 5)$ 18.47 $3.69r2¢ **31.** 3)$ 36.49 $12.16r1¢ **32.** 5)$ 48.75 $9.75

33. 2)$ 1.78 $0.89 **34.** 7)$ 14.09 $2.01r2¢ **35.** 5)$ 16.00 $3.20 **36.** 9)$ 297.50 $33.05r5¢

37. 8)$ 25.92 $3.24 **38.** 6)$ 53.74 $8.95r4¢ **39.** 4)$ 95.12 $23.78 **40.** 7)$ 306.22 $43.74r4¢

More Practice, page 126

Objectives 45 and 46

2-DIGIT DIVISORS

It is a good idea to save trial calculations made to determine the correct digit in a quotient. In a longer problem the same calculation may be required a second time.

Divide. $792 \div 21$

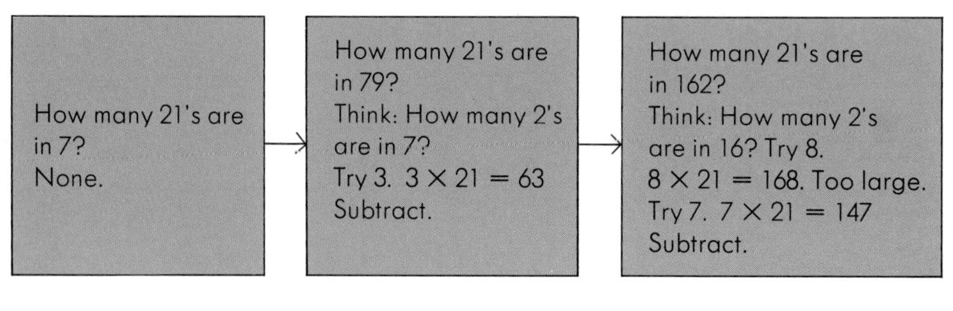


```
21)792
```

```
     3
21)792
   63
   16
```

```
     37 r 15
21)792
   63↓
   162
   147
    15
```

EXAMPLE 1. Divide. $187 \div 25$

```
      7 r 12
25)187
   175
    12
```

How many 25's are in 1? None.
How many 25's are in 18? None.
How many 25's are in 187?
Think: How many 2's are in 18?
Try 9. 9 × 25 = 225. Too large.
Try 8. 8 × 25 = 200. Still too large.
Try 7. 7 × 25 = 175. Subtract.

EXAMPLE 2. Divide. $458 \div 24$

```
     19 r 2
24)458
   24
   218
   216
     2
```

How many 24's are in 4? None.
How many 24's are in 45?
Think: How many 2's are in 4?
Try 2. 2 × 24 = 48. Too large.
Try 1. 1 × 24 = 24. Subtract.
How many 24's are in 218?
Think: How many 2's are in 21?
9 is the largest number to try.
9 × 24 = 216. Subtract.

TRY THIS

Divide.

1. $22\overline{)78}$ 3r12
2. $27\overline{)80}$ 2r26
3. $21\overline{)798}$ 38
4. $68\overline{)1427}$ 20r67
5. $56\overline{)48,832}$ 872
6. $72\overline{)\$96.48}$ \$1.34

PRACTICE

For extra practice, use Making Practice Fun 23 with this lesson.
Use Quiz 8 after this Practice.

Divide.

1. 96 ÷ 31 3r3
2. 88 ÷ 42 2r4
3. 648 ÷ 80 8r8
4. 400 ÷ 63 6r22
5. 310 ÷ 53 5r45
6. 200 ÷ 47 4r12
7. 90 ÷ 23 3r21
8. 81 ÷ 42 1r39
9. $27\overline{)348}$ 12r24
10. $31\overline{)738}$ 23r25
11. $54\overline{)892}$ 16r28
12. $87\overline{)986}$ 11r29
13. $43\overline{)1246}$ 28r42
14. $73\overline{)2654}$ 36r26
15. $92\overline{)3176}$ 34r48
16. $67\overline{)3825}$ 57r6
17. $31\overline{)5624}$ 181r13
18. $43\overline{)7396}$ 172
19. $71\overline{)9407}$ 132r35
20. $52\overline{)7000}$ 134r32
21. $63\overline{)13,545}$ 215
22. $47\overline{)32,641}$ 694r23
23. $84\overline{)62,412}$ 743
24. $72\overline{)41,641}$ 578r25
25. $23\overline{)9223}$ 401
26. $32\overline{)2249}$ 70r9
27. $44\overline{)26,620}$ 605
28. $56\overline{)1690}$ 30r10
29. $62\overline{)33,480}$ 540
30. $75\overline{)22,565}$ 300r65
31. $87\overline{)60,900}$ 700
32. $94\overline{)34,800}$ 370r20
33. $31\overline{)\$19.46}$ \$0.62r24¢
34. $52\overline{)\$102.96}$ \$1.98
35. $74\overline{)\$173.16}$ \$2.34
36. $45\overline{)\$55.50}$ \$1.23r15¢
37. $12\overline{)\$89.76}$ \$7.48
38. $21\overline{)\$22.89}$ \$1.09
39. $33\overline{)\$237.60}$ \$7.20
40. $54\overline{)\$432.07}$ \$8.00r7¢

PROBLEM CORNER

1. Jake has 500 dollars in 20-dollar bills. How many 20-dollar bills does he have? 25
2. There are 16 ounces in a pound. How many pounds are there in 528 ounces? 33

Objective 47

3-DIGIT DIVISORS

Divide. 19,674 ÷ 438

How many 438's are in 1967?
Think: How many 4's are in 19?
Try 4. 4 × 438 = 1752
Subtract.

→

How many 438's are in 2154?
Think: How many 4's are in 21?
Try 5. 5 × 438 = 2190. Too large.
Try 4. 4 × 438 = 1752
Subtract.

```
        4
438)19,674
    1752
     215
```

```
        44 r 402
438)19,674
    1752↓
     2154
     1752
      402
```

EXAMPLE 1. Divide. 2,357 ÷ 386

```
       6 r 41
386)2357
    2316
      41
```

How many 386's are in 2357?
Think: How many 3's are in 23?
Try 7. 7 × 386 = 2702. Too large.
Try 6. 6 × 386 = 2316. Subtract.

EXAMPLE 2. Divide. 42,497 ÷ 728

```
       58 r 273
728)42,497
    3640
     6097
     5824
      273
```

How many 728's are in 4249?
Think: How many 7's are in 42?
Try 6. 6 × 728 = 4368. Too large.
Try 5. 5 × 728 = 3640. Subtract.
How many 728's are in 6097?
Think: How many 7's are in 60?
Try 8. 8 × 728 = 5824. Subtract.

TRY THIS

Divide.

1. 221)895 4r11 **2.** 187)748 4 **3.** 826)7248 8r640

4. 827)9479 11r382 **5.** 741)63,874 86r148 **6.** 568)17,040 30

PRACTICE

For extra practice, use Making Practice Fun 24 with this lesson.

Divide.

1. 846 ÷ 326 2r194 **2.** 890 ÷ 212 4r42 **3.** 270 ÷ 125 2r20 **4.** 983 ÷ 640 1r343

5. 956 ÷ 407 2r142 **6.** 548 ÷ 113 4r96 **7.** 650 ÷ 386 1r264 **8.** 907 ÷ 287 3r46

9. 368)2944 8 **10.** 524)3700 7r32 **11.** 750)4500 6 **12.** 941)3774 4r10

13. 815)4095 5r20 **14.** 688)6292 9r100 **15.** 425)3000 7r25 **16.** 250)2000 8

17. 146)3289 22r77 **18.** 395)5545 14r15 **19.** 684)8308 12r100 **20.** 136)4624 34

21. 248)9425 38r1 **22.** 310)9641 31r31 **23.** 405)9750 24r30 **24.** 190)8360 44

25. 525)45,150 86 **26.** 736)31,000 42r88 **27.** 999)65,934 66 **28.** 888)78,144 88

29. 333)30,700 92r64 **30.** 140)10,640 76 **31.** 291)13,100 45r5 **32.** 407)19,129 47

33. 874)43,700 50 **34.** 736)30,000 40r560 **35.** 381)4000 10r190 **36.** 459)9200 20r20

37. 187)17,000 90r170 **38.** 512)40,960 80 **39.** 245)17,250 70r100 **40.** 650)39,000 60

PROBLEM CORNER

1. There are 144 items in a gross. How many gross are there in 1152 items? 8 gross

2. A plane averaged 328 miles per hour. It traveled 4592 miles. How many hours did the plane travel? 14 hours

More Practice, page 126

Objectives 157, 158, and 159

APPLICATION

Students might report on the ways in which their parents are paid.

RATES AND WAGES

One way to be paid is by the length of time worked. The time period can be the hour, the week, the month, or even the year.

The minimum wage law was first passed in 1938. This law is based on an hourly rate. The rate in 1938 was 25¢ per hour. Compare this to some recent rates in the graph.

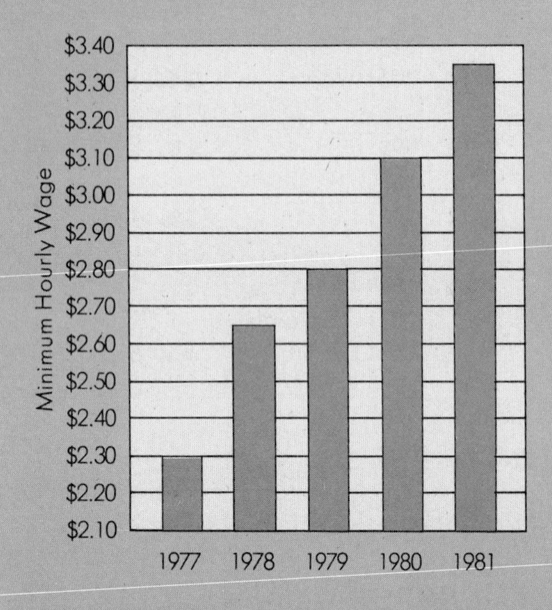

1. Find the minimum wage for 1980. $3.10
2. Find the minimum wage for 1981. $3.35
3. Find the daily minimum wage for an eight-hour day in:
   a. 1977 $18.40 b. 1979 $22.40 c. 1981 $26.80 d. 1938 $2.00
4. Find the weekly minimum wage for a forty-hour week in:
   a. 1977 $92.00 b. 1979 $112.00 c. 1981 $134.00 d. 1938 $10.00
5. A store clerk earned $ 10 for 40 hours at the minimum wage. What year was it? 1938
6. A secretary earned $ 18.40 for 8 hours at the minimum wage. What year was it? 1977

Try to have some examples of pay stubs available.

A pay stub is usually found attached to a paycheck.

Period Ending	Hours	Rate	Earnings: Amount Earned Regular Rate	Earnings: Overtime and Other	Gross Earnings	Deductions: FICA	Deductions: Federal Withholding Tax	Deductions: State Withholding Tax	Deductions: Others	Net Pay
3-4-82	40	4.50	180.00	68.00	248.00	14.72	35.18	9.67	Insurance 21.20	167.23

Statement of Earnings and Deductions for Employee's Record—Detach Before Cashing Check

Hours Worked (↑ Hours); Hourly Wage (↑ Rate); Overtime, Travel Allowance, Meal Allowance (↑ Overtime and Other); Social Security (↑ FICA); Pay After Deductions (↑ Net Pay)

7. How much was earned at the regular rate? $180.00
8. What were the overtime and other earnings? $68.00
9. What were the gross earnings? $248.00
10. What was the net pay? $167.23

Period Ending	Hours	Rate	Earnings: Amount Earned Regular Rate	Earnings: Overtime and Other	Gross Earnings	Deductions: FICA	Deductions: Federal Withholding Tax	Deductions: State Withholding Tax	Deductions: Others	Net Pay
4-7-82	48	4.00		73.00		15.10	37.15	9.88	Insurance 22.50	

Statement of Earnings and Deductions for Employee's Record—Detach Before Cashing Check

11. Find the amount earned at the regular rate. (Multiply hours by rate.) $192.00
12. Find the gross earnings. (Add overtime to amount earned at regular rate.) $265.00
13. Find the total deductions. $84.63
14. Find the net pay. (Subtract deductions from gross earnings.) $180.37

Period Ending	Hours	Rate	Earnings: Amount Earned Regular Rate	Earnings: Overtime and Other	Gross Earnings	Deductions: FICA	Deductions: Federal Withholding Tax	Deductions: State Withholding Tax	Deductions: Others	Net Pay
8-4-82	40	7.25		56.75		25.50	47.75	18.50	Ins. 23.00 Sav. 50.00	

Statement of Earnings and Deductions for Employee's Record—Detach Before Cashing Check

15. Find the gross earnings. $346.75
16. Find the net pay. $182.00

Objectives 48 and 49

DIVIDING A DECIMAL BY A WHOLE NUMBER

Divide. $17.421 \div 3$

Place the decimal point in the quotient.	→	Divide as with whole numbers.

```
3)17.421
```

```
    5.807
3)17.421
  15
   24
   24
    02
     0
     21
     21
      0
```

EXAMPLE 1. Divide. $171.36 \div 36$

```
       4.76
36)171.36
   144
    273
    252
     216
     216
```

Place the decimal point in the quotient.
Divide as with whole numbers.

TRY THIS

Divide.

1. $2\overline{)8.6}$ 4.3 **2.** $99\overline{)227.7}$ 2.3 **3.** $21\overline{)70.14}$ 3.34

EXAMPLES. Divide. $0.154 \div 2$

```
      0.077
2. 2)0.154
      0
      15
      14
       14
       14
```

0's must be placed in the quotient.

Divide. $15 \div 2$

```
       7.5
3.  2)15.0
      14
       10
       10
```

In problems like Example 3 a decimal point and 0's are written in the dividend. The division may then be carried to several places.

TRY THIS

Divide.

4. $5\overline{)2.5}$ 0.5
5. $9\overline{)0.18}$ 0.02
6. $8\overline{)0.016}$ 0.002
7. $5\overline{)12}$ 2.4
8. $6\overline{)3.9}$ 0.65
9. $8\overline{)3}$ 0.375

PRACTICE

For extra practice, use Making Practice Fun 25 with this lesson.

Divide.

1. $9.3 \div 3$ 3.1
2. $20.8 \div 4$ 5.2
3. $31.44 \div 6$ 5.24
4. $7.38 \div 2$ 3.69
5. $9.144 \div 8$ 1.143
6. $16.731 \div 9$ 1.859
7. $9.828 \div 4$ 2.457
8. $12.123 \div 3$ 4.041
9. $23\overline{)89.7}$ 3.9
10. $12\overline{)104.4}$ 8.7
11. $42\overline{)176.4}$ 4.2
12. $68\overline{)360.4}$ 5.3
13. $25\overline{)206.00}$ 8.24
14. $82\overline{)140.22}$ 1.71
15. $79\overline{)255.96}$ 3.24
16. $93\overline{)516.15}$ 5.55
17. $3\overline{)1.5}$ 0.5
18. $4\overline{)3.6}$ 0.9
19. $8\overline{)6.4}$ 0.8
20. $6\overline{)5.4}$ 0.9
21. $2\overline{)1.04}$ 0.52
22. $6\overline{)0.36}$ 0.06
23. $7\overline{)1.33}$ 0.19
24. $5\overline{)0.35}$ 0.07
25. $8\overline{)1.128}$ 0.141
26. $9\overline{)0.018}$ 0.002
27. $4\overline{)0.076}$ 0.019
28. $3\overline{)1.212}$ 0.404
29. $4\overline{)6}$ 1.5
30. $6\overline{)15}$ 2.5
31. $5\overline{)11}$ 2.2
32. $2\overline{)1}$ 0.5
33. $4\overline{)7}$ 1.75
34. $8\overline{)10}$ 1.25
35. $4\overline{)1.4}$ 0.35
36. $6\overline{)4.5}$ 0.75
37. $8\overline{)1}$ 0.125
38. $8\overline{)5}$ 0.625
39. $8\overline{)7}$ 0.875
40. $2\overline{)1.87}$ 0.935

PROBLEM CORNER

1. James had 15.3 meters of rope. He wants to cut it into 3 equal pieces. How long should each piece be? 5.1 meters
2. Six people are to work 9 pages of problems. How many pages should each do to share the work equally? 1.5 pages

Objective 50

SPECIAL QUOTIENTS

To divide by 10, 100, or 1000, we can use a short method.

Divide. $7000 \div 100$

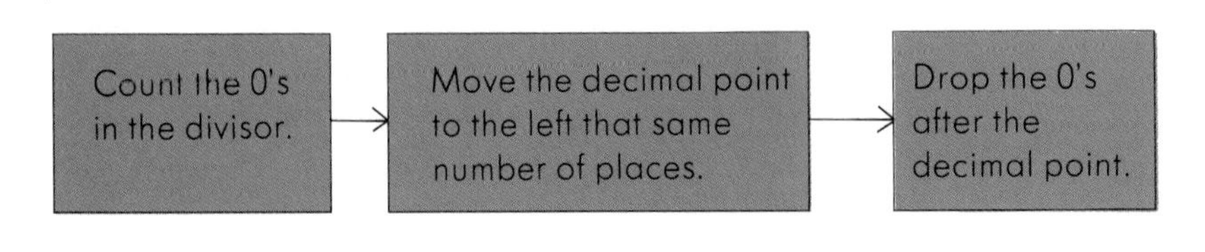

$7000 \div 100$ $\qquad$ $7000. \div 100$ $\qquad = \qquad 70$

EXAMPLES. Divide.

1. $500 \div 10 = 500. \div 10 = 50$

2. $732 \div 100 = 732. \div 100 = 7.32$

TRY THIS Divide.

1. $600 \div 100$ 6 **2.** $30{,}000 \div 1000$ 30 **3.** $52 \div 10$ 5.2

EXAMPLES. Divide.

3. $43.5 \div 10 = 4.35$ **4.** $527.45 \div 100 = 5.2745$

We sometimes need to write 0's.

EXAMPLES. Divide.

5. $8.7 \div 100 = 08.7 \div 100 = 0.087$

6. $9 \div 1000 = 009. \div 1000 = 0.009$

TRY THIS Divide.

4. $68.74 \div 10$ 6.874
5. $6 \div 100$ 0.06
6. $0.7 \div 1000$ 0.0007

PRACTICE Use Quiz 9 after this Practice.

Divide.

1. 80 ÷ 10 8 **2.** 700 ÷ 10 70 **3.** 8000 ÷ 10 800

4. 38 ÷ 10 3.8 **5.** 4748 ÷ 10 474.8 **6.** 386 ÷ 10 38.6

7. 600 ÷ 100 6 **8.** 4000 ÷ 100 40 **9.** 900 ÷ 100 9

10. 326 ÷ 100 3.26 **11.** 1348 ÷ 100 13.48 **12.** 248 ÷ 100 2.48

13. 8000 ÷ 1000 8 **14.** 22,000 ÷ 1000 22 **15.** 40,000 ÷ 1000 40

16. 3278 ÷ 1000 3.278 **17.** 4219 ÷ 1000 4.219 **18.** 73,489 ÷ 1000 73.489

19. 82.5 ÷ 10 8.25 **20.** 37.86 ÷ 10 3.786 **21.** 672.5 ÷ 10 67.25

22. 276.3 ÷ 100 2.763 **23.** 427.86 ÷ 100 4.2786 **24.** 425.4 ÷ 100 4.254

25. 8276.4 ÷ 1000 8.2764 **26.** 9387.6 ÷ 1000 9.3876 **27.** 18,427.3 ÷ 1000 18.4273

28. 9.3 ÷ 10 0.93 **29.** 0.9 ÷ 10 0.09 **30.** 6 ÷ 10 0.6

31. 8 ÷ 10 0.8 **32.** 4.7 ÷ 10 0.47 **33.** 0.3 ÷ 10 0.03

34. 3.4 ÷ 100 0.034 **35.** 0.1 ÷ 100 0.001 **36.** 2 ÷ 100 0.02

37. 3 ÷ 100 0.03 **38.** 3.8 ÷ 100 0.038 **39.** 0.5 ÷ 100 0.005

40. 4 ÷ 1000 0.004 **41.** 12 ÷ 1000 0.012 **42.** 86 ÷ 1000 0.086

PROBLEM CORNER

1. There are 10 millimeters in a centimeter. How many centimeters are there in 376 millimeters? 37.6 cm

2. There are 1000 kilograms in a metric ton. How many metric tons are there in 4236 kilograms? 4.236 metric tons

Objective 51

ESTIMATING QUOTIENTS

1-digit divisor: Estimate. 376 ÷ 6

> In finding exact quotients we always seek the number that is *less*. In Example 1 we would choose 5 × 7, because 5 × 8 is too large. But for an estimate, 5 × 8 is *closer* to 39.

Think: Is 6 × 6 or 6 × 7 closer to 37? Use 6. → Write 0's in the rest of the quotient.

$$6\overline{)376} \quad \text{quotient } 6 \qquad\qquad 6\overline{)376} \quad \text{quotient } 60$$

EXAMPLE 1. Estimate. 3974 ÷ 5

$$5\overline{)3974} \quad \text{quotient } 800$$

Think: Is 5 × 7 or 5 × 8 closer to 39? Use 8. Write 0's in the rest of the quotient.

EXAMPLE 2. Estimate. $18.78 ÷ 4

$$4\overline{)\$18.78} \longrightarrow 4\overline{)\$19} \longrightarrow 4\overline{)\$19} \quad \text{quotient } \$5$$

Round to the nearest dollar. 5 × 4 is closest to 19.

TRY THIS

Estimate.

1. $4\overline{)131}$ 30
2. $7\overline{)4128}$ 600
3. $6\overline{)\$41.47}$ $7

2-digit divisor: Estimate. 3764 ÷ 63

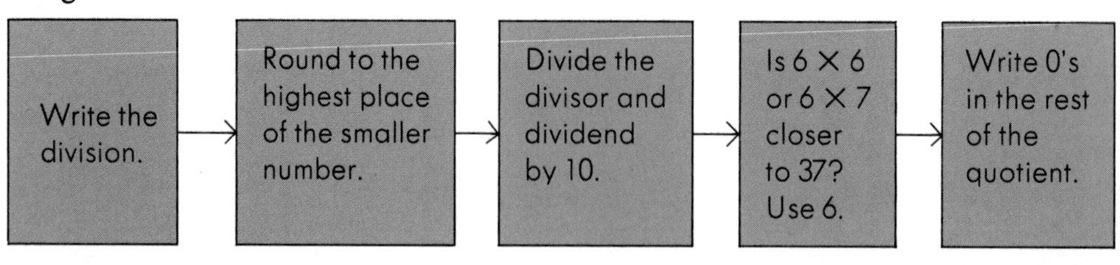

$$63\overline{)3764} \longrightarrow 60\overline{)3760} \longrightarrow 6\not{0}\overline{)376\not{0}} \longrightarrow 6\overline{)376} \;(6) \longrightarrow 6\overline{)376} \;(60)$$

EXAMPLE 3. Estimate. 5481 ÷ 69

$$69\overline{)5481} \longrightarrow 70\overline{)5480} \longrightarrow 7\overline{)548} \;(8) \longrightarrow 7\overline{)548} \;(80)$$

EXAMPLE 4. Estimate. $29.2 \div 3.89$

$3.89\overline{)29.2}$ $4\overline{)29}$ $\begin{array}{r}7\\4\overline{)29}\end{array}$

Round to the nearest whole numbers.
7×4 is closest to 29.

TRY THIS

Estimate.

4. $59\overline{)248}$ 4   **5.** $42\overline{)1931}$ 50   **6.** $9.12\overline{)44.129}$ 5

PRACTICE

For extra practice, use Making Practice Fun 26 with this lesson.

Estimate.

1. $8\overline{)576}$ 70	**2.** $4\overline{)312}$ 80	**3.** $5\overline{)398}$ 80	**4.** $6\overline{)435}$ 70
5. $4\overline{)2519}$ 600	**6.** $7\overline{)3486}$ 500	**7.** $9\overline{)8091}$ 900	**8.** $2\overline{)1234}$ 600
9. $4\overline{)31,497}$ 8000	**10.** $8\overline{)65,497}$ 8000	**11.** $5\overline{)44,987}$ 9000	**12.** $3\overline{)22,015}$ 7000
13. $9\overline{)\$36.70}$ \$4	**14.** $6\overline{)\$41.80}$ \$7	**15.** $3\overline{)\$20.80}$ \$7	**16.** $7\overline{)\$40.74}$ \$6
17. $6.2\overline{)18.13}$ 3	**18.** $8.78\overline{)26.1}$ 3	**19.** $3.124\overline{)16.01}$ 5	**20.** $7.8\overline{)41.8174}$ 5
21. $9.1\overline{)26.7}$ 3	**22.** $4.63\overline{)37.2}$ 7	**23.** $5.4\overline{)59.93}$ 12	**24.** $6.6\overline{)41.17}$ 6
25. $72\overline{)369}$ 5	**26.** $83\overline{)631}$ 8	**27.** $79\overline{)547}$ 7	**28.** $38\overline{)178}$ 4
29. $37\overline{)2756}$ 70	**30.** $93\overline{)4632}$ 50	**31.** $42\overline{)1943}$ 50	**32.** $49\overline{)3721}$ 70

PROBLEM CORNER

1. Six records cost \$ 41.98. Estimate the cost of one record. \$7

2. A car traveled 710 kilometers in 8 hours. Estimate how many kilometers it traveled in one hour. 90

3. Vito bought 3.3 kilograms of fish for \$ 11.37. Estimate how much the fish cost per kilogram. \$4

Objective 160

APPLICATION

Students should be asked to suggest other examples of jobs in which the worker is paid by the item handled or produced.

JOB AND FLAT RATES

Some income is based on the number of items handled or produced.

1. Sylvia delivered circulars. She was paid 3¢ per circular. Find how much she earned each day.

Day	Monday	Tuesday	Wednesday	Thursday	Friday
No. of Circulars	252	176	295	93	184
	$7.56	$5.28	$8.85	$2.79	$5.52

2. Sharon charges $ 3.50 to mow a lawn. How many lawns does she have to mow to earn:
   a. $ 63 18 b. $ 45.50 13 c. $ 87.50 25 d. $ 42 12
3. Some wallpaper hangers charge by the roll. Some charge by the hour. Who earned the most? Mr. Morales

   Mr. Morales hung 49 rolls in 7 hours. He charged $ 2.30 per roll.

   Mr. Yoshida hung 49 rolls in 7 hours. He charged $ 16.00 per hour.

The Quality Maintenance Service Co. pays employees flat rates for cleaning apartment units. Complete the following weekly payroll register by finding each employee's total pay.

	Employee	No. of Units Cleaned						Rate per Unit	
		M	T	W	Th	F	S		
4.	H. H. Elliott	10	13	8	15	7	9	$4.00	$248.00
5.	R. Salla	9	10	14	8	8	10	$3.50	$206.50
6.	T. M. Gill	11	12	12	6	18	6	$4.50	$292.50
7.	S. Blumenthal	14	5	12	9	12	11	$3.25	$204.75
8.	B. Vantucci	6	9	14	9	13	14	$4.75	$308.75
9.	N. Love	11	12	9	11	14	16	$5.00	$365.00

10. What was the total payroll? $1625.50
11. Suggest some reasons why the rate per unit varies. Answers will vary.

Burt Taylor makes armatures for the Electro Corporation. Use the graph to find the number of armatures Burt made each month from July to December.

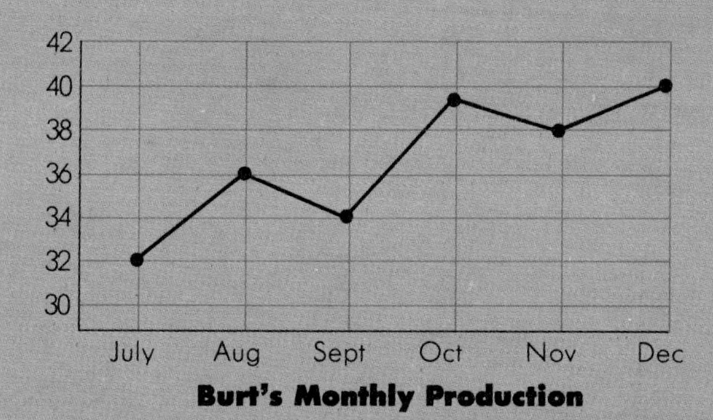

Burt's Monthly Production

12. Burt worked 22 days in July and 19 days in December. Find the average number of armatures produced each day during these months. Round to the nearest tenth. July: 1.5; December: 2.1
13. Burt earned a flat rate of $ 21.65 per armature. Find his earnings for August and November. August: $779.40; November: $822.70
14. How many more armatures did Burt produce in October than in September? 5

Objectives 52 and 53

DIVIDING A DECIMAL BY A DECIMAL

Divide. $8.73 \div 0.9$

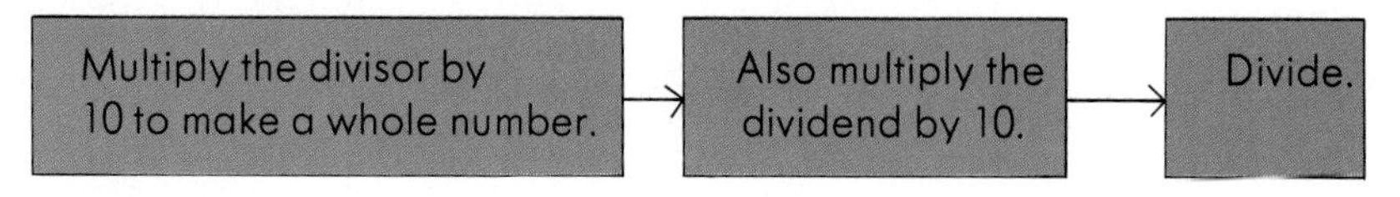

$0.9\overline{)8.73}$ $\quad$ $0.9\overline{)8.73}$ $\quad$ 9.7 / $0.9\overline{)8.73}$ / 81 / 63 / 63

The same multiplier is used in both places. The goal is to make the divisor a whole number. This method is equivalent to moving the decimal points in divisor and dividend the same number of places.

EXAMPLE 1. Divide. $22.63 \div 0.31$

$$\begin{array}{r} 73. \\ 0.31\overline{)22.63} \\ 217 \\ 93 \\ 93 \end{array}$$

Multiply the divisor by 100 to make a whole number.
Also multiply the dividend by 100.
Divide.

TRY THIS

Divide.

1. $0.3\overline{)7.44}$ 24.8 **2.** $0.23\overline{)9.43}$ 41 **3.** $0.002\overline{)8.642}$ 4321

EXAMPLE 2. Divide. $9.8 \div 0.002$

$$\begin{array}{r} 4900. \\ 0.002\overline{)9.800} \\ 8 \\ 18 \\ 18 \\ 00 \\ 0 \\ 00 \\ 0 \end{array}$$

Multiply the divisor and the dividend by 1000.
Write needed 0's.
Divide.

EXAMPLE 3. Divide. $8 \div 0.02$

$$\begin{array}{r} 400. \\ 0.02\overline{)8.00} \end{array}$$

Multiply the divisor and the dividend by 100.
Write needed 0's.

TRY THIS

Divide.

4. $0.08\overline{)6.4}$ 80 5. $0.012\overline{)7.2}$ 600 6. $0.71\overline{)355}$ 500

PRACTICE

For extra practice, use Making Practice Fun 27 with this lesson.

1. $4.5 \div 0.3$ 15	**2.** $67.4 \div 0.2$ 337	**3.** $82.17 \div 0.9$ 91.3	**4.** $4.26 \div 0.6$ 7.1
5. $8.4 \div 1.2$ 7	**6.** $220.1 \div 3.1$ 71	**7.** $78.3 \div 8.7$ 9	**8.** $4.32 \div 3.6$ 1.2
9. $0.04\overline{)1.68}$ 42	**10.** $0.01\overline{)7.692}$ 769.2	**11.** $0.02\overline{)41.28}$ 2064	**12.** $0.05\overline{)0.845}$ 16.9
13. $0.12\overline{)8.40}$ 70	**14.** $0.25\overline{)2.25}$ 9	**15.** $0.36\overline{)2.88}$ 8	**16.** $0.93\overline{)5.673}$ 6.1
17. $0.001\overline{)7.426}$ 7426	**18.** $0.002\overline{)7.426}$ 3713	**19.** $0.003\overline{)0.933}$ 311	**20.** $0.004\overline{)3.2848}$ 821.2
21. $0.005\overline{)8.7420}$ 1748.4	**22.** $0.012\overline{)0.072}$ 6	**23.** $0.123\overline{)0.984}$ 8	**24.** $0.423\overline{)2.538}$ 6
25. $0.05\overline{)1.5}$ 30	**26.** $0.13\overline{)5.2}$ 40	**27.** $0.34\overline{)27.2}$ 80	**28.** $0.08\overline{)17.6}$ 220
29. $0.004\overline{)1.6}$ 400	**30.** $0.003\overline{)6.36}$ 2120	**31.** $0.012\overline{)1.2}$ 100	**32.** $0.041\overline{)0.82}$ 20
33. $1.2\overline{)6}$ 5	**34.** $3.4\overline{)68}$ 20	**35.** $0.12\overline{)6}$ 50	**36.** $0.25\overline{)5}$ 20

PROBLEM CORNER

1. Mr. Harris has 6 liters of honey. He wants to put it in jars that hold 0.25 liters each. How many jars does he need? 24
2. If your heart beats once every 0.8 seconds, how many times will it beat in a minute? 75

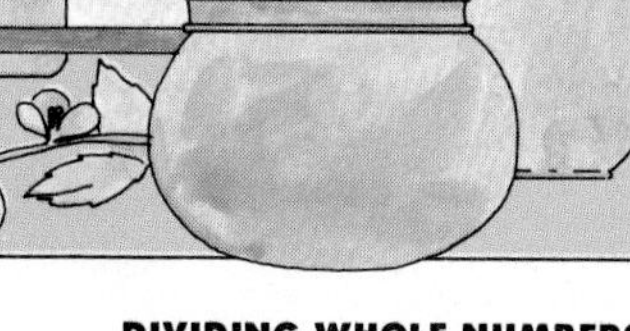

Objective 54

ROUNDING QUOTIENTS

Find the quotient to the nearest tenth. $5.3 \div 0.7$

Divide to 1 more place than to be rounded.	→	Round to the given place.

```
      7.5 7
0.7)5.3 0 0
    4 9
      4 0
      3 5
        5 0
        4 9
          1
```

$7.57 \approx 7.6$

$\approx$ *means is approximately equal to.*

EXAMPLE 1. Find the quotient to the nearest hundredth. $7 \div 0.9$

```
      7.7 7 7 ≈ 7.7 8
0.9)7.0 0 0 0
    6 3
      7 0
      6 3
        7 0
        6 3
          7 0
          6 3
            7
```

Divide to thousandths.
Round to hundredths.

EXAMPLE 2. Find the quotient to the nearest thousandth. $1 \div 3$

```
  0.3 3 3 3 ≈ 0.333
3)1.0 0 0 0
    9
    1 0
      9
      1 0
        9
        1 0
          9
            1
```

Divide to ten-thousandths.
Round to thousandths.

TRY THIS

Find each quotient to the nearest tenth.

1. $\overset{0.8}{8\overline{)6}}$ **2.** $\overset{23.5}{0.43\overline{)10.1}}$

Find each quotient to the nearest hundredth.

3. $\overset{6.67}{0.3\overline{)2}}$ **4.** $\overset{50.71}{0.14\overline{)7.1}}$

Find each quotient to the nearest thousandth.

5. $\overset{1.143}{7\overline{)8}}$ **6.** $\overset{13.286}{0.7\overline{)9.3}}$

PRACTICE

For extra practice, use Making Practice Fun 28 with this lesson.

Find each quotient to the nearest tenth.

1. $\overset{1.3}{3\overline{)4}}$ **2.** $\overset{1.3}{4\overline{)5}}$ **3.** $\overset{0.9}{7\overline{)6}}$ **4.** $\overset{0.9}{8\overline{)7}}$

5. $\overset{2.7}{3.2\overline{)8.7}}$ **6.** $\overset{2.1}{4.4\overline{)9.3}}$ **7.** $\overset{157.1}{0.07\overline{)11}}$ **8.** $\overset{88.9}{0.09\overline{)8}}$

Find each quotient to the nearest hundredth.

9. $\overset{0.89}{9\overline{)8}}$ **10.** $\overset{0.64}{11\overline{)7}}$ **11.** $\overset{1.22}{9\overline{)11}}$ **12.** $\overset{3.67}{3\overline{)11}}$

13. $\overset{3.33}{0.03\overline{)0.1}}$ **14.** $\overset{66.67}{0.03\overline{)2}}$ **15.** $\overset{4.30}{9.2\overline{)39.6}}$ **16.** $\overset{4.82}{1.7\overline{)8.2}}$

Find each quotient to the nearest thousandth.

17. $\overset{3.333}{3\overline{)10}}$ **18.** $\overset{1.429}{7\overline{)10}}$ **19.** $\overset{0.429}{7\overline{)3}}$ **20.** $\overset{0.111}{9\overline{)1}}$

21. $\overset{3.333}{0.3\overline{)1}}$ **22.** $\overset{2.857}{0.7\overline{)2}}$ **23.** $\overset{4.444}{0.9\overline{)4}}$ **24.** $\overset{7.778}{0.9\overline{)7}}$

PROBLEM CORNER

1. If you do 15 push ups in 11 seconds, how many push ups is that per second? Find the answer to the nearest hundredth. 1.36

2. If you do 20 sit ups in 11 seconds, how many sit ups is that per second? Find the answer to the nearest tenth. 1.8

More Practice, page 127

Objective 196

SOLVING PROBLEMS

Discuss this important idea with your class. Have students list some examples showing when this is true.

Approximate answers are often all that is needed for a problem.

EXAMPLE. An accounting firm paid the following annual salaries to its office employees.

Stenographer	$ 11,018
Typist	$ 9,276
File Clerk	$ 6,621
Secretary	$ 14,435

About how much money should be budgeted for office employees salaries?

For budget purposes, an approximate answer is sufficient. Round each salary and add to get an approximate total.

Stenographer	$ 11,000	
Typist	9,000	
File Clerk	7,000	
Secretary	+ 14,000	
	$ 41,000	Approximate Total

TRY THIS The accounting firm pays its chief accountant $ 39,895 annually. The two junior accountants are paid $ 18,115 annually. About how much should be budgeted for accountants' salaries? $76,000

PRACTICE

For extra practice, use Making Practice Fun 29 with this lesson. Use Quiz 10 after this Practice.

1. Fred Wise bought 600 shares of stock. The annual dividend on the stock was $ 2.10. About how much income can Fred expect yearly from the dividends? $1200
2. A population study showed that a large city had 577,642 people living in an area of just 47 square miles. About how many people live in each square mile? 11,552

3. In 1936, a 37-year-old man walked from Calcutta to Constantinople and back. He made the 5589 mile trip in just 59 days. About how many miles did he average per day? 90

4. One of the first automobile races took place in Chicago in 1895. The winner covered the 54.36 mile course in seven hours and 17 minutes. Estimate the average speed. 8 mph

5. The eruption of the volcano on Krakatoa in 1883 sent a tidal wave halfway around the world. The wave reached heights of 120 feet. It traveled 5450 miles to the South African coast in 12 hours. Estimate its average speed. 545 mph

6. Alaska has an area of 586,412 square miles. Rhode Island has an area of 1214 square miles. About how many times larger is Alaska than Rhode Island? 586

7. The distance from San Francisco, California to New York City is 3045 miles. About how many hours would it take to drive from San Francisco to New York City at an average speed of 55 miles per hour? 50

Terry Willis owns the Quality Maintenance Service. He provides janitorial and maintenance service for factories, offices, apartments, and homes.

Terry's business employs many people. People are employed to clean, supervise, and run his office. Their pay depends on the type of job they do.

1. Terry Willis draws a yearly salary of $ 22,500. How much is this monthly? $1875
2. Job supervisors earn $ 15,000 per year. What is their monthly salary? $1250
3. With overtime pay, full-time skilled employees may make $ 8800 annually. They are usually given 52 weeks pay for 50 weeks work. What is their weekly salary? $169.23
4. Shirley King is paid at the rate of $ 2.80 per hour. Her overtime rate is 1.5 times the regular rate. During a busy season, Shirley works a full 40-hour week plus 12 hours overtime. What does she earn that week? $162.40

Quality Maintenance uses a sliding scale for the cost of window washing. The more windows washed, the less the cost per window.

Number of Windows	Cost
1–20	$ 60.00
21–30	$ 85.00
31–50	$132.50
51–75	$180.00
76–100	$205.00

How much does it cost per window for the following jobs?

5. 20 windows $3.00
6. 22 windows $3.86
7. 29 windows $2.93
8. 37 windows $3.58
9. 53 windows $3.40
10. 72 windows $2.50
11. 93 windows $2.20
12. 15 windows $4.00

Terry Willis studied the graph of the gross monthly income.

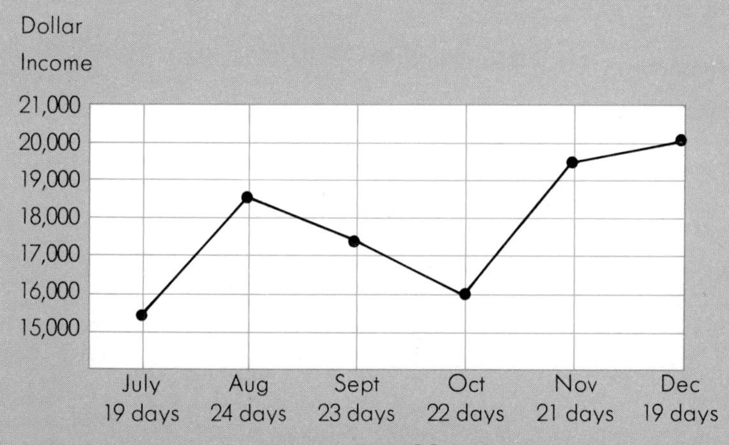

Gross Monthly Income

13. In which month was the income lowest? highest? July; December

Terry computed the daily gross income by considering the number of work days per month.

14. To the nearest dollar, what was the daily gross income per month? July: $816; Aug: $771; Sept: $761; Oct: $727; Nov: $929; Dec: $1053
15. Estimate the average daily gross income for the six months. $843

CHAPTER 4 REVIEW

For extra practice, use Making Practice Fun 30 with this lesson.

Divide. p. 98

1. $41 \div 7$ 5r6 **2.** $72 \div 3$ 24 **3.** $713 \div 8$ 89r1 **4.** $243 \div 6$ 40r3

5. $4\overline{)732}$ 183 **6.** $5\overline{)36{,}498}$ 7299r3 **7.** $9\overline{)4511}$ 501r2 **8.** $7\overline{)\$31.44}$ $4.49r1¢

9. $34\overline{)87}$ 2r19 **10.** $82\overline{)2641}$ 32r17 **11.** $65\overline{)52{,}455}$ 807 **12.** $43\overline{)\$355.64}$ $8.27r3¢

13. $512\overline{)4096}$ 8 **14.** $147\overline{)4716}$ 32r12 **15.** $947\overline{)82{,}389}$ 87 **16.** $747\overline{)29{,}993}$ 40r113

17. $4\overline{)8.712}$ 2.178 **18.** $56\overline{)268.8}$ 4.8 **19.** $3\overline{)0.012}$ 0.004 **20.** $5\overline{)4}$ 0.8

21. $3.6\overline{)34.92}$ 9.7 **22.** $0.05\overline{)3.75}$ 75 **23.** $0.2\overline{)8}$ 40 **24.** $0.25\overline{)5}$ 20

25. $700 \div 100$ 7 **26.** $48.7 \div 10$ 4.87 **27.** $7348.2 \div 1000$ 7.3482 **28.** $9 \div 100$ 0.09

Estimate. p. 112

29. $7\overline{)3480}$ 500 **30.** $47\overline{)2436}$ 50 **31.** $8\overline{)\$47.78}$ $6 **32.** $3.12\overline{)13.721}$ 4

Find the quotient to the nearest tenth. p. 118

33. $4\overline{)7}$ 1.8 **34.** $7\overline{)4}$ 0.6 **35.** $4.7\overline{)8.9}$ 1.9 **36.** $0.07\overline{)4}$ 57.1

Application p. 106

Period Ending			Earnings		Gross Earnings	Deductions				Net Pay
	Hours	Rate	Amount Earned Regular Rate	Overtime and Other		FICA	Federal Withholding Tax	State Withholding Tax	Others	
11-4-80	48	5.00		13.50		15.54	44.00	9.60	19.60	

Statement of Earnings and Deductions for Employee's Record—Detach Before Cashing Check

37. Find the gross earnings. $253.50

38. Find the net pay. $164.76

CHAPTER 4 TEST

Divide.

1. $35 \div 8$ 4r3 **2.** $72 \div 4$ 18 **3.** $473 \div 9$ 52r5 **4.** $284 \div 7$ 40r4

5. $5\overline{)620}$ 124 **6.** $6\overline{)37{,}492}$ 6248r4 **7.** $3\overline{)1514}$ 504r2 **8.** $8\overline{)\$25.83}$ \$3.22r7¢

9. $45\overline{)96}$ 2r6 **10.** $93\overline{)1302}$ 14 **11.** $76\overline{)30{,}786}$ 405r6 **12.** $54\overline{)\$511.38}$ \$9.47

13. $623\overline{)4463}$ 7r102 **14.** $258\overline{)5934}$ 23 **15.** $369\overline{)35{,}893}$ 97r100 **16.** $858\overline{)51{,}480}$ 60

17. $5\overline{)7.315}$ 1.463 **18.** $67\overline{)288.1}$ 4.3 **19.** $4\overline{)0.024}$ 0.006 **20.** $4\overline{)3}$ 0.75

21. $4.7\overline{)40.42}$ 8.6 **22.** $0.06\overline{)3.12}$ 52 **23.** $0.3\overline{)9}$ 30 **24.** $0.35\overline{)7}$ 20

25. $8000 \div 100$ 80 **26.** $538.3 \div 10$ 53.83 **27.** $4298.7 \div 1000$ 4.2987 **28.** $3 \div 10$ 0.3

Estimate.

29. $8\overline{)3142}$ 400 **30.** $58\overline{)3547}$ 60 **31.** $9\overline{)\$62.50}$ \$7 **32.** $4.9\overline{)24.73}$ 5

Find the quotient to the nearest hundredth.

33. $3\overline{)8}$ 2.67 **34.** $8\overline{)3}$ 0.38 **35.** $5.8\overline{)7.3}$ 1.26 **36.** $0.09\overline{)5}$ 55.56

Application

Period Ending	Hours	Rate	Earnings: Amount Earned Regular Rate	Earnings: Overtime and Other	Gross Earnings	Deductions: FICA	Deductions: Federal Withholding Tax	Deductions: State Withholding Tax	Deductions: Others	Net Pay
9-21-82	40	6.50		40.25		18.41	57.50	14.30	29.10	

Statement of Earnings and Deductions for Employee's Record—Detach Before Cashing Check

37. Find the gross earnings. \$300.25

38. Find the net pay. \$180.94

MORE PRACTICE

Divide. Use after page 99.

1. 60 ÷ 8 7r4	**2.** 52 ÷ 7 7r3	**3.** 15 ÷ 6 2r3	**4.** 80 ÷ 9 8r8
5. 2)87 43r1	**6.** 6)96 16	**7.** 4)174 43r2	**8.** 6)438 73
9. 7)74 10r4	**10.** 8)643 80r3	**11.** 4)160 40	**12.** 5)353 70r3

Divide. Use after page 101.

13. 897 ÷ 6 149r3	**14.** 873 ÷ 3 291	**15.** 4398 ÷ 7 628r2	**16.** 3920 ÷ 5 784
17. 6)7352 1225r2	**18.** 4)9287 2321r3	**19.** 7)68,749 9821r2	**20.** 5)47,845 9569
21. 3)313 104r1	**22.** 7)4256 608	**23.** 4)24,024 6006	**24.** 8)64,053 8006r5
25. 9)8209 912r1	**26.** 8)9832 1229	**27.** 6)4490 748r2	**28.** 4)33,907 8476r3

Divide. Use after page 103.

29. 70 ÷ 33 2r4	**30.** 90 ÷ 68 1r22	**31.** 440 ÷ 87 5r5	**32.** 400 ÷ 56 7r8
33. 24)300 12r12	**34.** 83)913 11	**35.** 74)1702 23	**36.** 67)3179 47r30
37. 61)6832 112	**38.** 48)6336 132	**39.** 78)47,156 604r44	**40.** 19)14,825 780r5
41. 17)2079 122r5	**42.** 24)9816 409	**43.** 36)83,495 2319r11	**44.** 59)11,992 203r15

Divide. Use after page 105.

45. 846 ÷ 731 1r115	**46.** 4992 ÷ 624 8	**47.** 2292 ÷ 326 7r10	**48.** 5200 ÷ 850 6r100
49. 912)11,967 13r111	**50.** 444)24,879 56r15	**51.** 399)16,758 42	**52.** 479)28,439 59r178
53. 248)17,460 70r100	**54.** 821)24,630 30	**55.** 596)53,765 90r125	**56.** 199)16,120 81r1
57. 123)4786 38r112	**58.** 487)60,529 124r141	**59.** 706)35,490 50r190	**60.** 263)19,248 73r49
61. 504)8964 17r396	**62.** 613)35,560 58r6	**63.** 345)29,824 86r154	**64.** 759)80,468 106r14
65. 699)49,629 71	**66.** 936)75,438 80r558	**67.** 288)47,735 165r215	**68.** 177)14,356 81r19

Divide. Use after page 109.

1. $26.88 \div 7$ 3.84 **2.** $46.4 \div 8$ 5.8 **3.** $14.73 \div 3$ 4.91 **4.** $2.490 \div 6$ 0.415

5. $82\overline{)787.2}$ 9.6 **6.** $31\overline{)107.88}$ 3.48 **7.** $43\overline{)73.1}$ 1.7 **8.** $62\overline{)7.626}$ 0.123

9. $6\overline{)0.18}$ 0.03 **10.** $4\overline{)0.008}$ 0.002 **11.** $8\overline{)4}$ 0.5 **12.** $10\overline{)2}$ 0.2

Divide. Use after page 111.

13. $46.4 \div 10$ 4.64 **14.** $3.8 \div 10$ 0.38 **15.** $6 \div 10$ 0.6 **16.** $0.9 \div 10$ 0.09

17. $7.8 \div 100$ 0.078 **18.** $3.76 \div 100$ 0.0376 **19.** $4 \div 100$ 0.04 **20.** $71 \div 100$ 0.71

21. $9.7 \div 1000$ 0.0097 **22.** $8.64 \div 1000$ 0.00864 **23.** $87 \div 1000$ 0.087 **24.** $3 \div 1000$ 0.003

Estimate. Use after page 113.

25. $212 \div 4$ 50 **26.** $6214 \div 7$ 900 **27.** $31{,}814 \div 5$ 6000 **28.** $15{,}749 \div 8$ 2000

29. $73\overline{)849}$ 10 **30.** $41\overline{)3349}$ 80 **31.** $79\overline{)33{,}124}$ 400 **32.** $67\overline{)36{,}841}$ 500

33. $4.3\overline{)21.8}$ 5 **34.** $8.17\overline{)42.722}$ 5 **35.** $6.2\overline{)23.916}$ 4 **36.** $3.914\overline{)25.72}$ 6

Divide. Use after page 117.

37. $264.6 \div 6.3$ 42 **38.** $34.08 \div 7.1$ 4.8 **39.** $2.448 \div 7.2$ 0.34 **40.** $3.43 \div 4.9$ 0.7

41. $0.03\overline{)2.943}$ 98.1 **42.** $7.3\overline{)0.438}$ 0.06 **43.** $0.28\overline{)95.2}$ 340 **44.** $0.004\overline{)0.7}$ 175

45. $0.3\overline{)9}$ 30 **46.** $0.7\overline{)14}$ 20 **47.** $0.02\overline{)14}$ 700 **48.** $0.08\overline{)64}$ 800

49. $0.24\overline{)48}$ 200 **50.** $0.12\overline{)6}$ 50 **51.** $0.12\overline{)24}$ 200 **52.** $0.013\overline{)52}$ 4000

Find each quotient to the nearest tenth. Use after page 119.

53. $7\overline{)9}$ 1.3 **54.** $4\overline{)13}$ 3.3 **55.** $8\overline{)11}$ 1.4 **56.** $3\overline{)4}$ 1.3

57. $4\overline{)3}$ 0.8 **58.** $7\overline{)1}$ 0.1 **59.** $9\overline{)7}$ 0.8 **60.** $7\overline{)6}$ 0.9

61. $3.2\overline{)7.3}$ 2.3 **62.** $9.6\overline{)20.15}$ 2.1 **63.** $0.3\overline{)13}$ 43.3 **64.** $0.09\overline{)4}$ 44.4

65. $2.8\overline{)16}$ 5.7 **66.** $0.9\overline{)8.25}$ 9.2 **67.** $7.1\overline{)25.8}$ 3.6 **68.** $0.24\overline{)16}$ 66.7

69. $0.8\overline{)29}$ 36.3 **70.** $5.8\overline{)22}$ 3.8 **71.** $0.52\overline{)9}$ 17.3 **72.** $8.3\overline{)34.67}$ 4.2

SKILLS REVIEW

Add. Refer to page 14.

1. 49 + 76 125

2. 94 + 36 130

3. 38 + 26 + 54 118

4. 428 + 62 490

5.
37
49
63
+ 47
196

6.
728
479
280
+ 146
1633

7.
$ 4.17
13.29
+ 6.54
$24.00

8.
$ 5.87
28.61
2.99
+ 0.73
$38.20

Add. Refer to page 16.

9.
9428
+ 6759
16,187

10.
7284
+ 6985
14,269

11.
37,487
+ 28,643
66,130

12.
328,412
+ 149,386
477,798

13.
72,497
65,388
+ 9,246
147,131

14.
329,421
658,187
+ 86,294
1,073,902

15.
37,285
49,784
65,118
+ 97,091
249,278

16.
427,286
698,487
407,759
+ 65,287
1,598,819

Add. Refer to page 22.

17. 8.7 + 3.9 12.6

18. 7.32 + 9.48 16.80

19. 8.746 + 0.943 9.689

20. 8.69 + 6.738 15.428

21.
7.63
9.87
+ 1.4
18.90

22.
9.72
8.746
+ 8.9
27.366

23.
9.87
3.86
1.98
+ 0.7
16.41

24.
6.4
4.295
3.87
+ 9.9
24.465

25. 2.3 + 0.8 + 19.46 22.56

26. 12.4 + 6.78 + 0.39 + 0.068 19.638

Estimate the sums. Refer to page 28.

27.
49
+ 87
140

28.
3297
+ 8746
12,000

29.
4293
+ 415
4700

30.
276
49
+ 62
390

31.
3.738
+ 1.9
6

32.
4.25
+ 6.704
11

33.
4.92
6.104
+ 8.6
20

34.
9.04
0.921
+ 8.1
18

Subtract. Refer to page 40.

1. 97 − 28 = 69	**2.** 64 − 37 = 27	**3.** 80 − 16 = 64	**4.** 42 − 37 = 5
5. 824 − 673 = 151	**6.** 920 − 637 = 283	**7.** 824 − 197 = 627	**8.** 835 − 444 = 391
9. 725 − 68 = 657	**10.** 479 − 85 = 394	**11.** 741 − 8 = 733	**12.** 627 − 9 = 618

Subtract. Refer to page 42.

13. 9487 − 2688 = 6799	**14.** 3241 − 1276 = 1965	**15.** 3498 − 2719 = 779	**16.** 7348 − 2619 = 4729
17. 28,741 − 19,198 = 9543	**18.** 36,287 − 27,499 = 8788	**19.** 326,145 − 198,087 = 128,058	**20.** 628,324 − 597,698 = 30,626

Subtract. Refer to page 44.

21. 500 − 146 = 354	**22.** 701 − 387 = 314	**23.** 5000 − 3298 = 1702	**24.** 8002 − 498 = 7504
25. 26,000 − 19,487 = 6513	**26.** 5007 − 2681 = 2326	**27.** 8040 − 1246 = 6794	**28.** 30,091 − 28,135 = 1956

Subtract. Refer to page 50.

29. 3.4 − 1.7 = 1.7	**30.** 8.75 − 4.69 = 4.06	**31.** 9.348 − 6.889 = 2.459	**32.** 8.764 − 3.871 = 4.893
33. 4.7 − 3.98 = 0.72	**34.** 8.15 − 6.9 = 1.25	**35.** 9.781 − 4.99 = 4.791	**36.** 9.214 − 6.5 = 2.714
37. 4.4 − 3.76 = 0.64	**38.** 8.74 − 3.987 = 4.753	**39.** 8.7 − 6.812 = 1.888	**40.** 9 − 3.47 = 5.53

Estimate the differences. Refer to page 54.

41. 781 − 93 ≈ 690	**42.** 872 − 181 ≈ 700	**43.** 3281 − 878 ≈ 2400	**44.** 3984 − 1299 ≈ 3000
45. 7.4 − 2.349 ≈ 5	**46.** 3.76 − 1.879 ≈ 2	**47.** 8.941 − 3.199 ≈ 6	**48.** 4.187 − 1.99 ≈ 2

Multiply. Refer to page 68.

1. 89×7 — 623
2. 46×5 — 230
3. 898×4 — 3592
4. $\$7.49 \times 7$ — \$52.43
5. 4287×9 — 38,583
6. 7307×8 — 58,456
7. 8291×7 — 58,037
8. 6489×5 — 32,445

Multiply. Refer to page 70.

9. 56×47 — 2632
10. 88×24 — 2112
11. 64×28 — 1792
12. 48×75 — 3600
13. 927×42 — 38,934
14. 729×67 — 48,843
15. $\$8.40 \times 98$ — \$823.20
16. $\$8.26 \times 54$ — \$446.04
17. 3784×86 — 325,424
18. 4281×81 — 346,761
19. 7285×18 — 131,130
20. 4784×36 — 172,224

Multiply. Refer to page 72.

21. 798×642 — 512,316
22. 786×424 — 333,264
23. 8749×183 — 1,601,067
24. $\$92.84 \times 366$ — \$33,979.44
25. 376×404 — 151,904
26. 928×480 — 445,440
27. 1376×604 — 831,104
28. $\$42.98 \times 730$ — \$31,375.40

Multiply. Refer to page 76.

29. 70×80 — 5600
30. 60×40 — 2400
31. 30×70 — 2100
32. 900×60 — 54,000
33. 300×700 — 210,000
34. 9000×80 — 720,000
35. 8000×300 — 2,400,000
36. 4000×6000 — 24,000,000

Multiply. Refer to page 82.

37. 93.6×7 — 655.2
38. 89.43×6 — 536.58
39. 4.287×35 — 150.045
40. 6.34×12 — 76.08
41. 8.7×3.1 — 26.97
42. 9.82×3.3 — 32.406
43. 87.1×0.32 — 27.872
44. 9.62×8.78 — 84.4636
45. 0.3×0.3 — 0.09
46. 0.7×0.06 — 0.042
47. 0.1×0.08 — 0.008
48. 0.03×0.2 — 0.006

Multiply. Refer to page 84.

1. 10×7.6 76 **2.** 10×9.87 98.7 **3.** 10×8.746 87.46 **4.** 10×3.9 39

5. 100×3.7 370 **6.** 100×4.28 428 **7.** 1000×3.7 3700 **8.** 1000×9.42 9420

9. 0.1×7.4 0.74 **10.** 0.1×6 0.6 **11.** 0.1×9.28 0.928 **12.** 0.1×0.4 0.04

13. 0.01×8.7 0.087 **14.** 0.01×8.73 0.0873 **15.** 0.001×8.76 0.00876 **16.** 0.001×8.5 0.0085

Estimate the products. Refer to page 76.

17. 87×9 810 **18.** 79×48 4000 **19.** 386×41 16,000 **20.** 925×648 540,000

21. 827×650 560,000 **22.** 3698×879 3,600,000 **23.** 4219×327 1,200,000 **24.** 7414×8785 63,000,000

Estimate the products. Refer to page 78.

25. 6.82×7 49 **26.** 8.75×6.24 54 **27.** 1.947×8.7 18 **28.** 3.682×4.79 20

29. $\$6.85 \times 7$ \$49 **30.** $\$9.15 \times 4$ \$36 **31.** $\$38.76 \times 29$ \$1200 **32.** $\$42.16 \times 58$ \$2400

Divide. Refer to page 98.

33. $36 \div 7$ 5r1 **34.** $71 \div 3$ 23r2 **35.** $92 \div 4$ 23 **36.** $186 \div 7$ 26r4

37. $5\overline{)328}$ 65r3 **38.** $4\overline{)321}$ 80r1 **39.** $6\overline{)62}$ 10r2 **40.** $7\overline{)352}$ 50r2

Divide. Refer to page 100.

41. $876 \div 4$ 219 **42.** $947 \div 5$ 189r2 **43.** $3724 \div 8$ 465r4 **44.** $4914 \div 6$ 819

45. $7\overline{)8241}$ 1177r2 **46.** $6\overline{)9487}$ 1581r1 **47.** $8\overline{)35,487}$ 4435r7 **48.** $5\overline{)36,491}$ 7298r1

49. $8\overline{)4012}$ 501r4 **50.** $6\overline{)3685}$ 614r1 **51.** $7\overline{)42,017}$ 6002r3 **52.** $9\overline{)87,319}$ 9702r1

Divide. Refer to page 102.

53. $89 \div 22$ 4r1 **54.** $98 \div 47$ 2r4 **55.** $228 \div 65$ 3r33 **56.** $897 \div 74$ 12r9

57. $62\overline{)3241}$ 52r17 **58.** $63\overline{)9487}$ 150r37 **59.** $87\overline{)52,431}$ 602r57 **60.** $18\overline{)11,294}$ 627r8

Divide. Refer to page 104.

1. 879 ÷ 624 1r255 **2.** 4149 ÷ 827 5r14 **3.** 1374 ÷ 225 6r24 **4.** 20,930 ÷ 645 32r290

5. $497\overline{)23{,}860}$ 48r4 **6.** $429\overline{)34{,}420}$ 80r100 **7.** $576\overline{)52{,}000}$ 90r160 **8.** $610\overline{)24{,}626}$ 40r226

Divide. Refer to page 108.

9. 9.45 ÷ 5 1.89 **10.** 9.6 ÷ 6 1.6 **11.** 14.72 ÷ 8 1.84 **12.** 1.323 ÷ 3 0.441

13. $71\overline{)610.6}$ 8.6 **14.** $84\overline{)261.24}$ 3.11 **15.** $34\overline{)214.2}$ 6.3 **16.** $51\overline{)16.983}$ 0.333

17. $4\overline{)0.12}$ 0.03 **18.** $7\overline{)0.007}$ 0.001 **19.** $6\overline{)3}$ 0.5 **20.** $5\overline{)2}$ 0.4

Divide. Refer to page 110.

21. 78.2 ÷ 10 7.82 **22.** 9.7 ÷ 10 0.97 **23.** 3 ÷ 10 0.3 **24.** 0.4 ÷ 10 0.04

25. 4.7 ÷ 100 0.047 **26.** 5.27 ÷ 100 0.0527 **27.** 8 ÷ 100 0.08 **28.** 42 ÷ 100 0.42

29. 6.3 ÷ 1000 0.0063 **30.** 7.42 ÷ 1000 0.00742 **31.** 94 ÷ 1000 0.094 **32.** 8 ÷ 1000 0.008

Estimate. Refer to page 112.

33. $8\overline{)650}$ 80 **34.** $9\overline{)7147}$ 800 **35.** $6\overline{)41{,}821}$ 7000 **36.** $9\overline{)19{,}247}$ 2000

37. $42\overline{)948}$ 20 **38.** $73\overline{)6241}$ 90 **39.** $87\overline{)33{,}421}$ 400 **40.** $81\overline{)40{,}628}$ 500

41. $3.7\overline{)16.5}$ 4 **42.** $9.32\overline{)27.491}$ 3 **43.** $3.3\overline{)23.84}$ 8 **44.** $4.376\overline{)27.98}$ 7

Divide. Refer to page 116.

45. 98.7 ÷ 4.7 21 **46.** 24.57 ÷ 6.3 3.9 **47.** 1.025 ÷ 4.1 0.25 **48.** 7.76 ÷ 9.7 0.8

49. $0.2\overline{)6}$ 30 **50.** $0.4\overline{)16}$ 40 **51.** $0.03\overline{)15}$ 500 **52.** $0.09\overline{)81}$ 900

53. $0.36\overline{)72}$ 200 **54.** $0.25\overline{)6}$ 24 **55.** $0.15\overline{)6}$ 40 **56.** $0.025\overline{)4}$ 160

Find each quotient to the nearest tenth. Refer to page 118.

57. $4\overline{)9}$ 2.3 **58.** $7\overline{)12}$ 1.7 **59.** $6\overline{)7}$ 1.2 **60.** $8\overline{)9}$ 1.1

61. $7\overline{)4}$ 0.6 **62.** $9\overline{)1}$ 0.1 **63.** $7\overline{)3}$ 0.4 **64.** $6\overline{)5}$ 0.8

65. $5.1\overline{)6.7}$ 1.3 **66.** $8.7\overline{)19.42}$ 2.2 **67.** $0.6\overline{)11}$ 18.3 **68.** $0.03\overline{)4}$ 133.3

CHAPTER 5 PRE-TEST

1. What part is shaded? p. 136 $\frac{5}{8}$

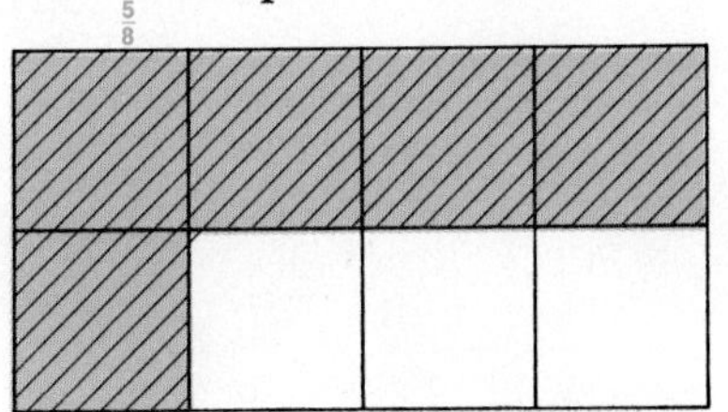

2. What part is shaded? $\frac{2}{3}$

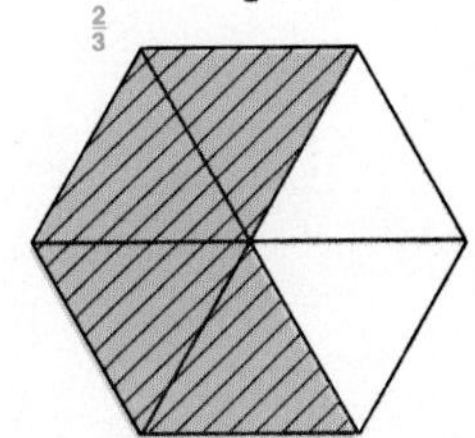

Divide, if possible. p. 138

3. $\frac{26}{1}$ 26 **4.** $\frac{26}{0}$ Not possible **5.** $\frac{12}{12}$ 1 **6.** $\frac{0}{16}$ 0

Are the fractions equal? p. 140

7. $\frac{3}{4}, \frac{6}{8}$ = yes **8.** $\frac{5}{4}, \frac{9}{7}$ no

Compare. p. 142

9. $\frac{11}{20}$ and $\frac{2}{3}$ $<$ **10.** $\frac{2}{3}$ and $\frac{9}{14}$ $>$

Convert to a fraction. pp. 144, 150

11. 1.11 $\frac{111}{100}$ **12.** 0.005 $\frac{5}{1000}$ **13.** $3\frac{1}{2}$ $\frac{7}{2}$ **14.** $9\frac{7}{8}$ $\frac{79}{8}$

Convert to a decimal. p. 146

15. $\frac{25}{16}$ 1.5625 **16.** $\frac{7}{8}$ 0.875 **17.** $\frac{947}{100}$ 9.47 **18.** $\frac{7813}{10{,}000}$ 0.7813

Convert to a decimal. Round to the nearest thousandth, when necessary. pp. 146, 154

19. $\frac{49}{12}$ 4.083 **20.** $\frac{5}{6}$ 0.833 **21.** $4\frac{3}{10}$ 4.3 **22.** $677\frac{4}{5}$ 677.8

Convert to a mixed number. pp. 152, 154

23. $\frac{9}{2}$ $4\frac{1}{2}$ **24.** $\frac{74}{9}$ $8\frac{2}{9}$ **25.** 2.8 $2\frac{8}{10}$ **26.** 112.73 $112\frac{73}{100}$

Application pp. 148, 156

27. Last week Miguel played his guitar for 4 hr, 45 min. Rita played her guitar for 2 hr, 55 min. How much longer did Miguel play than Rita? 1 hr, 50 min

28. When it is 8 PM in Los Angeles, what time is it in New York? 11 PM

5
FRACTIONS

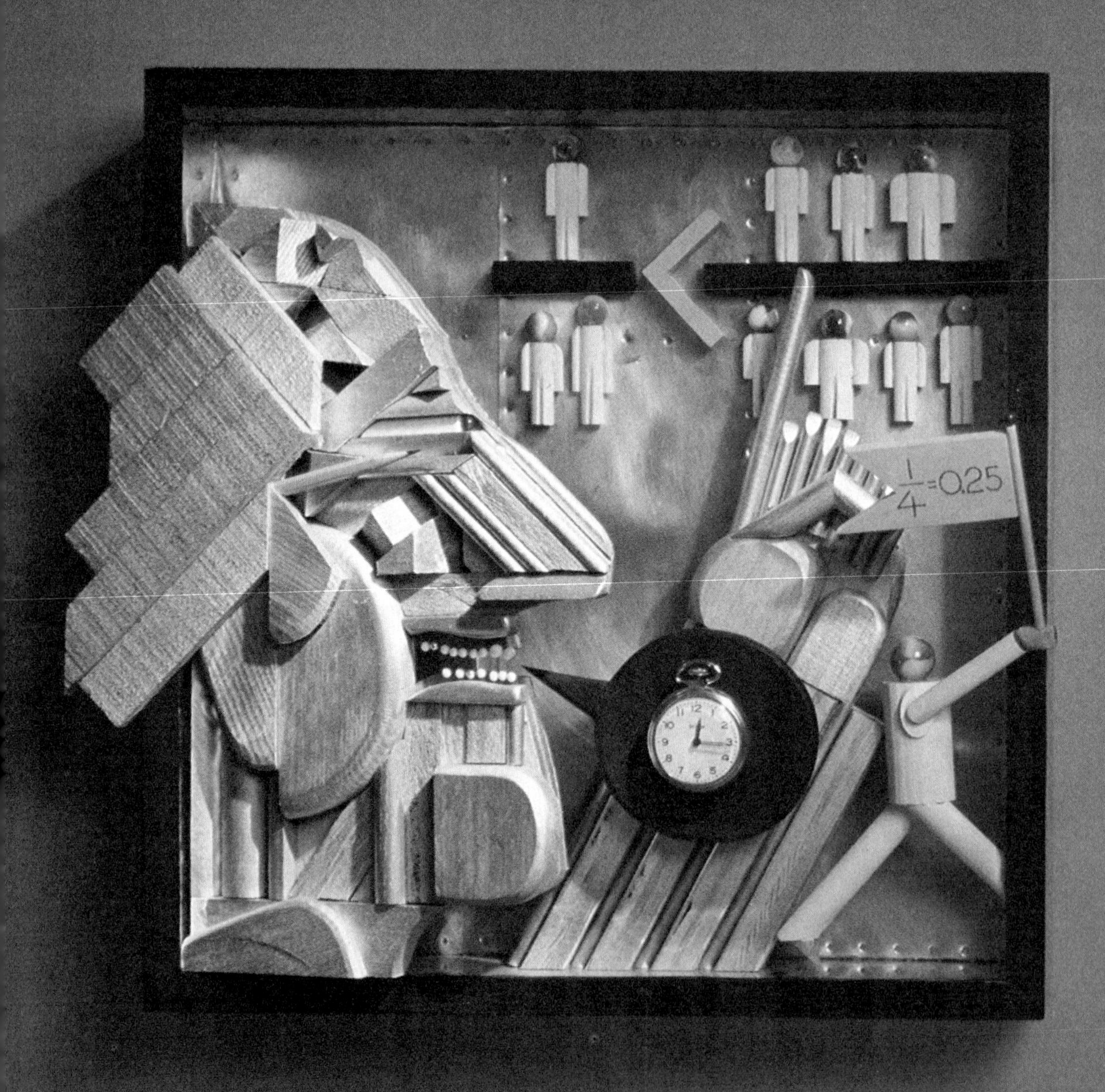

GETTING STARTED

FRACTION EQUO

WHAT'S NEEDED The 30 squares of paper with digits marked on them as used in GOTCHA, Chapter 1.

THE RULES

1. On a sheet of paper each player writes this:

   $\frac{[\ ]}{[\ ]} = \frac{[\ ]}{[\ ]} = \frac{[\ ]}{[\ ]}$

2. To play, a leader draws digits from the box, one at a time.
3. Each player decides whether or not to use the digit, but once the digit has been chosen it must be kept. The digit can be written in any space. Once chosen, the digit can be moved from space to space.
4. The first player to get three equal fractions wins.

Sample

Some students may be used to saying *equivalent* fractions instead of *equal* fractions. As used here, the terms mean the same.

Jacky: $\frac{[6]}{[9]} = \frac{[2]}{[3]} = \frac{[4]}{[6]}$ Yes, a winner!

Fred: $\frac{[3]}{[5]} = \frac{[6]}{[\ ]} = \frac{[2]}{[\ ]}$ No.

Jacky wins because she has three equal fractions.

Objective 55

USING FRACTIONS

Although there are many interpretations for fractions, the rules of computation are the same.

We use fractions to describe some of the objects in a set.

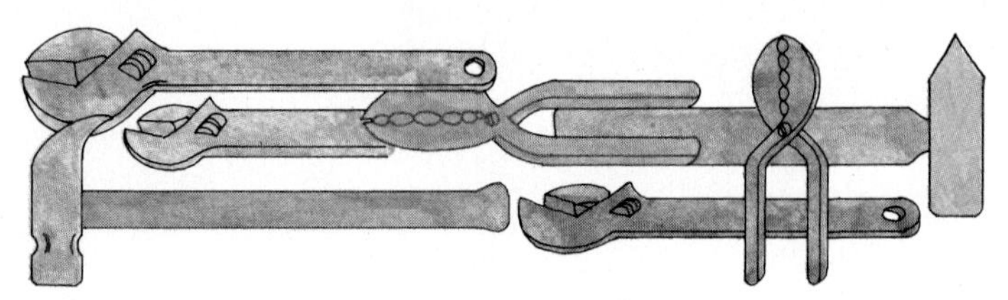

EXAMPLE 1. What part of the set of tools are hammers?

2 out of 7 tools are hammers. We write this as $\frac{2}{7}$.

TRY THIS

1. What part of the set of tools above are wrenches? $\frac{3}{7}$
2. What part of the set of tools are pliers? $\frac{2}{7}$

We call $\frac{2}{7}$ a fraction. $\frac{\text{Number of objects considered}}{\text{Number of objects in the set}}$ $\frac{2}{7}$

The top number, 2, is the numerator.
The bottom number, 7, is the denominator.

We also use a fraction to describe part of an object which has been divided into equal parts.

EXAMPLE 2. What part is shaded?

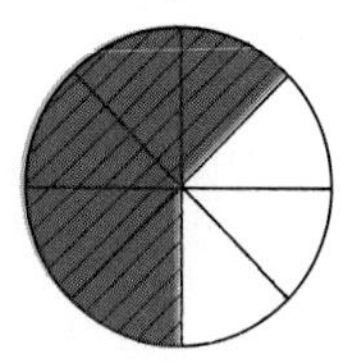

5 of the 8 equal parts are shaded.

We say $\frac{5}{8}$ is shaded.

TRY THIS

What part is shaded?

3. $\frac{5}{6}$

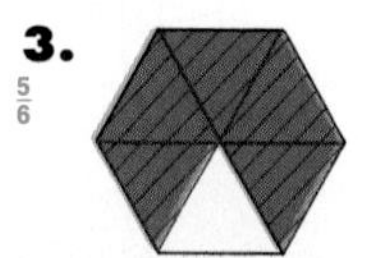

4. $\frac{1}{4}$

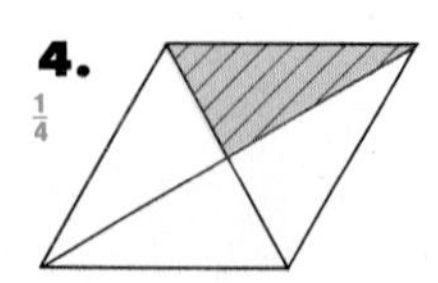

PRACTICE

For extra practice, use Making Practice Fun 31 with this lesson.

1. What part are women? $\frac{3}{5}$

2. What part are men? $\frac{2}{5}$

What part of each set is shaded?

3. $\frac{5}{8}$ **4.** $\frac{3}{5}$

5. $\frac{3}{5}$ **6.** $\frac{4}{7}$

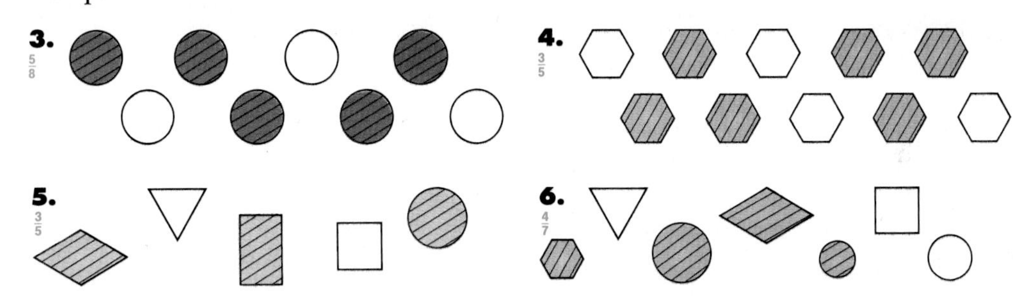

What part is shaded?

7. $\frac{3}{4}$

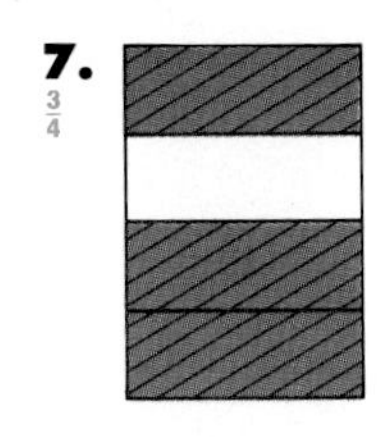

8. $\frac{1}{3}$

9. $\frac{1}{3}$

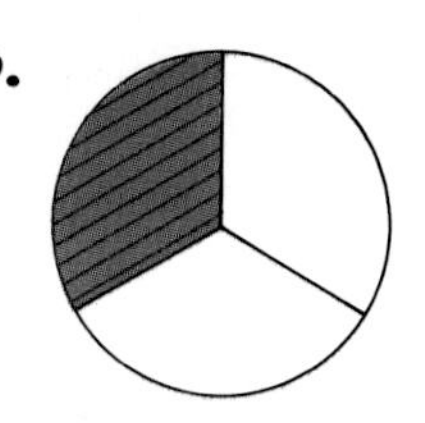

10. $\frac{1}{4}$

Identify the numerator and denominator of each fraction.

11. $\frac{11}{20}$ N: 11 D: 20 **12.** $\frac{14}{25}$ N: 14 D: 25 **13.** $\frac{19}{20}$ N: 19 D: 20 **14.** $\frac{18}{25}$ N: 18 D: 25 **15.** $\frac{51}{100}$ N: 51 D: 100 **16.** $\frac{71}{100}$ N: 71 D: 100

PROBLEM CORNER

The surface of the Earth is 3 parts water and 1 part land.

1. What fraction of the Earth is water? $\frac{3}{4}$

2. What fraction on the Earth is land? $\frac{1}{4}$

Objective 56

FRACTIONS MEAN DIVISION

We know that $\frac{8}{8} = 1$ because $8 \div 8 = 1$.
This is also true because $8 \times 1 = 8$.
Any nonzero number divided by itself is 1.

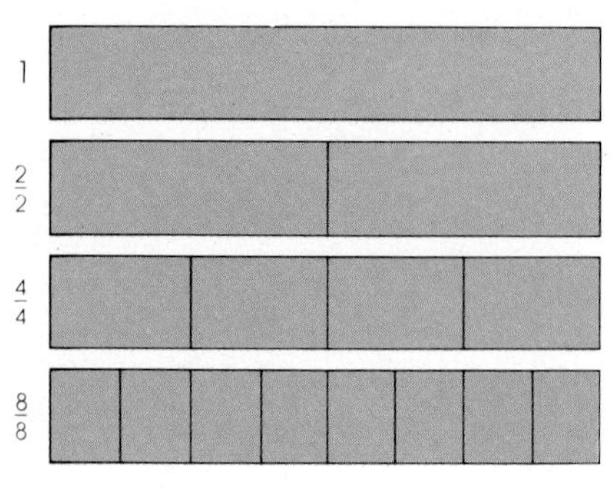

EXAMPLES. Divide.

1. $\frac{6}{6} = 1$ **2.** $\frac{9}{9} = 1$ **3.** $\frac{27}{27} = 1$

We know that $\frac{0}{4} = 0$ because $0 \div 4 = 0$. This is also true because $4 \times 0 = 0$.

0 divided by any nonzero number is 0.

EXAMPLES. Divide.

4. $\frac{0}{3} = 0$ **5.** $\frac{0}{7} = 0$ **6.** $\frac{0}{36} = 0$

TRY THIS

Divide.

1. $\frac{5}{5}$ 1 **2.** $\frac{42}{42}$ 1 **3.** $\frac{0}{8}$ 0 **4.** $\frac{0}{10}$ 0

We know that $\frac{8}{1} = 8$ because $8 \div 1 = 8$. This is also true because $1 \times 8 = 8$.

Any number divided by 1 is that number.

EXAMPLES. Divide.

7. $\frac{2}{1} = 2$ **8.** $\frac{6}{1} = 6$ **9.** $\frac{24}{1} = 24$

Division by 0 is not possible.

EXAMPLES. Divide, if possible.

In order for Example 10 to have an answer, 0 multiplied by some number must equal 14. But since $0 \times$ any number $= 0$, division by 0 is not possible.

10. $\frac{14}{0}$ Not possible **11.** $\frac{0}{6} = 0$ **12.** $\frac{0}{0}$ Not possible

TRY THIS

Divide, if possible.

5. $\frac{9}{1}$ 9 **6.** $\frac{36}{1}$ 36

7. $\frac{0}{12}$ 0 **8.** $\frac{19}{0}$ Not possible

9. $\frac{0}{47}$ 0 **10.** $\frac{24}{0}$ Not possible

PRACTICE

For extra practice, use Making Practice Fun 32 with this lesson.

Divide, if possible.

1. $\frac{8}{8}$ 1	**2.** $\frac{10}{10}$ 1	**3.** $\frac{9}{9}$ 1	**4.** $\frac{12}{12}$ 1
5. $\frac{6}{1}$ 6	**6.** $\frac{9}{1}$ 9	**7.** $\frac{8}{1}$ 8	**8.** $\frac{7}{1}$ 7
9. $\frac{0}{6}$ 0	**10.** $\frac{0}{8}$ 0	**11.** $\frac{0}{5}$ 0	**12.** $\frac{0}{9}$ 0
13. $\frac{24}{24}$ 1	**14.** $\frac{24}{1}$ 24	**15.** $\frac{0}{24}$ 0	**16.** $\frac{14}{14}$ 1
17. $\frac{0}{22}$ 0	**18.** $\frac{22}{22}$ 1	**19.** $\frac{22}{1}$ 22	**20.** $\frac{0}{15}$ 0
21. $\frac{16}{1}$ 16	**22.** $\frac{0}{16}$ 0	**23.** $\frac{18}{18}$ 1	**24.** $\frac{21}{1}$ 21
25. $\frac{0}{0}$ Not possible	**26.** $\frac{19}{0}$ Not possible	**27.** $\frac{18}{0}$ Not possible	**28.** $\frac{42}{0}$ Not possible
29. $\frac{13}{13}$ 1	**30.** $\frac{23}{1}$ 23	**31.** $\frac{25}{25}$ 1	**32.** $\frac{0}{42}$ 0
33. $\frac{124}{1}$ 124	**34.** $\frac{0}{256}$ 0	**35.** $\frac{0}{357}$ 0	**36.** $\frac{562}{562}$ 1
37. $\frac{0}{120}$ 0	**38.** $\frac{164}{164}$ 1	**39.** $\frac{1253}{1253}$ 1	**40.** $\frac{789}{1}$ 789
41. $\frac{45}{0}$ Not possible	**42.** $\frac{0}{0}$ Not possible	**43.** $\frac{32}{32}$ 1	**44.** $\frac{56}{0}$ Not possible
45. $\frac{0}{0}$ Not possible	**46.** $\frac{13}{1}$ 13	**47.** $\frac{28}{0}$ Not possible	**48.** $\frac{45}{0}$ Not possible

More Practice, page 162

Objective 57

EQUALITY

Compare $\frac{2}{4}$ and $\frac{3}{6}$.

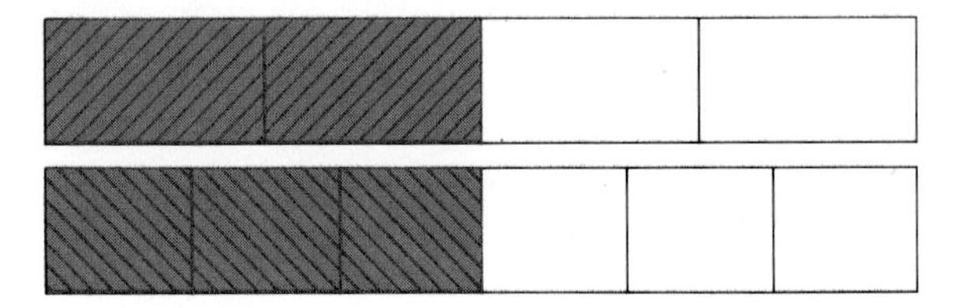

We see that $\frac{2}{4} = \frac{3}{6}$.

Tell whether the fractions are equal. $\frac{2}{4}, \frac{3}{6}$

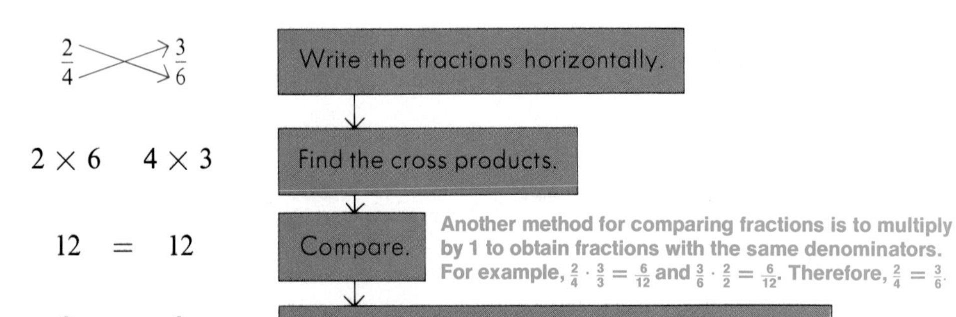

$\frac{2}{4} \quad \frac{3}{6}$

$2 \times 6 \quad 4 \times 3$

$12 = 12$

$\frac{2}{4} = \frac{3}{6}$

Another method for comparing fractions is to multiply by 1 to obtain fractions with the same denominators. For example, $\frac{2}{4} \cdot \frac{3}{3} = \frac{6}{12}$ and $\frac{3}{6} \cdot \frac{2}{2} = \frac{6}{12}$. Therefore, $\frac{2}{4} = \frac{3}{6}$.

We call 2×6 and 4×3 cross products.

EXAMPLES. Tell whether the fractions are equal.

1. $\frac{6}{10} \quad \frac{3}{5}$

$6 \times 5 \quad 10 \times 3$

$30 = 30$

$\frac{6}{10} = \frac{3}{5}$

2. $\frac{6}{7} \quad \frac{7}{8}$

$6 \times 8 \quad 7 \times 7$

$48 \neq 49$

$\frac{6}{7} \neq \frac{7}{8}$

$\neq$ *means "is not equal to."*

TRY THIS

Tell whether the fractions are equal.

1. $\frac{2}{6}, \frac{3}{9}$ yes **2.** $\frac{3}{4}, \frac{5}{7}$ no **3.** $\frac{8}{10}, \frac{4}{5}$ yes

PRACTICE

Tell whether the fractions are equal.

1. $\frac{3}{4}, \frac{9}{12}$ yes
2. $\frac{2}{4}, \frac{4}{6}$ no
3. $\frac{4}{5}, \frac{2}{3}$ no
4. $\frac{4}{8}, \frac{3}{6}$ yes

5. $\frac{1}{2}, \frac{2}{4}$ yes
6. $\frac{1}{3}, \frac{2}{6}$ yes
7. $\frac{1}{5}, \frac{2}{9}$ no
8. $\frac{1}{4}, \frac{3}{12}$ yes

9. $\frac{0}{3}, \frac{0}{4}$ yes
10. $\frac{0}{1}, \frac{0}{3}$ yes
11. $\frac{0}{5}, \frac{0}{7}$ yes
12. $\frac{0}{8}, \frac{0}{7}$ yes

13. $\frac{4}{5}, \frac{2}{3}$ no
14. $\frac{6}{8}, \frac{9}{12}$ yes
15. $\frac{3}{8}, \frac{6}{16}$ yes
16. $\frac{2}{6}, \frac{6}{18}$ yes

17. $\frac{1}{2}, \frac{1}{3}$ no
18. $\frac{1}{3}, \frac{1}{4}$ no
19. $\frac{1}{4}, \frac{1}{2}$ no
20. $\frac{1}{5}, \frac{1}{3}$ no

21. $\frac{2}{2}, \frac{5}{5}$ yes
22. $\frac{3}{3}, \frac{4}{4}$ yes
23. $\frac{5}{5}, \frac{3}{3}$ yes
24. $\frac{4}{4}, \frac{2}{2}$ yes

25. $\frac{2}{5}, \frac{3}{7}$ no
26. $\frac{3}{5}, \frac{5}{10}$ no
27. $\frac{4}{7}, \frac{5}{11}$ no
28. $\frac{12}{16}, \frac{3}{4}$ yes

29. $\frac{3}{2}, \frac{9}{6}$ yes
30. $\frac{16}{14}, \frac{8}{7}$ yes
31. $\frac{9}{3}, \frac{3}{1}$ yes
32. $\frac{12}{9}, \frac{8}{6}$ yes

33. $\frac{3}{9}, \frac{1}{2}$ no
34. $\frac{5}{2}, \frac{17}{7}$ no
35. $\frac{4}{6}, \frac{17}{7}$ no
36. $\frac{3}{2}, \frac{5}{2}$ no

37. $\frac{3}{10}, \frac{30}{100}$ yes
38. $\frac{70}{100}, \frac{700}{1000}$ yes
39. $\frac{5}{10}, \frac{520}{1000}$ no
40. $\frac{49}{100}, \frac{50}{1000}$ no

PROBLEM CORNER

No. The cross products are not equal.

1. Is $\frac{3}{4}$ gallon the same as $\frac{4}{5}$ gallon? Why?

2. A student got 63 out of 78 questions correct on one test. On another test the student got 75 out of 100 questions correct. Did the student get the same fraction correct on each test? Why?

No. The cross products are not equal.

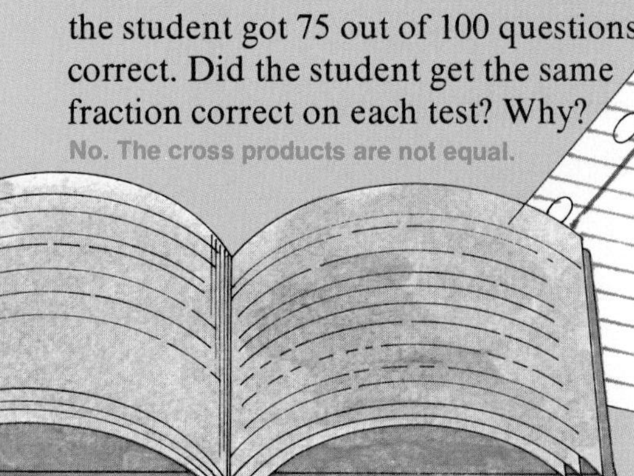

Objective 58

COMPARING FRACTIONS

Notice that $\frac{3}{4}$ is greater than $\frac{2}{5}$. $\frac{3}{4} > \frac{2}{5}$

Also see that $\frac{2}{5}$ is less than $\frac{3}{4}$. $\frac{2}{5} < \frac{3}{4}$

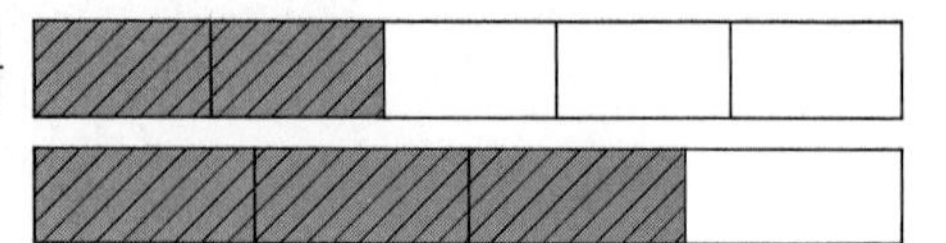

Compare. $\frac{2}{5}$ and $\frac{3}{4}$

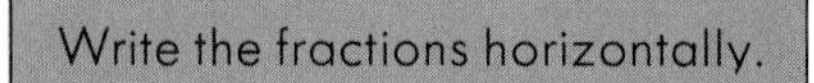

$2 \times 4 \quad 5 \times 3$ — Find the cross products.

$8 < 15$ — Compare.

$\frac{2}{5} < \frac{3}{4}$ — Write the same symbol between the fractions.

Always begin with the numerator of the first fraction to be sure the inequality symbol will be correct.

EXAMPLE 1. Compare. $\frac{3}{5}$ and $\frac{1}{2}$

$$\frac{3}{5} \quad \frac{1}{2}$$

$3 \times 2 \quad 5 \times 1$ *Find the cross products.*

$6 > 5$ *Compare.*

$\frac{3}{5} > \frac{1}{2}$ *Write the same symbol between the fractions.*

When the denominators are the same, we simply compare the numerators.

EXAMPLE 2. Compare. $\frac{4}{7}$ and $\frac{3}{7}$

$$\frac{4}{7} \quad \frac{3}{7}$$

$4 > 3$ *Compare the numerators.*

$\frac{4}{7} > \frac{3}{7}$ *Write the same symbol between the fractions.*

TRY THIS

Compare.

1. $\frac{2}{3}$ and $\frac{5}{8}$ $>$

2. $\frac{8}{12}$ and $\frac{3}{4}$ $<$

3. $\frac{4}{9}$ and $\frac{5}{9}$ $<$

PRACTICE

For extra practice, use Making Practice Fun 33 with this lesson.
Use Quiz 11 after this Practice.

Compare.

1. $\frac{5}{8}$ and $\frac{6}{8}$ $<$ **2.** $\frac{7}{6}$ and $\frac{5}{6}$ $>$ **3.** $\frac{9}{10}$ and $\frac{7}{10}$ $>$ **4.** $\frac{4}{9}$ and $\frac{8}{9}$ $<$

5. $\frac{1}{3}$ and $\frac{1}{4}$ $>$ **6.** $\frac{1}{5}$ and $\frac{1}{6}$ $>$ **7.** $\frac{1}{6}$ and $\frac{1}{8}$ $>$ **8.** $\frac{1}{5}$ and $\frac{1}{4}$ $<$

9. $\frac{2}{3}$ and $\frac{5}{7}$ $<$ **10.** $\frac{5}{8}$ and $\frac{3}{4}$ $<$ **11.** $\frac{3}{5}$ and $\frac{4}{7}$ $>$ **12.** $\frac{3}{8}$ and $\frac{4}{9}$ $<$

13. $\frac{3}{4}$ and $\frac{7}{9}$ $<$ **14.** $\frac{5}{6}$ and $\frac{4}{5}$ $>$ **15.** $\frac{3}{4}$ and $\frac{5}{7}$ $>$ **16.** $\frac{3}{2}$ and $\frac{7}{5}$ $>$

17. $\frac{11}{16}$ and $\frac{2}{3}$ $>$ **18.** $\frac{11}{12}$ and $\frac{12}{13}$ $<$ **19.** $\frac{5}{6}$ and $\frac{13}{16}$ $>$ **20.** $\frac{11}{15}$ and $\frac{3}{4}$ $<$

21. $\frac{8}{10}$ and $\frac{79}{100}$ $>$ **22.** $\frac{7}{10}$ and $\frac{73}{100}$ $<$ **23.** $\frac{5}{10}$ and $\frac{49}{100}$ $>$ **24.** $\frac{3}{10}$ and $\frac{31}{100}$ $<$

25. $\frac{6}{100}$ and $\frac{8}{1000}$ $>$ **26.** $\frac{9}{1000}$ and $\frac{1}{100}$ $<$ **27.** $\frac{2109}{1000}$ and $\frac{21}{10}$ $>$ **28.** $\frac{21{,}006}{1000}$ and $\frac{2105}{100}$ $<$

29. $\frac{5}{9}$ and $\frac{7}{20}$ $>$ **30.** $\frac{8}{25}$ and $\frac{3}{9}$ $<$ **31.** $\frac{3}{25}$ and $\frac{2}{17}$ $>$ **32.** $\frac{4}{25}$ and $\frac{3}{19}$ $>$

33. $\frac{20}{21}$ and $\frac{18}{19}$ $>$ **34.** $\frac{23}{25}$ and $\frac{25}{26}$ $<$ **35.** $\frac{18}{19}$ and $\frac{13}{15}$ $>$ **36.** $\frac{17}{18}$ and $\frac{19}{20}$ $<$

37. $\frac{10}{9}$ and $\frac{10}{8}$ $<$ **38.** $\frac{20}{7}$ and $\frac{20}{8}$ $>$ **39.** $\frac{40}{3}$ and $\frac{40}{7}$ $>$ **40.** $\frac{30}{4}$ and $\frac{30}{5}$ $>$

PROBLEM CORNER

Andy got 16 out of 19 questions correct on a test. On the next test he got 18 out of 21 correct. On which test did Andy score higher?
the second test

Objective 59

WRITING DECIMALS AS FRACTIONS

To convert 9.87 to a fraction, we can multiply by 100, then divide by 100.

$9.87 \times 100 = 987$ $\qquad 987 \div 100 = \frac{987}{100}$

Multiplying by 100 and then dividing by 100 is the same as multiplying by 1, which leaves the value of the number unchanged.

Convert to a fraction. 4.867

Count the number of decimal places. → Move the decimal point that many places to the right. → Write the whole number over 10, 100, 1000, etc. depending on the number of 0's needed.

4.867 — 3 places

4.867 — 3 moves

$\frac{4867}{1000}$ — three 0's

EXAMPLE 1. Convert to a fraction. 0.675

0.675 — *3 places*

0.675 — *3 moves*

$0.675 = \frac{675}{1000}$ — *three 0's*

EXAMPLE 2. Convert to a fraction. 1.6

1.6 — *1 place*

1.6 — *1 move*

$1.6 = \frac{16}{10}$ — *one 0*

EXAMPLE 3. Convert to a fraction. 21.43

21.43 — *2 places*

21.43 — *2 moves*

$\frac{2143}{100}$ — *two 0's*

TRY THIS

Convert to a fraction.

1. 0.679 $\frac{679}{1000}$ **2.** 3.7 $\frac{37}{10}$ **3.** 17.95 $\frac{1795}{100}$

4. 4.7732 $\frac{47,332}{10,000}$ **5.** 0.32 $\frac{32}{100}$ **6.** 1.05 $\frac{105}{100}$

PRACTICE

Convert to a fraction.

1. 0.1 $\frac{1}{10}$	**2.** 0.3 $\frac{3}{10}$	**3.** 0.4 $\frac{4}{10}$	**4.** 0.8 $\frac{8}{10}$
5. 1.3 $\frac{13}{10}$	**6.** 2.7 $\frac{27}{10}$	**7.** 1.4 $\frac{14}{10}$	**8.** 3.9 $\frac{39}{10}$
9. 6.8 $\frac{68}{10}$	**10.** 7.2 $\frac{72}{10}$	**11.** 3.1 $\frac{31}{10}$	**12.** 9.7 $\frac{97}{10}$
13. 0.17 $\frac{17}{100}$	**14.** 0.89 $\frac{89}{100}$	**15.** 0.27 $\frac{27}{100}$	**16.** 0.32 $\frac{32}{100}$
17. 0.03 $\frac{3}{100}$	**18.** 0.01 $\frac{1}{100}$	**19.** 0.07 $\frac{7}{100}$	**20.** 0.09 $\frac{9}{100}$
21. 17.98 $\frac{1798}{100}$	**22.** 14.95 $\frac{1495}{100}$	**23.** 21.18 $\frac{2118}{100}$	**24.** 43.57 $\frac{4357}{100}$
25. 3.167 $\frac{3167}{1000}$	**26.** 1.732 $\frac{1732}{1000}$	**27.** 46.032 $\frac{46,032}{1000}$	**28.** 45.916 $\frac{45,916}{1000}$
29. 1.007 $\frac{1007}{1000}$	**30.** 2.0118 $\frac{20,118}{10,000}$	**31.** 34.1212 $\frac{341,212}{10,000}$	**32.** 78.9011 $\frac{789,011}{10,000}$
33. 299.49 $\frac{29,949}{100}$	**34.** 191.3 $\frac{1913}{10}$	**35.** 234.019 $\frac{234,019}{1000}$	**36.** 0.0005 $\frac{5}{10,000}$
37. 10.4 $\frac{104}{10}$	**38.** 10.5 $\frac{105}{10}$	**39.** 2.05 $\frac{205}{100}$	**40.** 1.15 $\frac{115}{100}$
41. 4.95 $\frac{495}{100}$	**42.** 7.96 $\frac{796}{100}$	**43.** 1.004 $\frac{1004}{1000}$	**44.** 2.015 $\frac{2015}{1000}$

PROBLEM CORNER

1. Beth bought 8.5 gallons of gasoline. Convert the number of gallons to a fraction. $\frac{85}{10}$

2. Kevin wants to paint a wall in his bedroom. The wall is 2.7 meters high and 3.9 meters long. Convert each measurement to a fraction. $\frac{27}{10}, \frac{39}{10}$

More Practice, page 162

Objective 60

WRITING FRACTIONS AS DECIMALS

Convert to a decimal. $\frac{7}{20}$

Divide the numerator by the denominator. → Write the quotient.

$$20\overline{)7.00} = 0.35$$

$$\frac{7}{20} = 0.35$$

EXAMPLE 1. Convert to a decimal. $\frac{5}{8}$

$$8\overline{)5.000} = 0.625 \qquad \frac{5}{8} = 0.625$$

When a repeating decimal occurs, we divide to a certain decimal place and round. The stopping place is a matter of choice, but we will use ten-thousandths.

EXAMPLE 2. Convert to a decimal. $\frac{9}{11}$ Round to the nearest thousandth.

$$11\overline{)9.0000} = 0.8181 \approx 0.818$$

Divide to the ten-thousandths' place.
Round to the nearest thousandth.

TRY THIS

Convert to a decimal. Round to the nearest thousandth.

1. $\frac{3}{8}$ 0.375 **2.** $\frac{1}{16}$ 0.063 **3.** $\frac{7}{12}$ 0.583 **4.** $\frac{59}{12}$ 4.917

When the denominator is 10, 100, 1000, etc., we use a shorter method.

Convert to a decimal. $\frac{693}{100}$

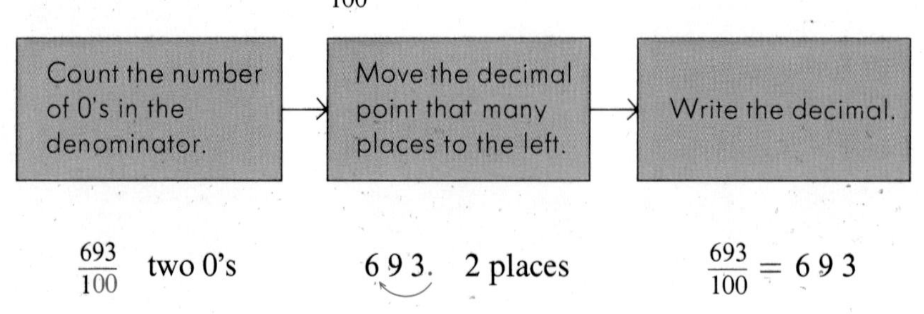

EXAMPLE 3. Convert to a decimal. $\frac{17}{1000}$

$\frac{17}{1000} \longrightarrow 0\,0\,1\,7. \longrightarrow 0.0\,1\,7$ *Move the decimal point 3 places to the left. Write necessary 0's.*

TRY THIS

Convert to a decimal.

5. $\frac{57}{100}$ 0.57 **6.** $\frac{17}{10}$ 1.7 **7.** $\frac{987}{10,000}$ 0.0987

PRACTICE

For extra practice, use Making Practice Fun 34 with this lesson.
Use Quiz 12 after this Practice.

Convert to a decimal.

1. $\frac{1}{5}$ 0.2 **2.** $\frac{7}{4}$ 1.75 **3.** $\frac{11}{4}$ 2.75 **4.** $\frac{9}{5}$ 1.8

5. $\frac{11}{8}$ 1.375 **6.** $\frac{13}{8}$ 1.625 **7.** $\frac{1}{8}$ 0.125 **8.** $\frac{53}{8}$ 6.625

9. $\frac{17}{16}$ 1.0625 **10.** $\frac{15}{16}$ 0.9375 **11.** $\frac{3}{16}$ 0.1875 **12.** $\frac{35}{16}$ 2.1875

13. $\frac{21}{125}$ 0.168 **14.** $\frac{17}{40}$ 0.425 **15.** $\frac{123}{200}$ 0.615 **16.** $\frac{5}{64}$ 0.078125

Convert to a decimal. Round to the nearest thousandth.

17. $\frac{1}{3}$ 0.333 **18.** $\frac{2}{3}$ 0.667 **19.** $\frac{5}{6}$ 0.833 **20.** $\frac{1}{6}$ 0.167

21. $\frac{2}{9}$ 0.222 **22.** $\frac{4}{9}$ 0.444 **23.** $\frac{4}{11}$ 0.364 **24.** $\frac{3}{11}$ 0.273

25. $\frac{25}{12}$ 2.083 **26.** $\frac{17}{12}$ 1.417 **27.** $\frac{19}{9}$ 2.111 **28.** $\frac{8}{3}$ 2.667

Convert to a decimal.

29. $\frac{8}{10}$ 0.8 **30.** $\frac{3}{10}$ 0.3 **31.** $\frac{29}{10}$ 2.9 **32.** $\frac{51}{10}$ 5.1

33. $\frac{99}{100}$ 0.99 **34.** $\frac{9}{100}$ 0.09 **35.** $\frac{246}{100}$ 2.46 **36.** $\frac{1567}{100}$ 15.67

37. $\frac{2}{1000}$ 0.002 **38.** $\frac{13}{1000}$ 0.013 **39.** $\frac{6789}{1000}$ 6.789 **40.** $\frac{14,012}{1000}$ 14.012

More Practice, pages 162 and 163

Objectives 161, 162, and 163

APPLICATION

MEASURING TIME

Time is given in many different measures. Sometimes we need to change from one measure to another.

1 minute (min) = 60 seconds (sec)	1 month (mo) = 30 days
1 hour (hr) = 60 minutes	1 year (yr) = 365 days
1 day = 24 hours	1 year = 52 weeks
1 week (wk) = 7 days	1 year = 12 months

To change from a larger to a smaller unit, we multiply.

EXAMPLE 1. How many minutes are there in 8 hours?

1 hour = 60 min
$8 \times 60 = 480$
8 hr = 480 min

This could also be calculated:
8 hr = 8×1 hr = 8×60 min = 480 min
Think: larger to smaller—multiply.

How many are there in each of the following?

1. hours in 4 days 96 hr
2. weeks in 3 years 156 wk
3. months in 4 years 48 mo
4. Mel played soccer for 4 hours. How many minutes was that? 240 min
5. Jeremy bought a stereo and will make monthly payments for 2 years. How many payments will he make? 24

To change from a smaller unit to a larger, we divide.

EXAMPLE 2. How many hours are there in 180 minutes?

1 hour = 60 min
$180 \div 60 = 3$
180 min = 3 hr

Think: smaller to larger—divide.

How many are there in each of the following?

6. hours in 300 min 5 hr
7. minutes in 420 sec 7 min
8. years in 1248 weeks 24 yr
9. Vicki rode her bicycle to work for 28 days in a row. How many weeks did she ride her bike? 4 wk
10. To train for the track team Audrey plans to run 5 miles each day for 90 days. How many months does she plan to run? 3 mo

Sometimes we express time in two measures.

```
WHOLE WHEAT BREAD

2 pkg active dry yeast
¼ c butter
2¼ c warm water
8 c whole wheat flour
Bake 80 minutes
```

EXAMPLE 3. How many hours and minutes should the bread bake?

1 hr = 60 min
80 min ÷ 60 min = 1 hr, 20 min

11. Steven's younger sister is 30 months old. How old is his sister in years and months? **2 yr, 6 mo**
12. Stacey can run the mile in 310 seconds. What is her speed in minutes and seconds? **5 min, 10 sec**

Often we need to add time.

EXAMPLE 4. Add. 3 hr, 25 min and 2 hr, 55 min

$$\begin{array}{r} 3 \text{ hr}, 25 \text{ min} \\ +\ 2 \text{ hr}, 55 \text{ min} \\ \hline 5 \text{ hr}, 80 \text{ min} \end{array} \begin{array}{l} \\ \\ = 5 \text{ hr} + 80 \text{ min} \\ = 5 \text{ hr} + 1 \text{ hr} + 20 \text{ min} \\ = 6 \text{ hr}, 20 \text{ min} \end{array}$$

13. Jack practiced the piano 5 hr, 29 min one week. The next week he practiced 6 hr, 45 min. How long did he practice in all? **12 hr, 14 min**
14. Sandy drove for 4 hr, 45 min. Martha drove for 3 hr, 35 min. How long did they drive altogether? **8 hr, 20 min**

We can also subtract time.

EXAMPLE 5. Subtract. 3 min, 25 sec and 1 min, 40 sec

$$\begin{array}{r} 3 \text{ min}, 25 \text{ sec} \\ -\ 1 \text{ min}, 40 \text{ sec} \\ \hline \end{array} = 2 \text{ min} + 60 \text{ sec} + 25 \text{ sec} = \begin{array}{r} 2 \text{ min}, 85 \text{ sec} \\ -\ 1 \text{ min}, 40 \text{ sec} \\ \hline 1 \text{ min}, 45 \text{ sec} \end{array}$$

15. Ralph has been playing the drums for 6 yr, 3 mo. Tony has been playing for 1 yr, 6 mo. How much longer has Ralph been playing? **4 yr, 9 mo**
16. In the 1920 Olympics Italy won the 4000 meter team cycling race with a time of 5 min, 20 sec. In 1960 they won it with a time of 4 min, 30 sec. By how much less time did they win in 1960? **50 sec**

Objective 61

WRITING MIXED NUMBERS AS FRACTIONS

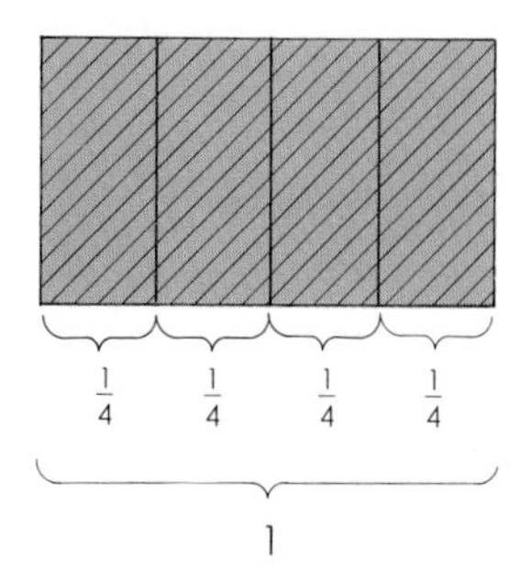

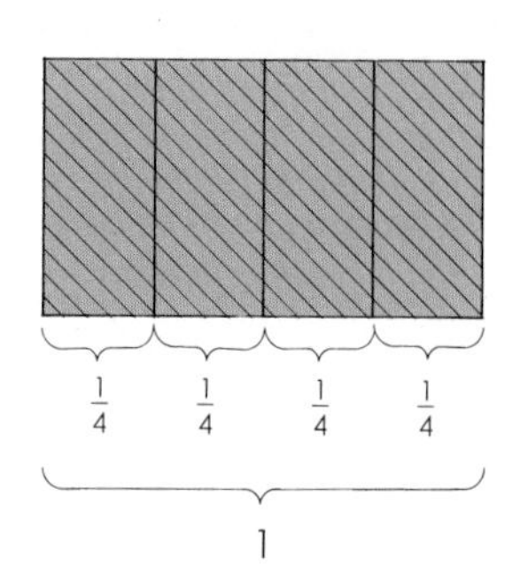

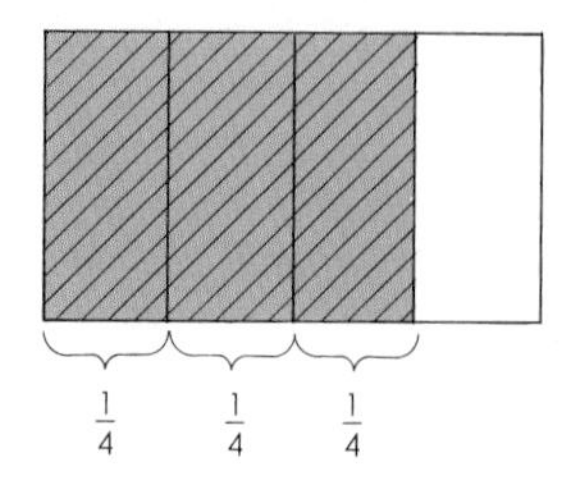

We can think of $2\frac{3}{4}$ as 2 whole units plus $\frac{3}{4}$ of a unit.

$2\frac{3}{4}$ is an example of a mixed number. It means $2 + \frac{3}{4}$.

EXAMPLES. Write as a mixed number.

1. $7 + \frac{2}{5} = 7\frac{2}{5}$ **2.** $\frac{3}{10} + 4 = 4\frac{3}{10}$

TRY THIS Write as a mixed number.

1. $8 + \frac{3}{4}$ $8\frac{3}{4}$ **2.** $\frac{2}{3} + 12$ $12\frac{2}{3}$

Convert to a fraction. $4\frac{3}{10}$

Multiply the denominator by the whole number. → Add the numerator. → Write the sum over the denominator.

$4\frac{3}{10}$ $10 \times 4 = 40$

$4\frac{3}{10}$ $40 + 3 = 43$

$\frac{43}{10}$

To see why this method works, remember that $4 = \frac{40}{10}$.
Then $4\frac{3}{10} = 4 + \frac{3}{10} = \frac{40}{10} + \frac{3}{10} = \frac{43}{10}$.

EXAMPLES. Convert to a fraction.

Add. *Add.*

3. $6\frac{2}{3} = \frac{20}{3}$ **4.** $10\frac{7}{8} = \frac{87}{8}$

Multiply. *Multiply.*

TRY THIS

Convert to a fraction.

3. $4\frac{2}{5}$ $\frac{22}{5}$ **4.** $10\frac{1}{6}$ $\frac{61}{6}$ **5.** $9\frac{3}{10}$ $\frac{93}{10}$

PRACTICE

Write as a mixed number.

1. $2 + \frac{1}{3}$ $2\frac{1}{3}$ **2.** $3 + \frac{1}{4}$ $3\frac{1}{4}$ **3.** $6 + \frac{3}{8}$ $6\frac{3}{8}$ **4.** $5 + \frac{2}{3}$ $5\frac{2}{3}$

5. $\frac{2}{3} + 9$ $9\frac{2}{3}$ **6.** $\frac{8}{9} + 7$ $7\frac{8}{9}$ **7.** $\frac{4}{5} + 10$ $10\frac{4}{5}$ **8.** $\frac{7}{8} + 12$ $12\frac{7}{8}$

Convert to a fraction.

9. $5\frac{2}{3}$ $\frac{17}{3}$ **10.** $4\frac{3}{5}$ $\frac{23}{5}$ **11.** $3\frac{5}{6}$ $\frac{23}{6}$ **12.** $7\frac{3}{4}$ $\frac{31}{4}$

13. $8\frac{1}{2}$ $\frac{17}{2}$ **14.** $5\frac{1}{3}$ $\frac{16}{3}$ **15.** $6\frac{1}{4}$ $\frac{25}{4}$ **16.** $9\frac{1}{5}$ $\frac{46}{5}$

17. $20\frac{2}{5}$ $\frac{102}{5}$ **18.** $30\frac{1}{4}$ $\frac{121}{4}$ **19.** $40\frac{1}{2}$ $\frac{81}{2}$ **20.** $50\frac{3}{4}$ $\frac{203}{4}$

21. $7\frac{3}{10}$ $\frac{73}{10}$ **22.** $6\frac{9}{10}$ $\frac{69}{10}$ **23.** $5\frac{7}{10}$ $\frac{57}{10}$ **24.** $8\frac{3}{10}$ $\frac{83}{10}$

25. $1\frac{5}{8}$ $\frac{13}{8}$ **26.** $1\frac{3}{5}$ $\frac{8}{5}$ **27.** $1\frac{5}{6}$ $\frac{11}{6}$ **28.** $1\frac{2}{3}$ $\frac{5}{3}$

29. $12\frac{3}{4}$ $\frac{51}{4}$ **30.** $15\frac{2}{3}$ $\frac{47}{3}$ **31.** $16\frac{7}{8}$ $\frac{135}{8}$ **32.** $18\frac{8}{9}$ $\frac{170}{9}$

33. $4\frac{8}{10}$ $\frac{48}{10}$ **34.** $5\frac{4}{10}$ $\frac{54}{10}$ **35.** $6\frac{2}{10}$ $\frac{62}{10}$ **36.** $8\frac{9}{10}$ $\frac{89}{10}$

37. $2\frac{3}{100}$ $\frac{203}{100}$ **38.** $5\frac{1}{100}$ $\frac{501}{100}$ **39.** $14\frac{57}{100}$ $\frac{1457}{100}$ **40.** $16\frac{95}{100}$ $\frac{1695}{100}$

More Practice, page 163

Objective 62

WRITING FRACTIONS AS MIXED NUMBERS

These circles are divided into thirds. Each part is $\frac{1}{3}$.

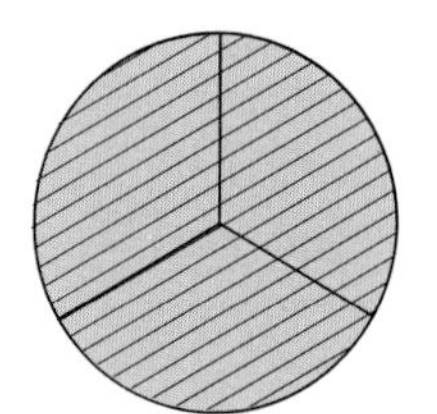

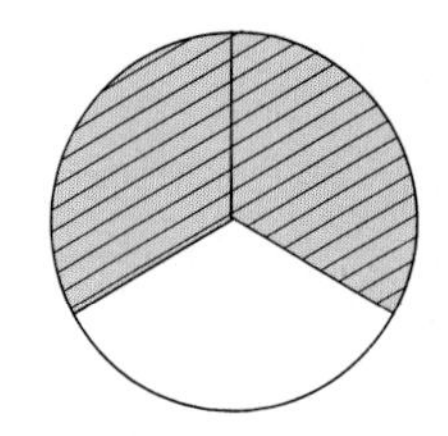

$$\frac{5}{3} = \underbrace{\frac{1}{3} + \frac{1}{3} + \frac{1}{3}}_{1} + \underbrace{\frac{1}{3} + \frac{1}{3}}_{\frac{2}{3}} = 1\frac{2}{3}$$

Convert to a mixed number. $\frac{74}{4}$

Divide. → The quotient is the whole number. The remainder over the divisor is the fraction.

Some students may not know how to simplify fractions, so it is not necessary that all answers be in simplest form.

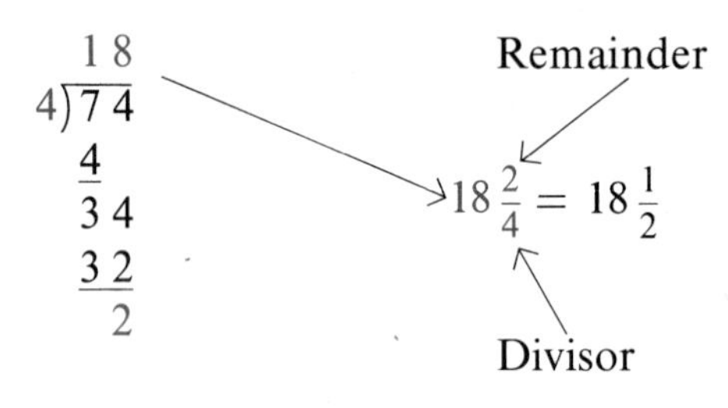

EXAMPLES. Convert to a mixed number.

1. $\frac{8}{5}$

$$\begin{array}{r} 1 \\ 5\overline{)8} \\ \underline{5} \\ 3 \end{array}$$

$\frac{8}{5} = 1\frac{3}{5}$

2. $\frac{122}{8}$

$$\begin{array}{r} 15 \\ 8\overline{)122} \\ \underline{120} \\ 2 \end{array}$$

$\frac{122}{8} = 15\frac{2}{8} = 15\frac{1}{4}$

TRY THIS

Convert to a mixed number.

1. $\frac{7}{3}$ $2\frac{1}{3}$ **2.** $\frac{11}{10}$ $1\frac{1}{10}$ **3.** $\frac{110}{6}$ $18\frac{2}{6}$ or $18\frac{1}{3}$

PRACTICE

For extra practice, use Making Practice Fun 35 with this lesson.

Convert to a mixed number.

1. $\frac{7}{5}$ $1\frac{2}{5}$
2. $\frac{13}{4}$ $3\frac{1}{4}$
3. $\frac{14}{3}$ $4\frac{2}{3}$
4. $\frac{7}{6}$ $1\frac{1}{6}$
5. $\frac{13}{4}$ $3\frac{1}{4}$
6. $\frac{19}{8}$ $2\frac{3}{8}$
7. $\frac{19}{5}$ $3\frac{4}{5}$
8. $\frac{13}{3}$ $4\frac{1}{3}$
9. $\frac{26}{4}$ $6\frac{2}{4}$ or $6\frac{1}{2}$
10. $\frac{27}{6}$ $4\frac{3}{6}$ or $4\frac{1}{2}$
11. $\frac{25}{2}$ $12\frac{1}{2}$
12. $\frac{28}{8}$ $3\frac{4}{8}$ or $3\frac{1}{2}$
13. $\frac{57}{10}$ $5\frac{7}{10}$
14. $\frac{69}{10}$ $6\frac{9}{10}$
15. $\frac{73}{10}$ $7\frac{3}{10}$
16. $\frac{41}{10}$ $4\frac{1}{10}$
17. $\frac{65}{10}$ $6\frac{5}{10}$ or $6\frac{1}{2}$
18. $\frac{54}{10}$ $5\frac{4}{10}$ or $5\frac{2}{5}$
19. $\frac{46}{10}$ $4\frac{6}{10}$ or $4\frac{3}{5}$
20. $\frac{78}{10}$ $7\frac{8}{10}$ or $7\frac{4}{5}$
21. $\frac{53}{7}$ $7\frac{4}{7}$
22. $\frac{62}{5}$ $12\frac{2}{5}$
23. $\frac{53}{8}$ $6\frac{5}{8}$
24. $\frac{49}{4}$ $12\frac{1}{4}$
25. $\frac{45}{6}$ $7\frac{3}{6}$ or $7\frac{1}{2}$
26. $\frac{50}{8}$ $6\frac{2}{8}$ or $6\frac{1}{4}$
27. $\frac{46}{4}$ $11\frac{2}{4}$ or $11\frac{1}{2}$
28. $\frac{50}{6}$ $8\frac{2}{6}$ or $8\frac{1}{3}$
29. $\frac{12}{8}$ $1\frac{4}{8}$ or $1\frac{1}{2}$
30. $\frac{18}{4}$ $4\frac{2}{4}$ or $4\frac{1}{2}$
31. $\frac{28}{6}$ $4\frac{4}{6}$ or $4\frac{2}{3}$
32. $\frac{22}{4}$ $5\frac{2}{4}$ or $5\frac{1}{2}$
33. $\frac{213}{100}$ $2\frac{13}{100}$
34. $\frac{467}{100}$ $4\frac{67}{100}$
35. $\frac{757}{100}$ $7\frac{57}{100}$
36. $\frac{321}{100}$ $3\frac{21}{100}$
37. $\frac{345}{8}$ $43\frac{1}{8}$
38. $\frac{223}{4}$ $55\frac{3}{4}$
39. $\frac{135}{6}$ $22\frac{3}{6}$ or $22\frac{1}{2}$
40. $\frac{185}{3}$ $61\frac{2}{3}$
41. $\frac{456}{16}$ $28\frac{8}{16}$ or $28\frac{1}{2}$
42. $\frac{554}{12}$ $46\frac{2}{12}$ or $46\frac{1}{6}$
43. $\frac{678}{15}$ $45\frac{3}{15}$ or $45\frac{1}{5}$
44. $\frac{567}{31}$ $18\frac{9}{31}$

PROBLEM CORNER

1. There are $\frac{365}{7}$ weeks in a year. Convert the number of weeks to a mixed number. $52\frac{1}{7}$ weeks
2. The Jefferson twins are making nut bread. Ada measures $\frac{7}{4}$ cups of sugar. Arthur measures $\frac{9}{2}$ cups of flour. Convert each measurement to a mixed number. $1\frac{3}{4}$ c; $4\frac{1}{2}$ c

Objectives 63 and 64

MIXED NUMBERS AND DECIMALS

Convert to a mixed number. 8.95

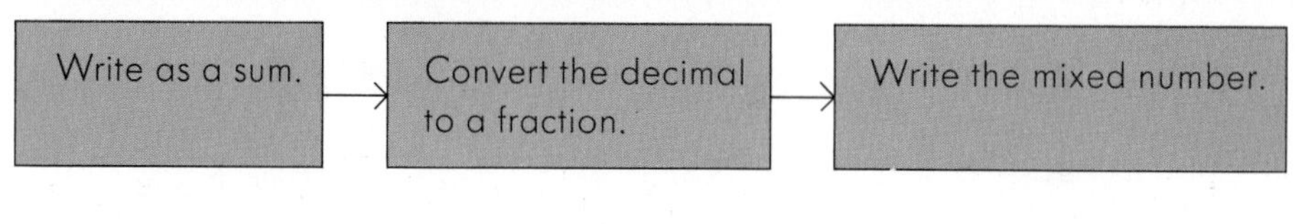

$8.95 = 8 + 0.95 \quad = \quad 8 + \frac{95}{100} \quad = \quad 8\frac{95}{100}$

EXAMPLES. Convert to a mixed number.

1. $6.3 = 6 + 0.3 = 6 + \frac{3}{10} = 6\frac{3}{10}$

2. $12.007 = 12\frac{7}{1000}$

TRY THIS

Convert to a mixed number.

1. 4.17 $4\frac{17}{100}$ **2.** 13.018 $13\frac{18}{1000}$ **3.** 1.9 $1\frac{9}{10}$

Convert to a decimal. $17\frac{3}{4}$

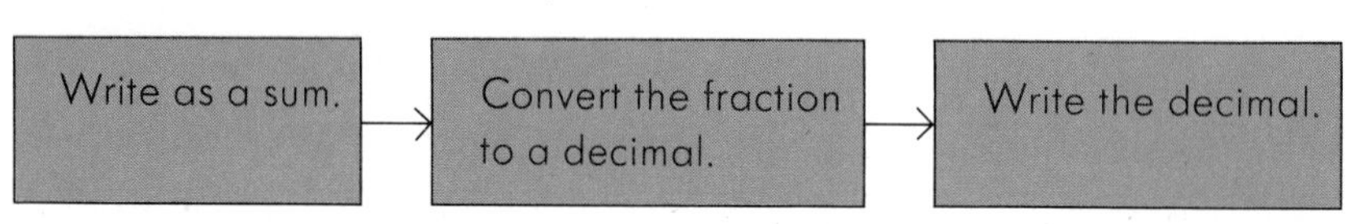

$17\frac{3}{4} = 17 + \frac{3}{4} \quad = \quad 17 + 0.75 \quad = \quad 17.75$

EXAMPLES. Convert to a decimal.

3. $9\frac{1}{2} = 9 + \frac{1}{2} = 9 + 0.5 = 9.5$

4. $28\frac{97}{100} = 28.97$

TRY THIS

Convert to a decimal.

4. $5\frac{6}{10}$ 5.6 **5.** $22\frac{4}{5}$ 22.8 **6.** $2\frac{33}{1000}$ 2.033

PRACTICE

Use Quiz 13 after this Practice.

Convert to a mixed number.

1. 1.2 $1\frac{2}{10}$ **2.** 3.7 $3\frac{7}{10}$ **3.** 11.1 $11\frac{1}{10}$ **4.** 99.7 $99\frac{7}{10}$

5. 8.98 $8\frac{98}{100}$ **6.** 1.23 $1\frac{23}{100}$ **7.** 24.76 $24\frac{76}{100}$ **8.** 32.22 $32\frac{22}{100}$

9. 1.113 $1\frac{113}{1000}$ **10.** 7.097 $7\frac{97}{1000}$ **11.** 89.001 $89\frac{1}{1000}$ **12.** 12.999 $12\frac{999}{1000}$

13. 4.7878 $4\frac{7878}{10,000}$ **14.** 5.6177 $5\frac{6177}{10,000}$ **15.** 11.4562 $11\frac{4562}{10,000}$ **16.** 13.1188 $13\frac{1188}{10,000}$

17. 124.9 $124\frac{9}{10}$ **18.** 621.7 $621\frac{7}{10}$ **19.** 134.56 $134\frac{56}{100}$ **20.** 221.08 $221\frac{8}{100}$

Convert to a decimal.

21. $1\frac{9}{10}$ 1.9 **22.** $3\frac{8}{10}$ 3.8 **23.** $18\frac{1}{10}$ 18.1 **24.** $23\frac{5}{10}$ 23.5

25. $1\frac{17}{100}$ 1.17 **26.** $7\frac{59}{100}$ 7.59 **27.** $13\frac{99}{100}$ 13.99 **28.** $14\frac{23}{100}$ 14.23

29. $4\frac{3}{1000}$ 4.003 **30.** $8\frac{1}{1000}$ 8.001 **31.** $166\frac{213}{1000}$ 166.213 **32.** $249\frac{188}{1000}$ 249.188

33. $8\frac{1}{5}$ 8.2 **34.** $7\frac{3}{5}$ 7.6 **35.** $9\frac{2}{5}$ 9.4 **36.** $6\frac{4}{5}$ 6.8

37. $13\frac{3}{8}$ 13.375 **38.** $14\frac{5}{8}$ 14.625 **39.** $19\frac{1}{8}$ 19.125 **40.** $20\frac{7}{8}$ 20.875

41. $10\frac{1}{2}$ 10.5 **42.** $17\frac{1}{4}$ 17.25 **43.** $35\frac{3}{4}$ 35.75 **44.** $29\frac{1}{4}$ 29.25

45. $23\frac{13}{16}$ 23.8125 **46.** $9\frac{11}{20}$ 9.55 **47.** $103\frac{14}{25}$ 103.56 **48.** $128\frac{23}{40}$ 128.575

PROBLEM CORNER

At service stations, pumps measure gasoline by one-tenth gallons. What will a pump read after filling an empty $16\frac{1}{5}$ gallon tank? 16.2

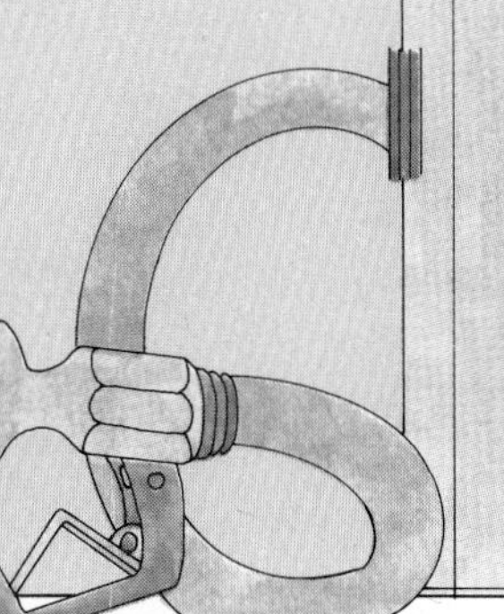

More Practice, page 163

Objectives 164 and 165

APPLICATION

TIME ZONES

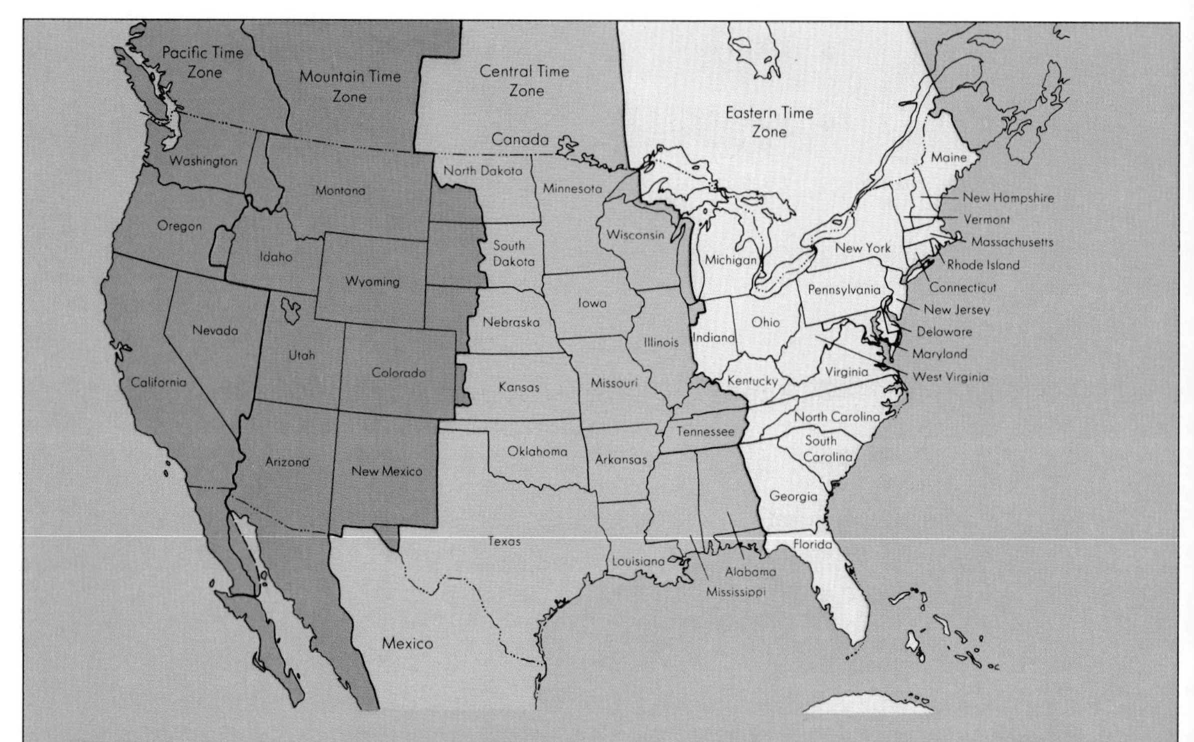

Most of the U.S. is divided into 4 time zones. It takes about 1 hour for the sun to pass over a zone.

A TV newscaster announces, "The tennis match will begin at 2 PM Eastern Standard Time (EST)." What time will this be in each zone?

1. Central (CST) 1 PM
2. Mountain (MST) 12 PM
3. Pacific (PST) 11 AM

Almost half the states have a city called Springfield. Use the time zone map to find what time it is in each city when it is 12:00 PM (noon) in Springfield, Illinois.

4. Springfield, Oregon 10 AM 5. Springfield, Maine 1 PM
6. Springfield, Colorado 11 AM

We can subtract to find how much time has passed.

EXAMPLE 1. School starts at 8:00 AM and ends at 2:30 PM. How long is the school day?

$$\begin{array}{r} 2{:}30\ \text{PM} \\ -\ 8{:}00\ \text{AM} \\ \hline \end{array} \longrightarrow \begin{array}{r} 14{:}30 \\ -\ 8{:}00 \\ \hline 6{:}30 \end{array} = 6\ \text{hr},\ 30\ \text{min}$$

We passed from AM to PM, so add 12 to 2:30.

7. Soccer practice begins at 10:00 AM and ends at 1:15 PM. How long is practice? 3 hr, 15 min
8. The concert started at 11:30 AM and ended at 2:00 PM. How long was the concert? 2 hr, 30 min

Airlines give the time of departure and the time of arrival in local time.

EXAMPLE 2. A plane departs Boston, Massachusetts at 10:30 AM and arrives in Los Angeles, California at 1:20 PM. How long does the trip take?

$$\begin{array}{r} 1{:}20\ \text{PM (PST)} \\ -\ 10{:}30\ \text{AM (EST)} \\ \hline \end{array} \longrightarrow \begin{array}{r} 1{:}20\ \text{PM (PST)} \\ -\ 7{:}30\ \text{AM (PST)} \\ \hline \end{array}$$

Subtract departure from arrival. Change to the same time zone.

$$\begin{array}{r} 1{:}20\ \text{PM} \\ -\ 7{:}30\ \text{AM} \\ \hline \end{array} \longrightarrow \begin{array}{r} 13{:}20 \\ -\ 7{:}30 \\ \hline \end{array} \longrightarrow \begin{array}{r} 12{:}80 \\ -\ 7{:}30 \\ \hline 5{:}50 \end{array} = 5\ \text{hr},\ 50\ \text{min}$$

Rename to subtract.

How long is each trip?

		Depart		Arrive	
1:52	9.	Chicago, Ill.	11:45 PM	New York City, N.Y.	2:37 AM
4:20	10.	Miami, Fla.	10:50 AM	Denver, Colo.	1:10 PM
4:05	11.	Philadelphia, Penn.	11:15 AM	Kansas City, Mo.	2:20 PM
4:25	12.	Seattle, Wash.	11:00 PM	Detroit, Mich.	6:25 AM

Miguel Romero is the operations manager for the Arden Plastic Company. He plans truck deliveries. He must think about loads, routes, time, and cost. The maximum legal weight of a loaded truck is 73,280 pounds.

1. Truck number 82 weighs 34,306 pounds. What is the heaviest load it can carry? 38,974 lb
2. Truck number 103 weighs 32,495 pounds. What is the heaviest load it can carry? 40,785 lb
3. Truck number 18 weighs 38,782 pounds. Can it carry a load which weighs 35,598 pounds? no
4. How many more pounds can truck number 103 carry than truck number 18? 6287 lb

Miguel records the time it takes to load the trucks. How long does it take to load each truck?

Truck	Start Loading	Finish Loading
23	6:15 AM	8:32 AM
184	7:20 AM	10:15 AM
73	9:13 AM	11:27 AM
87	10:06 AM	12:30 PM
159	11:50 AM	2:17 PM
41	1:45 PM	4:22 PM

5. truck 23 2:17
6. truck 184 2:55
7. truck 73 2:14
8. truck 87 2:24
9. truck 159 2:27
10. truck 41 2:37

Small trucks are used to make local deliveries. The drivers sign out and sign in to keep a record of hours driven. How much time did each driver take?

Name	Time Out	Time In
Sonny	6:35 AM	11:50 AM
Rhonda	7:10 AM	2:40 PM
Jacques	7:45 AM	1:45 PM
Mike	7:40 AM	1:30 PM
Red Eye	8:00 AM	2:45 PM
Buck	8:09 AM	3:12 PM

11. Sonny 5:15
12. Rhonda 7:30
13. Jacques 6:00
14. Mike 5:50
15. Red Eye 6:45
16. Buck 7:03

Arden Plastic Company ships products of many weights and sizes. Miguel must consider these measurements in deciding how to load the trucks, because there are weight limits for each truck.

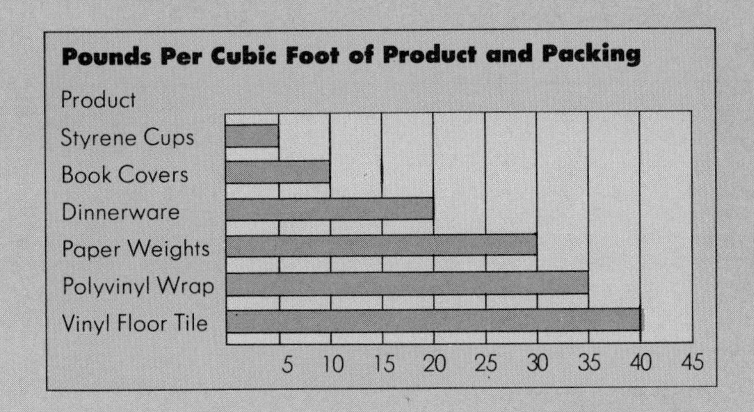

How many pounds are there in one cubic foot of each of the following?

17. polyvinyl wrap 35 lb
18. styrene cups 5 lb

How many pounds are there in 250 cubic feet of each?

19. book covers 2500 lb
20. dinnerware 5000 lb
21. paper weights 7500 lb
22. floor tile 10,000 lb

How many cubic feet are needed for 100 pounds of each?

23. styrene cups 20 cu ft
24. dinnerware 5 cu ft
25. polyvinyl wrap $2\frac{6}{7}$ cu ft
26. floor tile $2\frac{1}{2}$ cu ft

CHAPTER 5 REVIEW

For extra practice, use Making Practice Fun 36 with this lesson.

1. What part is shaded? p. 136 $\frac{3}{4}$

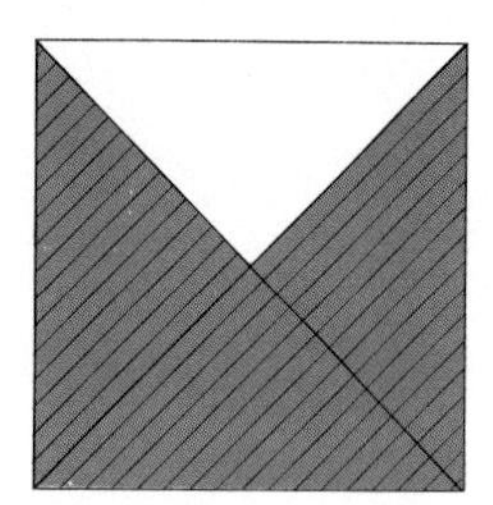

2. What part is shaded? $\frac{1}{3}$

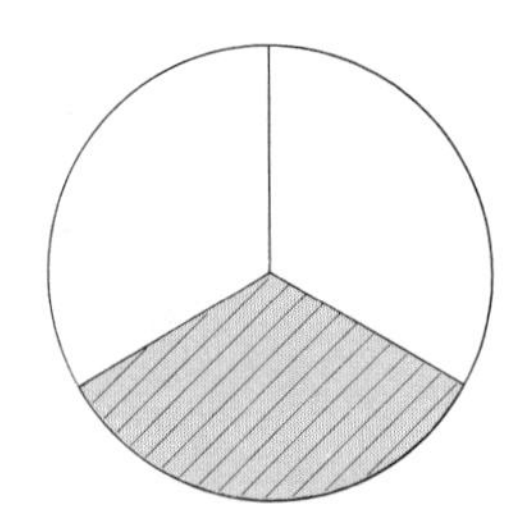

Divide, if possible. p. 138

3. $\frac{15}{1}$ 15 **4.** $\frac{0}{7}$ 0 **5.** $\frac{7}{7}$ 1 **6.** $\frac{12}{0}$ Not possible

Are the fractions equal? p. 140

7. $\frac{2}{4}, \frac{3}{6}$ yes **8.** $\frac{3}{5}, \frac{3}{7}$ no

Compare. p. 142

9. $\frac{6}{7}$ and $\frac{7}{8}$ $<$ **10.** $\frac{9}{11}$ and $\frac{4}{5}$ $>$

Convert to a fraction. pp. 144, 150

11. 1.093 $\frac{1093}{1000}$ **12.** 0.07 $\frac{7}{100}$ **13.** $4\frac{1}{3}$ $\frac{13}{3}$ **14.** $10\frac{5}{8}$ $\frac{85}{8}$

Convert to a decimal. p. 146

15. $\frac{9}{8}$ 1.125 **16.** $\frac{37}{40}$ 0.925 **17.** $\frac{13}{100}$ 0.13 **18.** $\frac{9898}{1000}$ 9.898

Convert to a decimal. Round to the nearest thousandth, if necessary. pp. 146, 154

19. $\frac{86}{7}$ 12.286 **20.** $\frac{4}{3}$ 1.333 **21.** $8\frac{1}{10}$ 8.1 **22.** $118\frac{3}{5}$ 118.6

Convert to a mixed number. pp. 152, 154

23. $\frac{19}{3}$ $6\frac{1}{3}$ **24.** $\frac{69}{7}$ $9\frac{6}{7}$ **25.** 4.9 $4\frac{9}{10}$ **26.** 8.013 $8\frac{13}{1000}$

Application pp. 148, 156

27. Jerry was on the ski slopes for 4 hr, 15 min. Sally was there for 2 hr, 50 min. How much longer was Jerry on the ski slopes than Sally? 1 hr, 25 min

28. When it is 7 PM in New York, what time is it in Los Angeles? 4 PM

CHAPTER 5 TEST

1. What part is shaded? $\frac{5}{6}$

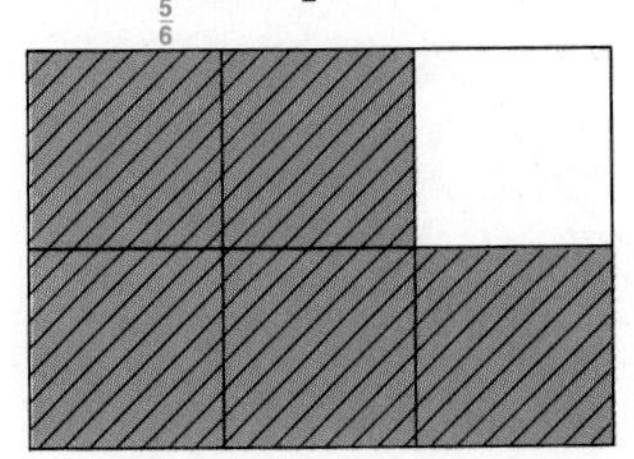

2. What part is shaded? $\frac{2}{5}$

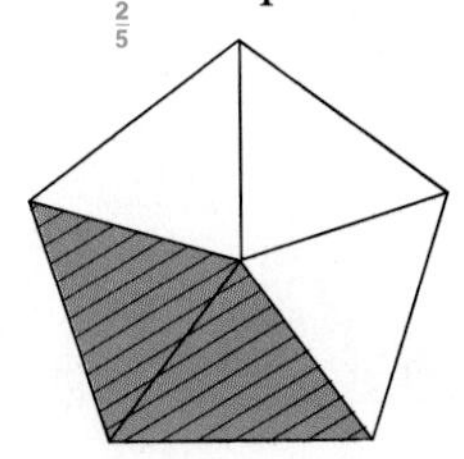

Divide, if possible.

3. $\frac{0}{23}$ 0 **4.** $\frac{27}{1}$ 27 **5.** $\frac{27}{0}$ Not possible **6.** $\frac{38}{38}$ 1

Are the fractions equal?

7. $\frac{2}{6}, \frac{6}{18}$ yes **8.** $\frac{7}{8}, \frac{8}{9}$ no

Compare.

9. $\frac{7}{6}$ and $\frac{8}{7}$ > **10.** $\frac{5}{12}$ and $\frac{3}{8}$ >

Convert to a fraction.

11. 1.237 $\frac{1237}{1000}$ **12.** 0.3 $\frac{3}{10}$ **13.** $2\frac{1}{8}$ $\frac{17}{8}$ **14.** $11\frac{4}{5}$ $\frac{59}{5}$

Convert to a decimal.

15. $\frac{11}{25}$ 0.44 **16.** $\frac{2}{5}$ 0.4 **17.** $\frac{7}{10}$ 0.7 **18.** $\frac{7784}{1000}$ 7.784

Convert to a decimal. Round to the nearest thousandth, if necessary.

19. $\frac{56}{13}$ 4.308 **20.** $\frac{5}{11}$ 0.455 **21.** $4\frac{9}{10}$ 4.9 **22.** $29\frac{3}{4}$ 29.75

Convert to a mixed number.

23. $\frac{25}{4}$ $6\frac{1}{4}$ **24.** $\frac{125}{9}$ $13\frac{8}{9}$ **25.** 2.8 $2\frac{4}{5}$ **26.** 120.67 $120\frac{67}{100}$

Application

27. Kevin spent 2 hr, 35 min mowing the lawn and 1 hr, 40 min trimming it. How much time did he take in all? 4 hr, 15 min

28. When it is 6 PM in Chicago, what time is it in New York? 7 PM

MORE PRACTICE

What part is shaded? Use after page 137.

1. $\frac{4}{5}$

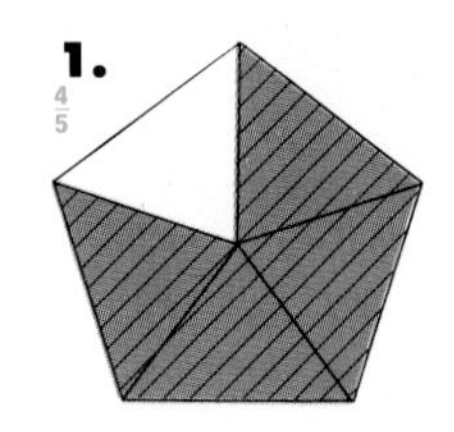

2. $\frac{3}{4}$

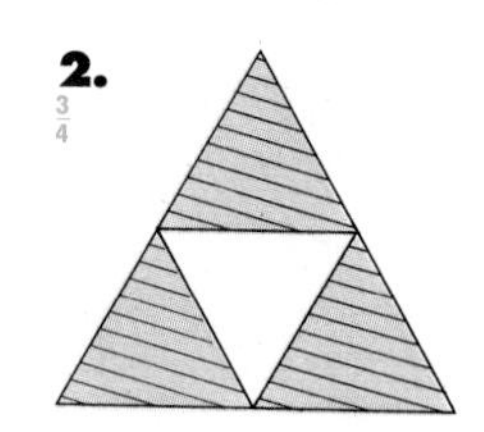

3. $\frac{1}{2}$

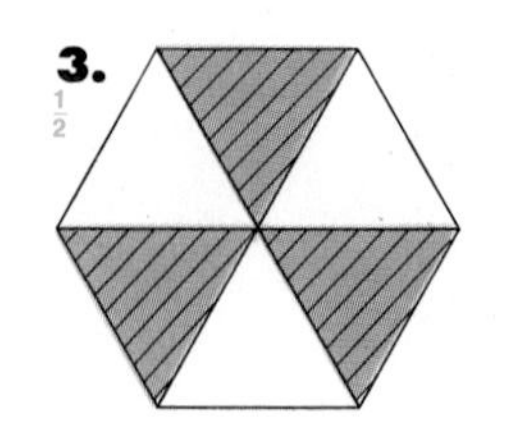

4. $\frac{7}{8}$

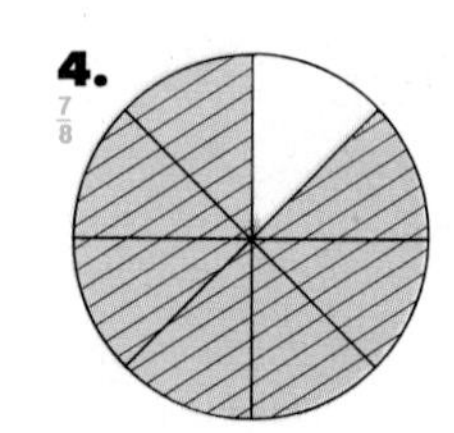

Divide, if possible. Use after page 139.

5. $\frac{39}{1}$ 39 **6.** $\frac{39}{39}$ 1 **7.** $\frac{43}{43}$ 1 **8.** $\frac{43}{0}$ Not possible

9. $\frac{17}{0}$ Not possible **10.** $\frac{17}{1}$ 17 **11.** $\frac{0}{38}$ 0 **12.** $\frac{0}{100}$ 0

Tell whether the fractions are equal. Use after page 141.

13. $\frac{1}{4}, \frac{1}{3}$ no **14.** $\frac{2}{3}, \frac{5}{8}$ no **15.** $\frac{3}{2}, \frac{6}{4}$ yes **16.** $\frac{2}{3}, \frac{5}{7}$ no

17. $\frac{6}{16}, \frac{3}{8}$ yes **18.** $\frac{2}{3}, \frac{8}{12}$ yes **19.** $\frac{7}{25}, \frac{2}{7}$ no **20.** $\frac{8}{40}, \frac{6}{30}$ yes

Compare. Use after page 143.

21. $\frac{1}{4}$ and $\frac{1}{3}$ $<$ **22.** $\frac{2}{3}$ and $\frac{5}{8}$ $>$ **23.** $\frac{5}{7}$ and $\frac{2}{3}$ $>$ **24.** $\frac{9}{10}$ and $\frac{89}{100}$ $>$

25. $\frac{5}{16}$ and $\frac{1}{4}$ $>$ **26.** $\frac{2}{5}$ and $\frac{1}{3}$ $>$ **27.** $\frac{7}{12}$ and $\frac{5}{6}$ $<$ **28.** $\frac{3}{4}$ and $\frac{7}{10}$ $>$

Convert to a fraction. Use after page 145.

29. 9.3 $\frac{93}{10}$ **30.** 1.07 $\frac{107}{100}$ **31.** 0.013 $\frac{13}{1000}$ **32.** 1.9132 $\frac{19,132}{10,000}$

33. 0.89 $\frac{89}{100}$ **34.** 0.9 $\frac{9}{10}$ **35.** 0.4133 $\frac{4133}{10,000}$ **36.** 7.567 $\frac{7567}{1000}$

Convert to a decimal. Use after page 147.

37. $\frac{4}{5}$ 0.8 **38.** $\frac{5}{4}$ 1.25 **39.** $\frac{7}{8}$ 0.875 **40.** $\frac{53}{20}$ 2.65

41. $\frac{13}{16}$ 0.8125 **42.** $\frac{32}{25}$ 1.28 **43.** $\frac{49}{40}$ 1.225 **44.** $\frac{64}{125}$ 0.512

Convert to a decimal. Round to the nearest thousandth. Use after page 147.

1. $\frac{1}{6}$ 0.167 **2.** $\frac{8}{3}$ 2.667 **3.** $\frac{4}{7}$ 0.571 **4.** $\frac{10}{9}$ 1.111

5. $\frac{12}{13}$ 0.923 **6.** $\frac{19}{12}$ 1.583 **7.** $\frac{14}{9}$ 1.556 **8.** $\frac{127}{11}$ 11.545

Convert to a decimal. Use after page 147.

9. $\frac{6}{10}$ 0.6 **10.** $\frac{17}{10}$ 1.7 **11.** $\frac{11}{1000}$ 0.011 **12.** $\frac{5673}{1000}$ 5.673

13. $\frac{23}{100}$ 0.23 **14.** $\frac{997}{100}$ 9.97 **15.** $\frac{569}{10,000}$ 0.0569 **16.** $\frac{18,977}{10,000}$ 1.8977

Convert to a fraction. Use after page 151.

17. $9\frac{1}{4}$ $\frac{37}{4}$ **18.** $7\frac{2}{3}$ $\frac{23}{3}$ **19.** $4\frac{5}{8}$ $\frac{37}{8}$ **20.** $10\frac{1}{2}$ $\frac{21}{2}$

21. $14\frac{2}{7}$ $\frac{100}{7}$ **22.** $8\frac{3}{10}$ $\frac{83}{10}$ **23.** $6\frac{3}{100}$ $\frac{603}{100}$ **24.** $17\frac{8}{9}$ $\frac{161}{9}$

25. $8\frac{7}{8}$ $\frac{71}{8}$ **26.** $13\frac{2}{5}$ $\frac{67}{5}$ **27.** $7\frac{5}{6}$ $\frac{47}{6}$ **28.** $19\frac{2}{7}$ $\frac{135}{7}$

Convert to a mixed number. Use after page 153.

29. $\frac{11}{3}$ $3\frac{2}{3}$ **30.** $\frac{9}{4}$ $2\frac{1}{4}$ **31.** $\frac{29}{5}$ $5\frac{4}{5}$ **32.** $\frac{35}{6}$ $5\frac{5}{6}$

33. $\frac{79}{10}$ $7\frac{9}{10}$ **34.** $\frac{457}{2}$ $228\frac{1}{2}$ **35.** $\frac{577}{12}$ $48\frac{1}{12}$ **36.** $\frac{492}{49}$ $10\frac{2}{49}$

37. $\frac{15}{7}$ $2\frac{1}{7}$ **38.** $\frac{98}{5}$ $19\frac{3}{5}$ **39.** $\frac{29}{3}$ $9\frac{2}{3}$ **40.** $\frac{137}{9}$ $15\frac{2}{9}$

Convert to a mixed number. Use after page 155.

41. 5.3 $5\frac{3}{10}$ **42.** 5.37 $5\frac{37}{100}$ **43.** 1.012 $1\frac{6}{500}$ **44.** 14.7 $14\frac{7}{10}$

45. 5.1137 $5\frac{1137}{10,000}$ **46.** 29.5142 $29\frac{5142}{10,000}$ **47.** 109.01 $109\frac{1}{100}$ **48.** 123.789 $123\frac{789}{1000}$

Convert to a decimal. Use after page 155.

49. $5\frac{9}{10}$ 5.9 **50.** $8\frac{41}{100}$ 8.41 **51.** $78\frac{7883}{10,000}$ 78.7883 **52.** $14\frac{1}{1000}$ 14.001

53. $9\frac{3}{5}$ 9.6 **54.** $12\frac{3}{8}$ 12.375 **55.** $10\frac{3}{4}$ 10.75 **56.** $8\frac{17}{20}$ 8.85

CALCULATOR CORNER

MULTIPLICATION "IN BETWEEN"

This activity emphasizes estimating and approximating. Discuss the size of the number used as a multiplier to obtain the desired result.

Players two

Materials two calculators, paper and pencils

Sample Aaron gives Jodi a calculator with a number, such as 212, on the display. He tells Jodi that by multiplying 212 by a number, she must get a number on the display which is between 120 and 130.

Jodi gives Aaron a calculator with a number, such as 183, on the display. She tells Aaron that by multiplying 183 by a number, he must get a number on the display which is between 260 and 270.

JODI Jodi writes down 212 and the interval 120–130.

Play 1 Jodi multiplies 212 by 0.8. She records 0.8.
Result: 169.5–too large

Play 2 Jodi multiplies 212 by 0.7. She records 0.7.
Result: 148.4–too large

Play 3 Jodi multiplies 212 by 0.6. She records 0.6.
Result: 127.2–OK

It took Jodi 3 plays, so her score is 3.

AARON Aaron writes down 183 and the interval 260–270.

Play 1 Aaron multiplies 183 by 2. He records 2.
Result: 366–too large

Play 2 Aaron multiplies 183 by 1.5. He records 1.5.
Result: 274.5–too large

Play 3 Aaron multiplies 183 by 1.4. He records 1.4.
Result: 256.2–too small

Play 4 Aaron multiplies 183 by 1.45. He records 1.45.
Result: 265.35–OK

It took Aaron 4 plays, so his score is 4.

Play continues for 5 rounds. The player with the lowest score wins.

CHAPTER 6 PRE-TEST

Multiply. p. 168

1. $\frac{1}{2} \cdot \frac{1}{5}$ $\frac{1}{10}$ **2.** $\frac{2}{3} \cdot \frac{4}{5}$ $\frac{8}{15}$ **3.** $6 \times \frac{2}{7}$ $1\frac{5}{7}$

4. Find a fraction equivalent to $\frac{3}{4}$ with a denominator of 12. p. 170 $\frac{9}{12}$

All the fractions are equivalent. Which is simplest? p. 172

5. $\frac{6}{16}, \frac{3}{8}, \frac{12}{32}, \frac{15}{40}, \frac{18}{48}$ $\frac{3}{8}$ **6.** $\frac{8}{28}, \frac{6}{21}, \frac{2}{7}, \frac{4}{14}$ $\frac{2}{7}$

Simplify. p. 172

7. $\frac{12}{18}$ $\frac{2}{3}$ **8.** $\frac{8}{20}$ $\frac{2}{5}$ **9.** $\frac{30}{6}$ 5

Multiply and simplify. pp. 174–176

10. $\frac{3}{8} \times \frac{4}{9}$ $\frac{1}{6}$ **11.** $\frac{6}{7} \cdot \frac{5}{3}$ $1\frac{3}{7}$ **12.** $\frac{4}{5} \cdot 20$ 16

13. $9 \times 4\frac{1}{3}$ 39 **14.** $6\frac{3}{4} \cdot \frac{2}{3}$ $4\frac{1}{2}$ **15.** $3\frac{1}{3} \cdot 1\frac{3}{4}$ $5\frac{5}{6}$

Find the reciprocal. p. 180

16. $\frac{3}{4}$ $\frac{4}{3}$ **17.** 8 $\frac{1}{8}$ **18.** $\frac{1}{10}$ 10

Divide and simplify. pp. 180–182

19. $\frac{5}{6} \div \frac{2}{3}$ $1\frac{1}{4}$ **20.** $\frac{3}{5} \div \frac{1}{2}$ $1\frac{1}{5}$ **21.** $\frac{2}{5} \div 6$ $\frac{1}{15}$

22. $33 \div 5\frac{1}{2}$ 6 **23.** $2\frac{1}{3} \div 1\frac{1}{6}$ 2 **24.** $2\frac{1}{12} \div 75$ $\frac{1}{36}$

Application pp. 178, 184

25. Rachel budgets $960 for transportation. By taking the bus to work, she can save $\frac{1}{3}$ of this expense. How much can she save? $320

26. For breakfast Alex eats the following. How many calories are in Alex's breakfast? 848

Food	Calories
1 cup of cereal	400 per cup
$\frac{1}{2}$ cup of milk	160 per cup
2 pieces of toast	100 per piece
$1\frac{1}{2}$ cups of orange juice	112 per cup

6
MULTIPLYING AND DIVIDING FRACTIONS

GETTING STARTED

FRACTION MULTO

WHAT'S NEEDED The 30 squares of paper with digits marked on them as in GOTCHA, Chapter 1.

THE RULES

1. On a sheet of paper each player writes this:

   $\frac{[\ ]}{[\ ]} \cdot \frac{[\ ]}{[\ ]}$

2. To play, a leader draws digits from the box, one at a time.
3. Each player writes the digit in any one of the four spaces. Once it has been written, it cannot be changed.
4. The leader continues to draw digits until four digits have been drawn.
5. Each player multiplies the resulting fractions. The winner is the player who has the largest product.

SAMPLE

Louis	**Chris**	**Sally**
$\frac{[2]}{[3]} \cdot \frac{[5]}{[4]} = \frac{10}{12}$	$\frac{[4]}{[5]} \cdot \frac{[3]}{[2]} = \frac{12}{10}$	$\frac{[4]}{[3]} \cdot \frac{[5]}{[2]} = \frac{20}{6}$

Sally wins because she has the largest product.

Objective 65

MULTIPLYING

We will simplify fractions later, so answers need not be written in simplest form.

Multiply. $\frac{2}{3} \times \frac{5}{7}$

Multiply the numerators. → Multiply the denominators.

$$\frac{2}{3} \times \frac{5}{7} = \frac{2 \times 5}{3 \times 7} = \frac{10}{21}$$

EXAMPLES. Multiply.

1. $\frac{3}{5} \times \frac{7}{8} = \frac{3 \times 7}{5 \times 8} = \frac{21}{40}$

2. $\frac{2}{5} \times \frac{4}{4} = \frac{8}{20}$ *Try to multiply mentally.*

3. $\frac{1}{3} \cdot \frac{1}{4} = \frac{1 \cdot 1}{3 \cdot 4} = \frac{1}{12}$ *We can also use a dot · for ×.*

TRY THIS

Multiply.

1. $\frac{2}{3} \times \frac{4}{5}$ $\frac{8}{15}$ **2.** $\frac{2}{3} \cdot \frac{2}{2}$ $\frac{4}{6}$

3. $\frac{1}{2} \times \frac{1}{5}$ $\frac{1}{10}$ **4.** $\frac{7}{1} \cdot \frac{5}{6}$ $5\frac{5}{6}$

EXAMPLE 4. Multiply. $6 \times \frac{4}{5}$

$$6 \times \frac{4}{5} = \frac{6}{1} \times \frac{4}{5} = \frac{6 \times 4}{1 \times 5} = \frac{24}{5} = 4\frac{4}{5}$$

EXAMPLE 5. Multiply. $\frac{3}{8} \cdot 11$

$$\frac{3}{8} \cdot 11 = \frac{3}{8} \cdot \frac{11}{1} = \frac{33}{8} = 4\frac{1}{8}$$

TRY THIS

Multiply.

5. $10 \cdot \frac{2}{3}$ $6\frac{2}{3}$ **6.** $\frac{2}{5} \times 16$ $6\frac{2}{5}$ **7.** $5 \cdot \frac{1}{8}$ $\frac{5}{8}$

PRACTICE

For extra practice, use Making Practice Fun 37 with this lesson.

Multiply.

1. $\frac{1}{4} \cdot \frac{1}{5}$ $\frac{1}{20}$ **2.** $\frac{1}{10} \cdot \frac{1}{4}$ $\frac{1}{40}$ **3.** $\frac{1}{5} \cdot \frac{1}{10}$ $\frac{1}{50}$ **4.** $\frac{1}{6} \cdot \frac{1}{8}$ $\frac{1}{48}$

5. $\frac{2}{3} \cdot \frac{1}{5}$ $\frac{2}{15}$ **6.** $\frac{3}{4} \cdot \frac{1}{3}$ $\frac{3}{12}$ **7.** $\frac{2}{5} \cdot \frac{1}{7}$ $\frac{2}{35}$ **8.** $\frac{3}{5} \cdot \frac{1}{2}$ $\frac{3}{10}$

9. $\frac{3}{4} \times \frac{2}{2}$ $\frac{6}{8}$ **10.** $\frac{4}{5} \times \frac{3}{3}$ $\frac{12}{15}$ **11.** $\frac{4}{4} \times \frac{2}{3}$ $\frac{8}{12}$ **12.** $\frac{8}{8} \times \frac{5}{6}$ $\frac{40}{48}$

13. $\frac{2}{5} \times \frac{5}{6}$ $\frac{10}{30}$ **14.** $\frac{2}{3} \times \frac{7}{8}$ $\frac{14}{24}$ **15.** $\frac{3}{8} \times \frac{2}{3}$ $\frac{6}{24}$ **16.** $\frac{4}{5} \times \frac{2}{6}$ $\frac{8}{30}$

17. $\frac{4}{9} \cdot \frac{5}{6}$ $\frac{20}{54}$ **18.** $\frac{2}{9} \cdot \frac{3}{4}$ $\frac{6}{36}$ **19.** $\frac{5}{6} \cdot \frac{3}{4}$ $\frac{15}{24}$ **20.** $\frac{2}{9} \cdot \frac{4}{9}$ $\frac{8}{81}$

21. $\frac{1}{5} \cdot 3$ $\frac{3}{5}$ **22.** $\frac{1}{3} \cdot 2$ $\frac{2}{3}$ **23.** $\frac{1}{6} \cdot 5$ $\frac{5}{6}$ **24.** $\frac{1}{7} \cdot 4$ $\frac{4}{7}$

25. $8 \times \frac{3}{5}$ $4\frac{4}{5}$ **26.** $1 \times \frac{2}{7}$ $\frac{2}{7}$ **27.** $8 \times \frac{2}{3}$ $5\frac{1}{3}$ **28.** $9 \times \frac{3}{5}$ $5\frac{2}{5}$

29. $\frac{7}{8} \cdot \frac{7}{8}$ $\frac{49}{64}$ **30.** $\frac{4}{5} \cdot \frac{4}{5}$ $\frac{16}{25}$ **31.** $\frac{5}{6} \cdot \frac{5}{6}$ $\frac{25}{36}$ **32.** $\frac{5}{8} \cdot \frac{5}{8}$ $\frac{25}{64}$

33. $\frac{2}{3} \cdot 23$ $15\frac{1}{3}$ **34.** $17 \cdot \frac{5}{6}$ $14\frac{1}{6}$ **35.** $\frac{3}{4} \cdot 19$ $14\frac{1}{4}$ **36.** $20 \cdot \frac{5}{3}$ $33\frac{1}{3}$

37. $\frac{10}{11} \cdot 32$ $29\frac{1}{11}$ **38.** $25 \cdot \frac{12}{13}$ $23\frac{1}{13}$ **39.** $20 \cdot \frac{14}{15}$ $18\frac{10}{15}$ **40.** $\frac{11}{12} \cdot 23$ $21\frac{1}{12}$

PROBLEM CORNER

1. It takes $\frac{3}{4}$ of a yard of ribbon to make a bow. How much ribbon is needed to make 7 bows? $5\frac{1}{4}$ yd

2. A granola recipe calls for $\frac{3}{4}$ of a cup of honey. How much honey is needed to make $\frac{1}{2}$ of a recipe. $\frac{3}{8}$ c

Objectives 66 and 67

EQUIVALENT FRACTIONS

Students should be reminded that *equivalent* means the same as *equal.*

When we multiply a number by 1, we get the same number. When we multiply a fraction by 1, we get an equivalent fraction.

Find a fraction equivalent to $\frac{3}{4}$.

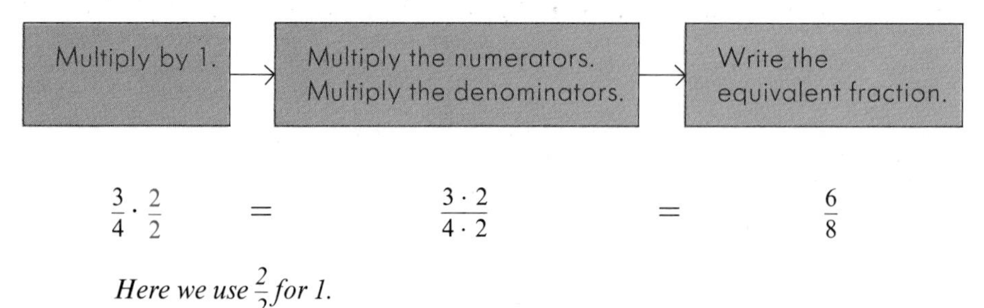

$$\frac{3}{4}\cdot\frac{2}{2} = \frac{3\cdot 2}{4\cdot 2} = \frac{6}{8}$$

Here we use $\frac{2}{2}$ *for 1.*

EXAMPLE 1. Find two other fractions equivalent to $\frac{3}{4}$.

$$\frac{3}{4} = \frac{3}{4}\cdot\frac{3}{3} = \frac{3\cdot 3}{4\cdot 3} = \frac{9}{12} \qquad \frac{3}{4} = \frac{3}{4}\cdot\frac{4}{4} = \frac{3\cdot 4}{4\cdot 4} = \frac{12}{16}$$

TRY THIS

Find three fractions equivalent to the given fraction.

1. $\frac{2}{3}$ $\frac{4}{6}, \frac{6}{9}, \frac{10}{15}$ **2.** $\frac{7}{8}$ $\frac{14}{16}, \frac{21}{24}, \frac{28}{32}$

Find a fraction equivalent to $\frac{5}{6}$ with a denominator of 24.

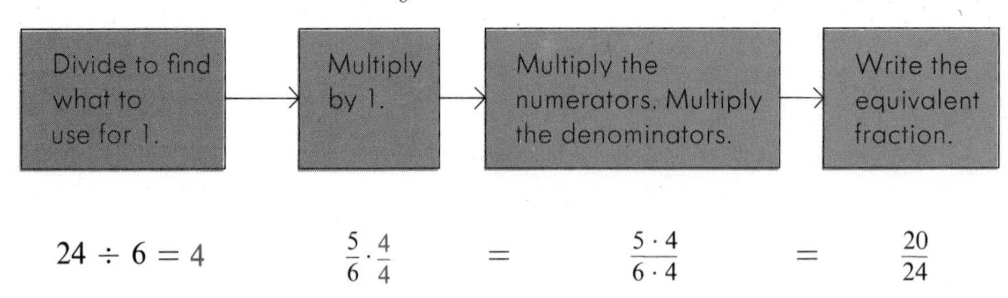

$$24 \div 6 = 4 \qquad \frac{5}{6}\cdot\frac{4}{4} = \frac{5\cdot 4}{6\cdot 4} = \frac{20}{24}$$

EXAMPLE 2. Find a fraction equivalent to $\frac{2}{5}$ with a denominator of 35.

$$\frac{2}{5} = \frac{2}{5}\cdot\frac{7}{7} = \frac{2\cdot 7}{5\cdot 7} = \frac{14}{35}$$

Since $35 \div 5 = 7$, *we multiply by* $\frac{7}{7}$.

TRY THIS

Find a fraction equivalent to the given one, but with the indicated denominator.

3. $\frac{2}{3}$ [9] $\frac{6}{9}$ **4.** $\frac{9}{10}$ [20] $\frac{18}{20}$ **5.** $\frac{1}{4}$ [32] $\frac{8}{32}$

PRACTICE

For extra practice, use Making Practice Fun 38 with this lesson.

Find a fraction equivalent to the given one, but with the indicated denominator.

1. $\frac{1}{2}$ [10] $\frac{5}{10}$	**2.** $\frac{1}{3}$ [6] $\frac{2}{6}$	**3.** $\frac{1}{2}$ [6] $\frac{3}{6}$	**4.** $\frac{1}{4}$ [12] $\frac{3}{12}$
5. $\frac{3}{4}$ [24] $\frac{18}{24}$	**6.** $\frac{2}{9}$ [18] $\frac{4}{18}$	**7.** $\frac{2}{3}$ [12] $\frac{8}{12}$	**8.** $\frac{2}{5}$ [10] $\frac{4}{10}$
9. $\frac{5}{6}$ [18] $\frac{15}{18}$	**10.** $\frac{7}{8}$ [16] $\frac{14}{16}$	**11.** $\frac{3}{5}$ [20] $\frac{12}{20}$	**12.** $\frac{2}{9}$ [27] $\frac{6}{27}$
13. $\frac{2}{3}$ [12] $\frac{8}{12}$	**14.** $\frac{2}{3}$ [15] $\frac{10}{15}$	**15.** $\frac{2}{3}$ [27] $\frac{18}{27}$	**16.** $\frac{2}{3}$ [30] $\frac{20}{30}$
17. $\frac{3}{4}$ [20] $\frac{15}{20}$	**18.** $\frac{3}{4}$ [12] $\frac{9}{12}$	**19.** $\frac{3}{4}$ [8] $\frac{6}{8}$	**20.** $\frac{3}{4}$ [36] $\frac{27}{36}$
21. $\frac{3}{8}$ [48] $\frac{18}{48}$	**22.** $\frac{7}{8}$ [32] $\frac{28}{32}$	**23.** $\frac{1}{6}$ [18] $\frac{3}{18}$	**24.** $\frac{7}{18}$ [36] $\frac{14}{36}$
25. $\frac{4}{5}$ [50] $\frac{40}{50}$	**26.** $\frac{1}{6}$ [24] $\frac{4}{24}$	**27.** $\frac{5}{8}$ [24] $\frac{15}{24}$	**28.** $\frac{1}{9}$ [18] $\frac{2}{18}$
29. $\frac{3}{8}$ [24] $\frac{9}{24}$	**30.** $\frac{5}{6}$ [24] $\frac{20}{24}$	**31.** $\frac{7}{22}$ [132] $\frac{42}{132}$	**32.** $\frac{10}{21}$ [126] $\frac{60}{126}$

PROBLEM CORNER

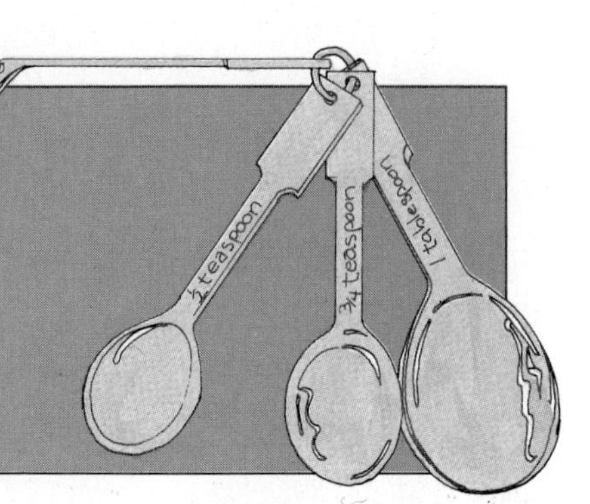

1. Tomas needs to measure $\frac{1}{2}$ c milk to make cereal. He only has a $\frac{1}{4}$ c measure. How many quarter cups should he use to get $\frac{1}{2}$ c? 2

2. Olivia needs to measure $\frac{3}{4}$ tsp salt to make soup. She only has a $\frac{1}{8}$ tsp measure. How many eighth teaspoons should she use to get $\frac{3}{4}$ tsp? 6

More Practice, page 190

Objectives 68 and 69

SIMPLIFYING

All the following are equivalent to $\frac{3}{4}$. $\frac{3}{4} = \frac{6}{8} = \frac{9}{12} = \frac{12}{16} = \frac{15}{20} = \frac{18}{24}$

$\frac{3}{4}$ is simplest because:

1. All the other fractions have larger numerators and denominators.
2. 1 is the only number that divides evenly into both the numerator and denominator.

EXAMPLE 1. All the fractions are equivalent. Which is simplest?

$\frac{2}{4}, \frac{3}{6}, \frac{1}{2}, \frac{6}{12}, \frac{5}{10}$ $\frac{1}{2}$ is simplest.

TRY THIS

All the fractions are equivalent. Which is simplest?

1. $\frac{6}{10}, \frac{15}{25}, \frac{18}{30}, \frac{3}{5}, \frac{12}{20}$ $\frac{3}{5}$ **2.** $\frac{15}{40}, \frac{9}{24}, \frac{12}{32}, \frac{6}{16}, \frac{3}{8}$ $\frac{3}{8}$

Simplify. $\frac{18}{24}$

Divide the numerator and denominator by the same number. → Continue dividing until 1 is the only number that divides evenly into both the numerator and denominator. → The fraction is simplified.

$\frac{18 \div 3}{24 \div 3} = \frac{6}{8}$ $\frac{6 \div 2}{8 \div 2} = \frac{3}{4}$ $\frac{18}{24} = \frac{3}{4}$

To check we multiply by 1: $\frac{3}{4} = \frac{3}{4} \cdot \frac{2}{2} = \frac{6}{8} \cdot \frac{3}{3} = \frac{18}{24}$

We do not recommend canceling as a means of simplifying. Sometimes students do not understand the process and cancel when they should not.

EXAMPLES. Simplify.

2. $\frac{28}{16} = \frac{28 \div 2}{16 \div 2} = \frac{14 \div 2}{8 \div 2} = \frac{7}{4} = 1\frac{3}{4}$

3. $\frac{2}{6} = \frac{2 \div 2}{6 \div 2} = \frac{1}{3}$

4. $\frac{8}{20} = \frac{8 \div 4}{20 \div 4} = \frac{2}{5}$

By choosing the largest number, we only need to divide once.

5. $\frac{30}{6} = \frac{30 \div 6}{6 \div 6} = \frac{5}{1} = 5$

TRY THIS

Simplify.

3. $\frac{10}{12}$ $\frac{5}{6}$

4. $\frac{32}{20}$ $1\frac{3}{5}$

5. $\frac{54}{6}$ 9

PRACTICE

All the fractions are equivalent. Which is simplest?

1. $\frac{10}{15}, \frac{2}{3}, \frac{8}{12}, \frac{6}{9}, \frac{14}{21}$ $\frac{2}{3}$

2. $\frac{8}{10}, \frac{12}{15}, \frac{24}{30}, \frac{28}{35}, \frac{4}{5}$ $\frac{4}{5}$

3. $\frac{12}{40}, \frac{15}{50}, \frac{3}{10}, \frac{6}{20}, \frac{9}{30}$ $\frac{3}{10}$

4. $\frac{1}{3}, \frac{2}{6}, \frac{3}{9}, \frac{4}{12}, \frac{5}{15}$ $\frac{1}{3}$

5. $\frac{5}{40}, \frac{4}{32}, \frac{3}{24}, \frac{2}{16}, \frac{1}{8}$ $\frac{1}{8}$

6. $\frac{20}{24}, \frac{15}{18}, \frac{35}{42}, \frac{5}{6}, \frac{10}{12}$ $\frac{5}{6}$

Simplify.

7. $\frac{2}{4}$ $\frac{1}{2}$ **8.** $\frac{4}{12}$ $\frac{1}{3}$ **9.** $\frac{3}{6}$ $\frac{1}{2}$ **10.** $\frac{2}{10}$ $\frac{1}{5}$

11. $\frac{6}{8}$ $\frac{3}{4}$ **12.** $\frac{9}{12}$ $\frac{3}{4}$ **13.** $\frac{6}{12}$ $\frac{1}{2}$ **14.** $\frac{6}{10}$ $\frac{3}{5}$

15. $\frac{10}{6}$ $1\frac{2}{3}$ **16.** $\frac{12}{8}$ $1\frac{1}{2}$ **17.** $\frac{12}{10}$ $1\frac{1}{5}$ **18.** $\frac{15}{6}$ $2\frac{1}{2}$

19. $\frac{12}{3}$ 4 **20.** $\frac{24}{8}$ 3 **21.** $\frac{25}{5}$ 5 **22.** $\frac{36}{4}$ 9

23. $\frac{18}{24}$ $\frac{3}{4}$ **24.** $\frac{42}{48}$ $\frac{7}{8}$ **25.** $\frac{20}{50}$ $\frac{2}{5}$ **26.** $\frac{40}{50}$ $\frac{4}{5}$

27. $\frac{14}{24}$ $\frac{7}{12}$ **28.** $\frac{14}{16}$ $\frac{7}{8}$ **29.** $\frac{14}{32}$ $\frac{7}{16}$ **30.** $\frac{15}{25}$ $\frac{3}{5}$

31. $\frac{8}{6}$ $1\frac{1}{3}$ **32.** $\frac{12}{10}$ $1\frac{1}{5}$ **33.** $\frac{16}{14}$ $1\frac{1}{7}$ **34.** $\frac{32}{14}$ $2\frac{2}{7}$

35. $\frac{6}{16}$ $\frac{3}{8}$ **36.** $\frac{4}{24}$ $\frac{1}{6}$ **37.** $\frac{16}{48}$ $\frac{1}{3}$ **38.** $\frac{15}{24}$ $\frac{5}{8}$

39. $\frac{100}{20}$ 5 **40.** $\frac{56}{7}$ 8 **41.** $\frac{42}{6}$ 7 **42.** $\frac{125}{25}$ 5

43. $\frac{35}{40}$ $\frac{7}{8}$ **44.** $\frac{25}{60}$ $\frac{5}{12}$ **45.** $\frac{30}{45}$ $\frac{2}{3}$ **46.** $\frac{40}{55}$ $\frac{8}{11}$

47. $\frac{10}{8}$ $1\frac{1}{4}$ **48.** $\frac{10}{4}$ $2\frac{1}{2}$ **49.** $\frac{20}{6}$ $3\frac{1}{3}$ **50.** $\frac{20}{8}$ $2\frac{1}{2}$

More Practice, page 190

Objective 70

MULTIPLYING AND SIMPLIFYING

Multiply and simplify. $\frac{3}{8} \times \frac{4}{9}$

Multiply the numerators.
Multiply the denominators. → Simplify.

$\frac{3}{8} \times \frac{4}{9} = \frac{12}{72}$ $\qquad$ $\frac{12}{72} = \frac{12 \div 12}{72 \div 12} = \frac{1}{6}$

EXAMPLE 1. Multiply and simplify. $\frac{6}{7} \times \frac{5}{3}$

$\frac{6}{7} \times \frac{5}{3} = \frac{30}{21}$ *Multiply numerators: $6 \times 5 = 30$*
Multiply denominators: $7 \times 3 = 21$

$= \frac{30 \div 3}{21 \div 3}$ *Simplify.*

$= \frac{10}{7}$

$= 1\frac{3}{7}$

TRY THIS

Multiply and simplify.

1. $\frac{2}{3} \times \frac{7}{8}$ $\frac{7}{12}$ **2.** $\frac{4}{5} \cdot \frac{5}{12}$ $\frac{1}{3}$ **3.** $\frac{10}{3} \cdot \frac{4}{5}$ $2\frac{2}{3}$

EXAMPLES. Multiply and simplify.

2. $24 \cdot \frac{3}{8} = \frac{24}{1} \cdot \frac{3}{8}$

$= \frac{72}{8}$

$= \frac{72 \div 8}{8 \div 8}$

$= \frac{9}{1} = 9$

3. $\frac{5}{6} \cdot 26 = \frac{5}{6} \cdot \frac{26}{1}$

$= \frac{130}{6}$

$= \frac{130 \div 2}{6 \div 2}$

$= \frac{65}{3} = 21\frac{2}{3}$

TRY THIS

Multiply and simplify.

4. $12 \cdot \frac{2}{3}$ 8 **5.** $\frac{3}{4} \times 14$ $10\frac{1}{2}$

PRACTICE

For extra practice, use Making Practice Fun 39 with this lesson.
Use Quiz 14 after this Practice.

Multiply and simplify.

1. $\frac{2}{3} \times \frac{1}{2}$ $\frac{1}{3}$ **2.** $\frac{4}{5} \times \frac{1}{4}$ $\frac{1}{5}$ **3.** $\frac{7}{8} \times \frac{1}{7}$ $\frac{1}{8}$ **4.** $\frac{5}{6} \times \frac{1}{5}$ $\frac{1}{6}$

5. $\frac{1}{8} \times \frac{4}{5}$ $\frac{1}{10}$ **6.** $\frac{2}{5} \times \frac{1}{6}$ $\frac{1}{15}$ **7.** $\frac{1}{4} \times \frac{2}{3}$ $\frac{1}{6}$ **8.** $\frac{3}{6} \times \frac{1}{6}$ $\frac{1}{12}$

9. $\frac{12}{5} \times \frac{9}{8}$ $2\frac{7}{10}$ **10.** $\frac{16}{15} \times \frac{5}{4}$ $1\frac{1}{3}$ **11.** $\frac{10}{9} \times \frac{7}{5}$ $1\frac{5}{9}$ **12.** $\frac{25}{12} \times \frac{4}{3}$ $2\frac{7}{9}$

13. $9 \cdot \frac{1}{9}$ 1 **14.** $4 \cdot \frac{1}{4}$ 1 **15.** $\frac{1}{3} \cdot 3$ 1 **16.** $\frac{1}{6} \cdot 6$ 1

17. $\frac{7}{10} \cdot \frac{10}{7}$ 1 **18.** $\frac{8}{9} \cdot \frac{9}{8}$ 1 **19.** $\frac{7}{5} \cdot \frac{5}{7}$ 1 **20.** $\frac{2}{11} \cdot \frac{11}{2}$ 1

21. $\frac{1}{4} \times 8$ 2 **22.** $\frac{1}{6} \times 12$ 2 **23.** $15 \times \frac{1}{3}$ 5 **24.** $14 \times \frac{1}{2}$ 7

25. $12 \times \frac{3}{4}$ 9 **26.** $18 \times \frac{5}{6}$ 15 **27.** $\frac{3}{8} \times 24$ 9 **28.** $\frac{2}{9} \times 36$ 8

29. $13 \times \frac{2}{5}$ $5\frac{1}{5}$ **30.** $15 \times \frac{1}{6}$ $2\frac{1}{2}$ **31.** $\frac{7}{10} \times 28$ $19\frac{3}{5}$ **32.** $\frac{5}{8} \times 34$ $21\frac{1}{4}$

33. $\frac{1}{6} \times 360$ 60 **34.** $\frac{1}{3} \times 120$ 40 **35.** $240 \times \frac{1}{8}$ 30 **36.** $150 \times \frac{1}{5}$ 30

37. $\frac{4}{10} \cdot \frac{5}{10}$ $\frac{1}{5}$ **38.** $\frac{7}{10} \cdot \frac{34}{150}$ $\frac{119}{750}$ **39.** $\frac{8}{10} \cdot \frac{45}{100}$ $\frac{9}{25}$ **40.** $\frac{3}{10} \cdot \frac{8}{10}$ $\frac{6}{25}$

41. $\frac{11}{24} \cdot \frac{3}{5}$ $\frac{11}{40}$ **42.** $\frac{15}{22} \cdot \frac{4}{7}$ $\frac{30}{77}$ **43.** $\frac{10}{21} \cdot \frac{3}{4}$ $\frac{5}{14}$ **44.** $\frac{17}{18} \cdot \frac{3}{5}$ $\frac{17}{30}$

PROBLEM CORNER

1. How much steak will be needed to serve 30 people if each person gets $\frac{2}{3}$ lb? 20 lbs

2. A gardener uses $\frac{2}{5}$ lb of peat moss for a rose bush. How much would be needed for 25 rose bushes? 10 lbs

Objective 71

MULTIPLYING WITH MIXED NUMBERS

Multiply and simplify. $6 \times 2\frac{1}{2}$

Convert to fractions. → Multiply. Simplify.

$$6 \times 2\frac{1}{2} = \frac{6}{1} \times \frac{5}{2} = \frac{30}{2} = 15$$

EXAMPLE 1. Multiply and simplify. $4\frac{2}{3} \cdot 8$

$4\frac{2}{3} \cdot 8 = \frac{14}{3} \cdot \frac{8}{1} = \frac{112}{3} = 37\frac{1}{3}$ *Convert to fractions. Multiply. Simplify.*

TRY THIS

Multiply and simplify.

1. $9 \times 4\frac{1}{3}$ 39 **2.** $6\frac{2}{5} \cdot 2$ $12\frac{4}{5}$

EXAMPLE 2. Multiply and simplify. $3\frac{1}{2} \cdot \frac{3}{4}$

$3\frac{1}{2} \cdot \frac{3}{4} = \frac{7}{2} \cdot \frac{3}{4} = \frac{21}{8} = 2\frac{5}{8}$ *Convert the mixed number to a fraction.*

TRY THIS

Multiply and simplify.

3. $4\frac{1}{8} \cdot \frac{2}{3}$ $2\frac{3}{4}$ **4.** $\frac{5}{6} \cdot 2\frac{1}{5}$ $1\frac{5}{6}$

EXAMPLE 3. Multiply and simplify. $2\frac{3}{4} \cdot 3\frac{2}{5}$

$2\frac{3}{4} \cdot 3\frac{2}{5} = \frac{11}{4} \cdot \frac{17}{5} = \frac{187}{20} = 9\frac{7}{20}$ *Convert the mixed numbers to fractions.*

TRY THIS

Multiply and simplify.

5. $3\frac{1}{3} \cdot 2\frac{1}{2}$ $8\frac{1}{3}$ **6.** $4\frac{1}{6} \cdot 2\frac{3}{5}$ $10\frac{5}{6}$

PRACTICE

For extra practice, use Making Practice Fun 40 with this lesson.

Multiply and simplify.

1. $9 \times 2\frac{1}{3}$ 21

2. $8 \times 3\frac{1}{5}$ $25\frac{3}{5}$

3. $4\frac{1}{2} \times 10$ 45

4. $5\frac{1}{7} \times 4$ $20\frac{4}{7}$

5. $3\frac{1}{5} \times \frac{2}{5}$ $1\frac{7}{25}$

6. $3\frac{1}{8} \times \frac{2}{3}$ $2\frac{1}{12}$

7. $1\frac{5}{8} \times \frac{2}{3}$ $1\frac{1}{12}$

8. $6\frac{2}{3} \times \frac{1}{4}$ $1\frac{2}{3}$

9. $6 \cdot 5\frac{2}{3}$ 34

10. $4 \cdot 3\frac{5}{8}$ $14\frac{1}{2}$

11. $2\frac{5}{6} \cdot 8$ $22\frac{2}{3}$

12. $5 \cdot 3\frac{3}{4}$ $18\frac{3}{4}$

13. $2\frac{1}{2} \cdot 3\frac{2}{3}$ $9\frac{1}{6}$

14. $3\frac{1}{2} \cdot 2\frac{1}{3}$ $8\frac{1}{6}$

15. $4\frac{1}{4} \cdot 3\frac{1}{2}$ $14\frac{7}{8}$

16. $1\frac{1}{4} \cdot 3\frac{1}{2}$ $4\frac{3}{8}$

17. $1\frac{3}{4} \cdot 1\frac{1}{2}$ $2\frac{5}{8}$

18. $1\frac{3}{4} \cdot 2\frac{1}{3}$ $4\frac{1}{12}$

19. $3\frac{2}{5} \cdot 5\frac{1}{3}$ $18\frac{2}{15}$

20. $3\frac{3}{4} \cdot 2\frac{1}{8}$ $7\frac{31}{32}$

21. $5\frac{3}{4} \times 1\frac{1}{3}$ $7\frac{2}{3}$

22. $3\frac{2}{5} \times 1\frac{1}{4}$ $4\frac{1}{4}$

23. $1\frac{3}{5} \times 3\frac{1}{3}$ $5\frac{1}{3}$

24. $3\frac{1}{6} \times 2\frac{3}{4}$ $8\frac{17}{24}$

25. $2\frac{2}{5} \cdot 2\frac{7}{8}$ $6\frac{9}{10}$

26. $2\frac{1}{10} \cdot 6\frac{2}{3}$ 14

27. $1\frac{3}{7} \cdot 2\frac{2}{5}$ $3\frac{3}{7}$

28. $2\frac{3}{10} \cdot 4\frac{3}{5}$ $10\frac{29}{50}$

29. $10\frac{1}{4} \cdot 50$ $512\frac{1}{2}$

30. $20\frac{1}{3} \cdot 60$ 1220

31. $40 \cdot 30\frac{1}{6}$ $1206\frac{2}{3}$

32. $20 \cdot 80\frac{1}{10}$ 1602

33. $5\frac{1}{4} \times 28$ 147

34. $40 \times 3\frac{5}{7}$ $148\frac{4}{7}$

35. $16 \times 4\frac{7}{8}$ 78

36. $20 \times 3\frac{2}{3}$ $73\frac{1}{3}$

37. $10\frac{1}{2} \cdot 10\frac{1}{3}$ $108\frac{1}{2}$

38. $10\frac{1}{4} \cdot 20\frac{1}{5}$ $207\frac{1}{20}$

39. $20\frac{1}{2} \cdot 20\frac{1}{4}$ $415\frac{1}{8}$

40. $20\frac{1}{2} \cdot 10\frac{1}{5}$ $209\frac{1}{10}$

PROBLEM CORNER

1. A long-playing record makes $33\frac{1}{3}$ revolutions per minute. It plays 12 minutes. How many revolutions does it make? 400
2. A car travels on an interstate highway at 55 miles per hour for $3\frac{1}{2}$ hr. How far does it go? $192\frac{1}{2}$ miles

Objective 166

APPLICATION

Discuss with your students the importance of budgeting.

BUDGETING

A budget is a plan which shows how money is to be spent and saved. Robert Carlson is planning his expenses. He has $ 2580 in his savings account and $ 868 in his checking account.

Income		Expenses	
Regular Job	$6500	Rent and Utilities	$3000
Part-time Job	$ 780	Food and Supplies	$2400
		Transportation	$1200
		Insurance	$ 500
		Personal	$1100

1. What is his total income? $7280
2. What are his total expenses? $8200
3. Will he have to use his savings account to pay his expenses? yes
4. What part of his expenses is for food? $\frac{12}{41}$
5. If Robert has a roommate, he can save $\frac{1}{2}$ of his expenses for rent and utilities. How much can he save? $1500
6. If Robert has a roommate, he can also save $\frac{3}{10}$ of his food and supplies expense. How much can he save? $720
7. How much in all can he save by having a roommate? $2220

The federal government budgets how it spends its money. The graph shows how each government dollar was spent in a previous year. What part of the dollar went for each expense?

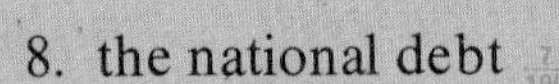

8. the national debt $\frac{7}{100}$
9. national defense $\frac{6}{25}$
10. human resources $\frac{1}{5}$
11. social and income security $\frac{17}{50}$
12. How much is spent annually by the government for natural resources?
not enough information

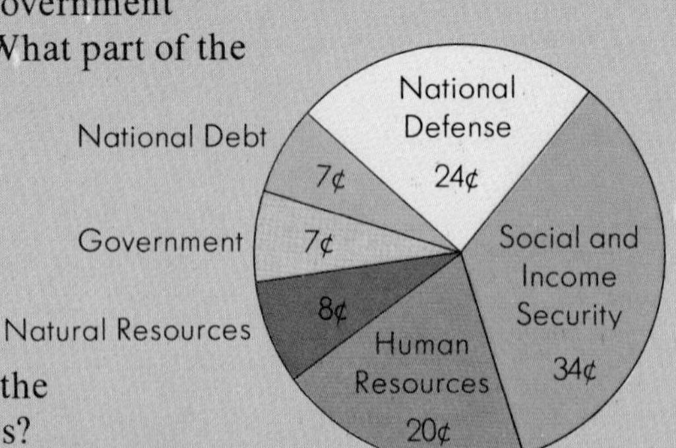

Janet Thompson's budget for volleyball was \$ 1000 for the season. The following expenses were estimated at the beginning of the year.

Equipment:	5 balls at \$10 each
Uniforms:	24 of each: T-shirts, \$3; shorts \$5; pair of socks, \$2
Towel Service:	10 weeks at \$7.50 per week
Transportation:	\$50 bus fee for each of 10 away games
Awards:	Remainder of budget money

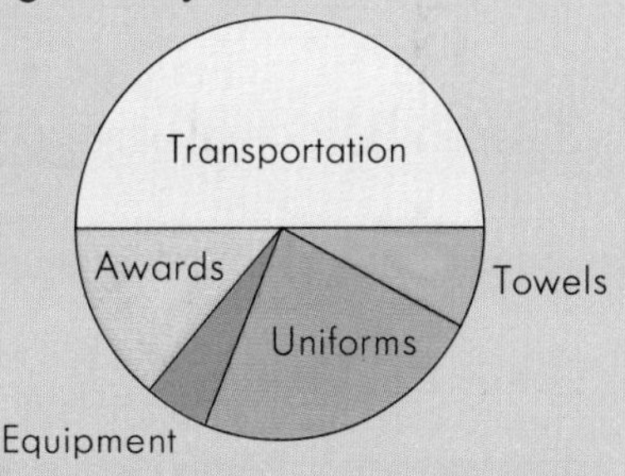

How much money is allowed for each budget item?

13. equipment \$50   14. uniforms \$240   15. towels \$75   16. transportation \$500

What part of the budget went for each of the following?

17. equipment $\frac{1}{20}$ 18. uniforms $\frac{6}{25}$ 19. towels $\frac{3}{40}$ 20. transportation $\frac{1}{2}$

21. How much money was left for awards? \$135
22. What part of the budget was left for awards? $\frac{27}{200}$

The income from 10 games is estimated to be \$ 184 per game. Of this, $\frac{2}{3}$ goes to the general athletic fund, $\frac{1}{4}$ goes for administration, and the rest goes into the volleyball award fund.

23. What is the total income from the games? \$1840
24. How much money goes to the general athletic fund? \$1226.67
25. How much money goes for administration? \$460
26. How much money goes into the volleyball award fund? \$153.33

Objectives 72 and 73

RECIPROCALS AND DIVIDING

To find the reciprocal of a fraction, we interchange the numerator and denominator.

Two numbers are also said to be reciprocals if their product is 1. Have your students check this with some products.

$\frac{2}{3} \to \frac{3}{2}$

EXAMPLES. Find the reciprocals.

1. $\frac{5}{4}$ The reciprocal of $\frac{5}{4}$ is $\frac{4}{5}$.

2. $\frac{1}{3}$ The reciprocal of $\frac{1}{3}$ is $\frac{3}{1}$, or 3.

3. 10. Think of 10 as $\frac{10}{1}$.
The reciprocal of 10 is $\frac{1}{10}$.

TRY THIS

Find the reciprocal.

1. $\frac{2}{5}$ $\frac{5}{2}$ **2.** $\frac{5}{2}$ $\frac{2}{5}$ **3.** $\frac{1}{5}$ 5 **4.** 24 $\frac{1}{24}$

Divide and simplify. $\frac{6}{5} \div \frac{3}{4}$

Find the reciprocal of the divisor. → Multiply by the reciprocal of the divisor. → Simplify.

The reciprocal of $\frac{3}{4}$ is $\frac{4}{3}$.

$$\frac{6}{5} \div \frac{3}{4} = \frac{6}{5} \cdot \frac{4}{3} = \frac{24}{15} = \frac{24 \div 3}{15 \div 3} = \frac{8}{5} = 1\frac{3}{5}$$

EXAMPLE 4. Divide and simplify. $\frac{5}{6} \div \frac{3}{2}$

$$\frac{5}{6} \div \frac{3}{2} = \frac{5}{6} \cdot \frac{2}{3} = \frac{10}{18} = \frac{10 \div 2}{18 \div 2} = \frac{5}{9}$$

The reciprocal of $\frac{3}{2}$ is $\frac{2}{3}$.

TRY THIS

Divide and simplify.

5. $\frac{6}{7} \div \frac{3}{4}$ $1\frac{1}{7}$ **6.** $\frac{2}{3} \div \frac{1}{4}$ $2\frac{2}{3}$ **7.** $\frac{4}{5} \div 8$ $\frac{1}{10}$

Marcus late to class 12:03

PRACTICE

Find the reciprocal.

1. $\frac{5}{6}$ $\frac{6}{5}$ **2.** $\frac{3}{5}$ $\frac{5}{3}$ **3.** $\frac{6}{9}$ $\frac{9}{6}$ **4.** $\frac{3}{8}$ $\frac{8}{3}$

5. 6 $\frac{1}{6}$ **6.** 2 $\frac{1}{2}$ **7.** 3 $\frac{1}{3}$ **8.** 5 $\frac{1}{5}$

9. $\frac{1}{3}$ 3 **10.** $\frac{1}{2}$ 2 **11.** $\frac{1}{6}$ 6 **12.** $\frac{1}{4}$ 4

Divide and simplify.

13. $\frac{3}{5} \div \frac{3}{4}$ $\frac{4}{5}$ **14.** $\frac{2}{3} \div \frac{3}{4}$ $\frac{8}{9}$ **15.** $\frac{3}{5} \div \frac{9}{4}$ $\frac{4}{15}$ **16.** $\frac{6}{7} \div \frac{3}{5}$ $1\frac{3}{7}$

17. $\frac{4}{3} \div \frac{1}{3}$ 4 **18.** $\frac{1}{3} \div \frac{1}{4}$ $1\frac{1}{3}$ **19.** $\frac{9}{8} \div \frac{1}{2}$ $2\frac{1}{4}$ **20.** $\frac{10}{9} \div \frac{1}{3}$ $3\frac{1}{3}$

21. $\frac{6}{5} \div 3$ $\frac{2}{5}$ **22.** $\frac{6}{5} \div 2$ $\frac{3}{5}$ **23.** $\frac{12}{7} \div 4$ $\frac{3}{7}$ **24.** $\frac{8}{7} \div 2$ $\frac{4}{7}$

25. $10 \div \frac{2}{3}$ 15 **26.** $8 \div \frac{5}{4}$ $6\frac{2}{5}$ **27.** $12 \div \frac{3}{2}$ 8 **28.** $24 \div \frac{3}{8}$ 64

29. $\frac{2}{3} \div \frac{2}{3}$ 1 **30.** $\frac{3}{4} \div \frac{3}{4}$ 1 **31.** $\frac{3}{8} \div \frac{3}{8}$ 1 **32.** $\frac{2}{5} \div \frac{2}{5}$ 1

33. $\frac{3}{4} \div 1$ $\frac{3}{4}$ **34.** $\frac{2}{5} \div 1$ $\frac{2}{5}$ **35.** $\frac{3}{10} \div 1$ $\frac{3}{10}$ **36.** $\frac{5}{12} \div 1$ $\frac{5}{12}$

37. $\frac{8}{15} \div \frac{4}{5}$ $\frac{2}{3}$ **38.** $\frac{6}{13} \div \frac{3}{26}$ 4 **39.** $\frac{9}{5} \div \frac{8}{10}$ $2\frac{1}{4}$ **40.** $\frac{101}{5} \div 10$ $2\frac{1}{50}$

PROBLEM CORNER

1. How many test tubes, each containing $\frac{3}{5}$ milliliter, can be filled from a container of 60 milliliters? 100

2. Each loop in a spring takes $\frac{3}{8}$ in. of wire. How many loops can be made from 120 in. of wire? 320

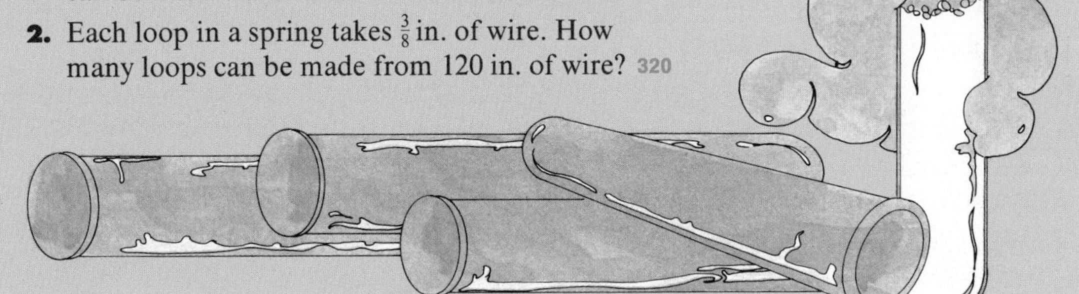

Objective 74

DIVIDING WITH MIXED NUMBERS

Divide. $1\frac{1}{2} \div \frac{1}{6}$

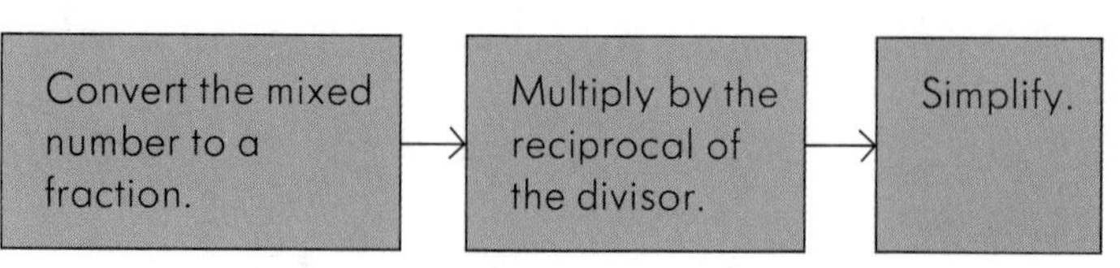

$$1\frac{1}{2} \div \frac{1}{6} = \frac{3}{2} \div \frac{1}{6} = \frac{3}{2} \cdot \frac{6}{1} = \frac{18}{2} = 9$$

EXAMPLE 1. Divide and simplify. $\frac{7}{8} \div 2\frac{1}{2}$

$$\frac{7}{8} \div 2\frac{1}{2} = \frac{7}{8} \div \frac{5}{2} = \frac{7}{8} \cdot \frac{2}{5} = \frac{14}{40} = \frac{7}{20}$$

Convert the mixed number to a fraction. Multiply by the reciprocal of the divisor. Simplify.

TRY THIS

Divide and simplify.

1. $2\frac{1}{3} \div \frac{3}{4}$ $3\frac{1}{9}$ **2.** $\frac{9}{10} \div 3\frac{2}{5}$ $\frac{9}{34}$

EXAMPLE 2. Divide and simplify. $35 \div 4\frac{1}{3}$

$$35 \div 4\frac{1}{3} = \frac{35}{1} \div \frac{13}{3} = \frac{35}{1} \cdot \frac{3}{13} = \frac{105}{13} = 8\frac{1}{13}$$

Convert each number to a fraction. Multiply by the reciprocal of the divisor. Simplify.

EXAMPLE 3. Divide and simplify. $2\frac{1}{3} \div 1\frac{3}{4}$

$$2\frac{1}{3} \div 1\frac{3}{4} = \frac{7}{3} \div \frac{7}{4} = \frac{7}{3} \cdot \frac{4}{7} = \frac{28}{21} = 1\frac{1}{3}$$

Convert each mixed number to a fraction. Multiply by the reciprocal of the divisor. Simplify.

TRY THIS

Divide and simplify.

3. $3\frac{1}{5} \div 32$ $\frac{1}{10}$ **4.** $1\frac{3}{5} \div 3\frac{1}{3}$ $\frac{12}{25}$

PRACTICE

For extra practice, use Making Practice Fun 41 with this lesson.
Use Quiz 15 after this Practice.

Divide and simplify.

1. $6 \div 4\frac{1}{2}$ $1\frac{1}{3}$

2. $20 \div 3\frac{1}{5}$ $6\frac{1}{4}$

3. $9 \div 2\frac{1}{4}$ 4

4. $18 \div 2\frac{1}{4}$ 8

5. $42 \div 3\frac{1}{5}$ $13\frac{1}{8}$

6. $34 \div 5\frac{1}{2}$ $6\frac{2}{11}$

7. $63 \div 2\frac{1}{3}$ 27

8. $52 \div 3\frac{1}{2}$ $14\frac{6}{7}$

9. $4\frac{3}{4} \div 1\frac{1}{3}$ $3\frac{9}{16}$

10. $2\frac{1}{2} \div 1\frac{1}{4}$ 2

11. $5\frac{4}{5} \div 2\frac{1}{2}$ $2\frac{8}{25}$

12. $3\frac{1}{2} \div 2\frac{2}{3}$ $1\frac{5}{16}$

13. $1\frac{7}{8} \div 1\frac{2}{3}$ $1\frac{1}{8}$

14. $7\frac{1}{3} \div 1\frac{1}{9}$ $6\frac{3}{5}$

15. $6\frac{7}{8} \div 1\frac{2}{3}$ $4\frac{1}{8}$

16. $4\frac{3}{8} \div 2\frac{5}{6}$ $1\frac{37}{68}$

17. $2\frac{2}{3} \div 3\frac{1}{5}$ $\frac{5}{6}$

18. $3\frac{1}{3} \div 5\frac{1}{2}$ $\frac{20}{33}$

19. $2\frac{1}{2} \div 2\frac{2}{3}$ $\frac{15}{16}$

20. $3\frac{1}{3} \div 5\frac{1}{2}$ $\frac{20}{33}$

21. $\frac{3}{5} \div 1\frac{1}{3}$ $\frac{9}{20}$

22. $\frac{2}{3} \div 3\frac{1}{4}$ $\frac{8}{39}$

23. $\frac{1}{2} \div 2\frac{1}{3}$ $\frac{3}{14}$

24. $\frac{5}{8} \div 3\frac{1}{3}$ $\frac{3}{16}$

25. $2\frac{1}{7} \div 5$ $\frac{3}{7}$

26. $5\frac{1}{4} \div 6$ $\frac{7}{8}$

27. $8\frac{2}{5} \div 7$ $1\frac{1}{5}$

28. $3\frac{3}{8} \div 3$ $1\frac{1}{8}$

29. $3\frac{1}{4} \div \frac{1}{2}$ $6\frac{1}{2}$

30. $4\frac{1}{5} \div 10$ $\frac{21}{50}$

31. $5\frac{1}{2} \div 22$ $\frac{1}{4}$

32. $4\frac{1}{3} \div 26$ $\frac{1}{6}$

33. $125 \div 12\frac{1}{2}$ 10

34. $86 \div 14\frac{1}{4}$ $6\frac{2}{57}$

35. $84 \div 8\frac{2}{5}$ 10

36. $130 \div 9\frac{3}{8}$ $13\frac{13}{15}$

37. $10\frac{1}{3} \div 3\frac{2}{5}$ $3\frac{2}{51}$

38. $3\frac{1}{5} \div 1\frac{1}{3}$ $2\frac{2}{5}$

39. $12\frac{1}{3} \div 6\frac{7}{8}$ $1\frac{131}{165}$

40. $11\frac{1}{4} \div 2\frac{1}{2}$ $4\frac{1}{2}$

PROBLEM CORNER

1. A truck carries $18\frac{1}{3}$ tons of sand. How many trips must the truck make to carry 550 tons? 30

2. The weight of water is $62\frac{1}{2}$ lb per cubic foot. How many cubic feet would be occupied by 250 lb of water? 4

Objectives 167 and 168

APPLICATION

FOOD AND NUTRITION

Our bodies need food for energy, growth, and maintenance of health. The following is a table showing the Recommended Daily Allowances (RDA) of calories and other nutrients.

RDA for a 15-Year-Old to 18-Year-Old Male

Calories	Protein	Niacin	Riboflavin	Thiamin	Magnesium	Iodine
3000	54 g	20 mg	1.8 mg	1.5 mg	400 mg	150 micrograms

The RDA for a 15-year-old to 18-year-old female is found by multiplying by the fraction shown after each of the items.

Tammy wants to find how many calories she should have in her diet.

calories $\left(\frac{7}{10}\right) \longrightarrow 3000 \times \frac{7}{10} = 2100$ calories

Find each RDA. The fractions are average figures and may vary depending on the individual.

1. Protein $\left(\frac{8}{9}\right)$ 48 g
2. Niacin $\left(\frac{7}{10}\right)$ 14 mg
3. Magnesium $\left(\frac{3}{4}\right)$ 300 mg
4. Iodine $\left(\frac{23}{30}\right)$ 115 micrograms
5. Riboflavin $\left(\frac{7}{9}\right)$ 1.4 mg
6. Thiamin $\left(\frac{11}{15}\right)$ 1.1 mg

Jesse compares some of the nutrients in his daily diet to the RDA.

	Calories	Protein	Vitamin C	Calcium	Iron
RDA	3000	54 g	45 mg	1200 mg	18 mg
Jesse's Diet	4200	46 g	20 mg	1600 mg	12 mg

He divides to find the part of the RDA in his diet.

$4200 \div 3000 = 1\frac{2}{5}$ of the RDA of calories

What part of the RDA of each nutrient does Jesse consume daily?

7. Protein $\frac{23}{27}$
8. Vitamin C $\frac{4}{9}$
9. Calcium $1\frac{1}{3}$
10. Iron $\frac{2}{3}$

Carbohydrates and calories are important factors in our diets.

Food	Unit of Measure	Carbohydrates	Calories
Broccoli	1 cup	8.2	44
Spaghetti	1 cup	60	308
Bread, whole wheat	1 slice	11.3	55
Chicken	5 oz	0	159
Brown rice	$\frac{1}{2}$ cup	26.9	100
Milk	1 cup	12	160
Apple juice	1 cup	34.4	124
Corn	1 cup	33.3	140
Ice cream	$\frac{1}{2}$ cup	14.6	145
Peaches	1 cup	12	46

11. Wednesday Verne ate 2 cups of spaghetti and Elise ate $1\frac{1}{2}$ cups. How many more calories did Verne eat than Elise? carbohydrates? 154; 30
12. Yoko had $1\frac{1}{2}$ cups of peaches for dessert, while Earl had $1\frac{1}{2}$ cups of ice cream. How many fewer calories did Yoko eat than Earl? 366 carbohydrates? 25.8

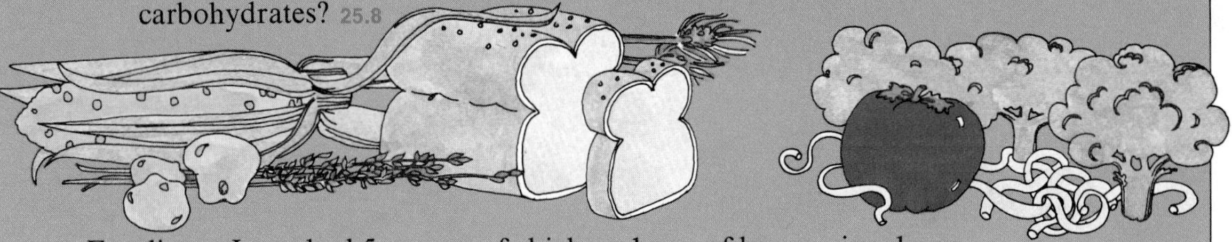

For dinner Irene had 5 ounces of chicken, 1 cup of brown rice, 1 cup of corn, and $1\frac{1}{2}$ cups of milk. Yuki had $1\frac{1}{2}$ cups of spaghetti, 2 slices of bread, 1 cup of broccoli, and 2 cups of apple juice.

13. How many calories were in Irene's dinner? Yuki's dinner? 739; 864
14. How many carbohydrates were in Irene's dinner? Yuki's dinner? 105.1; 189.6

Carbohydrates provide energy for the muscles in the body. The day before the track meet the track team members want to eat foods high in carbohydrates for energy.

15. Does 1 cup of brown rice or 2 slices of bread contain more carbohydrates? brown rice
16. To receive more carbohydrates should the members eat corn or broccoli? corn
17. Should the team members eat chicken or spaghetti? spaghetti
18. Does 1 cup of milk or 1 cup of apple juice contain more carbohydrates? How much more? apple juice; 22.4 carbohydrates

CAREER CAREER CAREER CAREER CAREER

HOME ECONOMIST

Lisa Hutchins is a home economist. She teaches day and evening classes at an adult education center.

There are many aspects of home economy. It is important that students realize we all practice home economy in our lives.

Travis and Emma Johnson need to cut expenses in order to save more money. Lisa shows the Johnsons how to cut out wasteful spending. She helps them plan a new budget for food, personal items, housing, and transportation.

How much will Travis and Emma save given the following monthly expenses?

1. food: \$ 234 \$29.25
2. personal items: \$ 116 \$34.80
3. housing: \$ 364 \$136.50
4. transportation: \$80 \$10

How much will Travis and Emma spend for the following expenses in their new budget?

5. food \$204.75
6. personal items \$81.20
7. housing \$227.50
8. transportation \$70

Food planning is an important part of home economy. Lisa helps Roger Sherman plan how much food he needs in order to serve a group of people. Roger plans to use $\frac{3}{8}$ pound of ground beef per serving. How many pounds of ground beef are needed to serve the following?

9. 4 people $1\frac{1}{2}$ pounds
10. 2 people $\frac{3}{4}$ pound
11. 10 people $3\frac{3}{4}$ pounds
12. 5 people $1\frac{7}{8}$ pounds

Sometimes it is necessary to change a recipe in order to serve more people or fewer people. Lisa's cooking class changes the Banana Nut Bread recipe. Then each person bakes a loaf of bread according to the new recipe.

Banana Nut Bread *Serves 6*

1 3/4 C sifted flour	2/3 C sugar
2 3/4 tsp baking powder	2 eggs
1/2 tsp salt	4 ripe bananas
1/3 C shortening	1/2 C nuts

13. How much flour is used to make Banana Nut Bread for 3 people? 8 people? $\frac{7}{8}$ cup; $2\frac{1}{3}$ cups
14. To make bread for 12 people, how much sugar is needed? How many bananas? $1\frac{1}{3}$ cups; 8 bananas
15. What amount of nuts is used to make bread for 4 people? $\frac{1}{3}$ cup

In a sewing class, Melba Cole wants to make pillows for her living room. Lisa helps Melba find how much one pillow will cost. Then Melba will decide how many pillows she can afford.

16. Find the total cost to make 1 pillow. $5.88
17. How much does it cost to make 4 pillows? $23.52
18. How many pillows can be made for $ 20? 3 pillows

	Amount per pillow	Unit Cost
Fabric	$\frac{1}{2}$ yd	$6.98 per yd
Trim	1 yd	$1.99 per yd
Filling	$\frac{1}{3}$ lb	$1.20 per lb

CHAPTER 6 REVIEW

For extra practice, use Making Practice Fun 42 with this lesson.

Multiply. p. 168

1. $\frac{1}{3} \cdot \frac{1}{4}$ $\frac{1}{12}$

2. $\frac{3}{5} \cdot \frac{2}{5}$ $\frac{6}{25}$

3. $5 \times \frac{4}{7}$ $2\frac{6}{7}$

4. Find a fraction equivalent to $\frac{7}{8}$ with a denominator of 40. p. 170 $\frac{35}{40}$

All the fractions are equivalent. Which is simplest? p. 172

5. $\frac{4}{10}, \frac{6}{15}, \frac{2}{5}, \frac{8}{20}, \frac{12}{30}$ $\frac{2}{5}$

6. $\frac{3}{18}, \frac{5}{30}, \frac{2}{12}, \frac{4}{24}, \frac{1}{6}$ $\frac{1}{6}$

Simplify. p. 172

7. $\frac{12}{24}$ $\frac{1}{2}$

8. $\frac{42}{7}$ 6

9. $\frac{2}{18}$ $\frac{1}{9}$

Multiply and simplify. pp. 174-176

10. $\frac{4}{3} \cdot 24$ 32

11. $\frac{3}{5} \cdot \frac{1}{6}$ $\frac{1}{10}$

12. $\frac{2}{3} \cdot \frac{15}{4}$ $2\frac{1}{2}$

13. $10 \times 8\frac{4}{5}$ 88

14. $7\frac{1}{2} \cdot \frac{4}{5}$ 6

15. $8\frac{1}{3} \cdot 5\frac{1}{4}$ $43\frac{3}{4}$

Find the reciprocal. p. 180

16. $\frac{1}{4}$ 4

17. 18 $\frac{1}{18}$

18. $\frac{5}{8}$ $\frac{8}{5}$

Divide and simplify. pp. 180-182

19. $\frac{3}{8} \div \frac{5}{4}$ $\frac{3}{10}$

20. $\frac{1}{5} \div \frac{1}{8}$ $1\frac{3}{5}$

21. $12 \div \frac{2}{3}$ 18

22. $20 \div 5\frac{1}{3}$ $3\frac{3}{4}$

23. $7\frac{1}{2} \div 5\frac{1}{4}$ $1\frac{3}{7}$

24. $7\frac{1}{2} \div 30$ $\frac{1}{4}$

Application pp. 178, 184

25. The government spends 7¢ of each dollar for the national debt. What part of the dollar is used to pay for the national debt? $\frac{7}{100}$

26. Susan needs $\frac{8}{9}$ as much protein in her diet as Jack. Jack should have 54 grams of protein each day. How much protein should Susan have? 48 g

CHAPTER 6 TEST

Multiply.

1. $\frac{1}{2} \cdot \frac{1}{6}$ $\frac{1}{12}$ **2.** $\frac{3}{5} \cdot \frac{4}{7}$ $\frac{12}{35}$ **3.** $8 \times \frac{2}{3}$ $5\frac{1}{3}$

4. Find a fraction equivalent to $\frac{5}{6}$ with a denominator of 12. $\frac{10}{12}$

All the fractions are equivalent. Which is simplest?

5. $\frac{4}{6}, \frac{6}{9}, \frac{8}{12}, \frac{2}{3}, \frac{10}{15}$ $\frac{2}{3}$ **6.** $\frac{9}{12}, \frac{3}{4}, \frac{6}{8}, \frac{15}{20}, \frac{12}{16}$ $\frac{3}{4}$

Simplify.

7. $\frac{12}{18}$ $\frac{2}{3}$ **8.** $\frac{3}{15}$ $\frac{1}{5}$ **9.** $\frac{54}{6}$ 9

Multiply and simplify.

10. $\frac{2}{5} \cdot \frac{5}{8}$ $\frac{1}{4}$ **11.** $\frac{3}{4} \times 36$ 27 **12.** $\frac{2}{3} \cdot \frac{9}{16}$ $\frac{3}{8}$

13. $20 \times 5\frac{1}{3}$ $106\frac{2}{3}$ **14.** $5\frac{1}{2} \cdot \frac{4}{5}$ $4\frac{2}{5}$ **15.** $9\frac{3}{5} \cdot 2\frac{1}{4}$ $21\frac{3}{5}$

Find the reciprocal.

16. 16 $\frac{1}{16}$ **17.** $\frac{7}{9}$ $\frac{9}{7}$ **18.** $\frac{1}{20}$ 20

Divide and simplify.

19. $36 \div \frac{1}{2}$ 72 **20.** $\frac{1}{4} \div \frac{2}{3}$ $\frac{3}{8}$ **21.** $\frac{1}{8} \div \frac{1}{6}$ $\frac{3}{4}$

22. $10 \div 8\frac{4}{5}$ $1\frac{3}{22}$ **23.** $4\frac{1}{3} \div 6\frac{1}{2}$ $\frac{2}{3}$ **24.** $7\frac{1}{3} \div 44$ $\frac{1}{6}$

Application

25. The Arrow Soccer Club spends $1000 for uniforms each year. Their total budget is $3500. What part of the total budget does the club spend for uniforms? $\frac{2}{7}$

26. There are 35 calories in $\frac{1}{2}$ cup of red raspberries. After school Janet ate $\frac{1}{2}$ cup of raspberries, and Jason ate $1\frac{1}{2}$ cups. How many more calories did Jason eat than Janet? 70

MORE PRACTICE

Multiply. Use after page 169.

1. $\frac{1}{6} \cdot \frac{1}{3}$ $\frac{1}{18}$
2. $\frac{2}{3} \cdot \frac{1}{5}$ $\frac{2}{15}$
3. $\frac{3}{4} \cdot \frac{5}{7}$ $\frac{15}{28}$
4. $9 \times \frac{5}{8}$ $5\frac{5}{8}$

5. $\frac{2}{5} \times \frac{4}{9}$ $\frac{8}{45}$
6. $\frac{7}{10} \times \frac{7}{8}$ $\frac{49}{80}$
7. $\frac{6}{11} \cdot 6$ $3\frac{3}{11}$
8. $3 \times \frac{9}{16}$ $1\frac{11}{16}$

Find a fraction equivalent to the given one, but with the indicated denominator. Use after page 171.

9. $\frac{1}{2}$ [8] $\frac{4}{8}$
10. $\frac{2}{3}$ [12] $\frac{8}{12}$
11. $\frac{2}{3}$ [15] $\frac{10}{15}$
12. $\frac{9}{10}$ [100] $\frac{90}{100}$

13. $\frac{7}{8}$ [48] $\frac{42}{48}$
14. $\frac{7}{12}$ [60] $\frac{35}{60}$
15. $\frac{13}{16}$ [80] $\frac{65}{80}$
16. $\frac{11}{32}$ [96] $\frac{33}{96}$

17. $\frac{3}{4}$ [28] $\frac{21}{28}$
18. $\frac{9}{16}$ [48] $\frac{27}{48}$
19. $\frac{5}{14}$ [42] $\frac{15}{42}$
20. $\frac{7}{8}$ [56] $\frac{49}{56}$

Simplify. Use after page 173.

21. $\frac{2}{4}$ $\frac{1}{2}$
22. $\frac{6}{18}$ $\frac{1}{3}$
23. $\frac{63}{72}$ $\frac{7}{8}$
24. $\frac{9}{12}$ $\frac{3}{4}$

25. $\frac{23}{46}$ $\frac{1}{2}$
26. $\frac{6}{54}$ $\frac{1}{9}$
27. $\frac{54}{81}$ $\frac{2}{3}$
28. $\frac{62}{93}$ $\frac{2}{3}$

29. $\frac{425}{575}$ $\frac{17}{23}$
30. $\frac{96}{84}$ $1\frac{1}{7}$
31. $\frac{32}{20}$ $1\frac{3}{5}$
32. $\frac{48}{36}$ $1\frac{1}{3}$

Multiply and simplify. Use after page 175.

33. $\frac{3}{4} \cdot \frac{1}{3}$ $\frac{1}{4}$
34. $\frac{2}{5} \cdot \frac{5}{2}$ 1
35. $\frac{3}{5} \times \frac{1}{6}$ $\frac{1}{10}$
36. $\frac{4}{5} \times \frac{15}{16}$ $\frac{3}{4}$

37. $6 \times \frac{7}{8}$ $5\frac{1}{4}$
38. $20 \times \frac{1}{5}$ 4
39. $40 \times \frac{3}{4}$ 30
40. $17 \times \frac{2}{5}$ $6\frac{4}{5}$

41. $9 \times \frac{1}{9}$ 1
42. $56 \cdot \frac{1}{8}$ 7
43. $\frac{7}{10} \cdot \frac{8}{10}$ $\frac{14}{25}$
44. $\frac{9}{10} \cdot \frac{2}{3}$ $\frac{3}{5}$

45. $\frac{16}{25} \times \frac{5}{8}$ $\frac{2}{5}$
46. $\frac{2}{5} \times 125$ 50
47. $\frac{7}{8} \times 64$ 56
48. $\frac{9}{7} \times \frac{2}{3}$ $\frac{6}{7}$

49. $\frac{15}{8} \cdot \frac{4}{3}$ $2\frac{1}{2}$
50. $\frac{5}{6} \times \frac{13}{5}$ $2\frac{1}{6}$
51. $\frac{8}{9} \cdot \frac{3}{5}$ $\frac{8}{15}$
52. $\frac{7}{2} \times \frac{5}{14}$ $1\frac{1}{4}$

Multiply and simplify. Use after page 177.

1. $8 \times 3\frac{1}{4}$ 26 **2.** $5 \times 2\frac{1}{2}$ $12\frac{1}{2}$ **3.** $3\frac{1}{4} \cdot \frac{2}{5}$ $1\frac{3}{10}$ **4.** $5\frac{2}{3} \cdot \frac{5}{6}$ $4\frac{13}{18}$

5. $4\frac{1}{3} \cdot 5\frac{1}{2}$ $23\frac{5}{6}$ **6.** $4\frac{1}{4} \cdot 6\frac{3}{8}$ $27\frac{3}{32}$ **7.** $9\frac{3}{10} \cdot 12$ $111\frac{3}{5}$ **8.** $1\frac{3}{7} \times 2\frac{5}{8}$ $3\frac{3}{4}$

9. $3\frac{1}{2} \cdot 3\frac{1}{2}$ $12\frac{1}{4}$ **10.** $2\frac{1}{3} \cdot 2\frac{1}{3}$ $5\frac{4}{9}$ **11.** $5\frac{3}{4} \times 5\frac{3}{4}$ $33\frac{1}{16}$ **12.** $3\frac{2}{5} \times \frac{2}{17}$ $\frac{2}{5}$

13. $5\frac{2}{3} \times 8$ $45\frac{1}{3}$ **14.** $3\frac{1}{2} \cdot 4\frac{2}{3}$ $16\frac{1}{3}$ **15.** $\frac{1}{2} \cdot 4\frac{5}{8}$ $2\frac{5}{16}$ **16.** $6\frac{1}{10} \cdot 1\frac{3}{8}$ $8\frac{31}{80}$

17. $\frac{3}{4} \cdot 2\frac{1}{2}$ $1\frac{7}{8}$ **18.** $1\frac{5}{7} \times 1\frac{5}{9}$ $2\frac{2}{3}$ **19.** $8\frac{3}{4} \times 10$ $87\frac{1}{2}$ **20.** $2\frac{1}{3} \times 3\frac{2}{5}$ $7\frac{14}{15}$

Find the reciprocal. Use after page 181.

21. $\frac{7}{8}$ $\frac{8}{7}$ **22.** $\frac{9}{10}$ $\frac{10}{9}$ **23.** $\frac{1}{12}$ 12 **24.** $\frac{11}{3}$ $\frac{3}{11}$

25. 23 $\frac{1}{23}$ **26.** $\frac{1}{37}$ 37 **27.** $\frac{23}{16}$ $\frac{16}{23}$ **28.** 135 $\frac{1}{135}$

Divide and simplify. Use after page 181.

29. $\frac{1}{2} \div \frac{1}{4}$ 2 **30.** $\frac{1}{3} \div 6$ $\frac{1}{18}$ **31.** $\frac{1}{3} \div \frac{1}{5}$ $1\frac{2}{3}$ **32.** $\frac{7}{8} \div 7$ $\frac{1}{8}$

33. $12 \div \frac{2}{3}$ 18 **34.** $\frac{7}{8} \div \frac{7}{8}$ 1 **35.** $\frac{3}{4} \div \frac{5}{6}$ $\frac{9}{10}$ **36.** $\frac{2}{3} \div \frac{4}{9}$ $1\frac{1}{2}$

37. $\frac{4}{5} \div \frac{5}{4}$ $\frac{16}{25}$ **38.** $8 \div \frac{2}{9}$ 36 **39.** $\frac{6}{5} \div \frac{7}{6}$ $1\frac{1}{35}$ **40.** $\frac{5}{8} \div \frac{3}{2}$ $\frac{5}{12}$

Divide and simplify. Use after page 183.

41. $1\frac{1}{2} \div 3$ $\frac{1}{2}$ **42.** $9 \div 4\frac{1}{2}$ 2 **43.** $1\frac{3}{5} \div \frac{2}{3}$ $2\frac{2}{5}$ **44.** $3\frac{1}{4} \div \frac{5}{6}$ $3\frac{9}{10}$

45. $16\frac{1}{2} \div 4\frac{1}{4}$ $3\frac{15}{17}$ **46.** $5\frac{1}{4} \div 10$ $\frac{21}{40}$ **47.** $2\frac{1}{10} \div 1\frac{1}{5}$ $1\frac{3}{4}$ **48.** $100 \div 33\frac{1}{3}$ 3

49. $10\frac{1}{2} \div 5\frac{1}{2}$ $1\frac{10}{11}$ **50.** $14\frac{1}{2} \div 3\frac{5}{8}$ 4 **51.** $30\frac{1}{4} \div 121$ $\frac{1}{4}$ **52.** $15\frac{3}{4} \div 9\frac{1}{8}$ $1\frac{53}{73}$

53. $6\frac{3}{5} \div 4$ $1\frac{13}{20}$ **54.** $6\frac{5}{6} \div \frac{3}{4}$ $9\frac{1}{9}$ **55.** $3\frac{2}{5} \div 2\frac{4}{7}$ $1\frac{29}{90}$ **56.** $\frac{1}{2} \div 2\frac{2}{3}$ $\frac{3}{16}$

CALCULATOR CORNER

GET 100

Players one or more
Materials one calculator

This activity stresses number relationships and allows for creativity in determining a variety of solutions. It can easily be adapted for use with an entire class, using only one calculator.

* Suppose a calculator has 127 on the display.
How many ways can we get 100 on the display?

Try these on your calculator.

1. Press [−] [2] [7] [=]
2. Press [−] [1] [0] [−] [1] [0] [−] [7] [=]
3. Press [−] [1] [2] [7] [+] [1] [0] [0] [=]

Start with 127. Find 5 more ways to get 100 on the display. Write down your steps.

**Suppose 127 is still on the display. This time we will get 100 on the display but must press [×] at least once.

Try these on your calculator.

4. Press [×] [2] [−] [1] [2] [7] [−] [2] [7] [=]
5. Press [×] [1] [−] [2] [7] [=]

Start with 127. Find 5 more ways to get 100 on the display. You must press [×] at least once. Write down your steps.

EXERCISES Write down the steps for each of the following.

1. Start with 154. Find 10 ways to get 100 on the display.
2. Start with 136. Find 10 ways to get 100 on the display. You must press [×] at least once.
3. Start with 182. Find 10 ways to get 100 on the display. You must press [÷] at least once.

CHAPTER 7 PRE-TEST

Add and simplify. p. 196

1. $\frac{1}{2} + \frac{5}{2}$ 3
2. $\frac{3}{8} + \frac{2}{8}$ $\frac{5}{8}$
3. $\frac{5}{12} + \frac{11}{12} + \frac{2}{12}$ $1\frac{1}{2}$

Find the LCD. p. 198

4. $\frac{5}{12}, \frac{7}{16}$ 48
5. $\frac{1}{3}, \frac{2}{5}$ 15
6. $\frac{1}{2}, \frac{1}{4}, \frac{1}{3}$ 12

Add and simplify. p. 200

7. $\frac{7}{8} + \frac{2}{3}$ $1\frac{13}{24}$
8. $\frac{7}{10} + \frac{9}{100}$ $\frac{79}{100}$
9. $\frac{1}{2} + \frac{3}{4} + \frac{5}{6}$ $2\frac{1}{12}$

Add and simplify. p. 202

10. $6\frac{2}{5} + 7\frac{4}{5}$ $14\frac{1}{5}$
11. $12\frac{5}{8} + 7$ $19\frac{5}{8}$
12. $9\frac{1}{4} + 5\frac{1}{6}$ $14\frac{5}{12}$
13. $10\frac{1}{8} + 5\frac{1}{2} + 7\frac{3}{4}$ $23\frac{3}{8}$
14. $\frac{3}{8} + 2\frac{5}{6}$ $3\frac{5}{24}$
15. $7\frac{1}{12} + 4\frac{3}{4}$ $11\frac{5}{6}$

Subtract and simplify. p. 206

16. $\frac{5}{6} - \frac{3}{6}$ $\frac{1}{3}$
17. $\frac{5}{6} - \frac{3}{4}$ $\frac{1}{12}$
18. $\frac{17}{24} - \frac{5}{8}$ $\frac{1}{12}$

Subtract and simplify. pp. 208–210

19. $7\frac{5}{8} - 5\frac{1}{8}$ $2\frac{1}{2}$
20. $15\frac{3}{4} - 9$ $6\frac{3}{4}$
21. $7\frac{5}{6} - 2\frac{1}{3}$ $5\frac{1}{2}$
22. $10\frac{1}{6} - 5\frac{7}{8}$ $4\frac{7}{24}$
23. $14 - 7\frac{5}{6}$ $6\frac{1}{6}$
24. $16\frac{1}{3} - 7\frac{2}{3}$ $8\frac{2}{3}$

Application pp. 204, 212

25. George drank the following amounts of water one day. How much did he drink in all? 7 cups

Home	School	Restaurant
$4\frac{1}{2}$ cups	$\frac{3}{4}$ cup	$1\frac{3}{4}$ cups

26. A carpenter cuts $2\frac{3}{4}$ ft off a board which is $7\frac{1}{2}$ ft long. How much is left? $4\frac{3}{4}$ ft

7

ADDING AND SUBTRACTING FRACTIONS

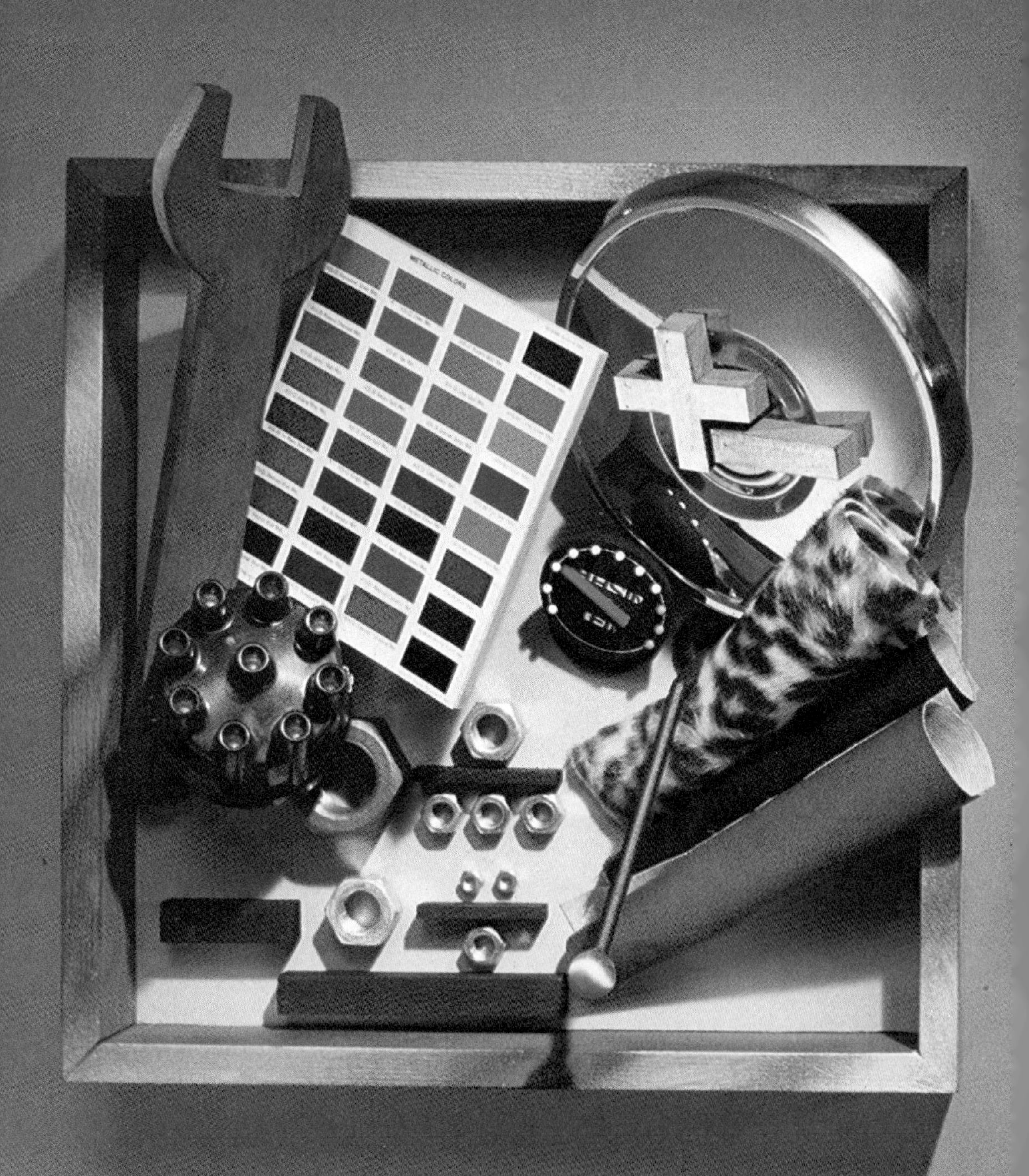

GETTING STARTED

FRACTION ADDO

WHAT'S NEEDED The 30 squares of paper with digits marked on them as in GOTCHA, Chapter 1.

THE RULES

1. On a sheet of paper each player writes this:

$$\frac{[\ ]}{[\ ]} + \frac{[\ ]}{[\ ]}$$

2. To play, a leader draws digits from the box, one at a time.
3. Each player writes the digit in any one of the four spaces. Once it has been written, it cannot be changed.
4. The leader continues to draw digits until four digits have been drawn.
5. Each player adds the resulting fractions. The winner is the player who has the largest sum.

Sample

Louis

$$\frac{[2]}{[3]} + \frac{[5]}{[4]} = \frac{23}{12}$$

Chris

$$\frac{[4]}{[5]} + \frac{[3]}{[2]} = \frac{23}{10}$$

Sally

$$\frac{[4]}{[3]} + \frac{[5]}{[2]} = \frac{23}{6}$$

Sally wins because she has the largest sum.

Objective 75

ADDING WITH THE SAME DENOMINATORS

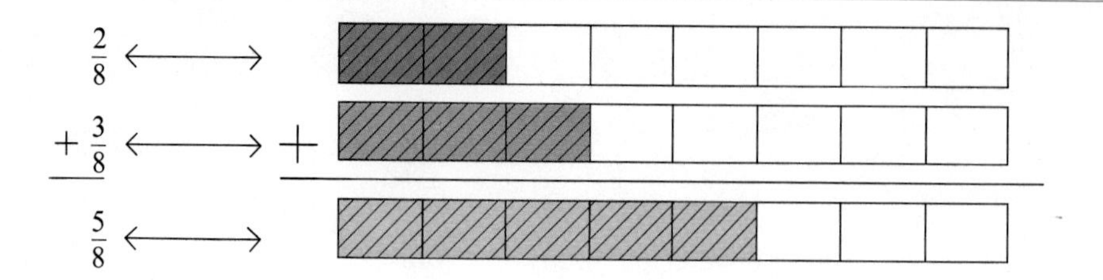

Add and simplify. $\frac{2}{3} + \frac{4}{3}$

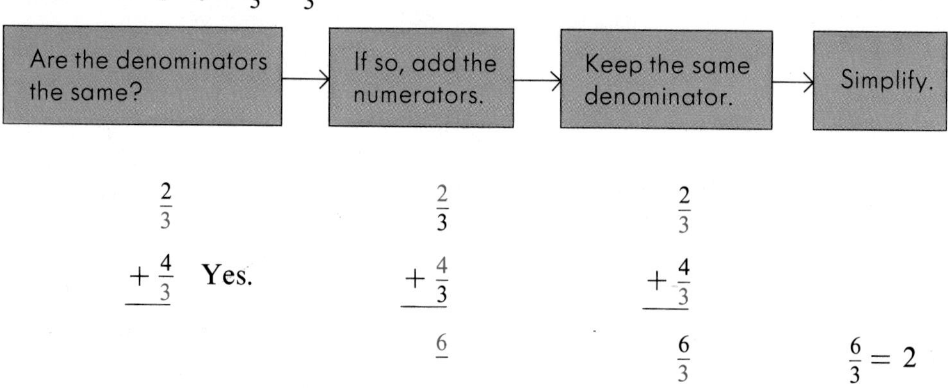

EXAMPLE 1. Add and simplify. $\frac{2}{5} + \frac{1}{5}$

$\frac{2}{5}$ *Are the denominators the same? Yes.*

$+\frac{1}{5}$ *Add the numerators.*

$\frac{3}{5}$ *Keep the same denominator.*

EXAMPLE 2. Add and simplify. $\frac{3}{12} + \frac{5}{12} + \frac{1}{12}$

$\frac{3}{12}$ *Are the denominators the same? Yes.*

$\frac{5}{12}$ *Add the numerators.*

$+\frac{1}{12}$ *Keep the same denominator.*

$\frac{9}{12} = \frac{3}{4}$ *Simplify.*

TRY THIS

Add and simplify.

1. $\frac{1}{3}+\frac{2}{3}$ 1
2. $\frac{5}{12}+\frac{2}{12}$ $\frac{7}{12}$
3. $\frac{9}{16}+\frac{3}{16}+\frac{2}{16}$ $\frac{7}{8}$

PRACTICE

For extra practice, use Making Practice Fun 43 with this lesson.

Add and simplify.

1. $\frac{4}{5}+\frac{1}{5}$ 1
2. $\frac{1}{4}+\frac{3}{4}$ 1
3. $\frac{7}{8}+\frac{1}{8}$ 1
4. $\frac{1}{6}+\frac{5}{6}$ 1
5. $\frac{3}{8}+\frac{4}{8}$ $\frac{7}{8}$
6. $\frac{1}{5}+\frac{2}{5}$ $\frac{3}{5}$
7. $\frac{3}{10}+\frac{4}{10}$ $\frac{7}{10}$
8. $\frac{3}{12}+\frac{2}{12}$ $\frac{5}{12}$
9. $\frac{2}{9}+\frac{1}{9}+\frac{5}{9}$ $\frac{8}{9}$
10. $\frac{1}{10}+\frac{3}{10}+\frac{4}{10}$ $\frac{4}{5}$
11. $\frac{5}{16}+\frac{7}{16}+\frac{3}{16}$ $\frac{15}{16}$
12. $\frac{3}{8}+\frac{1}{8}+\frac{2}{8}$ $\frac{3}{4}$
13. $\frac{3}{2}+\frac{5}{2}$ 4
14. $\frac{5}{3}+\frac{4}{3}$ 3
15. $\frac{11}{10}+\frac{9}{10}$ 2
16. $\frac{11}{12}+\frac{13}{12}$ 2
17. $\frac{5}{12}+\frac{9}{12}$ $1\frac{1}{6}$
18. $\frac{11}{10}+\frac{1}{10}$ $1\frac{1}{5}$
19. $\frac{13}{16}+\frac{5}{16}$ $1\frac{1}{8}$
20. $\frac{11}{8}+\frac{3}{8}$ $1\frac{3}{4}$
21. $\frac{3}{10}+\frac{1}{10}+\frac{9}{10}$ $1\frac{3}{10}$
22. $\frac{5}{6}+\frac{11}{6}+\frac{1}{6}$ $2\frac{5}{6}$
23. $\frac{3}{5}+\frac{2}{5}+\frac{4}{5}$ $1\frac{4}{5}$
24. $\frac{7}{8}+\frac{5}{8}+\frac{3}{8}$ $1\frac{7}{8}$
25. $\frac{13}{15}+\frac{12}{15}$ $1\frac{2}{3}$
26. $\frac{13}{14}+\frac{15}{14}$ 2
27. $\frac{11}{12}+\frac{11}{12}$ $1\frac{5}{6}$
28. $\frac{21}{16}+\frac{21}{16}$ $2\frac{5}{8}$
29. $\frac{15}{100}+\frac{31}{100}$ $\frac{23}{50}$
30. $\frac{24}{100}+\frac{22}{100}$ $\frac{23}{50}$
31. $\frac{51}{100}+\frac{14}{100}$ $\frac{13}{20}$
32. $\frac{71}{100}+\frac{14}{100}$ $\frac{17}{20}$
33. $\frac{21}{50}+\frac{9}{50}$ $\frac{3}{5}$
34. $\frac{7}{60}+\frac{5}{60}$ $\frac{1}{5}$
35. $\frac{48}{80}+\frac{12}{80}$ $\frac{3}{4}$
36. $\frac{45}{90}+\frac{15}{90}$ $\frac{2}{3}$

PROBLEM CORNER

1. Marti ran $\frac{5}{10}$ of a mile and then $\frac{3}{10}$ of a mile. How far did she run in all? $\frac{4}{5}$ mile
2. Lou bought $\frac{1}{3}$ lb of peanuts and $\frac{2}{3}$ lb of cashews. How many lbs of nuts did he buy in all? 1 lb

Objective 76

LEAST COMMON DENOMINATOR

The LCD is the smallest number that is divisible by the denominators.

What is the LCD of $\frac{3}{4}$ and $\frac{1}{8}$? This method of finding the LCD can easily be done mentally.

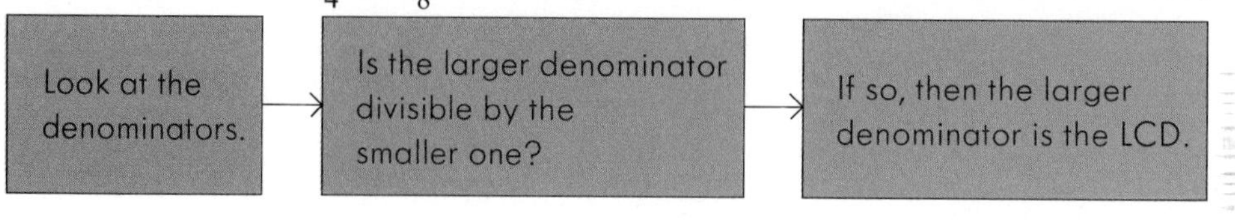

4 and 8 Is 8 divisible by 4? Yes. The LCD is 8.

EXAMPLE 1. What is the LCD of $\frac{2}{3}$, $\frac{5}{6}$, and $\frac{1}{2}$?

The denominators are 3, 6, and 2.
Is 6 divisible by 3 and divisible by 2? Yes. The LCD is 6.

TRY THIS

Find the LCD.

1. $\frac{5}{6}$, $\frac{7}{18}$ 18 **2.** $\frac{2}{3}$, $\frac{3}{4}$, $\frac{7}{12}$ 12

What is the LCD of $\frac{1}{12}$ and $\frac{2}{15}$?

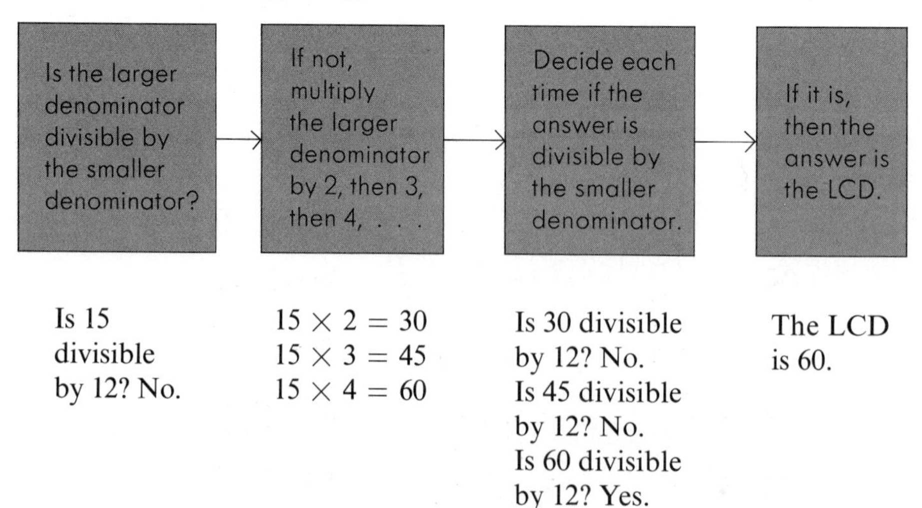

EXAMPLE 2. What is the LCD of $\frac{1}{6}$, $\frac{1}{3}$, and $\frac{3}{4}$?

The denominators are 6, 3, and 4.
Is 6 divisible by 3 and divisible by 4? No.
$6 \times 2 = 12$ Is 12 divisible by 3 and divisible by 4? Yes.
The LCD is 12.

TRY THIS

Find the LCD.

3. $\frac{3}{8}, \frac{5}{6}$ 24

4. $\frac{1}{2}, \frac{1}{5}, \frac{1}{4}$ 20

PRACTICE

For extra practice, use Making Practice Fun 44 with this lesson.

Find the LCD.

1. $\frac{2}{3}, \frac{5}{6}$ 6
2. $\frac{2}{3}, \frac{1}{9}$ 9
3. $\frac{1}{2}, \frac{5}{6}$ 6
4. $\frac{1}{4}, \frac{3}{8}$ 8
5. $\frac{1}{8}, \frac{1}{6}$ 24
6. $\frac{1}{4}, \frac{1}{6}$ 12
7. $\frac{1}{9}, \frac{1}{6}$ 18
8. $\frac{1}{8}, \frac{1}{12}$ 24
9. $\frac{2}{5}, \frac{1}{10}$ 10
10. $\frac{2}{3}, \frac{1}{12}$ 12
11. $\frac{4}{5}, \frac{7}{10}$ 10
12. $\frac{3}{4}, \frac{1}{12}$ 12
13. $\frac{5}{12}, \frac{3}{8}, \frac{1}{4}$ 24
14. $\frac{4}{3}, \frac{5}{6}, \frac{1}{8}$ 24
15. $\frac{7}{8}, \frac{1}{16}, \frac{1}{12}$ 48
16. $\frac{2}{9}, \frac{7}{12}, \frac{1}{6}$ 36
17. $\frac{3}{20}, \frac{3}{4}$ 20
18. $\frac{2}{15}, \frac{2}{5}$ 15
19. $\frac{1}{18}, \frac{2}{3}$ 18
20. $\frac{3}{4}, \frac{3}{16}$ 16
21. $\frac{1}{8}, \frac{7}{12}$ 24
22. $\frac{3}{8}, \frac{5}{6}$ 24
23. $\frac{3}{4}, \frac{5}{6}$ 12
24. $\frac{5}{6}, \frac{7}{9}$ 18
25. $\frac{5}{12}, \frac{2}{15}$ 60
26. $\frac{3}{10}, \frac{5}{12}$ 60
27. $\frac{3}{16}, \frac{1}{12}$ 48
28. $\frac{9}{10}, \frac{11}{16}$ 80
29. $\frac{3}{4}, \frac{2}{5}, \frac{11}{20}$ 20
30. $\frac{1}{8}, \frac{2}{3}, \frac{15}{24}$ 24
31. $\frac{2}{3}, \frac{3}{4}, \frac{5}{12}$ 12
32. $\frac{1}{5}, \frac{1}{3}, \frac{4}{15}$ 15
33. $\frac{4}{5}, \frac{2}{3}, \frac{1}{6}$ 30
34. $\frac{4}{5}, \frac{3}{10}, \frac{1}{4}$ 20
35. $\frac{3}{8}, \frac{5}{6}, \frac{1}{2}$ 24
36. $\frac{3}{4}, \frac{1}{6}, \frac{5}{8}$ 24
37. $\frac{3}{8}, \frac{7}{10}$ 40
38. $\frac{3}{10}, \frac{1}{15}$ 30
39. $\frac{5}{12}, \frac{11}{18}$ 36
40. $\frac{11}{12}, \frac{3}{16}$ 48
41. $\frac{5}{16}, \frac{3}{24}$ 48
42. $\frac{13}{18}, \frac{1}{24}$ 72
43. $\frac{11}{24}, \frac{15}{36}$ 72
44. $\frac{12}{15}, \frac{9}{20}$ 60

More Practice, page 218

Objective 77

ADDING WITH DIFFERENT DENOMINATORS

Add and simplify. $\frac{1}{6} + \frac{3}{4}$

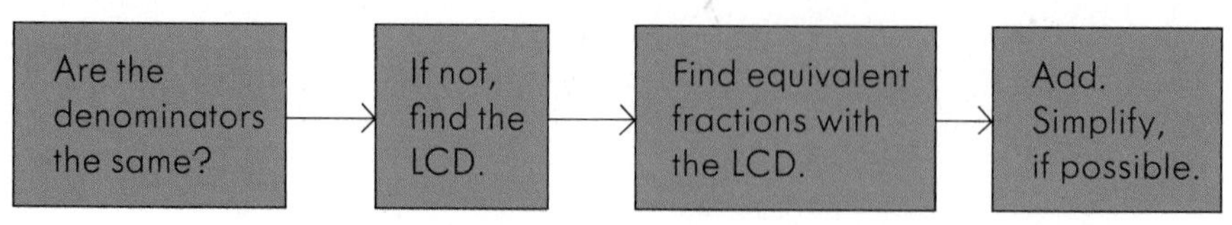

$\frac{1}{6} + \frac{3}{4}$ No.

The LCD is 12.

$\frac{1}{6} = \frac{2}{12}$, $+\frac{3}{4} = \frac{9}{12}$

$\frac{1}{6} = \frac{2}{12}$, $+\frac{3}{4} = \frac{9}{12}$, $\frac{11}{12}$

EXAMPLES

1. Add and simplify. $\frac{4}{5} + \frac{7}{10}$

$$\frac{4}{5} = \frac{8}{10}$$
$$+\frac{7}{10} = \frac{7}{10}$$
$$\frac{15}{10} = \frac{3}{2} = 1\frac{1}{2}$$

Are the denominators the same? No.
If not, find the LCD. The LCD is 10.
Find equivalent fractions with the LCD.
Add. Simplify.

2. Add and simplify. $\frac{5}{12} + \frac{7}{8}$

$$\frac{5}{12} = \frac{10}{24}$$
$$+\frac{7}{8} = \frac{21}{24}$$
$$\frac{31}{24} = 1\frac{7}{24}$$

3. Add. $\frac{5}{9} + \frac{7}{18} + \frac{5}{6}$

$$\frac{5}{9} = \frac{10}{18}$$
$$\frac{7}{18} = \frac{7}{18}$$
$$+\frac{5}{6} = \frac{15}{18}$$
$$\frac{32}{18} = \frac{16}{9}$$
$$= 1\frac{7}{9}$$

TRY THIS

Add and simplify.

1. $\frac{3}{8} + \frac{5}{6}$ $1\frac{5}{24}$ **2.** $\frac{1}{6} + \frac{7}{18}$ $\frac{5}{9}$ **3.** $\frac{1}{2} + \frac{2}{5} + \frac{6}{10}$ $1\frac{1}{2}$

PRACTICE

For extra practice, use Making Practice Fun 45 with this lesson.

Add and simplify.

1. $\frac{2}{3} + \frac{5}{6}$ $1\frac{1}{2}$
2. $\frac{2}{3} + \frac{1}{9}$ $\frac{7}{9}$
3. $\frac{1}{2} + \frac{5}{6}$ $1\frac{1}{3}$
4. $\frac{1}{4} + \frac{3}{8}$ $\frac{5}{8}$
5. $\frac{2}{5} + \frac{1}{10}$ $\frac{1}{2}$
6. $\frac{2}{3} + \frac{1}{12}$ $\frac{3}{4}$
7. $\frac{4}{5} + \frac{7}{10}$ $1\frac{1}{2}$
8. $\frac{3}{4} + \frac{1}{12}$ $\frac{5}{6}$
9. $\frac{5}{12} + \frac{3}{8}$ $\frac{19}{24}$
10. $\frac{4}{3} + \frac{5}{6}$ $2\frac{1}{6}$
11. $\frac{7}{8} + \frac{1}{16}$ $\frac{15}{16}$
12. $\frac{2}{9} + \frac{7}{12}$ $\frac{29}{36}$
13. $\frac{1}{8} + \frac{7}{12}$ $\frac{17}{24}$
14. $\frac{3}{8} + \frac{5}{6}$ $1\frac{5}{24}$
15. $\frac{3}{4} + \frac{5}{6}$ $1\frac{7}{12}$
16. $\frac{5}{6} + \frac{7}{9}$ $1\frac{11}{18}$
17. $\frac{3}{10} + \frac{1}{100}$ $\frac{31}{100}$
18. $\frac{1}{10} + \frac{7}{100}$ $\frac{17}{100}$
19. $\frac{2}{10} + \frac{6}{100}$ $\frac{13}{50}$
20. $\frac{9}{10} + \frac{3}{100}$ $\frac{93}{100}$
21. $\frac{3}{10} + \frac{7}{40}$ $\frac{19}{40}$
22. $\frac{4}{10} + \frac{3}{20}$ $\frac{11}{20}$
23. $\frac{9}{10} + \frac{11}{30}$ $1\frac{4}{15}$
24. $\frac{1}{10} + \frac{3}{50}$ $\frac{4}{25}$
25. $\frac{5}{12} + \frac{2}{15}$ $\frac{11}{20}$
26. $\frac{3}{10} + \frac{5}{12}$ $\frac{43}{60}$
27. $\frac{3}{16} + \frac{1}{12}$ $\frac{13}{48}$
28. $\frac{9}{10} + \frac{11}{16}$ $1\frac{47}{80}$
29. $\frac{4}{5} + \frac{3}{25}$ $\frac{23}{25}$
30. $\frac{7}{20} + \frac{4}{5}$ $1\frac{3}{20}$
31. $\frac{3}{8} + \frac{15}{16}$ $1\frac{5}{16}$
32. $\frac{3}{4} + \frac{1}{20}$ $\frac{4}{5}$
33. $\frac{1}{8} + \frac{1}{6} + \frac{1}{2}$ $\frac{19}{24}$
34. $\frac{1}{4} + \frac{1}{6} + \frac{1}{3}$ $\frac{3}{4}$
35. $\frac{1}{9} + \frac{1}{6} + \frac{1}{3}$ $\frac{11}{18}$
36. $\frac{3}{20} + \frac{3}{4} + \frac{1}{5}$ $1\frac{1}{10}$
37. $\frac{2}{15} + \frac{2}{5} + \frac{2}{3}$ $1\frac{1}{5}$
38. $\frac{1}{18} + \frac{2}{3} + \frac{1}{6}$ $\frac{8}{9}$
39. $\frac{3}{4} + \frac{2}{3} + \frac{1}{12}$ $1\frac{1}{2}$
40. $\frac{1}{8} + \frac{2}{3} + \frac{1}{24}$ $\frac{5}{6}$
41. $\frac{2}{3} + \frac{3}{4} + \frac{5}{12}$ $1\frac{5}{6}$
42. $\frac{5}{10} + \frac{31}{100} + \frac{1}{1000}$ $\frac{811}{1000}$
43. $\frac{3}{10} + \frac{27}{100} + \frac{5}{1000}$ $\frac{23}{40}$
44. $\frac{1}{10} + \frac{11}{100} + \frac{3}{1000}$ $\frac{213}{1000}$

PROBLEM CORNER

1. Sara walked $\frac{3}{4}$ of a mile to a friend's house, and then $\frac{7}{6}$ of a mile to class. How far did Sara walk? $1\frac{11}{12}$ miles
2. Allen bought $\frac{1}{4}$ lb of bean sprouts, $\frac{1}{2}$ lb of mushrooms, and $\frac{2}{3}$ lb of carrots. How many pounds of vegetables did Allen buy? $1\frac{5}{12}$ lbs

Objective 78

ADDING WITH MIXED NUMBERS

Add and simplify. $12\frac{1}{2} + 7\frac{3}{10}$

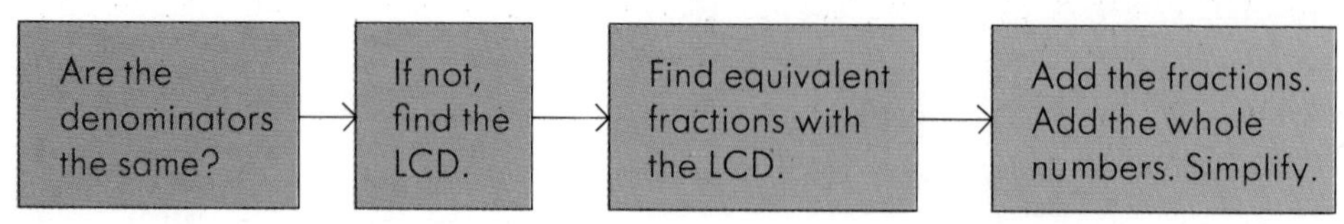

$$\begin{array}{r} 12\frac{1}{2} \\ +\ 7\frac{3}{10} \\ \hline \end{array} \text{ No.}$$

The LCD is 10.

$$\begin{array}{rcr} 12\frac{1}{2} & = & 12\frac{5}{10} \\ +\ 7\frac{3}{10} & = & 7\frac{3}{10} \\ \hline \end{array}$$

$$\begin{array}{r} 12\frac{5}{10} \\ +\ 7\frac{3}{10} \\ \hline 19\frac{8}{10} = 19\frac{4}{5} \end{array}$$

EXAMPLES. Add and simplify.

1. $$\begin{array}{rcr} 4\frac{1}{2} & = & 4\frac{3}{6} \\ +\ 7\frac{1}{3} & = & 7\frac{2}{6} \\ \hline & & 11\frac{5}{6} \end{array}$$

2. $$\begin{array}{r} 7\frac{5}{8} \\ +\ 3\frac{1}{8} \\ \hline 10\frac{6}{8} = 10\frac{3}{4} \end{array}$$

3. $$\begin{array}{r} 19\frac{3}{5} \\ +\ 6 \\ \hline 25\frac{3}{5} \end{array}$$ Think of 6 as $6\frac{0}{5}$.

TRY THIS

Add and simplify.

1. $8\frac{2}{5} + 3\frac{4}{10}$ $11\frac{4}{5}$ **2.** $16\frac{2}{9} + 7\frac{4}{9}$ $23\frac{2}{3}$ **3.** $12 + 9\frac{7}{8}$ $21\frac{7}{8}$

EXAMPLE 4. Add and simplify. $5\frac{2}{3} + 3\frac{5}{6} + 2\frac{1}{4}$

$$\begin{array}{rcr} 5\frac{2}{3} & = & 5\frac{8}{12} \\ 3\frac{5}{6} & = & 3\frac{10}{12} \\ +\ 2\frac{1}{4} & = & 2\frac{3}{12} \\ \hline \end{array}$$

$$10\frac{21}{12} = 10 + \frac{21}{12} = 10 + \frac{7}{4}$$
$$= 10 + 1\frac{3}{4} = 11\frac{3}{4}$$

Simplify. Convert the fraction to a mixed number.

Add and simplify.

TRY THIS **4.** $9\frac{3}{4} + 3\frac{5}{6}$ $13\frac{7}{12}$ **5.** $3\frac{1}{2} + 5\frac{1}{4} + 7\frac{3}{10}$ $16\frac{1}{20}$

PRACTICE

For extra practice, use Making Practice Fun 46 with this lesson.
Use Quiz 16 after this Practice.

Add and simplify.

1. $1\frac{2}{3} + 5\frac{1}{3}$ 7 **2.** $4\frac{5}{6} + 3\frac{5}{6}$ $8\frac{2}{3}$ **3.** $4\frac{1}{6} + 3\frac{5}{6}$ 8 **4.** $8\frac{2}{3} + 5\frac{2}{3}$ $14\frac{1}{3}$

5. $1\frac{1}{4} + 1\frac{2}{3}$ $2\frac{11}{12}$ **6.** $1\frac{2}{5} + 2\frac{1}{4}$ $3\frac{13}{20}$ **7.** $2\frac{3}{4} + 2\frac{2}{5}$ $5\frac{3}{20}$ **8.** $2\frac{1}{6} + 1\frac{1}{5}$ $3\frac{11}{30}$

9. $4\frac{3}{8} + 6\frac{5}{12}$ $10\frac{19}{24}$ **10.** $3\frac{2}{9} + 4\frac{1}{6}$ $7\frac{7}{18}$ **11.** $8\frac{3}{4} + 5\frac{1}{6}$ $13\frac{11}{12}$ **12.** $5\frac{3}{8} + 10\frac{5}{6}$ $16\frac{5}{24}$

13. $\frac{5}{8} + 1\frac{5}{6}$ $2\frac{11}{24}$ **14.** $14\frac{5}{8} + \frac{1}{4}$ $14\frac{7}{8}$ **15.** $5\frac{1}{4} + \frac{1}{10}$ $5\frac{7}{20}$ **16.** $2\frac{3}{16} + \frac{1}{2}$ $2\frac{11}{16}$

17. $5\frac{3}{8} + 12$ $17\frac{3}{8}$ **18.** $6\frac{2}{3} + 14$ $20\frac{2}{3}$ **19.** $15 + 4\frac{5}{6}$ $19\frac{5}{6}$ **20.** $18 + 2\frac{1}{8}$ $20\frac{1}{8}$

21. $16\frac{1}{4} + 25\frac{3}{10}$ $41\frac{11}{20}$ **22.** $14\frac{1}{6} + 12\frac{2}{3}$ $26\frac{5}{6}$ **23.** $24\frac{1}{3} + 12\frac{2}{9}$ $36\frac{5}{9}$ **24.** $14\frac{3}{5} + 16\frac{1}{10}$ $30\frac{7}{10}$

25. $\frac{3}{4} + 2\frac{1}{4} + 3\frac{1}{4}$ $6\frac{1}{4}$ **26.** $\frac{2}{3} + 5\frac{1}{3} + 6\frac{2}{3}$ $12\frac{2}{3}$ **27.** $9\frac{4}{5} + \frac{1}{5} + 1\frac{2}{5}$ $11\frac{2}{5}$

28. $3\frac{5}{16} + 6\frac{1}{4} + 2\frac{1}{2}$ $12\frac{1}{16}$ **29.** $3\frac{5}{6} + 2\frac{5}{12} + 5\frac{1}{3}$ $11\frac{7}{12}$ **30.** $4\frac{1}{3} + 2\frac{5}{8} + 3\frac{1}{4}$ $10\frac{5}{24}$

31. $4\frac{7}{8} + \frac{1}{2} + 5\frac{3}{4}$ $11\frac{1}{8}$ **32.** $4\frac{1}{6} + 2\frac{2}{3} + 7\frac{1}{2}$ $14\frac{1}{3}$ **33.** $2\frac{5}{6} + 10\frac{4}{9} + 9\frac{1}{3}$ $22\frac{11}{18}$

PROBLEM CORNER

1. One day a stock opened at $63\frac{7}{8}$ and gained $1\frac{3}{4}$. What was the closing price? $\$65\frac{5}{8}$
2. A fabric store sold three pieces of burlap. One was $6\frac{1}{4}$ yd, one was $4\frac{1}{2}$ yd, and one was $10\frac{5}{6}$ yd long. What was the total length of the burlap? $21\frac{7}{12}$ yd

Objective 169

APPLICATION

FOOD VALUES

Nutrients in food promote growth and repair tissues of the body.

Food	Serving Size	Calories	Protein (gm)	Calcium (mg)	Iron (mg)	Vitamin C (mg)
Whole Milk	1 cup	160	9	288	0.2	3
Skimmed Milk	1 cup	90	9	303	0.2	3
Whole Wheat Bread	1 slice	55	2	22	0.5	0
Peanut Butter	1 tbs	90	4	12	0.3	0
Swiss Cheese	1 oz	105	8	262	0.3	0
Spinach (cooked)	1 cup	45	6	223	3.6	54
Vanilla Ice Cream	$\frac{1}{7}$ qt	165	3	100	0.1	1

1. Compare the Vitamin C in whole and skimmed milk. Which has more Vitamin C? How much more? neither
2. How many tablespoons of peanut butter does it take to get the same quantity of protein as in 1 oz of Swiss cheese? 2 tbs
3. How many tablespoons of peanut butter does it take to get the same quantity of calcium as in 1 oz of Swiss cheese? 21.8
4. How many slices of whole wheat bread would it take to get the same amount of calcium as in 1 cup of whole milk? 13.1
5. How many slices of whole wheat bread would it take to get the same amount of calcium as in $\frac{1}{7}$ qt of vanilla ice cream? 4.5

Yesterday Barry and Joan drank these quantities of skimmed milk.

	Home	**School**	**Restaurant**
Barry	$3\frac{3}{8}$ cups	$3\frac{1}{2}$ cups	$1\frac{1}{8}$ cups
Joan	$3\frac{1}{2}$ cups	$2\frac{3}{4}$ cups	$1\frac{2}{3}$ cups

6. How much skimmed milk did each person drink? Barry: 8 cups; Joan: $7\frac{11}{12}$ cups
7. How many calories were in the milk that each person drank? Barry: 720; Joan: 712.5
8. How many grams of protein did Barry and Joan each receive? Barry: 72 g; Joan: 71.25 g
9. How many milligrams of calcium did each receive? Barry: 2424 mg; Joan: 2399 mg
10. How many milligrams of Vitamin C did each receive? Barry: 24 mg; Joan: 23.75 mg

On Tuesday Juanita had one serving of each food in the table. How much of each of the following did she receive?

11. calories 710
12. protein 41 gm
13. calcium 1210 mg
14. iron 5.2 mg
15. vitamin C 61 mg

The following show that some foods are high in certain nutrients.

See answer section for #16–19.

16. For each item in the table find the amount of iron divided by the number of calories. Which food is highest in iron? spinach
17. For each item in the table find the amount of protein divided by the number of calories. Which food is highest in protein? spinach
18. For each item in the table find the amount of calcium divided by the number of calories. Which food is highest in calcium? spinach
19. For each item in the table find the amount of Vitamin C divided by the number of calories. Which food is highest in Vitamin C? spinach

Objective 79

SUBTRACTING

Subtract and simplify. $\frac{7}{10} - \frac{3}{10}$

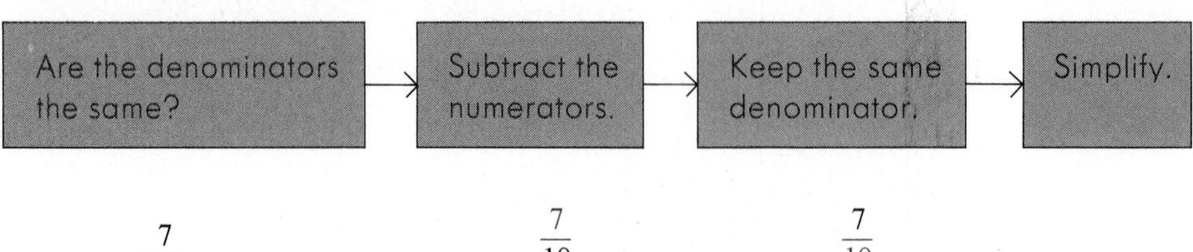

$$\begin{array}{r} \frac{7}{10} \\ -\frac{3}{10} \\ \hline \end{array} \quad \text{Yes.}$$

$$\begin{array}{r} \frac{7}{10} \\ -\frac{3}{10} \\ \hline 4 \end{array}$$

$$\begin{array}{r} \frac{7}{10} \\ -\frac{3}{10} \\ \hline \frac{4}{10} \end{array}$$

$$\frac{4}{10} = \frac{2}{5}$$

EXAMPLE 1. Subtract and simplify. $\frac{11}{12} - \frac{1}{12}$

$$\begin{array}{r} \frac{11}{12} \\ -\frac{1}{12} \\ \hline \frac{10}{12} = \frac{5}{6} \end{array}$$

Are the denominators the same? Yes.
Subtract the numerators.
Keep the same denominator.
Simplify.

TRY THIS

Subtract and simplify.

1. $\frac{7}{8} - \frac{5}{8}$ $\frac{1}{4}$ **2.** $\frac{12}{16} - \frac{4}{16}$ $\frac{1}{2}$

When the denominators are different, we find equivalent fractions with the LCD.

Subtract and simplify. $\frac{3}{4} - \frac{1}{6}$

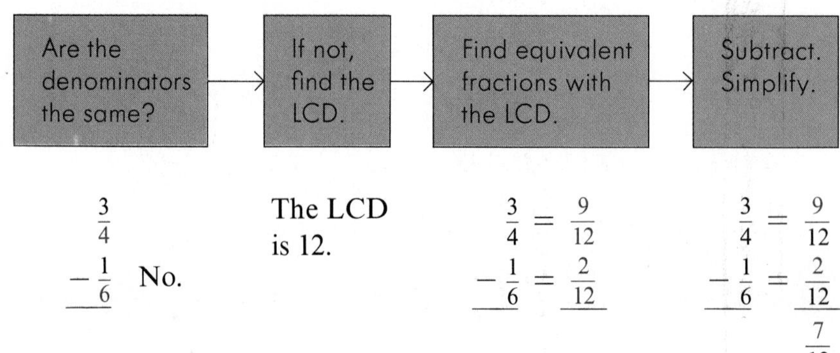

EXAMPLE 2. Subtract and simplify. $\frac{5}{6} - \frac{7}{12}$

$$\begin{array}{rl} \frac{5}{6} = \frac{10}{12} & \\ -\frac{7}{12} = \frac{7}{12} & \\ \hline \frac{3}{12} = \frac{1}{4} & \end{array}$$

Are the denominators the same? No.
The LCD is 12.
Find equivalent fractions with the LCD.

Subtract. Simplify.

TRY THIS

Subtract and simplify.

3. $\frac{4}{5} - \frac{3}{10}$ $\frac{1}{2}$ **4.** $\frac{5}{6} - \frac{1}{9}$ $\frac{13}{18}$

PRACTICE

Subtract and simplify.

1. $\frac{5}{6} - \frac{1}{6}$ $\frac{2}{3}$ **2.** $\frac{3}{4} - \frac{1}{4}$ $\frac{1}{2}$ **3.** $\frac{7}{5} - \frac{2}{5}$ 1 **4.** $\frac{4}{3} - \frac{2}{3}$ $\frac{2}{3}$

5. $\frac{11}{12} - \frac{2}{12}$ $\frac{3}{4}$ **6.** $\frac{9}{10} - \frac{3}{10}$ $\frac{3}{5}$ **7.** $\frac{15}{16} - \frac{1}{16}$ $\frac{7}{8}$ **8.** $\frac{17}{18} - \frac{5}{18}$ $\frac{2}{3}$

9. $\frac{3}{4} - \frac{1}{8}$ $\frac{5}{8}$ **10.** $\frac{2}{3} - \frac{1}{8}$ $\frac{13}{24}$ **11.** $\frac{3}{4} - \frac{1}{2}$ $\frac{1}{4}$ **12.** $\frac{3}{5} - \frac{1}{4}$ $\frac{7}{20}$

13. $\frac{2}{3} - \frac{1}{9}$ $\frac{5}{9}$ **14.** $\frac{5}{6} - \frac{2}{3}$ $\frac{1}{6}$ **15.** $\frac{3}{8} - \frac{1}{4}$ $\frac{1}{8}$ **16.** $\frac{5}{6} - \frac{1}{2}$ $\frac{1}{3}$

17. $\frac{4}{3} - \frac{5}{6}$ $\frac{1}{2}$ **18.** $\frac{5}{12} - \frac{3}{8}$ $\frac{1}{24}$ **19.** $\frac{7}{12} - \frac{2}{9}$ $\frac{13}{36}$ **20.** $\frac{7}{8} - \frac{1}{16}$ $\frac{13}{16}$

21. $\frac{2}{5} - \frac{2}{15}$ $\frac{4}{15}$ **22.** $\frac{3}{4} - \frac{3}{20}$ $\frac{3}{5}$ **23.** $\frac{3}{4} - \frac{4}{16}$ $\frac{1}{2}$ **24.** $\frac{2}{3} - \frac{1}{18}$ $\frac{11}{18}$

25. $\frac{5}{6} - \frac{3}{8}$ $\frac{11}{24}$ **26.** $\frac{7}{12} - \frac{1}{8}$ $\frac{11}{24}$ **27.** $\frac{5}{6} - \frac{7}{9}$ $\frac{1}{18}$ **28.** $\frac{5}{6} - \frac{3}{4}$ $\frac{1}{12}$

29. $\frac{5}{12} - \frac{3}{10}$ $\frac{7}{60}$ **30.** $\frac{5}{12} - \frac{2}{15}$ $\frac{17}{60}$ **31.** $\frac{9}{10} - \frac{11}{16}$ $\frac{17}{80}$ **32.** $\frac{3}{16} - \frac{1}{12}$ $\frac{5}{48}$

33. $\frac{2}{3} - \frac{1}{8}$ $\frac{13}{24}$ **34.** $\frac{3}{4} - \frac{2}{5}$ $\frac{7}{20}$ **35.** $\frac{1}{3} - \frac{1}{5}$ $\frac{2}{15}$ **36.** $\frac{3}{4} - \frac{2}{3}$ $\frac{1}{12}$

37. $\frac{4}{5} - \frac{7}{20}$ $\frac{9}{20}$ **38.** $\frac{4}{5} - \frac{3}{25}$ $\frac{17}{25}$ **39.** $\frac{3}{4} - \frac{1}{20}$ $\frac{7}{10}$ **40.** $\frac{15}{16} - \frac{3}{8}$ $\frac{9}{16}$

More Practice, page 219

Objective 80

SUBTRACTING WITH MIXED NUMBERS

Subtract and simplify. $9\frac{4}{5} - 3\frac{1}{2}$

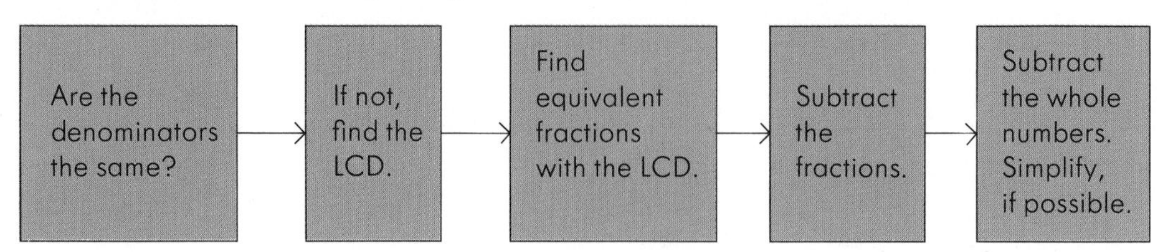

$\begin{array}{r} 9\frac{4}{5} \\ -\,3\frac{1}{2} \\ \hline \end{array}$ No.	The LCD is 10.	$\begin{array}{r} 9\frac{4}{5} = 9\frac{8}{10} \\ -\,3\frac{1}{2} = 3\frac{5}{10} \\ \hline \end{array}$	$\begin{array}{r} 9\frac{8}{10} \\ -\,3\frac{5}{10} \\ \hline \frac{3}{10} \end{array}$	$\begin{array}{r} 9\frac{8}{10} \\ -\,3\frac{5}{10} \\ \hline 6\frac{3}{10} \end{array}$

EXAMPLE 1. Subtract and simplify. $7\frac{3}{4} - 2\frac{1}{4}$

$$\begin{array}{r} 7\frac{3}{4} = 7\frac{3}{4} \\ -\,2\frac{1}{4} = 2\frac{1}{4} \\ \hline 5\frac{2}{4} = 5\frac{1}{2} \end{array}$$

Are the denominators the same? Yes.

Subtract the fractions.
Subtract the whole numbers.
Simplify.

EXAMPLE 2. Subtract and simplify. $16\frac{3}{7} - 9$

$$\begin{array}{r} 16\frac{3}{7} \\ -\quad 9 \\ \hline 7\frac{3}{7} \end{array}$$

Are the denominators the same. Yes.
Think of 9 as $9\frac{0}{7}$.

EXAMPLE 3. Subtract and simplify. $15\frac{5}{6} - 8\frac{1}{2}$

$$\begin{array}{r} 15\frac{5}{6} = 15\frac{5}{6} \\ -\quad 8\frac{1}{2} = \ \ 8\frac{3}{6} \\ \hline 7\frac{2}{6} = 7\frac{1}{3} \end{array}$$

Are the denominators the same? No.
The LCD is 6.
Find equivalent fractions with the LCD.
Subtract. Simplify.

TRY THIS

Subtract and simplify.

1. $11\frac{5}{8} - 6\frac{3}{8}$ $5\frac{1}{4}$ **2.** $21\frac{7}{8} - 9$ $12\frac{7}{8}$ **3.** $8\frac{2}{3} - 5\frac{1}{2}$ $3\frac{1}{6}$

PRACTICE

For extra practice, use Making Practice Fun 47 with this lesson.

Subtract and simplify.

1. $4\frac{3}{5} - 2\frac{1}{5}$ $2\frac{2}{5}$ **2.** $8\frac{7}{9} - 2\frac{1}{9}$ $6\frac{2}{3}$ **3.** $6\frac{5}{8} - 3\frac{1}{8}$ $3\frac{1}{2}$ **4.** $9\frac{7}{10} - 4\frac{1}{10}$ $5\frac{3}{5}$

5. $14\frac{5}{6} - 9\frac{5}{6}$ 5 **6.** $23\frac{7}{8} - 8\frac{7}{8}$ 15 **7.** $14\frac{3}{7} - 5\frac{3}{7}$ 9 **8.** $31\frac{2}{5} - 7\frac{2}{5}$ 24

9. $6\frac{3}{5} - 2\frac{1}{2}$ $4\frac{1}{10}$ **10.** $8\frac{2}{5} - 4\frac{1}{3}$ $4\frac{1}{15}$ **11.** $7\frac{2}{3} - 6\frac{1}{2}$ $1\frac{1}{6}$ **12.** $8\frac{4}{5} - 1\frac{1}{4}$ $7\frac{11}{20}$

13. $7\frac{1}{3} - 3\frac{1}{6}$ $4\frac{1}{6}$ **14.** $9\frac{1}{4} - 5\frac{1}{8}$ $4\frac{1}{8}$ **15.** $10\frac{1}{3} - 6\frac{1}{9}$ $4\frac{2}{9}$ **16.** $9\frac{1}{2} - 2\frac{1}{4}$ $7\frac{1}{4}$

17. $5\frac{7}{8} - 3$ $2\frac{7}{8}$ **18.** $9\frac{1}{4} - 2$ $7\frac{1}{4}$ **19.** $5\frac{5}{6} - 4$ $1\frac{5}{6}$ **20.** $9\frac{4}{5} - 7$ $2\frac{4}{5}$

21. $27\frac{4}{5} - 19$ $8\frac{4}{5}$ **22.** $26\frac{5}{8} - 13$ $13\frac{5}{8}$ **23.** $32\frac{3}{4} - 16$ $16\frac{3}{4}$ **24.** $29\frac{1}{6} - 11$ $18\frac{1}{6}$

25. $24\frac{5}{6} - 2\frac{5}{12}$ $22\frac{5}{12}$ **26.** $23\frac{5}{16} - 6\frac{1}{4}$ $17\frac{1}{16}$ **27.** $34\frac{5}{8} - 2\frac{1}{3}$ $32\frac{7}{24}$ **28.** $28\frac{2}{3} - 2\frac{1}{6}$ $26\frac{1}{2}$

29. $9\frac{5}{6} - \frac{1}{6}$ $9\frac{2}{3}$ **30.** $10\frac{3}{4} - \frac{1}{4}$ $10\frac{1}{2}$ **31.** $4\frac{9}{10} - \frac{7}{10}$ $4\frac{1}{5}$ **32.** $8\frac{9}{12} - \frac{5}{12}$ $8\frac{1}{3}$

33. $14\frac{3}{4} - \frac{1}{8}$ $14\frac{5}{8}$ **34.** $12\frac{1}{3} - \frac{1}{4}$ $12\frac{1}{12}$ **35.** $20\frac{5}{8} - \frac{1}{4}$ $20\frac{3}{8}$ **36.** $16\frac{5}{6} - \frac{1}{3}$ $16\frac{1}{2}$

PROBLEM CORNER

1. Joel bicycled $11\frac{7}{10}$ miles on Saturday and $4\frac{1}{10}$ miles on Monday. How much farther did he bicycle on Saturday than on Monday? $7\frac{3}{5}$ km

2. A pattern calls for $3\frac{3}{8}$ yards of fabric and $2\frac{1}{4}$ yards of elastic. How much more fabric is needed than elastic? $1\frac{1}{8}$ yd

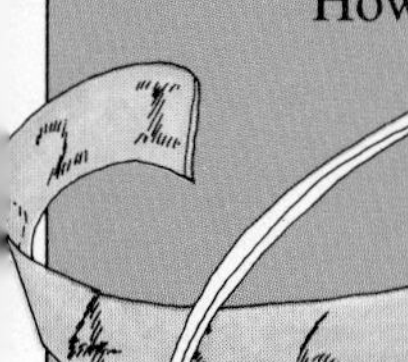

More Practice, page 219

Objective 81

SUBTRACTING WITH RENAMING

Subtract and simplify. $12 - 9\frac{3}{8}$

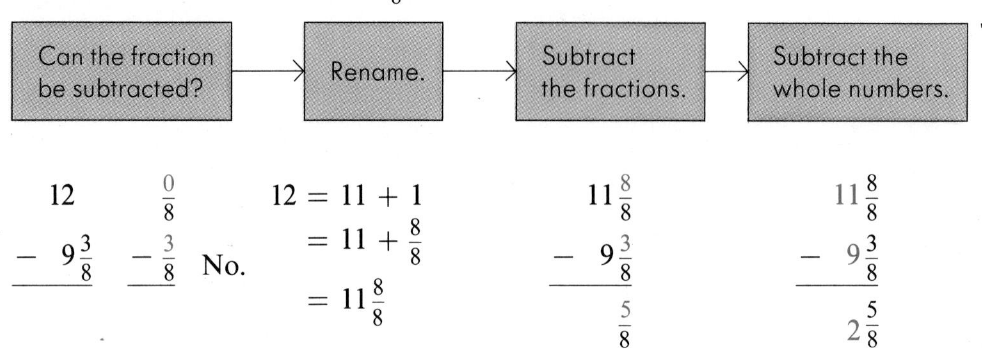

EXAMPLE 1. Subtract and simplify. $24 - 17\frac{2}{5}$

$$\begin{array}{rcl} 24 & = & 23\frac{5}{5} \\ -\ 17\frac{2}{5} & = & 17\frac{2}{5} \\ \hline & & 6\frac{3}{5} \end{array} \qquad \begin{array}{l} 24 = 23 + 1 \\ 24 = 23 + \frac{5}{5} \\ 24 = 23\frac{5}{5} \end{array}$$

TRY THIS

Subtract and simplify.

1. $10 - 3\frac{1}{6}$ $6\frac{5}{6}$ **2.** $21 - 14\frac{5}{8}$ $6\frac{3}{8}$

Subtract and simplify. $7\frac{1}{6} - 2\frac{1}{4}$

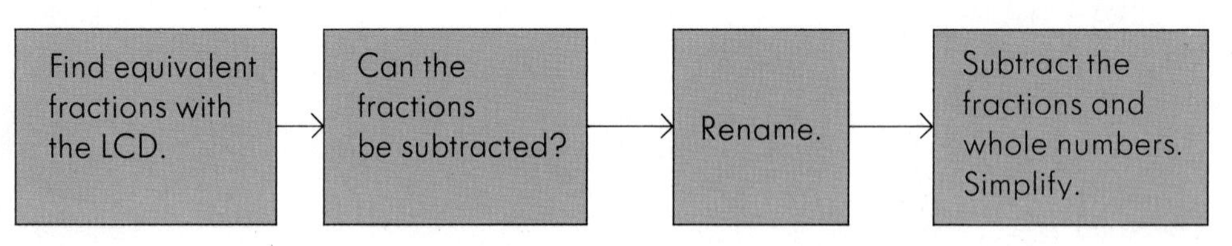

$$\begin{array}{rcl} 7\frac{1}{6} & = & 7\frac{2}{12} \\ -\ 2\frac{1}{4} & = & 2\frac{3}{12} \end{array}$$

The LCD is 12.

No, $\frac{2}{12}$ is smaller than $\frac{3}{12}$.

$$7\frac{2}{12} = 6 + \frac{12}{12} + \frac{2}{12} = 6\frac{14}{12}$$

$$\begin{array}{r} 6\frac{14}{12} \\ -\ 2\frac{3}{12} \\ \hline 4\frac{11}{12} \end{array}$$

EXAMPLE 2. Subtract and simplify. $10\frac{1}{3} - 7\frac{5}{6}$

$$\begin{array}{rcrcr} 10\frac{1}{3} & = & 10\frac{2}{6} & = & 9\frac{8}{6} \\ -\ \ 7\frac{5}{6} & = & 7\frac{5}{6} & = & 7\frac{5}{6} \\ \hline & & & & 2\frac{3}{6} = 2\frac{1}{2} \end{array}$$

Rename. $10\frac{2}{6} = 9 + \frac{6}{6} + \frac{2}{6} = 9\frac{8}{6}$

TRY THIS

Subtract.

3. $5\frac{1}{8} - 2\frac{1}{6}$ $2\frac{23}{24}$

4. $13\frac{1}{12} - 3\frac{3}{4}$ $9\frac{1}{3}$

PRACTICE

For extra practice, use Making Practice Fun 48 with this lesson.
Use Quiz 17 after this Practice.

Subtract and simplify.

1. $5 - 1\frac{1}{3}$ $3\frac{2}{3}$ **2.** $4 - 2\frac{1}{8}$ $1\frac{7}{8}$ **3.** $9 - 1\frac{1}{6}$ $7\frac{5}{6}$ **4.** $6 - 2\frac{1}{5}$ $3\frac{4}{5}$

5. $14 - 8\frac{5}{6}$ $5\frac{1}{6}$ **6.** $17 - 9\frac{4}{9}$ $7\frac{5}{9}$ **7.** $23 - 6\frac{3}{4}$ $16\frac{1}{4}$ **8.** $31 - 4\frac{7}{8}$ $26\frac{1}{8}$

9. $40 - 15\frac{3}{8}$ $24\frac{5}{8}$ **10.** $20 - 16\frac{3}{5}$ $3\frac{2}{5}$ **11.** $30 - 9\frac{7}{10}$ $20\frac{3}{10}$ **12.** $40 - 12\frac{3}{7}$ $27\frac{4}{7}$

13. $8\frac{2}{9} - 2\frac{7}{9}$ $5\frac{4}{9}$ **14.** $4\frac{3}{5} - 2\frac{4}{5}$ $1\frac{4}{5}$ **15.** $4\frac{1}{6} - 1\frac{5}{6}$ $2\frac{1}{3}$ **16.** $5\frac{1}{8} - 2\frac{3}{8}$ $2\frac{3}{4}$

17. $9\frac{1}{4} - 2\frac{1}{2}$ $6\frac{3}{4}$ **18.** $10\frac{2}{9} - 6\frac{1}{3}$ $3\frac{8}{9}$ **19.** $9\frac{1}{8} - 5\frac{1}{4}$ $3\frac{7}{8}$ **20.** $7\frac{1}{6} - 3\frac{5}{6}$ $3\frac{1}{3}$

21. $8\frac{1}{4} - 1\frac{4}{5}$ $6\frac{9}{20}$ **22.** $7\frac{1}{2} - 6\frac{2}{3}$ $\frac{5}{6}$ **23.** $18\frac{1}{6} - 5\frac{3}{4}$ $12\frac{5}{12}$ **24.** $10\frac{5}{6} - 5\frac{3}{8}$ $5\frac{11}{24}$

25. $31 - 19\frac{11}{12}$ $11\frac{1}{12}$ **26.** $18 - 16\frac{2}{9}$ $1\frac{7}{9}$ **27.** $22 - 14\frac{7}{8}$ $7\frac{1}{8}$ **28.** $24 - 13\frac{5}{6}$ $10\frac{1}{6}$

29. $14\frac{1}{8} - \frac{3}{4}$ $13\frac{3}{8}$ **30.** $12\frac{1}{4} - \frac{1}{3}$ $11\frac{11}{12}$ **31.** $20\frac{1}{4} - \frac{5}{8}$ $19\frac{5}{8}$ **32.** $16\frac{1}{3} - \frac{5}{6}$ $15\frac{1}{2}$

33. $28\frac{2}{3} - 2\frac{1}{6}$ $26\frac{1}{2}$ **34.** $34\frac{1}{3} - 2\frac{5}{8}$ $31\frac{17}{24}$ **35.** $23\frac{1}{4} - 7\frac{5}{16}$ $15\frac{15}{16}$ **36.** $20\frac{5}{12} - 2\frac{5}{6}$ $17\frac{7}{12}$

More Practice, page 219

Objective 170

APPLICATION Solving household problems is an extremely important application of mathematics. A great deal of time and money can be saved by being able to deal with such applications.

HOUSEHOLD PROBLEMS

Linda Torres plans to build a bookcase. She makes a sketch to use during the project.

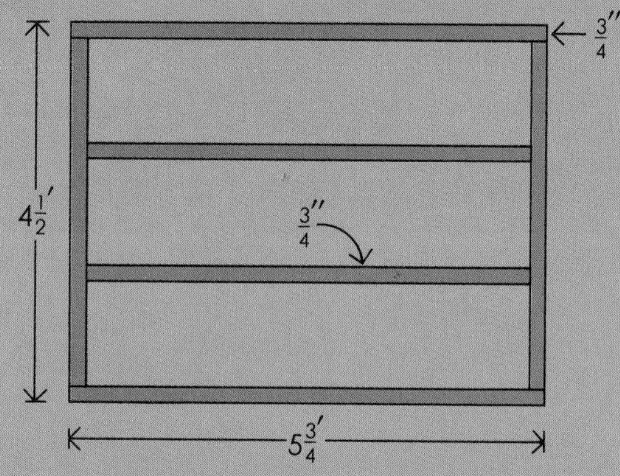

$4\frac{1}{2}'$ means $4\frac{1}{2}$ ft

$\frac{3}{4}''$ means $\frac{3}{4}$ in.

1. How many inches long is each side piece? $52\frac{1}{2}''$
2. How long is each middle shelf? $67\frac{1}{2}''$
3. How many feet of wood will be needed? $31\frac{1}{2}'$
4. How many inches are there between each shelf? $17''$
5. What important measure is left off the drawing? depth
6. To make the bookcase she buys 4 boards, each 10 ft long. They cost \$ 3.25 each. What is the total cost? \$13.00
7. Estimate how much lumber will be left over. $8\frac{1}{2}'$

Pete O'Brien wants to put up a clothesline to save electricity by not using the electric dryer.

How many feet of clothesline are used for each of the following?

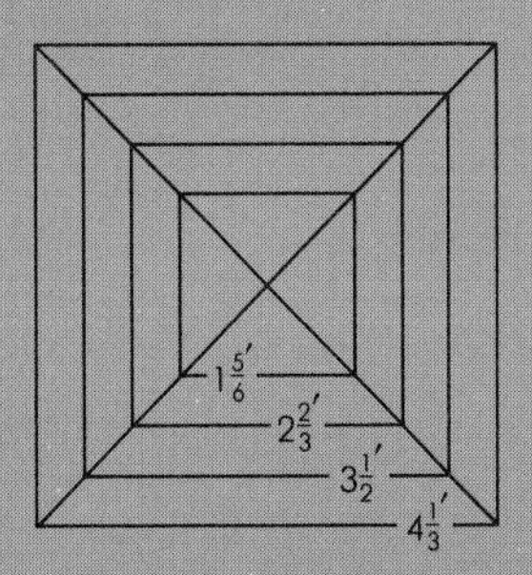

8. the $1\frac{5}{6}'$ square $7\frac{1}{3}'$
9. the $2\frac{2}{3}'$ square $10\frac{2}{3}'$
10. the $3\frac{1}{2}'$ square $14'$
11. the $4\frac{1}{3}'$ square $17\frac{1}{3}'$
12. How many feet of clothesline are used in all? $49\frac{1}{3}'$
13. Clothesline comes in a 50 ft roll. How much of the roll will not be used? $\frac{2}{3}$ ft

The Andersons want to refurnish a room.

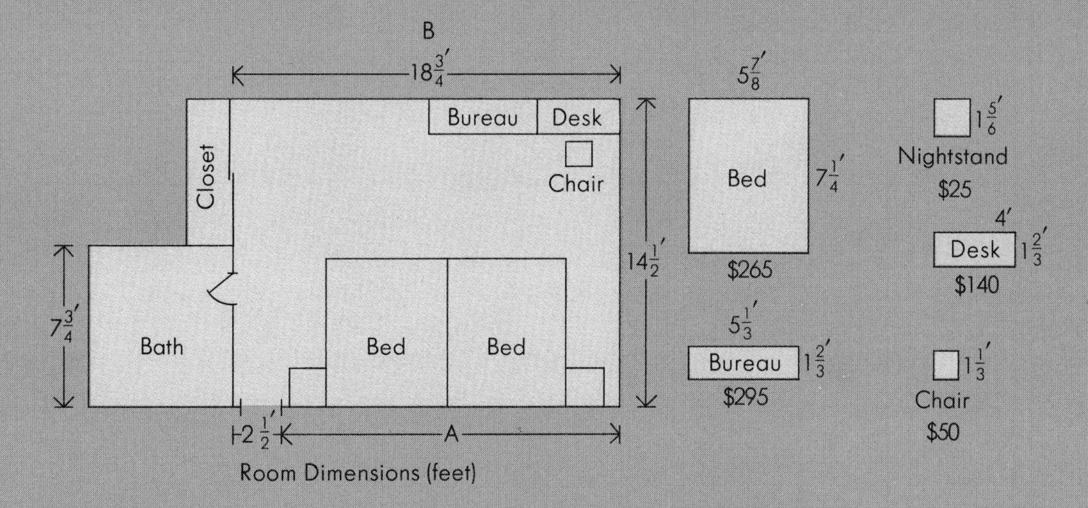

14. What is the length of wall A? $16\frac{1}{4}'$
15. How long is the closet? $6\frac{3}{4}'$
16. How much room do the two beds and nightstands take up along wall A? $15\frac{5}{12}'$
17. How much room is left on wall B after putting in the bureau and desk? $9\frac{5}{12}'$
18. Find the cost of furnishing the room. $1065
19. Curtains and lamps for the room cost $ 525. What is the cost of furnishing the room if these items are included? $1590
20. What is the cost of furnishing the room if only one bed and one nightstand are used? $775

CAREER CAREER CAREER CAREER CAREER

VAN CUSTOMIZING

Mark Ryan owns Custom Van Works. The employees specialize in upholstery, carpentry, auto painting, and body work.

The Wilsons want benches and a table built into their van. Mark's estimate must consider the following.

How Much Will It Cost?		
Materials:	Fabric and Foam	\$70
	Wood	\$20
	Laminated Plastic	\$10
	Trim and Hardware	\$16
Labor:	10 hours at \$15 per hour	

1. How much will the materials cost? \$116
2. How much will the labor cost? \$150
3. What is the total cost of the job to the Wilsons? \$266
4. Mark makes a profit of $\frac{1}{4}$ on the cost of the materials. How much is this? \$29
5. Mark pays his workers \$ 8 per hour. What profit does he make on the labor charges? \$70
6. What is Mark's total profit? \$99

The various types of furniture Mark puts in a van require different amounts of fabric.

Captain's Chair	$\frac{1}{5}$ bolt
Armless Chair	$\frac{1}{6}$ bolt
Double Sofa Seat	$\frac{5}{6}$ bolt
Sofa Bed	$\frac{7}{10}$ bolt

How many bolts of fabric would be needed for each job?

7. 1 captain's chair
   1 armless chair
   1 double sofa seat
   1 sofa bed $1\frac{9}{10}$

8. 2 captain's chairs
   2 sofa beds $1\frac{4}{5}$

9. 1 captain's chair
   2 armless chairs
   2 double sofa seats $1\frac{14}{15}$

10. 1 captain's chair
    3 armless chairs
    1 sofa bed $1\frac{2}{5}$

Mark charges $ 0.72 per square foot to put plain sun screen film on windows. How much would each of these jobs cost?

11. 1 rear window, 8.3 square feet $5.98
12. 2 rear windows, 3.4 square feet each $4.90
13. 2 side windows: 1 is 6.8 square feet, and the other is 9.4 square feet $11.66

Mark must know how long a job will take. Employees who refinish paneling work at different rates according to their skill.

Jan took 3 hours to refinish 216 square feet of paneling. She was paid $ 5 per hour. Chiang took 4 hours to refinish 384 square feet of paneling. He was paid $ 6.75 per hour.

14. How much did Mark pay each person? Jan: $15; Chiang: $27
15. At that rate, how many square feet can each person do in 1 hour? Jan: 72; Chiang: 96
16. To the nearest cent, what is Mark's cost per square foot to have each employee refinish paneling? Jan: 7¢; Chiang: 7¢

CHAPTER 7 REVIEW

For extra practice, use Making Practice Fun 49 with this lesson.

Add and simplify. p. 196

1. $\frac{1}{4} + \frac{7}{4}$ 2
2. $\frac{5}{8} + \frac{1}{8}$ $\frac{3}{4}$
3. $\frac{7}{10} + \frac{9}{10} + \frac{6}{10}$ $2\frac{1}{5}$

Find the LCD. p. 198

4. $\frac{4}{5}, \frac{7}{10}$ 10
5. $\frac{5}{6}, \frac{7}{8}$ 24
6. $\frac{4}{9}, \frac{2}{3}, \frac{1}{6}$ 18

Add and simplify. p. 200

7. $\frac{5}{6} + \frac{7}{8}$ $1\frac{17}{24}$
8. $\frac{4}{5} + \frac{1}{10}$ $\frac{9}{10}$
9. $\frac{2}{3} + \frac{3}{4} + \frac{1}{6}$ $1\frac{7}{12}$

Add and simplify. p. 202

10. $5\frac{1}{8} + 4\frac{1}{2}$ $9\frac{5}{8}$
11. $19\frac{3}{4} + 8$ $27\frac{3}{4}$
12. $16\frac{1}{4} + 7\frac{2}{3}$ $23\frac{11}{12}$
13. $5\frac{1}{3} + 18\frac{1}{2} + 4\frac{5}{6}$ $28\frac{2}{3}$
14. $\frac{5}{8} + 4\frac{1}{3}$ $4\frac{23}{24}$
15. $6\frac{3}{4} + 8\frac{3}{16}$ $14\frac{15}{16}$

Subtract and simplify. p. 206

16. $\frac{8}{5} - \frac{3}{5}$ 1
17. $\frac{7}{8} - \frac{1}{12}$ $\frac{19}{24}$
18. $\frac{87}{100} - \frac{3}{10}$ $\frac{57}{100}$

Subtract and simplify. pp. 208–210

19. $9\frac{7}{10} - 2\frac{1}{10}$ $7\frac{3}{5}$
20. $12\frac{7}{8} - 7$ $5\frac{7}{8}$
21. $8\frac{3}{4} - 7\frac{1}{2}$ $1\frac{1}{4}$
22. $15\frac{1}{4} - 3\frac{1}{2}$ $11\frac{3}{4}$
23. $32 - 8\frac{2}{5}$ $23\frac{3}{5}$
24. $17\frac{3}{8} - 9\frac{5}{8}$ $7\frac{3}{4}$

Application pp. 204, 212

25. Les drank the following amounts of milk one day. How much did he drink in all? $6\frac{11}{12}$ c

Home	School	Restaurant
$2\frac{3}{4}$ cups	$1\frac{2}{3}$ cups	$2\frac{1}{2}$ cups

26. How much clothesline would it take to make a rectangle $8\frac{1}{2}$ yds by $7\frac{3}{4}$ yds? $32\frac{1}{2}$ yds

CHAPTER 7 TEST

Add and simplify.

1. $\frac{2}{3} + \frac{7}{3}$ 3

2. $\frac{6}{12} + \frac{2}{12}$ $\frac{2}{3}$

3. $\frac{7}{8} + \frac{5}{8} + \frac{2}{8}$ $1\frac{3}{4}$

Find the LCD.

4. $\frac{5}{6}, \frac{2}{3}$ 6

5. $\frac{7}{12}, \frac{1}{8}$ 24

6. $\frac{3}{4}, \frac{1}{2}, \frac{5}{6}$ 12

Add and simplify.

7. $\frac{5}{6} + \frac{2}{3}$ $1\frac{1}{2}$

8. $\frac{7}{12} + \frac{1}{8}$ $\frac{17}{24}$

9. $\frac{1}{3} + \frac{3}{4} + \frac{1}{8}$ $1\frac{5}{24}$

Add and simplify.

10. $10\frac{1}{6} + 2\frac{1}{3}$ $12\frac{1}{2}$

11. $14\frac{2}{3} + 9$ $23\frac{2}{3}$

12. $4\frac{1}{6} + 6\frac{5}{8}$ $10\frac{19}{24}$

13. $7\frac{1}{5} + 4\frac{1}{2} + 8\frac{3}{10}$ 20

14. $\frac{3}{4} + 7\frac{1}{5}$ $7\frac{19}{20}$

15. $9\frac{2}{3} + 6\frac{5}{12}$ $16\frac{1}{12}$

Subtract and simplify.

16. $\frac{7}{8} - \frac{3}{8}$ $\frac{1}{2}$

17. $\frac{7}{8} - \frac{1}{4}$ $\frac{5}{8}$

18. $\frac{8}{9} - \frac{5}{6}$ $\frac{1}{18}$

Subtract and simplify.

19. $11\frac{5}{6} - 8\frac{1}{6}$ $3\frac{2}{3}$

20. $12\frac{3}{10} - 9$ $3\frac{3}{10}$

21. $7\frac{4}{9} - 2\frac{1}{3}$ $5\frac{1}{9}$

22. $10\frac{2}{3} - 8\frac{4}{5}$ $1\frac{13}{15}$

23. $29 - 5\frac{3}{8}$ $23\frac{5}{8}$

24. $14\frac{3}{10} - 6\frac{9}{10}$ $7\frac{2}{5}$

Application

25. Selma drank the following amounts of milk one day. How much did she drink in all? $7\frac{1}{8}$ c

Home	School	Restaurant
$3\frac{1}{8}$ cups	$2\frac{1}{4}$ cups	$1\frac{3}{4}$ cups

26. How much clothesline would it take to make a square $5\frac{2}{3}$ ft on a side? $22\frac{2}{3}$ ft

MORE PRACTICE

Add and simplify. Use after page 197.

1. $\frac{1}{8}+\frac{7}{8}$ 1 **2.** $\frac{5}{10}+\frac{2}{10}$ $\frac{7}{10}$ **3.** $\frac{2}{9}+\frac{4}{9}$ $\frac{2}{3}$ **4.** $\frac{9}{12}+\frac{15}{12}$ 2

5. $\frac{2}{7}+\frac{9}{7}$ $1\frac{4}{7}$ **6.** $\frac{3}{5}+\frac{9}{5}$ $2\frac{2}{5}$ **7.** $\frac{8}{3}+\frac{7}{3}$ 5 **8.** $\frac{5}{4}+\frac{1}{4}$ $1\frac{1}{2}$

Find the LCD. Use after page 199.

9. $\frac{1}{3}, \frac{1}{5}$ 15 **10.** $\frac{3}{4}, \frac{2}{5}$ 20 **11.** $\frac{5}{8}, \frac{2}{3}$ 24 **12.** $\frac{7}{10}, \frac{1}{3}$ 30

13. $\frac{1}{2}, \frac{3}{4}$ 4 **14.** $\frac{5}{6}, \frac{1}{3}$ 6 **15.** $\frac{3}{10}, \frac{2}{5}$ 10 **16.** $\frac{5}{6}, \frac{7}{12}$ 12

17. $\frac{3}{4}, \frac{1}{6}, \frac{2}{3}$ 12 **18.** $\frac{7}{8}, \frac{5}{6}, \frac{1}{2}$ 24 **19.** $\frac{3}{8}, \frac{5}{12}, \frac{2}{9}$ 72 **20.** $\frac{7}{10}, \frac{3}{100}, \frac{41}{50}$ 100

Add and simplify. Use after page 201.

21. $\frac{1}{3}+\frac{1}{5}$ $\frac{8}{15}$ **22.** $\frac{3}{4}+\frac{2}{5}$ $1\frac{3}{20}$ **23.** $\frac{5}{8}+\frac{2}{3}$ $1\frac{7}{24}$ **24.** $\frac{7}{10}+\frac{1}{3}$ $1\frac{1}{30}$

25. $\frac{1}{2}+\frac{3}{4}$ $1\frac{1}{4}$ **26.** $\frac{1}{6}+\frac{1}{3}$ $\frac{1}{2}$ **27.** $\frac{7}{10}+\frac{2}{5}$ $1\frac{1}{10}$ **28.** $\frac{5}{6}+\frac{7}{12}$ $1\frac{5}{12}$

29. $\frac{3}{4}+\frac{1}{6}$ $\frac{11}{12}$ **30.** $\frac{7}{8}+\frac{5}{6}$ $1\frac{17}{24}$ **31.** $\frac{3}{8}+\frac{5}{12}$ $\frac{19}{24}$ **32.** $\frac{7}{10}+\frac{5}{100}$ $\frac{3}{4}$

Add and simplify. Use after page 203.

33. $13\frac{2}{3}+9$ $22\frac{2}{3}$ **34.** $15\frac{2}{3}+7\frac{3}{4}$ $23\frac{5}{12}$ **35.** $8\frac{1}{2}+9\frac{1}{4}$ $17\frac{3}{4}$ **36.** $5\frac{3}{4}+7\frac{3}{10}$ $13\frac{1}{20}$

37. $12\frac{2}{5}+8\frac{1}{10}$ $20\frac{1}{2}$ **38.** $8\frac{1}{6}+9\frac{1}{8}$ $17\frac{7}{24}$ **39.** $5\frac{7}{8}+6\frac{3}{4}$ $12\frac{5}{8}$ **40.** $17\frac{7}{10}+6\frac{3}{10}$ 24

41. $18+5\frac{5}{9}$ $23\frac{5}{9}$ **42.** $6\frac{2}{3}+5\frac{2}{3}$ $12\frac{1}{3}$ **43.** $7\frac{1}{2}+9$ $16\frac{1}{2}$ **44.** $14\frac{4}{5}+8\frac{1}{3}$ $23\frac{2}{15}$

45. $1\frac{2}{3}+9\frac{5}{6}+7\frac{1}{2}$ 19 **46.** $4\frac{1}{5}+10\frac{3}{4}+9\frac{1}{2}$ $24\frac{9}{20}$ **47.** $7\frac{1}{10}+6\frac{2}{5}+14\frac{3}{4}$ $28\frac{1}{4}$

48. $21\frac{1}{3}+9\frac{5}{6}+12\frac{1}{2}$ $43\frac{2}{3}$ **49.** $14\frac{1}{6}+9\frac{2}{3}+6\frac{3}{4}$ $30\frac{7}{12}$ **50.** $16\frac{4}{5}+2\frac{1}{6}+8\frac{3}{10}$ $27\frac{4}{15}$

Subtract and simplify. Use after page 207.

1. $\frac{7}{8} - \frac{1}{8}$ $\frac{3}{4}$	**2.** $\frac{11}{12} - \frac{1}{12}$ $\frac{5}{6}$	**3.** $\frac{3}{5} - \frac{1}{2}$ $\frac{1}{10}$	**4.** $\frac{2}{3} - \frac{1}{4}$ $\frac{5}{12}$
5. $\frac{3}{4} - \frac{1}{6}$ $\frac{7}{12}$	**6.** $\frac{7}{10} - \frac{3}{5}$ $\frac{1}{10}$	**7.** $\frac{5}{12} - \frac{3}{8}$ $\frac{1}{24}$	**8.** $\frac{8}{9} - \frac{5}{6}$ $\frac{1}{18}$
9. $\frac{7}{12} - \frac{1}{10}$ $\frac{29}{60}$	**10.** $\frac{5}{6} - \frac{3}{8}$ $\frac{11}{24}$	**11.** $\frac{7}{10} - \frac{3}{100}$ $\frac{67}{100}$	**12.** $\frac{5}{16} - \frac{1}{12}$ $\frac{11}{48}$
13. $\frac{4}{5} - \frac{3}{4}$ $\frac{1}{20}$	**14.** $\frac{7}{9} - \frac{2}{3}$ $\frac{1}{9}$	**15.** $\frac{4}{7} - \frac{1}{2}$ $\frac{1}{14}$	**16.** $\frac{5}{8} - \frac{3}{5}$ $\frac{1}{40}$

Subtract and simplify. Use after page 209.

17. $8\frac{7}{8} - 5\frac{5}{8}$ $3\frac{1}{4}$	**18.** $11\frac{9}{10} - 7\frac{3}{10}$ $4\frac{3}{5}$	**19.** $15\frac{4}{5} - 8$ $7\frac{4}{5}$	**20.** $12\frac{5}{6} - 11$ $1\frac{5}{6}$
21. $13\frac{5}{6} - 7\frac{1}{3}$ $6\frac{1}{2}$	**22.** $12\frac{9}{10} - 4\frac{2}{5}$ $8\frac{1}{2}$	**23.** $11\frac{3}{8} - 4\frac{1}{6}$ $7\frac{5}{24}$	**24.** $15\frac{8}{9} - 11\frac{5}{6}$ $4\frac{1}{18}$
25. $14\frac{3}{4} - 6$ $8\frac{3}{4}$	**26.** $11\frac{2}{3} - 7$ $4\frac{2}{3}$	**27.** $12\frac{5}{6} - 5\frac{1}{6}$ $7\frac{2}{3}$	**28.** $16\frac{11}{12} - 9\frac{5}{12}$ $7\frac{1}{2}$
29. $21\frac{1}{2} - 18\frac{1}{3}$ $3\frac{1}{6}$	**30.** $17\frac{5}{6} - 9\frac{3}{4}$ $12\frac{1}{12}$	**31.** $8\frac{5}{9} - 3\frac{1}{4}$ $5\frac{11}{36}$	**32.** $23\frac{6}{7} - 15\frac{5}{14}$ $8\frac{1}{2}$
33. $32\frac{4}{5} - 6\frac{2}{15}$ $26\frac{2}{3}$	**34.** $15\frac{7}{8} - 4\frac{1}{2}$ $11\frac{3}{8}$	**35.** $29\frac{7}{10} - 17\frac{9}{20}$ $12\frac{1}{4}$	**36.** $10\frac{7}{9} - 4\frac{1}{6}$ $6\frac{11}{18}$

Subtract and simplify. Use after page 211.

37. $7 - 2\frac{1}{4}$ $4\frac{3}{4}$	**38.** $10 - 5\frac{1}{8}$ $4\frac{7}{8}$	**39.** $20 - 4\frac{7}{10}$ $15\frac{3}{10}$	**40.** $52 - 39\frac{5}{6}$ $12\frac{1}{6}$
41. $10\frac{1}{6} - 3\frac{5}{6}$ $6\frac{1}{3}$	**42.** $12\frac{3}{8} - 4\frac{5}{8}$ $7\frac{3}{4}$	**43.** $24\frac{1}{10} - 13\frac{9}{10}$ $10\frac{1}{5}$	**44.** $18\frac{5}{12} - 3\frac{7}{12}$ $14\frac{5}{6}$
45. $50\frac{1}{2} - 37\frac{3}{4}$ $12\frac{3}{4}$	**46.** $11\frac{1}{2} - 8\frac{3}{5}$ $2\frac{9}{10}$	**47.** $9\frac{1}{8} - 7\frac{1}{6}$ $1\frac{23}{24}$	**48.** $12\frac{1}{10} - 8\frac{3}{4}$ $3\frac{7}{20}$
49. $12 - 5\frac{2}{3}$ $6\frac{1}{3}$	**50.** $21 - 8\frac{1}{2}$ $12\frac{1}{2}$	**51.** $30 - 7\frac{3}{8}$ $22\frac{5}{8}$	**52.** $16 - 6\frac{7}{11}$ $9\frac{4}{11}$
53. $10\frac{2}{7} - 4\frac{3}{7}$ $5\frac{6}{7}$	**54.** $15\frac{1}{3} - 9\frac{2}{3}$ $5\frac{2}{3}$	**55.** $8\frac{1}{5} - 3\frac{4}{5}$ $4\frac{2}{5}$	**56.** $20\frac{2}{9} - 15\frac{8}{9}$ $4\frac{1}{3}$
57. $14\frac{2}{3} - 7\frac{7}{8}$ $6\frac{19}{24}$	**58.** $10\frac{1}{6} - 3\frac{7}{12}$ $6\frac{7}{12}$	**59.** $28\frac{1}{2} - 8\frac{2}{9}$ $20\frac{5}{18}$	**60.** $17\frac{3}{5} - 6\frac{2}{3}$ $10\frac{14}{15}$

CALCULATOR CORNER

A STRANGE CALCULATOR

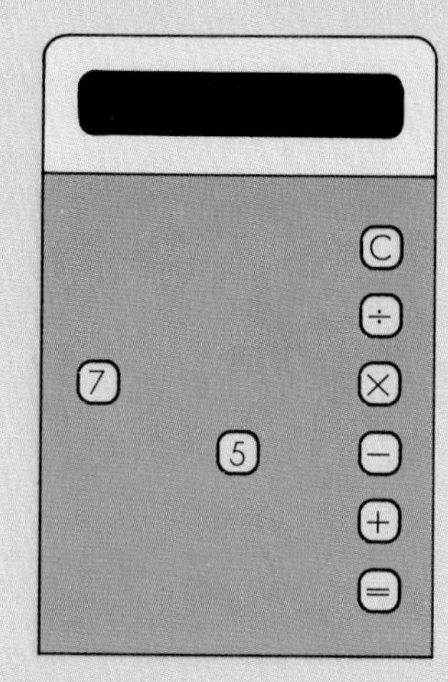

This strange calculator has only the digits [5] and [7]. It also has [C], [=], [+], [−], [×], and [÷].

For this exercise we are assuming that we have a calculator with a constant addition mode. That is, if we press [+] [5] [=] [=], the calculator continues to add 5 for each [=] pressed. If your students do not have such calculators, they will have to press [+] [5] [=] [+] [5] [=].

Try these on your calculator.

1. To get a 6 on the display press [7] [÷] [7] [+] [5] [=]
2. To get a 9 on the display press [5] [÷] [5] [=] [+] [=] [+] [7] [=]
3. To get an 80 on the display press [7] [×] [5] [=] [+] [=] [+] [5] [=] [=]

Use the strange calculator described above. Tell how to get these numbers on the display. Write down your steps.

1. 30 2. 16 3. 50 4. 90

Suppose you had a calculator with only [4], [6], [C], [+], [−], [×], [÷] and [=].

Try these on your calculator.

1. To get a 5 on the display press [6] [÷] [6] [+] [4] [=]
2. To get 0.5 on the display press [6] [÷] [6] [+] [=] [÷] [4] [=]
3. To get 100 on the display press [6] [+] [4] [=] [×] [=]

Use this calculator for the following. Tell how to get these numbers on the display. Write down your steps.

1. 22 2. 34 3. 94 4. 0.25

CHAPTER 8 PRE-TEST

Write the ratio. Simplify, if possible. p. 224

1. 6 problems in 4 minutes $\frac{3}{2}$

2. 4 green apples out of a total of 7 apples $\frac{4}{7}$

3. 2 blue marbles, 3 red marbles $\frac{2}{3}$

4. 9 hits in 15 times at bat $\frac{3}{5}$

Find the unit rate. p. 226

5. A man walks 8 miles in 2 hours. How many miles per hour is this? 4 mi/hr

6. If a car travels 96 kilometers on 8 liters of gas, how many kilometers per liter is this? 12 km/L

7. If 5 items cost 74¢, what is the cost per item? Round to tenths. 14.8¢

8. If a car travels 419 kilometers in 8 hours, how many kilometers per hour is this? Round to tenths. 52.4 km/hr

Tell whether each proportion is true. p. 230

9. $\frac{12}{15} = \frac{6}{7}$ no

10. $\frac{5}{6} = \frac{15}{18}$ yes

Solve. p. 232

11. $\frac{2}{3} = \frac{12}{x}$ 18

12. $\frac{9}{8} = \frac{x}{24}$ 27

13. $\frac{3}{7} = \frac{4}{x}$ $9\frac{1}{3}$

14. $\frac{7}{5} = \frac{x}{8}$ 11.2

15. If 5 items cost 85¢, how much would 9 items cost? $1.53

16. If a bicycle travels 45 kilometers in 9 hours, how many kilometers would be covered in 4 hours? 20 kilometers

The figures are similar. Find x. p. 234

13.5 cm

17.

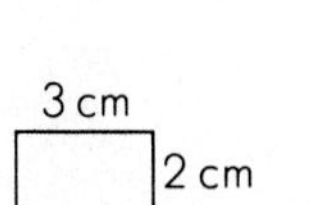

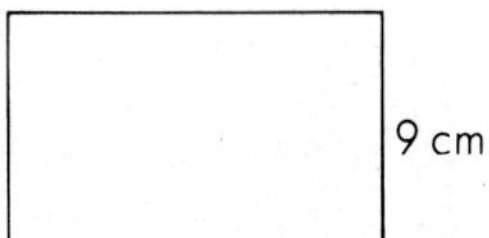

15 ft

18.

16 ft
20 ft
12 ft
x

Application pp. 228, 236

19. Two liters of milk cost $ 0.95. Four liters cost $ 1.83. Which is the more economical buy? 4 liters

20. A pack of 3 cassette tapes costs $ 8.99. The shipping cost is $ 1.08. What is the total cost? $10.07

8
RATIO AND PROPORTION

GETTING STARTED

THE GREATEST RATIO

WHAT'S NEEDED The 30 squares of paper with digits marked on them as in GOTCHA, Chapter 1. Take out the 0's. Leave the other 27 digits in the box.

THE RULES

1. On a sheet of paper each player writes the following. __ __ __ __ to __
2. A leader draws a digit from the box. Each player writes the digit on any one of the blanks. Once it is written it cannot be changed or moved.
3. The leader draws four more digits, one at a time. Write each digit on a blank as it is drawn.
4. Each player now has a ratio.
5. Who has the greatest ratio? Find out by dividing the last digit into the first four.

SAMPLE 7 6 4 5 to 3

```
     2548.3
  3)7645.0
    6
    16
    15
     14
     12
      25
      24
       10
        9
        1
```

The player with the largest quotient has the greatest ratio and is the winner.

Objectives 82 and 83

RATIOS

The ratio of wins to losses for San Francisco is 57 to 38. We write the ratio as $\frac{57}{38}$.

The ratio of losses to wins for Cincinnati is 39 to 54. We write the ratio as $\frac{39}{54}$.

The ratio of wins to games played (wins plus losses) for Los Angeles is 54 to 94. We write $\frac{54}{94}$.

National League West

	Won	Lost
San Francisco	57	38
Cincinnati	54	39
Los Angeles	54	40

EXAMPLES. Write the ratio.

1. pennies to dimes $\frac{4}{3}$ (4 to 3)

2. dimes to pennies $\frac{3}{4}$ (3 to 4)

3. pennies to coins $\frac{4}{7}$ (4 to 7)

TRY THIS

Write the ratio.

1. Circles to squares $\frac{2}{3}$

2. Squares to circles $\frac{3}{2}$

3. Squares to figures $\frac{3}{5}$

EXAMPLES. Write the ratio.

4. 7 problems in 5 minutes $\frac{7}{5}$

5. 17 breaths per 60 seconds $\frac{17}{60}$

6. 3 apples for 78¢ $\frac{3}{78} = \frac{1}{26}$

7. 210 people on 7 buses $\frac{210}{7} = \frac{30}{1}$ *Leave 1 as the denominator.*

TRY THIS

Write the ratio. Simplify, if possible.

4. 12 miles in 5 hours $\frac{12}{5}$

5. 72 heartbeats per 60 seconds $\frac{72}{60} = \frac{6}{5}$

6. 25 pennies for 5 nickels $\frac{5}{1}$

7. 8 hits in 32 times at bat $\frac{1}{4}$

PRACTICE

Write the ratio.

Boston: Won 62, Lost 28

1. Wins to losses $\frac{62}{28}$

2. Losses to wins $\frac{28}{62}$

3. Wins to games played $\frac{62}{90}$

New York: Won 48, Lost 38

4. Wins to losses $\frac{48}{38}$

5. Losses to wins $\frac{38}{48}$

6. Losses to games played $\frac{38}{86}$

4 dogs, 3 cats

7. Dogs to cats $\frac{4}{3}$

8. Cats to dogs $\frac{3}{4}$

9. Dogs to animals $\frac{4}{7}$

3 dimes, 5 nickels

10. Dimes to nickels $\frac{3}{5}$

11. Nickels to dimes $\frac{5}{3}$

12. Nickels to coins $\frac{5}{8}$

Write the ratio. Simplify, if possible.

13. 210 kilometers in 3 hours $\frac{70}{1}$

14. 2 parts hydrogen for 1 part oxygen $\frac{2}{1}$

15. 200 books on 5 shelves $\frac{40}{1}$

16. 2 eggs per 3 servings $\frac{2}{3}$

17. 8 hours sleep in 24 hours $\frac{1}{3}$

18. 2 teachers for 50 students $\frac{1}{25}$

19. 27 hits in 100 times at bat $\frac{27}{100}$

20. 2000 grams in 2 kilograms $\frac{1000}{1}$

21. 100 miles on 5 gallons of gas $\frac{20}{1}$

22. 10 nickels for 2 quarters $\frac{5}{1}$

23. 17 right out of 20 problems $\frac{17}{20}$

24. 3 wrong out of 20 problems $\frac{3}{20}$

More Practice, page 242

Objective 84

UNIT RATES

A **rate** is a ratio used to compare two measures. A **unit rate** is the rate for (per) one.

Maria drives 162 miles in 3 hours. How many miles per hour does she travel?

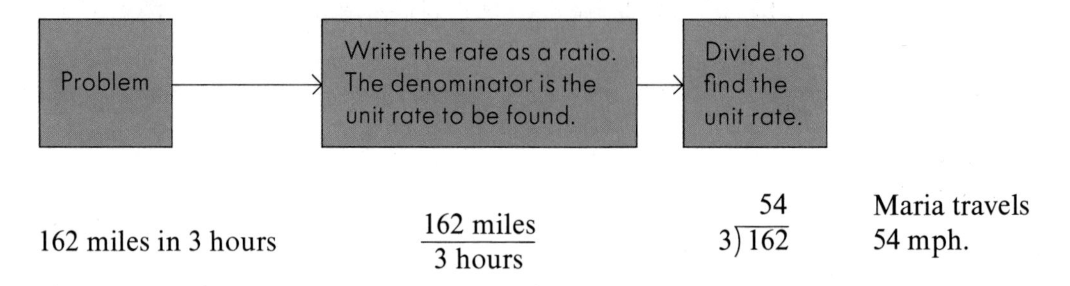

162 miles in 3 hours $\qquad \dfrac{162 \text{ miles}}{3 \text{ hours}} \qquad 3\overline{)162}$ = 54 $\qquad$ Maria travels 54 mph.

EXAMPLE 1. A car travels 272 kilometers on 34 liters of gas. How many kilometers per liter is this?

$\dfrac{272 \text{ km}}{34 \text{ L}} \qquad 34\overline{)272}$ = 8

Write the rate. Divide.
The unit rate is 8 kilometers per liter.

EXAMPLE 2. A car travels 272 kilometers on 34 liters of gas. How many liters per kilometer is this?

$\dfrac{34 \text{ L}}{272 \text{ km}}$

```
      0.1 2 ≈ 0.1
2 7 2)3 4.0 0
      2 7 2
        6 8 0
        5 4 4
        1 3 6
```

Write the rate.
Divide. Round to tenths.
The car uses 0.1 liter per kilometer.

The unit rate for money is often called the unit cost or unit price.

EXAMPLE 3. 30 tablets cost 80 cents. Find the cost per tablet.

$\dfrac{80¢}{30 \text{ tablets}}$

```
      2.6 6 ≈ 2.7
3 0)8 0.0 0
    6 0
    2 0 0
    1 8 0
      2 0 0
      1 8 0
        2 0
```

Write the rate.
Divide. Round to tenths.
The cost is 2.7¢ per tablet.

TRY THIS

Find the unit rate.

1. A car travels 98 miles on 5 gallons of gas. How many miles per gallon is this? Round to the nearest tenth. 19.6 miles/gallon
2. 7 oranges cost 84¢. What is the unit price? 12¢
3. Pierre receives $ 26.75 for 5 hours of work. What is his hourly wage? $5.35 per hour

PRACTICE

For extra practice, use Making Practice Fun 50 with this lesson.

Find the unit rate.

1. Kay drives 1000 kilometers in 20 hours. How many kilometers per hour does she drive? 50 km/hr
2. A car travels 96 miles on 6 gallons of gas. How many miles per gallon is this? 16 mi/gal
3. 10 pounds of grapes cost $ 3.75. What is the unit price? Round to the nearest tenth. $0.38
4. Charlie sprints 100 meters in 16 seconds. How many meters per second is this? Round to the nearest tenth. 6.3 m/sec
5. John sprints 100 yards in 10 seconds. How many yards per second is this? 10 yd/sec
6. John sprints 100 yards in 10 seconds. How many seconds per yard is this? Round to the nearest tenth. 0.1 sec/yd
7. Miko serves 5 kilograms of hamburger to 15 people. How many kilograms per person is this? Round to the nearest tenth. 0.3 kg per person
8. Miko serves 5 kilograms of hamburger to 15 people. How many people per kilogram is this? 3 people per kg
9. If 5 apples cost 94¢, how much does 1 apple cost? Round to the nearest tenth. 18.8¢
10. If 3 ties cost $ 28.74, how much does 1 tie cost? $9.58
11. Lori receives $ 29.95 for working 9 hours. What is her hourly wage? $3.33/hr
12. Jeff reads 102 pages in 4 hours. How many pages per hour is this? Round to the nearest tenth. 25.5 pages/hr

More Practice, page 242

Objective 171

APPLICATION

Discuss with your students the reasons why comparison shopping is an important skill. Have them suggest other ways in which items could be compared.

COMPARISON SHOPPING

We use unit prices to find the most economical buy.

Which package of rice is the more economical: the 2 kg package or the 0.5 kg package? To find the unit price, divide the price by the amount.

2 kg $\longrightarrow$ \$ 1.89 $\div$ 2 = \$ 0.95 per kg 0.5 kg $\longrightarrow$ \$ 0.49 $\div$ 0.5 = \$ 0.98 per kg

The 2 kg package is the more economical.

Which is the more economical buy on rice?

1. the 12.5 kg package or the 5 kg package 12.5 kg package
2. the 5 kg package or the 2 kg package 5 kg package
3. the 12.5 kg package or the 0.5 kg package 12.5 kg package

What is the unit price of each liquid detergent? Round to the nearest tenth.

4. 12 oz for 79¢ 6.6¢
5. 16 oz for $ 1.09 6.8¢
6. 22 oz for $ 1.39 6.3¢
7. 48 oz for $ 2.49 5.2¢
8. Which size is the most economical? 48 oz for $2.49
9. Which is the least economical? 16 oz for $1.09

Sometimes the unit price is the cost per unit of prepared product.

10. Find the cost per prepared quart of each lemon drink. 29¢; 13¢; 20¢
11. Assume the quality is equal. Which is the best buy? Lemon Drink Mix

12. Find the cost per liter for each batter. 89¢; 93¢
13. If the quality is equal, which is the better buy? Pancake Mix

14. Find the cost per ounce of each soup. 4¢; 5¢
15. How many ounces of soup can be made from the condensed soup? 22 ounces
16. From which can would you get the greatest number of ounces of prepared soup? Condensed Soup
17. Find the cost per prepared ounce of each soup. 2¢; 5¢
18. If the quality is equal, which is the better buy? Condensed Soup

Objectives 85 and 86

PROPORTIONS

A **proportion** is a sentence that says two ratios are equal. If the cross products are the same, the proportion is true.

Tell whether the proportion is true. $\frac{2}{3} = \frac{6}{9}$

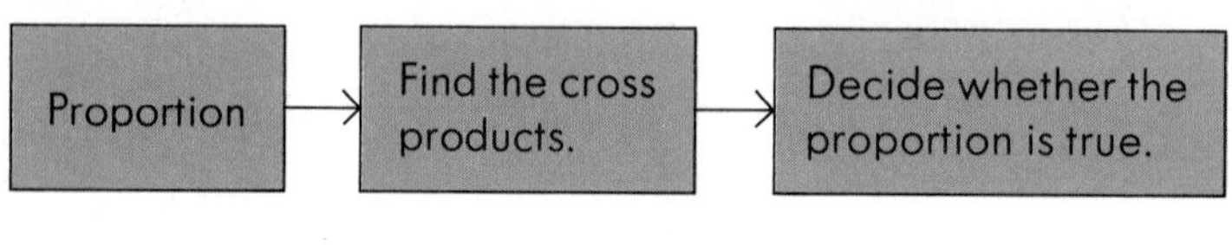

$\frac{2}{3} = \frac{6}{9}$ $\qquad$ $\frac{2}{3} = \frac{6}{9}$

$2 \cdot 9 = 18$
$3 \cdot 6 = 18$
$18 = 18$
The proportion is true.

The ratios 2 to 3 and 6 to 9 are equal. We say 2 is to 3 as 6 is to 9.

EXAMPLES. Tell whether each proportion is true.

1. $\frac{6}{8} = \frac{3}{4}$

Find the cross products.
$6 \cdot 4 = 24$
$8 \cdot 3 = 24$
$24 = 24$ *The proportion is true.*

2. $\frac{4}{6} = \frac{8}{10}$

$4 \cdot 10 = 40$
$6 \cdot 8 = 48$
$40 \neq 48$ *The proportion is not true.*

TRY THIS

Tell whether each proportion is true.

1. $\frac{1}{2} = \frac{2}{3}$ no **2.** $\frac{5}{10} = \frac{2}{4}$ yes

EXAMPLE 3. Solve. $\frac{5}{8} = \frac{10}{x}$

$5 \cdot x = 8 \cdot 10$ *Find the cross products.*
$5x = 80$
$x = \frac{80}{5}$ *Since 80 is 5 times x, we can find x by dividing by 5.*
$x = 16$ *Solve.*

EXAMPLES. Solve. You can write the solution as a mixed number or a decimal.

4. $\frac{x}{5} = \frac{2}{3}$

$x \cdot 3 = 5 \cdot 2$

$3x = 10$

$x = \frac{10}{3} = 3\frac{1}{3}$

5. $\frac{x}{7} = \frac{3}{4}$

$x \cdot 4 = 7 \cdot 3$

$4x = 21$

$x = \frac{21}{4} = 5\frac{1}{4}$ or 5.25

TRY THIS

Solve.

3. $\frac{3}{5} = \frac{12}{x}$ 20

4. $\frac{1}{2} = \frac{x}{5}$ $2\frac{1}{2}$

5. $\frac{x}{4} = \frac{7}{6}$ $4\frac{2}{3}$

PRACTICE

For extra practice, use Making Practice Fun 51 with this lesson.

Tell whether each proportion is true.

1. $\frac{1}{2} = \frac{5}{10}$ yes
2. $\frac{2}{3} = \frac{4}{6}$ yes
3. $\frac{5}{2} = \frac{10}{4}$ yes
4. $\frac{8}{6} = \frac{4}{3}$ yes

5. $\frac{4}{5} = \frac{20}{25}$ yes
6. $\frac{2}{7} = \frac{6}{21}$ yes
7. $\frac{75}{25} = \frac{1}{3}$ no
8. $\frac{50}{1} = \frac{100}{2}$ yes

Solve.

9. $\frac{1}{2} = \frac{x}{6}$ 3
10. $\frac{1}{2} = \frac{6}{x}$ 12
11. $\frac{1}{4} = \frac{16}{x}$ 64
12. $\frac{1}{4} = \frac{x}{16}$ 4

13. $\frac{x}{5} = \frac{2}{10}$ 1
14. $\frac{5}{x} = \frac{10}{20}$ 10
15. $\frac{8}{12} = \frac{4}{x}$ 6
16. $\frac{25}{75} = \frac{1}{x}$ 3

17. $\frac{1}{2} = \frac{x}{7}$ $3\frac{1}{2}$
18. $\frac{1}{3} = \frac{x}{7}$ $2\frac{1}{3}$
19. $\frac{1}{6} = \frac{7}{x}$ 42
20. $\frac{3}{8} = \frac{7}{x}$ $18\frac{2}{3}$

21. $\frac{5}{12} = \frac{x}{6}$ $2\frac{1}{2}$
22. $\frac{9}{10} = \frac{x}{5}$ $4\frac{1}{2}$
23. $\frac{5}{x} = \frac{9}{11}$ $6\frac{1}{9}$
24. $\frac{9}{x} = \frac{2}{3}$ $13\frac{1}{2}$

25. $\frac{1}{2} = \frac{x}{9}$ 4.5
26. $\frac{1}{4} = \frac{x}{9}$ 2.25
27. $\frac{8}{4} = \frac{5}{x}$ 2.5
28. $\frac{2}{7} = \frac{5}{x}$ 17.5

29. $\frac{x}{4} = \frac{5}{8}$ 2.5
30. $\frac{x}{5} = \frac{8}{6}$ 6.67
31. $\frac{7}{x} = \frac{4}{5}$ 8.75
32. $\frac{2}{x} = \frac{5}{9}$ 3.6

More Practice, page 242

Objective 87

SOLVING PROBLEMS USING PROPORTIONS

How far is it from Griffith to West Lafayette?

The legend on the map tells us the ratio: 0.5 centimeters to 25 kilometers.

We measure and find that on the map it is 3 centimeters between the cities.

Think: 0.5 cm is to 25 km as 3 cm is to x km.

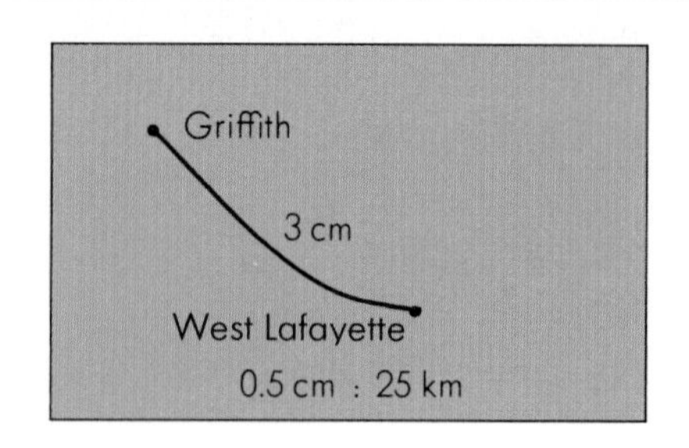

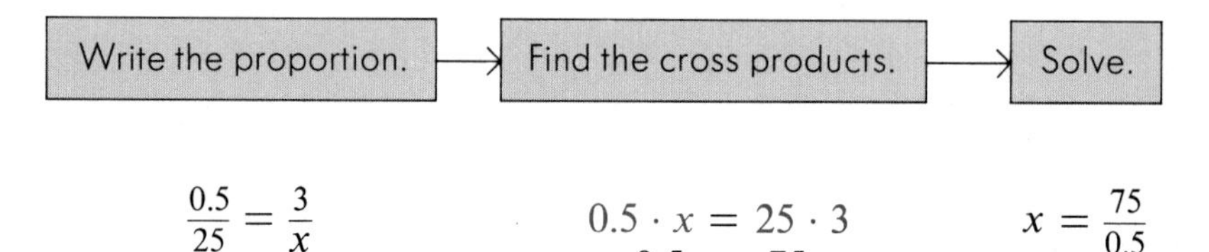

$$\frac{0.5}{25} = \frac{3}{x}$$

$$0.5 \cdot x = 25 \cdot 3$$
$$0.5x = 75$$

$$x = \frac{75}{0.5}$$
$$x = 150$$

It is 150 km from Griffith to West Lafayette.

EXAMPLE 1. A car can travel 100 kilometers on 8 liters of gas. How far can it travel on 10 liters?

Think: 100 km is to 8 L as x km is to 10 L.

$\frac{100}{8} = \frac{x}{10}$ *Write the proportion.*

$100 \cdot 10 = 8 \cdot x$ *Find the cross products.*

$1000 = 8x$

$\frac{1000}{8} = x$ *Solve.*

$125 = x$ The car can travel 125 km on 10 liters of gas.

EXAMPLE 2. Potatoes cost \$ 1.80 for 3 kilograms. How much did 7 kilograms cost?

Think: 3 kg is to \$ 1.80 as 7 kg is to x.

$\frac{3}{180} = \frac{7}{x}$ *Write the proportion.*

$3 \cdot x = 7 \cdot 180$ *Find the cross products.*

$3x = 1260$

$x = \frac{1260}{3} = 420$ 7 kg of potatoes cost \$ 4.20.

TRY THIS

1. Oranges were 8 for 96 cents. How many could you buy for 60 cents? 5 oranges
2. A car traveled 159 miles in 3 hours. At this rate, how far did it travel in 8 hours? 424 miles

PRACTICE

For extra practice, use Making Practice Fun 52 with this lesson.

Solve.

1. A football team has won 3 out of its first 4 games. At this rate how many will it win in 16 games? 12 games
2. On a map $\frac{1}{2}$ inch represents 100 miles. What distance does $2\frac{1}{2}$ inches represent? 500 miles
3. If you can do 3 problems in 15 seconds, how long will it take to do 10 problems? 50 seconds
4. If 8 tickets cost $ 6.00, how many can you buy for $ 9.00? 12 tickets
5. A pitcher struck out 3.4 batters per 9 innings. At this rate how many would be struck out in 297 innings? 112.2 batters
6. The property tax on a home is $ 5 per $ 1000 assessed evaluation. What is the tax on a $ 45,000 home? $225
7. If 9 pounds of hamburger cost $ 14.58, how much do 6 pounds cost? $9.72
8. A clock loses 3 minutes every 12 hours. How much will it lose in 72 hours? 18 minutes
9. A company car was driven 4200 miles in the first 4 months. At this rate, how far will it be driven in 12 months? 12,600 miles
10. Mary had 7 hits in her first 20 times at bat. At this rate how many hits will she get in 340 times at bat? 119 hits
11. Jerry earned $ 32.80 for 8 hours of work. If he earned $ 82.00, how many hours did he work? 20 hours
12. A car travels 150 kilometers on 10 liters of gas. How many liters of gas are needed to travel 500 kilometers? $33\frac{1}{3}$ liters
13. If 3 records cost $ 16.50, how much do 5 records cost? $27.50
14. If 6 boxes of raisins cost $ 2, how many boxes can you buy for $ 4.80? 14.4 boxes
15. If 2 cans of tennis balls cost $ 6.50, how much do 5 cans cost? $16.25
16. If you make 3 baskets in 5 tries, how many baskets can you make in 20 tries? 12 baskets

More Practice, page 243

Objective 88

SIMILAR FIGURES AND PROPORTIONS

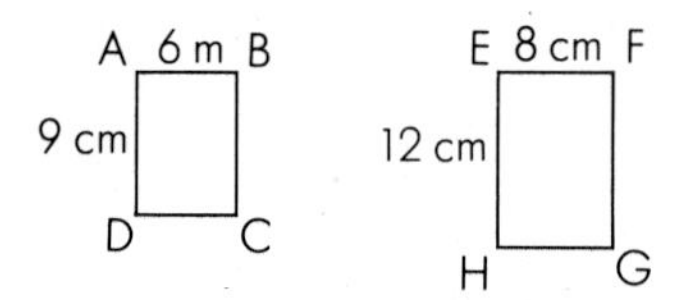

Figure $ABCD$ is similar to figure $EFGH$. Sides $\overline{AB}$ and $\overline{EF}$ are corresponding. Sides $\overline{AD}$ and $\overline{EH}$ are corresponding. In similar figures corresponding lengths are proportional. $\frac{6}{9} = \frac{8}{12}$

The figures are similar. Find x.

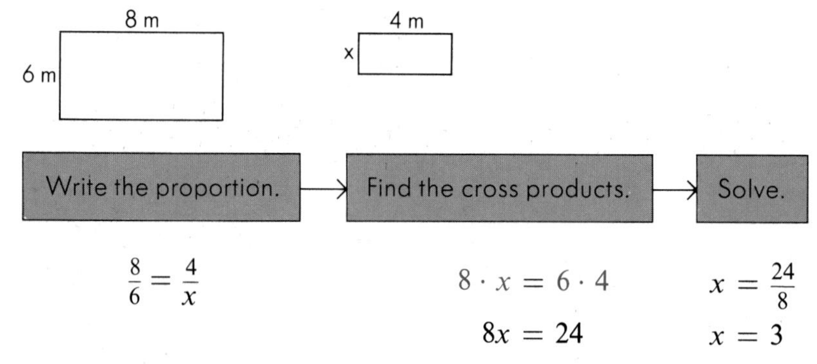

$$\frac{8}{6} = \frac{4}{x}$$

$$8 \cdot x = 6 \cdot 4$$
$$8x = 24$$

$$x = \frac{24}{8}$$
$$x = 3$$

The side is 3 m long.

EXAMPLE 1. The figures are similar. Find x.

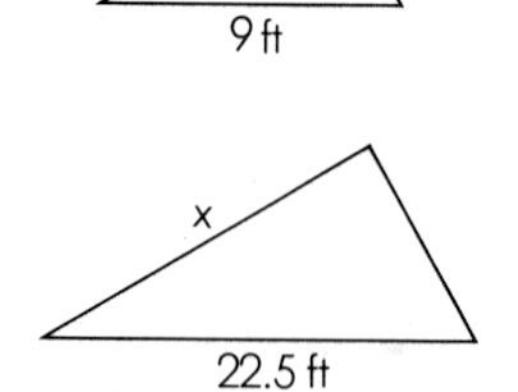

$\frac{4}{9} = \frac{x}{22.5}$ *Write the proportion.*

$4 \cdot 22.5 = 9 \cdot x$ *Find the cross products.*

$90 = 9x$

$\frac{90}{9} = x$ *Solve.*

$10 = x$ *The side is 10 feet long.*

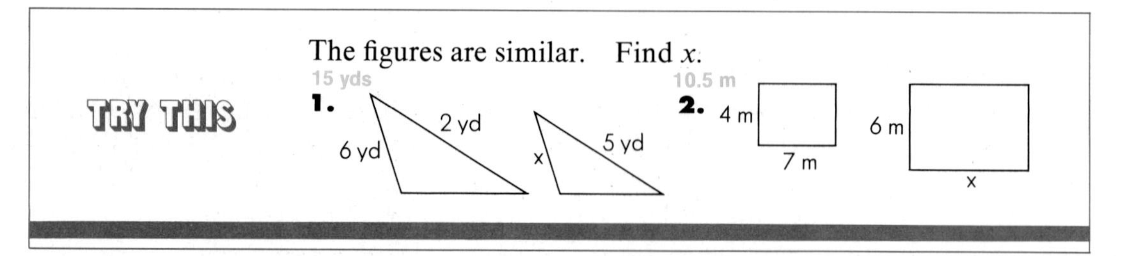

PRACTICE

Use Quiz 18 after this Practice.

The figures are similar. Find x.

8 mm

1.

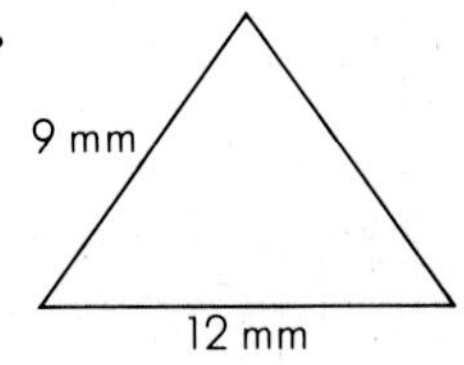

6 mm
x

18.75 in.

2.

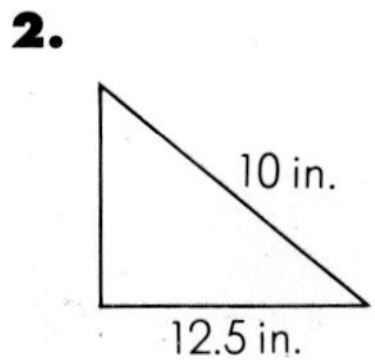

15 in.
x

6 m

3.

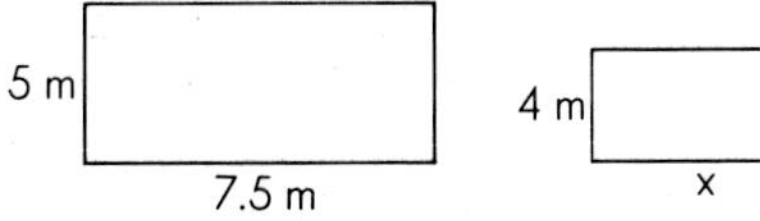

12 yd

4.

8 yd
6 yd
9 yd
x

20 km

5.

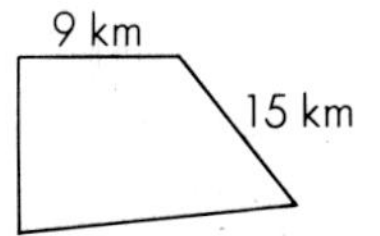

12 km
x

$5\frac{1}{2}$ ft

6.

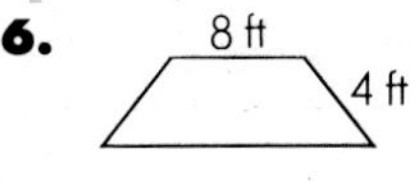

11 ft
x

2 cm

7.

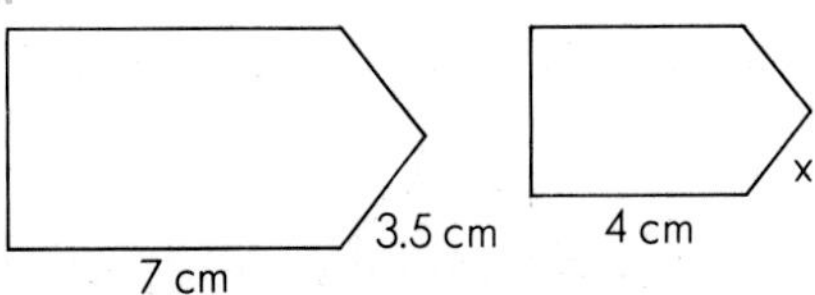

16 cm

8.

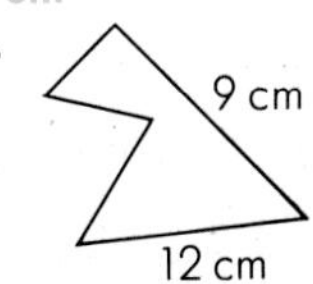

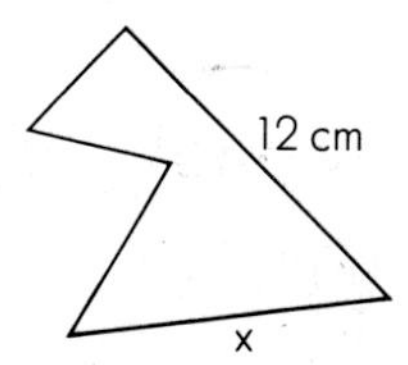

PROBLEM CORNER

How tall is the flagpole? 9 ft

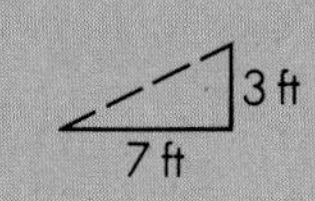

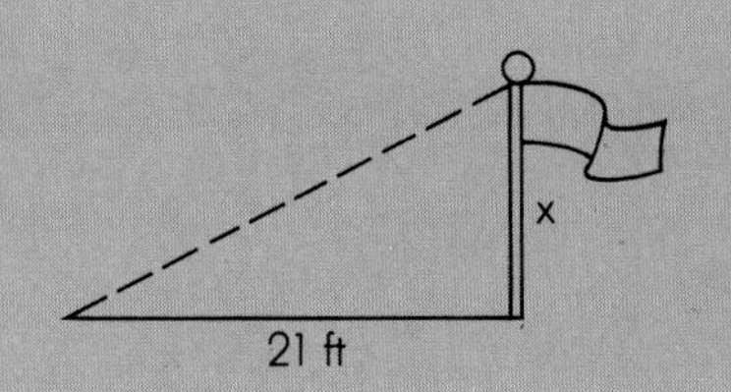

Objectives 172 and 173

APPLICATION

Discuss with your students the advantages and disadvantages of catalog shopping.

CATALOG SHOPPING

Many items can be purchased from a catalog. Product and shipping information is written in code.

KNIT SHIRTS **Polyester and Cotton**

Child sizes 8-10-12-14-16-18

A. 12D-3211-T **B.** 12D-3212-T **C.** 12D-3213-T

(3211: style ↗ ↖ color)

Shipping weight 7 oz $5.99 pack of 3 $16.95

Adult sizes S-M-L-XL-XXL

A. 12D-3611-T **B.** 12D-3612-T **C.** 12D-3613-T

Shipping weight 12 oz $6.99 pack of 3 $19.65

KNIT SHIRTS IN 3 COLORS

Polyester and Cotton

11 Red 12 Blue 13 Green
Machine wash and dry

Find the cost and shipping weight of the following.

1. 1 child's shirt, size 12 $5.99; 7 oz
2. 1 adult's shirt, size XL $6.99; 12 oz
3. 1 pack of 3 children's shirts, size 16 $16.95 21 oz
4. 1 pack of 3 adults' shirts, size M $19.65 36 oz

How much do you save per shirt by buying a pack of 3 shirts?

5. Child sizes $0.34
6. Adult sizes $0.44

Consumer Hint

The unit cost of shipping ($ per ounce) decreases as the weight increases. When possible plan ahead and purchase several items in one order.

Shipping charges are added to the cost of the items in the order. Find the shipping cost for each order.

APPROXIMATE DISTANCE FROM CATALOG DISTRIBUTION CENTER								
Shipping Weight	Local Zone	Zones 1 and 2 Not over 150 Miles	Zone 3 151 to 300 Miles	Zone 4 301 to 600 Miles	Zone 5 601 to 1000 Miles	Zone 6 1001 to 1400 Miles	Zone 7 1401 to 1800 Miles	Zone 8 Over 1800 Miles
1 oz to 8 oz	$0.28	$0.36	$0.36	$0.38	$0.43	$ 0.46	$ 0.51	$ 0.56
9 oz to 15 oz	0.68	0.74	0.74	0.74	0.81	0.90	0.99	1.08
1 lb to 2 lbs	0.75	0.85	0.90	1.10	1.20	1.30	1.40	1.50
2 lbs 1 oz to 3 lbs	0.80	1.00	1.05	1.20	1.30	1.40	1.60	1.75
3 lbs 1 oz to 5 lbs	0.90	1.15	1.25	1.35	1.50	1.75	1.95	2.05
5 lbs 1 oz to 10 lbs	1.10	1.40	1.50	1.75	2.10	2.45	2.85	3.00
10 lbs 1 oz to 15 lbs	1.30	1.80	2.00	2.25	2.85	3.30	3.75	4.20

7. 3 lb 10 oz, 200 miles $1.25
8. 10 lb 8 oz, 1700 miles $3.75
9. 14 oz, 10 miles $0.68
10. 1 lb 12 oz, New York to Los Angeles $1.50
11. 130 oz, 1000 miles (16 oz = 1 lb) $2.10

Find the cost and weight of each item.

COST = QUANTITY × UNIT PRICE   WEIGHT = QUANTITY × UNIT WEIGHT

	Quantity	Number	Item-Size	Unit Price	Unit Weight	Cost	Weight
12.	3	623BF	blouse-M	$4.95	9 oz	$14.85	1 lb 11 oz
13.	2	3794BR	pants-28	$9.95	1 lb 2 oz	$19.90	2 lb 4 oz
14.	1	5321SM	art brushes set of 6	$7.62	12 oz	$ 7.62	12 oz
15.	5	2634PX	tube of oil paint	$1.99	12 oz	$ 9.95	3 lb 12 oz
					Subtotal	$52.32	8 lb 7 oz
					Shipping Cost	$ 1.50	
					Total Cost	$53.82	

16. Find the subtotal cost and total weight of the four items.
17. Use the shipping chart to find the shipping cost to zone 3.
18. Find the total cost.

What is the actual unit cost of the following items? (Catalog cost + shipping cost)

19. 1 pair of pants sent to zone 1 $10.80
20. 1 tube of oil paint sent to zone 6 $2.89

CAREER CAREER CAREER CAREER CAREER

MERCHANDISING

Betsy Rathman owns the Melody Music Store. She sells records, musical instruments, and provides music lessons.

Betsy finds that about $\frac{1}{6}$ of all Americans play a musical instrument. Find the number of instrument players in these areas.

1. New England: population, 12,600,000 2,100,000
2. West Coast: population, 28,620,000 4,770,000
3. Southwest: population, 18,610,000 $3{,}101{,}666\frac{2}{3}$
4. Great Lakes: population, 42,250,000 $7{,}041{,}666\frac{2}{3}$

A cash discount is given for each item purchased at some sales at the Melody Music store.

For each of the following find the sale price. Round to the nearest cent.

5. $4.95 $3.71
6. $12.95 $9.71
7. $7.95 $5.96
8. $21.95 $16.46

Nov. 5, 1983 City Gazette

MELODY MUSIC SPECIAL
1/4 Off All Records, Tapes, Cassettes
1 WEEK ONLY NOV. 6–12

Melody Music employs teachers to give music lessons.

ORGAN AND PIANO
Private Lessons

\$9.50 -- $\frac{1}{2}$ hour \$17.50 -- 1 hour

50 $\frac{1}{2}$ - hour lessons -- \$425

GUITAR
Private Lessons

\$8.50 -- $\frac{1}{2}$ hour \$14 -- 1 hour

50 $\frac{1}{2}$ - hour lessons -- \$325

9. For each type of lesson which rate is the best value?
Organ and Piano: 50 $\frac{1}{2}$-hour lessons; Guitar: 50 $\frac{1}{2}$-hour lessons
10. For which set of 50 lessons is a larger discount given? How much would each of the 50 lessons cost? Guitar; \$6.50
11. Find the cost of the best value for $\frac{1}{2}$-hour piano lessons, 1 per week for 1 year. \$444
12. Teachers suggest practicing 15 minutes daily for each $\frac{1}{2}$-hour lesson per week. For a 1-hour lesson each week, how much practice time per week is suggested? 3 hours, 30 minutes

Summary Sheet			
Item	Quantity	Wholesale	Retail
LP Singles	1820	\$7134.40	\$15,690.50
LP Albums	365	\$2151.30	\$ 4,732.80
45 Singles	2360	\$1132.80	\$ 2,832.00
Classical Album Sets	20	\$ 532.60	\$ 1,270.75

For each department Betsy keeps inventory records she calls summary sheets. These tell her when to reorder. They also give the wholesale and retail value of the stock. The **markup** is the difference between the wholesale and retail costs. The **rate of markup** is the retail cost divided by the wholesale cost.

13. For each group of items find the markup.
14. For each item what is the rate of markup?
15. Which of the four items has the highest rate of markup? 45 singles

13. Singles: \$8556.10
Albums: \$2581.50
45 Singles: \$1699.20
Classical: \$738.15

14. Singles: 2.2
Albums: 2.2
45 Singles: 2.5
Classical: 2.4

CHAPTER 8 REVIEW

For extra practice, use Making Practice Fun 53 with this lesson.

Write the ratio. Simplify, if possible. p. 224

1. 4 peaches to 3 plums $\frac{4}{3}$

2. 9 questions in 12 minutes $\frac{3}{4}$

3. 4 cups per 2 servings $\frac{2}{1}$

4. 2 dollars for 8 quarters $\frac{1}{4}$

Find the unit rate. p. 226

5. If 3 items cost 93¢, what is the cost per item? 31¢

6. If 5 hours of lessons cost $ 16.25, what is the cost per hour? $3.25

7. If a person drives 187 kilometers in 4 hours, how many kilometers per hour is this? Round to tenths. 46.8 km/hr

8. If a car travels 386 miles on 20 gallons of gas, how many miles per gallon is this? Round to tenths. 19.3 mi/gal

Tell whether each proportion is true. p. 230

9. $\frac{7}{8} = \frac{6}{7}$ no

10. $\frac{15}{18} = \frac{5}{6}$ yes

Solve. p. 232

11. $\frac{5}{6} = \frac{15}{x}$ 18

12. $\frac{4}{3} = \frac{x}{15}$ 20

13. $\frac{3}{4} = \frac{7}{x}$ $9\frac{1}{3}$

14. $\frac{7}{5} = \frac{x}{9}$ 12.6

15. If there are 14 books for 21 students, how many are there for 15 students? 10

16. If 3 items cost $ 7.50, how much would 5 items cost? $12.50

The figures are similar. Find x. p. 234

17. 4 m

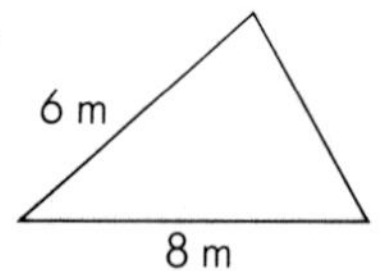

18. 12 yd

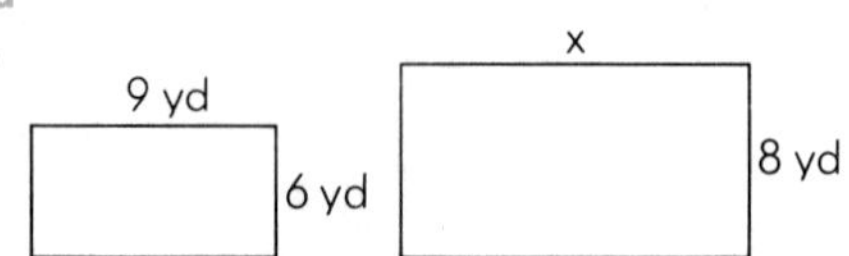

Application pp. 228, 236

19. A 15-ounce can of tomato sauce costs $ 0.47. A 29-ounce can costs $ 0.78. Which is the more economical buy? the 29-ounce can

20. A box of golf balls costs $ 15.39. The shipping cost is $ 0.90. What is the total cost? $16.29

CHAPTER 8 TEST

Write the ratio. Simplify, if possible.

1. 7 pages per 2 hours $\frac{7}{2}$

2. 9 chairs to 6 desks $\frac{3}{2}$

3. 12 pens to 9 pencils $\frac{4}{3}$

4. 8 hits in 23 times at bat $\frac{8}{23}$

Find the unit rate.

5. If a woman's wages are $ 18.75 for 3 hours of work, what is the amount per hour? $6.25

6. If a car travels 184 miles on 8 gallons of gas, how many miles per gallon is this? 23 mi/gal

7. If a man drives 798 kilometers in 4 hours, how many kilometers per hour is this? 199.5 km/hr

8. If 3 items cost 77¢, what is the cost per item? $25\frac{2}{3}$¢ or 26¢

Tell whether each proportion is true.

9. $\frac{9}{15} = \frac{3}{5}$ yes

10. $\frac{4}{3} = \frac{12}{18}$ no

Solve.

11. $\frac{6}{3} = \frac{12}{x}$ 6

12. $\frac{7}{8} = \frac{x}{24}$ 21

13. $\frac{5}{6} = \frac{9}{x}$ $10\frac{4}{5}$

14. $\frac{3}{8} = \frac{x}{20}$ 7.5

15. If 8 items cost $ 20.00, what is the cost for 6 items? $15.00

16. If Joe jogs 28 miles in 4 hours, how far will he jog in 9 hours? 63 miles

The figures are similar. Find x.

18 cm

17.

8 cm

16 cm

9 cm

x

5 in.

18.

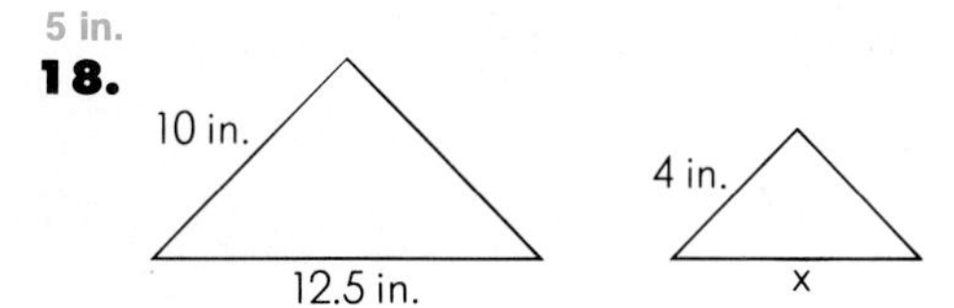

Application

19. A 12-ounce can of juice concentrate costs $ 1.39. The concentrate must be mixed with 3 cans of water. Find the cost per prepared ounce of juice. $0.03

20. A sleeping bag costs $ 32.98. The shipping cost is $ 1.95. What is the total cost? $34.93

MORE PRACTICE

Write the ratio. Simplify, if possible. Use after page 225.

1. 10 tires to 2 cars $\frac{5}{1}$

2. 8 wheel covers to 2 cars $\frac{4}{1}$

3. 9 stitches to 3 centimeters $\frac{3}{1}$

4. 3 dresses in 6 hours $\frac{1}{2}$

5. 8 carries for 40 yards $\frac{1}{5}$

6. 12 hits in 36 times at bat $\frac{1}{3}$

7. 56 goals in 8 games $\frac{7}{1}$

8. 5 kicks for 170 yards $\frac{1}{34}$

Find the unit rate. Use after page 227.

9. If a dozen eggs cost 96¢, how much does 1 egg cost? 8¢

10. 10 pounds of potatoes cost \$ 1.50. What is the price per pound? \$0.15

11. Ann receives \$ 23.75 for working 5 hours. How much is this per hour? \$4.75

12. Lester drives 458.4 miles in 8 hours. How many miles per hour is this? 57.3 mph

13. Nora drives 179 kilometers in 2 hours. How many kilometers per hour is this? Round to the nearest tenth. 89.5 km/hr

14. A car travels 77 miles on 3 gallons of gas. How many miles per gallon is this? Round to the nearest tenth. 25.7 mi/gal

15. Bonnie reads 185 pages in 3 hours. How many pages per hour does she read? Round to the nearest tenth. 61.7 pages

16. A van travels 148 kilometers on 9 liters of gas. How many kilometers per liter is this? Round to the nearest tenth. 16.4 km/L

Tell whether each proportion is true. Use after page 231.

17. $\frac{2}{3} = \frac{3}{4}$ no
18. $\frac{7}{8} = \frac{8}{7}$ no
19. $\frac{5}{6} = \frac{10}{12}$ yes
20. $\frac{1}{3} = \frac{3}{1}$ no

21. $\frac{4}{6} = \frac{6}{9}$ yes
22. $\frac{9}{12} = \frac{6}{9}$ no
23. $\frac{6}{7} = \frac{7}{8}$ no
24. $\frac{4}{8} = \frac{7}{14}$ yes

Solve. Use after page 231

25. $\frac{7}{8} = \frac{x}{16}$ 14
26. $\frac{5}{4} = \frac{25}{x}$ 20
27. $\frac{3}{9} = \frac{x}{3}$ 1
28. $\frac{12}{16} = \frac{6}{x}$ 8

29. $\frac{1}{2} = \frac{x}{5}$ $2\frac{1}{2}$
30. $\frac{3}{4} = \frac{10}{x}$ $13\frac{1}{3}$
31. $\frac{6}{x} = \frac{5}{7}$ $8\frac{2}{5}$
32. $\frac{x}{9} = \frac{8}{7}$ $10\frac{2}{7}$

Solve. Use after page 233.

1. If 8 pencils cost 48¢, how many pencils can be purchased for 72¢? 12 pencils

2. If 8 pencils cost 48¢, how much will 10 pencils cost? 60¢

3. Ken receives $ 11.25 for 3 hours of work. At that rate how much will he receive for working 2 hours? $7.50

4. Alice is paid $ 11.25 for working 3 hours. How many hours does she work to receive $ 18.75? 5 hours

5. There are 70 ounces of juice in 5 bottles. How many ounces are there in 7 bottles? 98 ounces

6. If 5 bottles contain 60 ounces, how many bottles contain 96 ounces? 8 bottles

7. There are 165 minutes in 3 class periods. How many minutes are there in 5 class periods? 275 minutes

8. If 3 class periods are 165 minutes long, how many class periods are there in 110 minutes? 2 periods

The figures are similar. Find x. Use after page 235.

12 cm
9.

5 cm
15 cm
x
36 cm

15 m
10.

4 m
5 m
12 m
x

8 ft
11.

9 ft
12 ft
6 ft
x

3 in.
12.

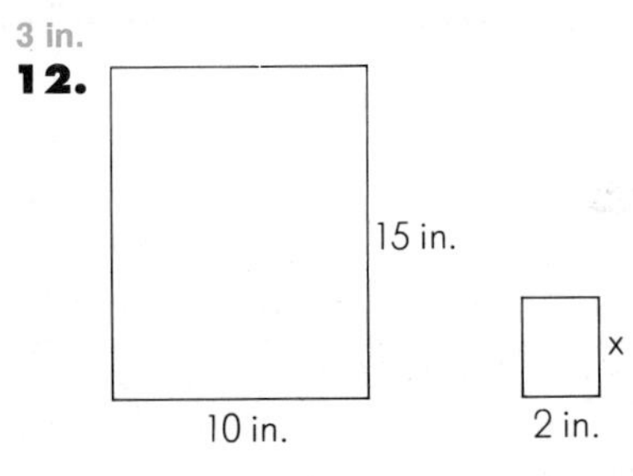

0.9 m
13.

3.6 m
1.8 m
1.8 m
x

$4\frac{1}{2}$ yd
14.

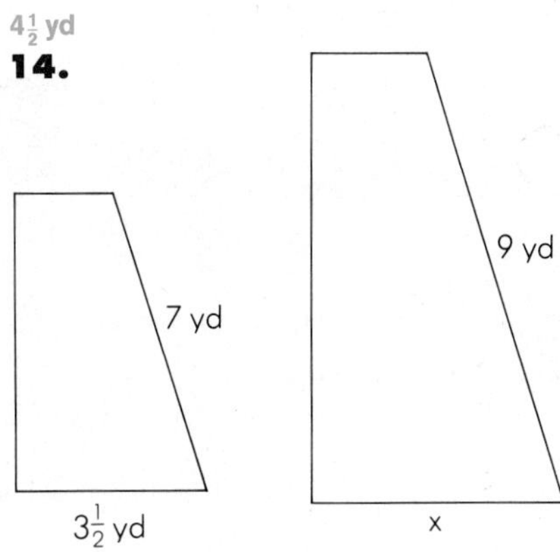

CALCULATOR CORNER

This activity is concerned primarily with number relationships and the effect that performing various operations have on a given number.

WIPE OUT

Players two
Materials two calculators, paper and pencils

SAMPLE Chin enters a five-digit whole number, such as 59,762, on the display. Kristen enters a five-digit whole number, such as 48,829, on the display.

They exchange calculators. Each player must now attempt to "wipe out" the number and get 0 on the display.

To do so they may add, subtract, multiply, or divide by any one or two-digit whole number or decimal except 0.

Kristen		**Chin**	
Play 1	59,762 −62 59,700	**Play 1**	48,829 −99 48,730
Play 2	×0.01 597	**Play 2**	−30 48,700
Play 3	−97 500	**Play 3**	×0.01 487
Play 4	×0.1 50	**Play 4**	−87 400
Play 5	−50 0	**Play 5**	×0.01 4
		Play 6	−4 0

The activity could be adjusted to a class activity by dividing the class into two teams. Various members of the team could suggest plays to reach 0.

Kristen's score is 5. Chin's score is 6.

Kristen wins the game, because she reaches 0 in fewer plays.

CHAPTER 9 PRE-TEST

Convert to a decimal. p. 248

1. 37% 0.37 **2.** 136% 1.36 **3.** 9% 0.09 **4.** 4.2% 0.042

Convert to a percent. p. 248

5. 0.17 17% **6.** 2.47 247% **7.** 0.03 3% **8.** 0.004 0.4%

Convert to a fraction, mixed number, or whole number. p. 250

9. 37% $\frac{37}{100}$ **10.** 45% $\frac{9}{20}$ **11.** 120% $1\frac{1}{5}$ **12.** $12\frac{1}{2}\%$ $\frac{1}{8}$

Convert to a percent. p. 252

13. $\frac{4}{5}$ 80% **14.** $\frac{19}{20}$ 95% **15.** $3\frac{1}{2}$ 350% **16.** 6 600%

17. $\frac{2}{3}$ $66\frac{2}{3}\%$ **18.** $\frac{1}{8}$ $12\frac{1}{2}\%$ **19.** $2\frac{1}{3}$ $233\frac{1}{3}\%$ **20.** $1\frac{3}{7}$ 143%

Translate. Do not solve. p. 254

21. 18% of 38 is what number? $18\% \times 38 = n$

22. 42 is 75% of what number? $42 = 75\% \times x$

23. 5 is what percent of 90? $5 = n \times 90$

24. What percent of 10 is 3? $x \times 10 = 3$

Solve. pp. 256–262

25. What is 25% of 40? 10

26. 20% of 32 is what number? 6.4

27. 8 is what percent of 12? $66\frac{2}{3}\%$

28. What percent of 9 is 3? $33\frac{1}{3}\%$

29. 4 is 20% of what number? 20

30. 36 is $16\frac{2}{3}\%$ of what number? 216

Find the percent of increase or decrease. p. 264

31. From 10 to 12 20%

32. From 12 to 10 $16\frac{2}{3}\%$

Application pp. 260, 266

33. Matt earns \$ 300 per month plus 9% of sales over \$ 1000. How much does he earn if his total sales are \$ 3300 in a month? \$507

34. A tennis racquet regularly sells for \$ 35. It is on sale for 20% off. What is the sale price? \$28

9
PERCENT

GETTING STARTED

MAXIMIZE

WHAT'S NEEDED The 30 squares of paper with digits marked on them as in GOTCHA, Chapter 1.

THE RULES

1. On a sheet of paper each player writes the following.
   ___ ___% of ___ ___ ___ is what number?
2. A leader draws a digit from the box. Each player writes the digit on any one of the blanks. Once it is written, it cannot be changed or moved.
3. The leader draws four more digits, one at a time. Write each digit on a blank as it is drawn.
4. Each player is now ready to solve the problem.

SAMPLE 3 8 % of 9 4 0 is what number? Remind students that *of* translates to x.

5. Who has the largest number? Find out by multiplying.

SAMPLE

```
    940
  × 0.38
   7520
  28200
  357.20
```

The player with the largest answer is the winner.

Objectives 89 and 90

CONVERTING PERCENTS AND DECIMALS

Percent means hundredths: $n\%$ means $n \times 0.01$ $87\% = 87 \times 0.01 = 0.87$
We replace % with $\times$ 0.01 and multiply.

Convert 87% to a decimal.

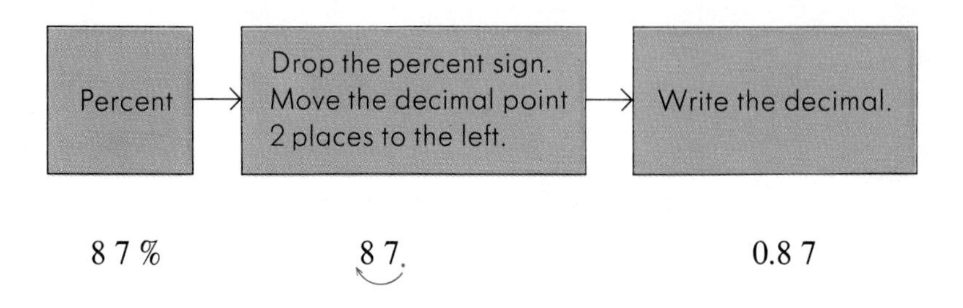

EXAMPLES. Convert to a decimal.

1. $247\% \longrightarrow 2.47. \longrightarrow 2.47$

2. $8\% \longrightarrow .08. \longrightarrow 0.08$

3. $7.6\% \longrightarrow .07.6 \longrightarrow 0.076$

4. $3\frac{1}{3}\% \longrightarrow .03.\frac{1}{3} \longrightarrow 0.03\frac{1}{3}$

TRY THIS

Convert to a decimal.

1. 25% 0.25 **2.** 100% 1.00 **3.** 4% 0.04 **4.** $4\frac{1}{2}\%$ 0.045

Decimals can be written as percents: $n \times 0.01$ means $n\%$ $0.79 = 79 \times 0.01 = 79\%$
We write the number as $n \times 0.01$.
Then replace $\times$ 0.01 with %.

Convert 0.79 to a percent.

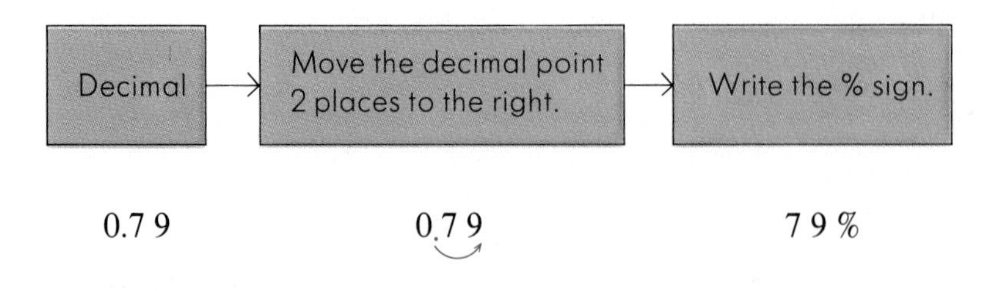

EXAMPLES. Convert to a percent.

5. $0.006 \longrightarrow 0.00.6 \longrightarrow 0.6\%$

6. $0.02\frac{1}{2} \longrightarrow 0.02\frac{1}{2} \longrightarrow 2\frac{1}{2}\%$

7. $0.3 \longrightarrow 0.30 \longrightarrow 30\%$

8. $6 \longrightarrow 6.00 \longrightarrow 600\%$

no book Shawn Phillips 1/20/86

TRY THIS

Convert to a percent.

5. 0.72 72% **6.** 7 700% **7.** 7.2 720% **8.** $0.66\frac{2}{3}$ $66\frac{2}{3}$%

PRACTICE

For extra practice, use Making Practice Fun 54 with this lesson.

Convert to a decimal.

1. 15% 0.15 **2.** 241% 2.41 **3.** 42% 0.42 **4.** 312% 3.12

5. 6% 0.06 **6.** 1% 0.01 **7.** 7% 0.07 **8.** 3% 0.03

9. 3.9% 0.039 **10.** 8.7% 0.087 **11.** 1.2% 0.012 **12.** 5.5% 0.055

13. 0.8% 0.008 **14.** 0.7% 0.007 **15.** 0.1% 0.001 **16.** 0.9% 0.009

17. 0.04% 0.0004 **18.** 0.18% 0.0018 **19.** 0.09% 0.0009 **20.** 0.42% 0.0042

21. $33\frac{1}{3}$% $0.33\frac{1}{3}$ **22.** $5\frac{1}{2}$% $0.05\frac{1}{2}$ **23.** $9\frac{2}{3}$% $0.09\frac{2}{3}$ **24.** $18\frac{3}{4}$% $0.18\frac{3}{4}$

Convert to a percent.

25. 0.99 99% **26.** 1.82 182% **27.** 0.71 71% **28.** 2.04 204%

29. 0.01 1% **30.** 0.09 9% **31.** 0.05 5% **32.** 0.04 4%

33. 0.001 0.1% **34.** 0.039 3.9% **35.** 0.008 0.8% **36.** 0.082 8.2%

37. $0.05\frac{1}{2}$ $5\frac{1}{2}$% **38.** $0.07\frac{2}{3}$ $7\frac{2}{3}$% **39.** $0.43\frac{1}{3}$ $43\frac{1}{3}$% **40.** $1.36\frac{3}{4}$ $136\frac{3}{4}$%

41. 0.5 50% **42.** 1.7 170% **43.** 0.1 10% **44.** 3.9 390%

PROBLEM CORNER

1. By the age of 14 (age 16 for boys), the average girl has attained 98.3% of her total height. Write the decimal for this percent. 0.983

2. Water makes up about 0.7 of a person's weight. Write the percent for this. 70%

More Practice, page 272

Objective 91

CONVERTING PERCENTS TO FRACTIONS

Percent means hundredths: $n\%$ means $n \times \frac{1}{100}$

Convert 25% to a fraction.

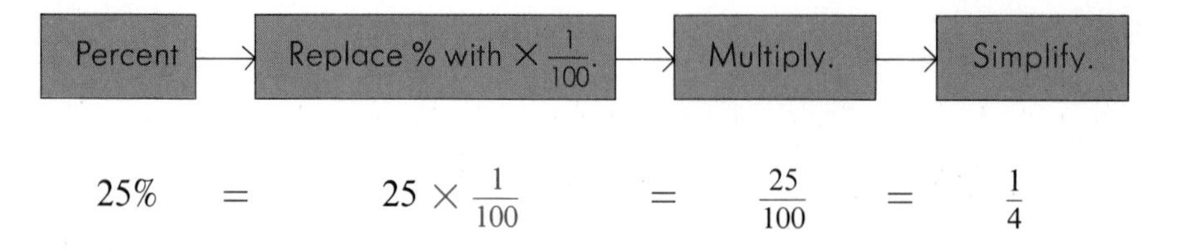

$$25\% = 25 \times \frac{1}{100} = \frac{25}{100} = \frac{1}{4}$$

EXAMPLES. Convert to a fraction, mixed number, or whole number.

1. $19\% = 19 \times \frac{1}{100} = \frac{19}{100}$

2. $125\% = 125 \times \frac{1}{100} = \frac{125}{100} = 1\frac{25}{100} = 1\frac{1}{4}$

TRY THIS Convert to a fraction, mixed number, or whole number.

1. 31% $\frac{31}{100}$ **2.** 75% $\frac{3}{4}$ **3.** 400% 4 **4.** 150% $1\frac{1}{2}$

Percents may contain fractions, mixed numbers, or decimals.

EXAMPLES. Convert to a fraction.

3. $\frac{2}{3}\% = \frac{2}{3} \times \frac{1}{100} = \frac{2}{300} = \frac{1}{150}$

4. $33\frac{1}{3}\% = \frac{100}{3}\% = \frac{100}{3} \times \frac{1}{100} = \frac{100}{300} = \frac{1}{3}$

Rename the mixed numbers as fractions. Replace % with $\times \frac{1}{100}$. Multiply. Simplify.

TRY THIS Convert to a fraction.

5. $\frac{1}{2}\%$ $\frac{1}{200}$ **6.** $16\frac{2}{3}\%$ $\frac{1}{6}$ **7.** $12\frac{1}{2}\%$ $\frac{1}{8}$

EXAMPLES. Convert to a fraction.

5. $13.9\% = 0.139 = \frac{139}{1000}$

6. $0.8\% = 0.008 = \frac{8}{1000} = \frac{1}{125}$

Convert the percent to a decimal. Convert the decimal to a fraction. Simplify.

TRY THIS

Convert to a fraction.

8. 10.3% $\frac{103}{1000}$ **9.** 0.4% $\frac{1}{250}$ **10.** 0.04% $\frac{1}{2500}$

PRACTICE

For extra practice, use Making Practice Fun 55 with this lesson.

Convert to a fraction, mixed number, or whole number.

1. 49% $\frac{49}{100}$	**2.** 3% $\frac{3}{100}$	**3.** 99% $\frac{99}{100}$	**4.** 9% $\frac{9}{100}$
5. 10% $\frac{1}{10}$	**6.** 20% $\frac{1}{5}$	**7.** 30% $\frac{3}{10}$	**8.** 90% $\frac{9}{10}$
9. 5% $\frac{1}{20}$	**10.** 15% $\frac{3}{20}$	**11.** 85% $\frac{17}{20}$	**12.** 95% $\frac{19}{20}$
13. 2% $\frac{1}{50}$	**14.** 14% $\frac{7}{50}$	**15.** 48% $\frac{12}{25}$	**16.** 98% $\frac{49}{50}$
17. 100% 1	**18.** 300% 3	**19.** 500% 5	**20.** 1000% 10
21. 130% $1\frac{3}{10}$	**22.** 175% $1\frac{3}{4}$	**23.** 180% $1\frac{4}{5}$	**24.** 225% $2\frac{1}{4}$

Convert to a fraction.

25. $\frac{1}{3}\%$ $\frac{1}{300}$	**26.** $\frac{1}{4}\%$ $\frac{1}{400}$	**27.** $\frac{3}{4}\%$ $\frac{3}{400}$	**28.** $\frac{4}{5}\%$ $\frac{1}{125}$
29. $\frac{3}{8}\%$ $\frac{3}{800}$	**30.** $\frac{7}{8}\%$ $\frac{7}{800}$	**31.** $\frac{2}{3}\%$ $\frac{1}{150}$	**32.** $\frac{7}{10}\%$ $\frac{7}{1000}$
33. $66\frac{2}{3}\%$ $\frac{2}{3}$	**34.** $83\frac{1}{3}\%$ $\frac{5}{6}$	**35.** $14\frac{2}{7}\%$ $\frac{1}{7}$	**36.** $28\frac{4}{7}\%$ $\frac{2}{7}$
37. $37\frac{1}{2}\%$ $\frac{3}{8}$	**38.** $62\frac{1}{2}\%$ $\frac{5}{8}$	**39.** $44\frac{4}{9}\%$ $\frac{4}{9}$	**40.** $77\frac{7}{9}\%$ $\frac{7}{9}$
41. 15.3% $\frac{153}{1000}$	**42.** 99.9% $\frac{999}{1000}$	**43.** 0.5% $\frac{1}{200}$	**44.** 5.7% $\frac{57}{1000}$
45. 7.75% $\frac{31}{400}$	**46.** 8.25% $\frac{33}{400}$	**47.** 0.05% $\frac{1}{2000}$	**48.** 9.50% $\frac{19}{200}$

PROBLEM CORNER

1. China has about 30% of the world's population. Write a fraction for this. $\frac{3}{10}$

2. About $66\frac{2}{3}\%$ of the Earth is covered with water. Write a fraction for this. $\frac{2}{3}$

More Practice, page 272

Objective 92

CONVERTING FRACTIONS TO PERCENTS

Convert $\frac{1}{2}$ to a percent.

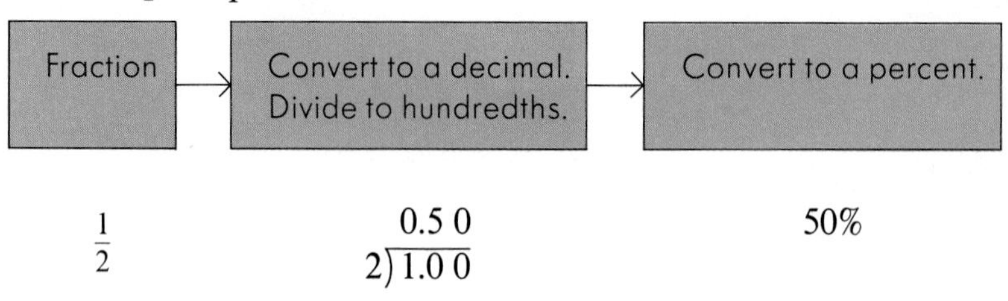

$\frac{1}{2}$ $\quad 2\overline{)1.00} = 0.50 \quad$ 50%

EXAMPLE. Convert to a percent.

1. $\frac{3}{4} \longrightarrow 4\overline{)3.00} = 0.75 \longrightarrow 75\%$
2. $1\frac{1}{2} \longrightarrow \frac{3}{2} \longrightarrow 2\overline{)3.00} = 1.50 \longrightarrow 150\%$

Convert to a decimal. Divide to hundredths. Convert to a percent.

For mixed numbers, change to a fraction.

TRY THIS Convert to a percent.

1. $\frac{1}{4}$ 25% **2.** $\frac{3}{10}$ 30% **3.** $1\frac{2}{5}$ 140% **4.** $2\frac{3}{10}$ 230%

When we do not have a 0 remainder, we write a decimal or a fraction.

EXAMPLES. Convert to a percent.

3. $\frac{1}{3} \longrightarrow 3\overline{)1.00} = 0.33\frac{1}{3} = 33\frac{1}{3}\%$

```
  0.33 1/3
3)1.00
   9
   10
    9
    1
```

4. $\frac{3}{8} \longrightarrow 8\overline{)3.000} = 0.375 = 37.5\%$

```
  0.375
8)3.000
  2 4
    60
    56
     40
     40
```

TRY THIS Convert to a percent.

5. $\frac{2}{3}$ $66\frac{2}{3}\%$ **6.** $\frac{5}{8}$ 62.5%

7. $1\frac{5}{6}$ $183\frac{1}{3}\%$ **8.** $2\frac{7}{8}$ $287\frac{1}{2}\%$

PRACTICE

For extra practice, use Making Practice Fun 56 with this lesson.

Convert to a percent.

1. $\frac{2}{5}$ 40% **2.** $\frac{3}{5}$ 60% **3.** $\frac{3}{10}$ 30% **4.** $\frac{7}{10}$ 70%

5. $\frac{17}{20}$ 85% **6.** $\frac{6}{25}$ 24% **7.** $\frac{3}{50}$ 6% **8.** $\frac{38}{100}$ 38%

9. $1\frac{1}{4}$ 125% **10.** $1\frac{3}{4}$ 175% **11.** $2\frac{1}{2}$ 250% **12.** $1\frac{1}{5}$ 120%

13. 3 300% **14.** 2 200% **15.** 1 100% **16.** 4 400%

17. $\frac{1}{6}$ $16\frac{2}{3}\%$ **18.** $\frac{2}{6}$ $33\frac{1}{3}\%$ **19.** $\frac{4}{6}$ $66\frac{2}{3}\%$ **20.** $\frac{5}{6}$ $83\frac{1}{3}\%$

21. $\frac{1}{8}$ 12.5% **22.** $\frac{3}{7}$ $42\frac{6}{7}\%$ **23.** $\frac{5}{7}$ $71\frac{3}{7}\%$ **24.** $\frac{7}{8}$ 87.5%

25. $1\frac{1}{6}$ $116\frac{2}{3}\%$ **26.** $1\frac{2}{3}$ $166\frac{2}{3}\%$ **27.** $2\frac{1}{9}$ $211\frac{1}{9}\%$ **28.** $3\frac{1}{3}$ $333\frac{1}{3}\%$

29. $1\frac{3}{7}$ $142\frac{6}{7}\%$ **30.** $2\frac{2}{3}$ $266\frac{2}{3}\%$ **31.** $1\frac{5}{9}$ $155\frac{5}{9}\%$ **32.** $1\frac{5}{6}$ $183\frac{1}{3}\%$

33. $\frac{1}{3}$ $33\frac{1}{3}\%$ **34.** $\frac{5}{8}$ 62.5% **35.** $\frac{2}{7}$ $28\frac{4}{7}\%$ **36.** $\frac{4}{7}$ $57\frac{1}{7}\%$

37. $\frac{1}{9}$ $11\frac{1}{9}\%$ **38.** $\frac{8}{9}$ $88\frac{8}{9}\%$ **39.** $\frac{1}{11}$ $9\frac{1}{11}\%$ **40.** $\frac{9}{11}$ $81\frac{9}{11}\%$

41. $1\frac{1}{3}$ $133\frac{1}{3}\%$ **42.** $3\frac{2}{3}$ $366\frac{2}{3}\%$ **43.** $2\frac{5}{6}$ $283\frac{1}{3}\%$ **44.** $1\frac{4}{6}$ $166\frac{2}{3}\%$

45. $2\frac{1}{6}$ $216\frac{2}{3}\%$ **46.** $1\frac{1}{8}$ 112.5% **47.** $1\frac{7}{8}$ 187.5% **48.** $2\frac{3}{8}$ 237.5%

PROBLEM CORNER

1. The human body is about $\frac{2}{3}$ water. Write a percent for this. $66\frac{2}{3}\%$

2. About $\frac{1}{3}$ of the human body is not water. Write a percent for this. $33\frac{1}{3}\%$

3. About $\frac{4}{5}$ of all households own one or more cars. Write a percent for this. 80%

Objectives 93 and 94

TRANSLATING TO NUMBER SENTENCES

To solve problems with percents, it is helpful to translate to number sentences.

Translate. 36% of 97 is what number?

$$36\% \times 97 = a$$

Of translates to ×. *Is* translates to =.
What or *what percent* or *what number* translates to any letter.

EXAMPLES. Translate.

1. What is 19% of 27?

$$n = 19\% \times 27$$

2. What percent of 76 is 8?

$$n \times 76 = 8$$

TRY THIS

Translate.

1. What is 60% of 70? $x = 60\% \times 70$

2. 76 is 19% of what number? $76 = 19\% \times n$

3. 18 is what percent of 36? $18 = b \times 36$

Students should realize that the decimal equivalents for many mixed number percents are repeating decimals. It is often easier to use the fraction equivalents for these percents.

We often use the following percent, decimal, and fraction equivalents. They should be memorized. Some decimals have been rounded to the nearest thousandth.

Percent	Decimal	Fraction	Percent	Decimal	Fraction
10%	0.1	$\frac{1}{10}$	50%	0.5	$\frac{1}{2}$
$12\frac{1}{2}\%$	0.125	$\frac{1}{8}$	$66\frac{2}{3}\%$	0.667	$\frac{2}{3}$
$16\frac{2}{3}\%$	0.167	$\frac{1}{6}$	75%	0.75	$\frac{3}{4}$
20%	0.2	$\frac{1}{5}$	$37\frac{1}{2}\%$	0.375	$\frac{3}{8}$
25%	0.25	$\frac{1}{4}$	$62\frac{1}{2}\%$	0.625	$\frac{5}{8}$
$33\frac{1}{3}\%$	0.333	$\frac{1}{3}$	$87\frac{1}{2}\%$	0.875	$\frac{7}{8}$

EXAMPLES. Translate.

3. Convert the percent to a decimal.
10% of 48 is what number?
$$\downarrow \quad \downarrow \quad \downarrow \quad \swarrow$$
$$0.1 \times 48 = n$$

4. Convert the percent to a fraction.
56 is $12\frac{1}{2}$% of what number?
$$\downarrow \quad \downarrow \quad \downarrow \quad \swarrow$$
$$56 = \frac{1}{8} \times n$$

TRY THIS

4. Translate. Convert the percent to a decimal.
What is 50% of 48? $x = 0.5 \times 48$

5. Translate. Convert the percent to a fraction.
$62\frac{1}{2}$% of what number is 32? $\frac{5}{8} \times a = 32$

PRACTICE

Use Quiz 19 after this Practice.

Translate.

1. 17% of 36 is what number? $17\% \times 36 = b$

2. What is 42% of 98? $a = 42\% \times 98$

3. 36 is $12\frac{1}{2}$% of what number? $36 = 12\frac{1}{2}\% \times x$

4. $66\frac{2}{3}$% of what number is 9? $66\frac{2}{3}\% \times b = 9$

5. 8.7 is what percent of 19.6? $8.7 = a \times 19.6$

6. What percent of $7\frac{1}{2}$ is $3\frac{1}{4}$? $x \times 7\frac{1}{2} = 3\frac{1}{4}$

Translate. Convert each percent to a decimal.

7. $87\frac{1}{2}$% of what number is 80? $0.875 \times n = 80$

8. 48 is 75% of what number? $48 = 0.75 \times b$

9. What is 70% of 98? $x = 0.7 \times 98$

10. 50% of 94.2 is what number? $0.5 \times 94.2 = n$

11. 10% of what number is 84? $0.1 \times a = 84$

12. 3.7 is 25% of what number? $3.7 = 0.25 \times y$

Translate. Convert each percent to a fraction.

13. What is 25% of 8? $x = \frac{1}{4} \times 8$

14. $16\frac{2}{3}$% of 18 is what number? $\frac{1}{6} \times 18 = y$

15. $12\frac{1}{2}$% of what number is 16? $\frac{1}{8} \times a = 16$

16. 40 is 10% of what number? $40 = \frac{1}{10} \times b$

17. What is $87\frac{1}{2}$% of 80? $n = \frac{7}{8} \times 80$

18. $62\frac{1}{2}$% of 48 is what number? $\frac{5}{8} \times 48 = c$

More Practice, page 272

Objective 95

FINDING A PERCENT OF A NUMBER

When we solve percent problems, we can write the percent either as a decimal or as a fraction.

USING A DECIMAL		USING A FRACTION
25% of 75 is what number?	Problem	25% of 75 is what number?
$0.25 \times 75 = n$	Translate.	$\frac{1}{4} \times 75 = n$
$0.25 \times 75 = 18.75$	Multiply.	$\frac{1}{4} \times \frac{75}{1} = \frac{75}{4} = 18\frac{3}{4}$
$18.75 = n$		$18\frac{3}{4} = n$
25% of 75 is 18.75.	Remind students that $18.75 = 18\frac{3}{4}$.	25% of 75 is $18\frac{3}{4}$.

EXAMPLES. Solve. Convert each percent to a decimal.

1. What is 38% of 95?
$n = 0.38 \times 95$
$n = 36.10$
36.10 is 38% of 95.

2. 3.8% of 9.5 is what number?
$0.038 \times 9.5 = n$
$0.3610 = n$
3.8% of 9.5 is 0.3610.

TRY THIS

Solve. Convert each percent to a decimal.

1. 8% of 159 is what number? 12.72
2. What is 0.5% of 37? 0.185

EXAMPLES. Solve. Convert each percent to a fraction.

3. $12\frac{1}{2}$% of 96 is what?

$\frac{1}{8} \times 96 = n$

$\frac{1}{8} \times \frac{96}{1} = \frac{96}{8} = 12$

$12 = n$

$12\frac{1}{2}$% of 96 is 12.

4. What is 75% of 40?

$n = \frac{3}{4} \times 40$

$\frac{3}{4} \times \frac{40}{1} = \frac{120}{4} = 30$

$n = 30$

30 is 75% of 40.

TRY THIS

Solve. Convert each percent to a fraction.

3. $16\frac{2}{3}\%$ of 50 is what number? $8\frac{1}{3}$

4. What is 50% of 36? 18

PRACTICE

For extra practice, use Making Practice Fun 57 with this lesson.

Solve. Convert each percent to a decimal.

1. 97% of 500 is what number? 485

2. 1% of 73 is what number? 0.73

3. What is 3% of 10? 0.3

4. What is 128% of 100? 128

5. 49% of 7 is what number? 3.43

6. 146% of 5 is what number? 7.3

7. 3.1% of 10 is what number? 0.31

8. 9.7% of 20 is what number? 1.94

9. What is 12.7% of 88? 11.176

10. What is 0.6% of 52? 0.312

11. What is 4.9% of 87.3? 4.2777

12. What is 59.7% of 34.8? 20.7756

Solve. Convert each percent to a fraction.

13. 10% of 98 is what number? $9\frac{4}{5}$

14. $87\frac{1}{2}\%$ of 16 is what number? 14

15. $12\frac{1}{2}\%$ of 40 is what number? 5

16. 25% of 86 is what number? $21\frac{1}{2}$

17. What is $62\frac{1}{2}\%$ of 10? $6\frac{1}{4}$

18. What is $16\frac{2}{3}\%$ of 100? $16\frac{2}{3}$

19. What is 50% of 84? 42

20. What is $37\frac{1}{2}\%$ of 120? 45

21. 20% of 8 is what number? $1\frac{3}{5}$

22. 75% of 218 is what number? $163\frac{1}{2}$

PROBLEM CORNER

1. About 55% of Americans wear glasses. In a group of 50 people, about how many would wear glasses? 28 people

2. About $66\frac{2}{3}\%$ of a person's weight is water. How much water is there in a person who weighs 120 pounds? 80 pounds

More Practice, page 273

Objective 96

WHAT PERCENT A NUMBER IS OF ANOTHER

Solve.

6 is what percent of 8?

$6 = n \times 8$

$\frac{6}{8} = n$

$8\overline{)6.00}$ = 0.75 = 75%

$75\% = n$

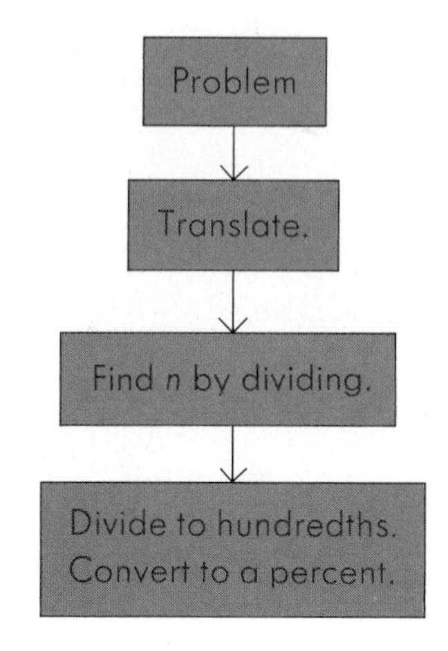

6 is 75% of 8.

EXAMPLES. Solve. Sometimes it is easier to reduce the fraction before dividing. Example: $\frac{6}{9} = \frac{2}{3}$ We know that the equivalent for $\frac{2}{3}$ is $66\frac{2}{3}\%$.

1. 6 is what percent of 9?

$6 = n \times 9$ *Translate.*

$\frac{6}{9} = n$ *Find n by dividing.*

$9\overline{)6.00}$ = $0.66\frac{6}{9} = 0.66\frac{2}{3} = 66\frac{2}{3}\%$ *Divide to hundredths. Convert to a percent.*

```
  54
   60
   54
    6
```

$66\frac{2}{3}\% = n$ 6 is $66\frac{2}{3}\%$ of 9.

2. 9 is what percent of 6?

$9 = n \times 6$ *Translate.*

$\frac{9}{6} = n$ *Find n by dividing.*

$6\overline{)9.00}$ = 1.50 = 150% *Divide to hundredths. Convert to a percent.*

$150\% = n$ 9 is 150% of 6.

3. What percent of 8 is 3?

$n \times 8 = 3$ *Translate.*

$n = \frac{3}{8}$ *Find n by dividing.*

$n = 37\frac{1}{2}\%$ $37\frac{1}{2}\%$ of 8 is 3.

TRY THIS

Solve.

1. 8 is what percent of 16? 50%
2. 10 is what percent of 6? $166\frac{2}{3}\%$
3. What percent of 9 is 7? $77\frac{7}{9}\%$

PRACTICE

For extra practice, use Making Practice Fun 58 with this lesson.

Solve.

1. 1 is what percent of 2? 50%
2. 3 is what percent of 4? 75%
3. 1 is what percent of 3? $33\frac{1}{3}\%$
4. 2 is what percent of 3? $66\frac{2}{3}\%$
5. 4 is what percent of 6? $66\frac{2}{3}\%$
6. 7 is what percent of 8? $87\frac{1}{2}\%$
7. 6 is what percent of 4? 150%
8. 8 is what percent of 7? $114\frac{2}{7}\%$
9. 10 is what percent of 2? 500%
10. 8 is what percent of 1? 800%
11. 5 is what percent of 5? 100%
12. 12 is what percent of 8? 150%
13. What percent of 6 is 2? $33\frac{1}{3}\%$
14. What percent of 6 is 6? 100%
15. What percent of 8 is 3? $37\frac{1}{2}\%$
16. What percent of 8 is 4? 50%
17. What percent of 20 is 16? 80%
18. What percent of 30 is 25? $83\frac{1}{3}\%$
19. What percent of 4 is 10? 250%
20. What percent of 4 is 12? 300%
21. What percent of 9 is 12? $133\frac{1}{3}\%$
22. What percent of 6 is 24? 400%
23. What percent of 100 is 63? 63%
24. What percent of 100 is 75? 75%

More Practice, page 273

Objectives 174 and 175

APPLICATION Have your students suggest advantages and disadvantages of working for commissions and tips.

WORKING FOR COMMISSIONS AND TIPS

Some workers are paid a commission rather than a fixed salary. Commission may be based on the number of items sold.

1. Rico Torrez receives 8¢ for each pack of raisins he sells. He ordered 500 packs and sold 80% of them. How much is his commission? $32.00

Commission is usually based on a percent of sale.

Nancy WhiteEagle earns 7% commission on her monthly sales. How much did she earn these months?

2. January—total sales = $ 8820.00 $617.40
3. June—total sales = $ 10,150.00 $710.50

Some salespeople are paid a salary plus a commission.

Martin Herbert earns \$ 500 per month plus 2% commission on sales over \$ 1000. How much will he earn these months?

4. January—total sales = \$ 2450.00 \$529.00
5. September—total sales = \$ 4165.00 \$563.30

Some workers are paid a graduated commission. The commission rate increases as sales increase. Each month Marion Reesey earns $12\frac{1}{2}\%$ of sales on the first \$ 500, 15% on sales up to \$ 1000 and 10% on sales over \$ 1000. Find each percentage of commission and add to find each total monthly salary.

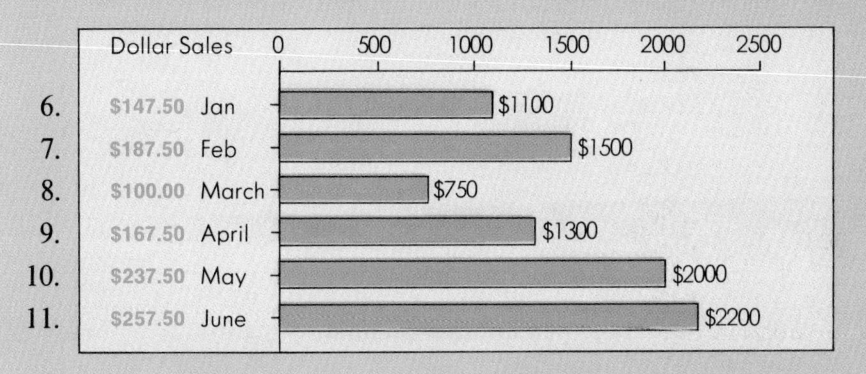

Marilyn Fisher earns \$ 3.50 an hour in salary as a waitress. She averages \$ 4.50 per hour in tips in an 8-hour day. Round each percent to the nearest tenth.

12. What are Marilyn's total daily wages? \$64.00
13. What percent of her daily wages is in tips? 56.3%
14. What percent of her daily wages is in salary? 43.8%

Last year Marilyn earned \$ 16,000 for the year. At the above rate find how much she earned in each of the following.

15. salary \$7008 16. tips \$9008 It is important for students to be able to estimate amounts such as tips. Have your students estimate the tips for these problems.

Restaurant tips are often based on the amount of the customer's check. Many diners tip 15% of the total check. Find how much that would be for the following checks.

17. \$ 20 \$3 18. \$ 46.50 \$7.00
19. \$ 9.45 \$1.42 20. \$ 58.20 \$8.73

Objective 97

FINDING A NUMBER GIVEN A PERCENT

USING A DECIMAL

6 is 75 % of what number?

$6 = 0.75 \times n$

$\frac{6}{0.75} = n$

$$0.75\overline{)6.00} = 8.$$

$$\begin{array}{r} 8. \\ 0.75\overline{)6.00} \\ \underline{600} \\ 0 \end{array}$$

$8 = n$

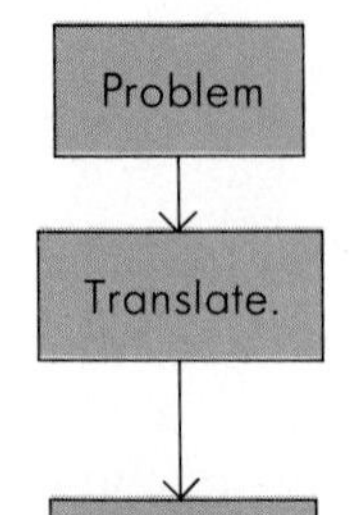

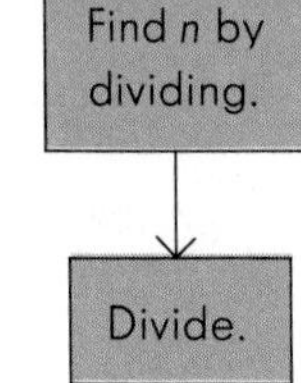

6 is 75 % of 8.

USING A FRACTION

6 is 75 % of what number?

$6 = \frac{3}{4} \times n$

Refer students to page 254 for the fraction equivalent to the percent.

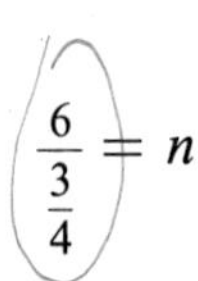

$\frac{6}{\frac{3}{4}} = n$

$6 \div \frac{3}{4} = \frac{6}{1} \times \frac{4}{3} = \frac{24}{3} = 8$

$8 = n$

EXAMPLE 1. Solve. Convert the percent to a decimal.

9 is 36 % of what number?

$9 = 0.36 \times n$ *Translate. 36% = 0.36*

$\frac{9}{0.36} = n$ *Divide.*

$$\begin{array}{r} 25. \\ 0.36\overline{)9.00} \end{array}$$

$25 = n$ 9 is 36% of 25.

Sometimes it is easier to use fractions.

EXAMPLE 2. Solve. Convert the percent to a fraction.

42 is $33\frac{1}{3}$% of what number?

$42 = \frac{1}{3} \times n$ *Translate.*

$\frac{42}{\frac{1}{3}} = n$ *Divide.*

$42 \div \frac{1}{3} = \frac{42}{1} \times \frac{3}{1} = 126$

$126 = n$ 42 is $33\frac{1}{3}$% of 126.

TRY THIS

Solve.

1. Convert the percent to a decimal.
   30 is 15% of what number? 200
2. Convert the percent to a fraction.
   12 is $66\frac{2}{3}$% of what number? 18

PRACTICE

For extra practice, use Making Practice Fun 59 with this lesson.

Solve. Convert each percent to a decimal.

1. 7 is 70% of what number? 10
2. 9 is 30% of what number? 30
3. 6 is 15% of what number? 40
4. 8 is 16% of what number? 50
5. 2 is 6% of what number? $33\frac{1}{3}$
6. 4 is 8% of what number? 50
7. 2 is 60% of what number? $3\frac{1}{3}$
8. 4 is 80% of what number? 5
9. 8 is 4% of what number? 200
10. 8 is 40% of what number? 20

Solve. Convert each percent to a fraction.

11. 4 is $12\frac{1}{2}$% of what number? 32
12. 21 is $16\frac{2}{3}$% of what number? 126
13. 8 is 20% of what number? 40
14. 9 is $12\frac{1}{2}$% of what number? 72
15. 10 is $16\frac{2}{3}$% of what number? 60
16. 4 is 50% of what number? 8
17. 9 is 25% of what number? 36
18. 72 is 20% of what number? 360
19. 1 is $33\frac{1}{3}$% of what number? 3
20. 48 is 75% of what number? 64

PROBLEM CORNER

1. Three students in the room have red hair. This is 10% of the total number of students. How many students are in the room? 30
2. 40% of a person's weight is muscle. A person has 80 pounds of muscle. What is the person's total weight? 200 pounds

More Practice, page 273

Objectives 98 and 99

PERCENT OF INCREASE OR DECREASE

Find the percent of increase: from 80 to 100

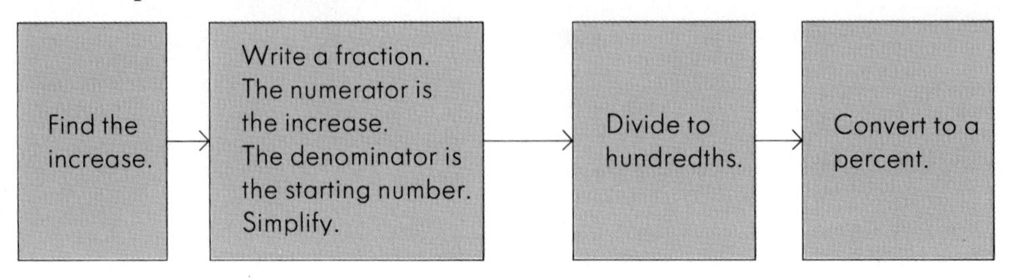

$$\begin{array}{r} 100 \\ -\ 80 \\ \hline 20 \end{array} \qquad \frac{\text{increase}}{\text{starting number}} \rightarrow \frac{20}{80} = \frac{1}{4} \qquad 4\overline{)1.00} = 0.25 \qquad 25\%$$

EXAMPLE 1. Find the percent of increase: from 30 to 40

$$\begin{array}{r} 40 \\ -\ 30 \\ \hline 10 \end{array}$$ *Find the increase.*

$$\frac{\text{increase}}{\text{starting number}} \rightarrow \frac{10}{30} = \frac{1}{3}$$ *Write a fraction. Simplify.*

$$3\overline{)1.00} = 0.33\tfrac{1}{3} = 33\tfrac{1}{3}\%$$ *Divide to hundredths. Convert to a percent.*

EXAMPLE 2. Find the percent of decrease: from 30 to 25

$$\begin{array}{r} 30 \\ -\ 25 \\ \hline 5 \end{array}$$ *Find the decrease.*

$$\frac{\text{decrease}}{\text{starting number}} \rightarrow \frac{5}{30} = \frac{1}{6}$$ *Write a fraction. Simplify.*

$$\begin{array}{r} 0.16\tfrac{4}{6} \\ 6\overline{)1.00} \\ \underline{6} \\ 40 \\ \underline{36} \\ 4 \end{array} \quad 0.16\tfrac{4}{6} = 0.16\tfrac{2}{3} = 16\tfrac{2}{3}\%$$ *Divide to hundredths. Convert to a percent.*

TRY THIS

Find the percent of increase or decrease.

1. From 12 to 20 $66\frac{2}{3}\%$ **2.** From 100 to 105 5%

3. From 15 to 12 20% **4.** From 100 to 60 40%

PRACTICE

Use Quiz 20 after this Practice.

Find the percent of increase or decrease.

1. From 8 to 10 25%	**2.** From 12 to 15 25%	**3.** From 20 to 25 25%
4. From 10 to 8 20%	**5.** From 24 to 18 25%	**6.** From 40 to 30 25%
7. From 10 to 15 50%	**8.** From 6 to 9 50%	**9.** From 5 to 7.5 50%
10. From 20 to 10 50%	**11.** From 34 to 17 50%	**12.** From 16 to 8 50%
13. From 5 to 10 100%	**14.** From 10 to 20 100%	**15.** From 50 to 100 100%
16. From 9 to 3 $66\frac{2}{3}\%$	**17.** From 42 to 35 $16\frac{2}{3}\%$	**18.** From 100 to 87 13%
19. From 6 to 8 $33\frac{1}{3}\%$	**20.** From 9 to 12 $33\frac{1}{3}\%$	**21.** From 12 to 16 $33\frac{1}{3}\%$
22. From 80 to 60 25%	**23.** From 45 to 35 $22\frac{2}{9}\%$	**24.** From 50 to 30 40%
25. From 10 to 11 10%	**26.** From 16 to 22 $37\frac{1}{2}\%$	**27.** From 45 to 50 $11\frac{1}{9}\%$
28. From 3 to 2 $33\frac{1}{3}\%$	**29.** From 4 to 3 25%	**30.** From 5 to 4 20%

PROBLEM CORNER

1. Jose weighed 100 pounds. A year later he weighed 120 pounds. What was his percent of increase in weight? 20%

2. Maria lowered the thermostat setting in her house. The heating bill was $ 80 per month. Now it is $ 75. What was the percent of decrease? $6\frac{1}{4}\%$

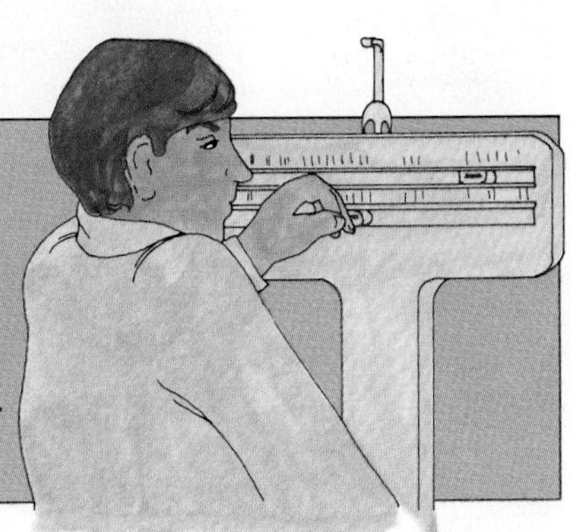

More Practice, page 273

Objectives 99, 176, 177, 178, and 179

APPLICATION

Students should realize that sales are used by the merchandiser for many reasons. Sale items and prices should be evaluated carefully.

BUYING ON SALE

Sporting Equipment on Sale—15% to 50% OFF the regular price.

Find the sale price. Find the amount of savings.

1. $12.76; $3.19
   BASEBALL GLOVES
   Regularly $15.95
   NOW 20% Off

2. $2.63; $2.62
   TENNIS BALLS
   50% Off Our Regular Price of $5.25

3. $8.21; $2.74
   25% Off
   PRO SKATEBOARDS
   Regularly $10.95

Find the percent of decrease.

4. 11.1%
   TENNIS RACQUETS
   Were $36
   NOW $32

5. 16.7%
   FOOTBALLS
   Formerly $12.00
   NOW $10

6. 40%
   SOCCER BALLS
   Were $8.95
   NOW $5.37

Find the percent off. Find the sale price.

7. $33\frac{1}{3}$%; $7.32
   1/3 OFF Were $10.98
   SIDEWALK SKATES
   NOW ______

8. 50%; $1.49
   1/2 OFF
   PING-PONG BALLS
   Were $2.98 per box
   NOW ______ box

9. 25%; $18.66
   1/4 Off
   SKI PANTS
   Were $24.88
   NOW ______

Find the original price.

10. $24.00
    FISHING RODS
    10% Off
    NOW $21.60

11. $39.00
    WARM-UP SUITS
    1/4 OFF
    NOW $29.25

12. $10.00
    SOCCER SHORTS
    30% Off
    NOW $7.00

A popular type of sale offer:

BUY 2 GET 1 FREE

You really pay for 2 items and get 3 items.

Each item costs $\frac{2}{3}$ of the regular price. The savings is $\frac{1}{3}$ or $33\frac{1}{3}\%$.

Find the sale price for each item.

BUY 2 GET 1 Free

13. Soap 29¢ each bar $19\frac{1}{3}$¢ or 19¢
14. Soup 55¢ per can $36\frac{2}{3}$¢ or 37¢
15. Dog Food 43¢ per can $28\frac{2}{3}$¢ or 29¢
16. Toothpaste 89¢ each tube $59\frac{1}{3}$¢ or 59¢

Read the ad carefully. Find the unit sale price. Find the percent of savings.

2 FOR 1 SALE
GET 2 FOR THE PRICE OF 1

17. Paper towels 66¢ each roll 33¢; 50%
18. Note pads 48¢ each 24¢; 50%
19. Auto oil $ 1.08 per qt 54¢; 50%
20. Soap powder $ 4.26 per box $2.13; 50%

Assume the quality is equal. Find the sale price for each size. Which is the most economical?

21.

SOAP
20 oz size, regularly 60¢, now 25% off
40 oz size, regularly $1.08, now 20% off
104 oz size, regularly $2.55, now 15% off

20 oz: 45¢; 40 oz: 86.4¢
104 oz: $2.17; 104 oz size

22.

APPLE JUICE
8 oz can, regularly 8/$1.00, now 15/$1.25
24 oz can, regularly 3/$1.00, now 4/$1.00
64 oz can, regularly 2/$1.59, now 2/$1.35

8 oz: $8\frac{1}{3}$¢ or 8¢; 24 oz: 25¢
64 oz: $67\frac{1}{2}$¢ or 68¢; 8 oz can

How much change would you receive from a $ 20 bill for each purchase?

23. Buy 6 pairs of socks at 25% off. The regular price is $ 1.78 per pair. $11.99
24. Buy 1 dozen drinking glasses. The regular cost is 69¢ each. On sale you can buy 3 for the price of 2. $14.48
25. Buy 15 rolls of paper towels at $\frac{1}{3}$ off. The regular price is 3 for $ 1.00. $16.67
26. Buy 2 pounds of meat at $\frac{2}{3}$ of the regular price of $ 1.47 per pound. $18.04
27. Buy 5 pounds of meat at $\frac{1}{3}$ off. The regular price is $ 1.98 per pound. $13.40

CAREER CAREER CAREER CAREER CAREER

BEAUTICIAN

Bill Oberle owns The Beauty Culture Studio. Among the specialists working for him are hair stylists, manicurists, and facial treatment operators.

Formula for Mrs. Darlene Spies	
95 mL of #1 red	25 mL of #4 drab
Mix with 200 mL peroxide	

Hair coloring requires knowledge of chemistry and proportions. Each customer has a special color formula.

1. What percent of the solution is #1 red? 30%
2. What percent of the solution is #4 drab? 8%
3. What is the ratio of red to drab?

Bill used only 10 mL of #4 drab to make a small amount of solution.

4. How much of #1 red should be used? 38 mL
5. How much peroxide should be used? 80 mL

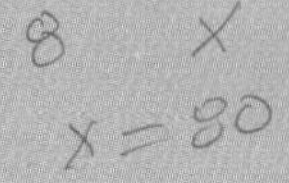

Bill pays $ 8.00 for a 5 L jug of shampoo concentrate. 60 mL of the concentrate is used for each customer's shampoo.

6. How many shampoos can be done from a jug of concentrate? 83
7. How much does the concentrate for one shampoo cost? 9.6¢

Bill's employees earn a weekly salary. They also get tips and a percentage of their weekly gross sales.

Income for One Week—5 Day Week				
Operator	Salary	Percentage	Gross Sales	Tips
Sally	$ 50	20%	$ 425	$ 65
Joel	$ 75	25%	$ 675	$ 95
Eiko	$100	$37\frac{1}{2}$%	$ 842	$120
Jorge	$125	15%	$1250	$175

Find the income for one week for each operator. (Income = salary + % of gross + tips)

8. Sally $200
9. Joel $338.75
10. Eiko $535.75
11. Jorge $487.50

Tips are related to gross sales. Many patrons tip according to their bill for services.
Express the ratio of tips to sales to the nearest 1%.

12. Sally 15%
13. Joel 14%
14. Eiko 14%
15. Jorge 14%

Joel's goal is to increase his gross sales to $ 783 per week.

16. What percent of increase in gross sales does he want? 16%
17. How much does he estimate his tips to be? (See problem 13.) $109.62
18. How much will Joel earn weekly? $380.37

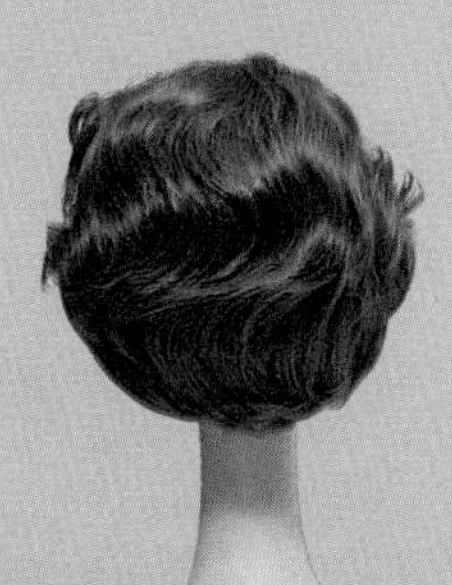
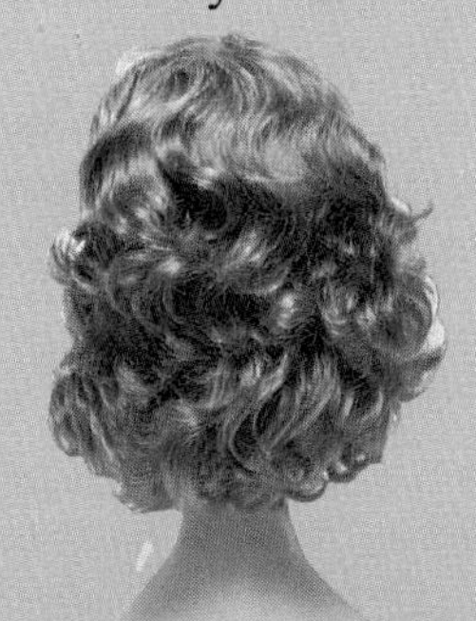

CHAPTER 9 REVIEW

For extra practice, use Making Practice Fun 60 with this lesson.

Convert to a decimal. p. 248

1. 43% 0.43 **2.** 120% 1.20 **3.** 8% 0.08 **4.** 6.3% 0.063

Convert to a percent. p. 248

5. 0.19 19% **6.** 1.78 178% **7.** 0.05 5% **8.** 0.004 0.4%

Convert to a fraction, mixed number, or whole number. p. 250

9. 49% $\frac{49}{100}$ **10.** 24% $\frac{6}{25}$ **11.** 145% $1\frac{9}{20}$ **12.** $33\frac{1}{3}\%$ $\frac{1}{3}$

Convert to a percent. p. 252

13. $\frac{3}{5}$ 60% **14.** $\frac{4}{25}$ 16% **15.** $2\frac{1}{4}$ 225% **16.** 5 500%

17. $\frac{1}{3}$ $33\frac{1}{3}\%$ **18.** $\frac{3}{8}$ $37\frac{1}{2}\%$ **19.** $2\frac{2}{3}$ $266\frac{2}{3}\%$ **20.** $1\frac{2}{7}$ 128.6%

Translate. Do not solve. p. 254

21. 17% of 35 is what number? $17\% \times 35 = n$

22. What percent of 20 is 4? $x \times 20 = 4$

23. 54 is what percent of 96? $54 = x \times 96$

24. 7 is $33\frac{1}{3}\%$ of what number? $7 = 33\frac{1}{3}\% \times n$

Solve. pp. 256–262

25. 36% of 50 is what number? 18

26. What is 90% of 120? 108

27. What percent of 12 is 6? 50%

28. 7 is what percent of 21? $33\frac{1}{3}\%$

29. 40 is $33\frac{1}{3}\%$ of what number? 120

30. 9 is 75% of what number? 12

Find the percent of increase or decrease. p. 264

31. From 15 to10 $33\frac{1}{3}\%$

32. From 10 to 15 50%

Application pp. 260, 266

33. Sandra earns 8% commission on her monthly sales. How much commission does she earn on total monthly sales of $ 8450? $676

34. Skateboards usually cost $ 24. On sale they cost $ 18. Find the percent of decrease in price. 25%

CHAPTER 9 TEST

Convert to a decimal.

1. 97% 0.97 **2.** 326% 3.26 **3.** 7% 0.07 **4.** 2.7% 0.027

Convert to a percent.

5. 0.37 37% **6.** 1.62 162% **7.** 0.03 3% **8.** 0.007 0.7%

Convert to a fraction, mixed number, or whole number.

9. 67% $\frac{67}{100}$ **10.** 15% $\frac{3}{20}$ **11.** 125% $1\frac{1}{4}$ **12.** $66\frac{2}{3}\%$ $\frac{2}{3}$

Convert to a percent.

13. $\frac{4}{5}$ 80% **14.** $\frac{7}{20}$ 35% **15.** $1\frac{3}{4}$ 175% **16.** 4 400%

17. $\frac{3}{8}$ $37\frac{1}{2}\%$ **18.** $\frac{3}{7}$ 42.9% **19.** $1\frac{1}{6}$ $116\frac{2}{3}\%$ **20.** $2\frac{5}{9}$ 255.6%

Translate. Do not solve.

21. 15 is what percent of 45? $15 = x \times 45$

22. 76% of 19 is what number? $76\% \times 19 = x$

23. 9 is 50% of what number? $9 = 50\% \times x$

24. What percent of 9 is 5? $n \times 9 = 5$

Solve.

25. What is 25% of 32? 8

26. $16\frac{2}{3}\%$ of 42 is what number? 7

27. 9 is what percent of 45? 20%

28. What percent of 18 is 12? $66\frac{2}{3}\%$

29. 50 is 25% of what number? 200

30. 7 is 75% of what number? $9\frac{1}{3}$

Find the percent of increase or decrease.

31. From 20 to 25 25%

32. From 25 to 20 20%

Application

33. Russ earns $ 3.25 an hour in salary. He averages $ 4.75 an hour in tips. How much does he earn in an 8-hour day? $64

34. Ski jackets are on sale for $\frac{1}{4}$ off. The regular price is $ 30. Find the sale price. $22.50

MORE PRACTICE

Convert to a decimal. Use after page 249.

1. 44% 0.44 **2.** 38% 0.38 **3.** 78% 0.78 **4.** 96% 0.96

5. 130% 1.30 **6.** 240% 2.40 **7.** 350% 3.50 **8.** 680% 6.80

9. 9% 0.09 **10.** 7% 0.07 **11.** 1% 0.01 **12.** 5% 0.05

13. 7.8% 0.078 **14.** 9.2% 0.092 **15.** 0.3% 0.003 **16.** 0.7% 0.007

Convert to a percent. Use after page 249.

17. 0.26 26% **18.** 0.98 98% **19.** 0.39 39% **20.** 0.46 46%

21. 1.12 112% **22.** 1.37 137% **23.** 2.85 285% **24.** 4.32 432%

25. 0.01 1% **26.** 0.07 7% **27.** 0.09 9% **28.** 0.06 6%

29. 0.003 0.3% **30.** 0.008 0.8% **31.** 0.001 0.1% **32.** 0.007 0.7%

Convert to a fraction, mixed number, or whole number. Use after page 251.

33. 77% $\frac{77}{100}$ **34.** 13% $\frac{13}{100}$ **35.** 45% $\frac{9}{20}$ **36.** 75% $\frac{3}{4}$

37. 175% $1\frac{3}{4}$ **38.** 250% $2\frac{1}{2}$ **39.** 400% 4 **40.** 700% 7

Convert to a percent. Use after page 253.

41. $\frac{4}{5}$ 80% **42.** $\frac{7}{10}$ 70% **43.** $\frac{7}{25}$ 28% **44.** $\frac{4}{50}$ 8%

45. $3\frac{3}{4}$ 375% **46.** $1\frac{1}{2}$ 150% **47.** 2 200% **48.** 6 600%

49. $\frac{2}{3}$ $66\frac{2}{3}\%$ **50.** $\frac{1}{8}$ $12\frac{1}{2}\%$ **51.** $\frac{5}{8}$ $62\frac{1}{2}\%$ **52.** $\frac{5}{6}$ $83\frac{1}{3}\%$

53. $1\frac{1}{3}$ $133\frac{1}{3}\%$ **54.** $3\frac{3}{8}$ $337\frac{1}{2}\%$ **55.** $2\frac{1}{6}$ $216\frac{2}{3}\%$ **56.** $4\frac{7}{8}$ $487\frac{1}{2}\%$

Translate. Do not solve. Use after page 255.

57. 18% of 36 is what number? $18\% \times 36 = n$

58. What is 19% of 76? $n = 19\% \times 76$

59. What percent of 40 is 8? $n \times 40 = 8$

60. What percent of 70 is 6? $n \times 70 = 6$

Solve. Use after page 257.

1. 37% of 10 is what number? 3.7

2. 75% of 12 is what number? 9

3. 90% of 20 is what number? 18

4. $12\frac{1}{2}$% of 16 is what number? 2

5. What is 80% of 70? 56

6. What is 25% of 20? 5

7. What is $87\frac{1}{2}$% of 24? 21

8. What is $66\frac{2}{3}$% of 30? 20

9. $33\frac{1}{3}$% of 60 is what number? 20

10. What is 50% of 70? 35

Solve. Use after page 259.

11. What percent of 20 is 10? 50%

12. What percent of 15 is 3? 20%

13. What percent of 100 is 39? 39%

14. What percent of 90 is 9? 10%

15. 6 is what percent of 30? 20%

16. 9 is what percent of 36? 25%

17. 12 is what percent of 36? $33\frac{1}{3}$%

18. 18 is what percent of 9? 200%

19. What percent of 50 is 25? 50%

20. 25 is what percent of 75? $33\frac{1}{3}$%

Solve. Use after page 263.

21. 6 is 10% of what number? 60

22. 7 is 20% of what number? 35

23. 8 is $33\frac{1}{3}$% of what number? 24

24. 32 is 50% of what number? 64

25. 30 is 75% of what number? 40

26. 16 is $66\frac{2}{3}$% of what number? 24

27. 9 is 100% of what number? 9

28. 50 is 100% of what number? 50

29. 6 is $12\frac{1}{2}$% of what number? 48

30. 50 is 50% of what number? 100

Find the percent of increase or decrease. Use after page 265.

31. From 18 to 25 38.9%

32. From 15 to 18 20%

33. From 35 to 30 14.3%

34. From 30 to 35 $16\frac{2}{3}$%

35. From 100 to 120 20%

36. From 120 to 100 $16\frac{2}{3}$%

SKILLS REVIEW

Add. Refer to pages 14–17.

1. $298 + 472$ — 770
2. $6497 + 8584$ — 15,081
3. $398{,}108 + 169{,}942$ — 568,050
4. $473{,}493 + 87{,}875$ — 561,368
5. $4298 + 7394 + 6285 + 4073$ — 22,050
6. $17{,}478 + 29{,}407 + 33{,}987 + 49{,}785$ — 130,657
7. $496{,}784 + 748{,}147 + 286{,}595 + 985{,}589$ — 2,517,115
8. $429{,}784 + 68{,}498 + 8{,}747 + 9{,}606$ — 516,635

Add. Refer to page 22.

9. $9.9 + 3.6$ — 13.5
10. $6.72 + 3.98$ — 10.70
11. $14.784 + 9.608$ — 24.392
12. $5.78 + 6.9$ — 12.68
13. $3.97 + 0.4 + 0.827$ — 5.197
14. $0.984 + 7.38 + 1.8$ — 10.164
15. $7.673 + 0.8 + 0.87 + 8.107$ — 17.450
16. $6.984 + 8.746 + 9.98 + 0.7$ — 26.410
17. $4.39 + 15.8$ — 20.19
18. $0.6 + 9.7$ — 10.3
19. $2.4 + 0.3 + 5.71$ — 8.41
20. $9.52 + 11.6$ — 21.12

Subtract. Refer to pages 40–45.

21. $749 - 298$ — 451
22. $37{,}385 - 19{,}496$ — 17,889
23. $742{,}781 - 647{,}398$ — 95,383
24. $725{,}684 - 39{,}708$ — 685,976
25. $800 - 147$ — 653
26. $3007 - 2988$ — 19
27. $74{,}000 - 28{,}146$ — 45,854
28. $90{,}085 - 18{,}198$ — 71,887

Subtract. Refer to page 50.

29. $8.7 - 1.9$ — 6.8
30. $9.37 - 8.19$ — 1.18
31. $6.784 - 0.928$ — 5.856
32. $6.085 - 4.276$ — 1.809
33. $9.742 - 1.8$ — 7.942
34. $8.072 - 3.86$ — 4.212
35. $7.4 - 1.874$ — 5.526
36. $6 - 1.72$ — 4.28
37. $13.68 - 6$ — 7.68
38. $22 - 8.4$ — 13.6
39. $5.6 - 0.89$ — 4.71
40. $24.1 - 7.93$ — 16.17
41. $18 - 9.3$ — 8.7
42. $16.2 - 6.43$ — 9.77
43. $19.71 - 4$ — 15.71
44. $20.3 - 7.28$ — 13.02

Multiply. Refer to pages 68–73.

1. 88×7 616 **2.** 427×8 3416 **3.** 7264×6 43,584 **4.** 87×25 2175

5. 426×37 15,762 **6.** 758×643 487,394 **7.** 4986×278 1,386,108 **8.** 6720×807 5,423,040

Multiply. Refer to page 82.

9. 9.8×6 58.8 **10.** 7.32×5 36.60 **11.** 9.376×4 37.504 **12.** 7.8×3.2 24.96

13. 8.32×6.9 57.408 **14.** 9.78×3.28 32.0784 **15.** 0.3×0.2 0.06 **16.** 0.04×0.1 0.004

17. 4.7×1.3 6.11 **18.** 2.9×0.9 2.61 **19.** 5.82×1.3 7.566 **20.** 12.7×2.86 36.322

Divide. Refer to pages 98–105.

21. $59 \div 8$ 7r3 **22.** $75 \div 4$ 18r3 **23.** $359 \div 6$ 59r5 **24.** $962 \div 7$ 137r3

25. $4\overline{)5732}$ 1433 **26.** $9\overline{)6354}$ 706 **27.** $21\overline{)75}$ 3r12 **28.** $84\overline{)19{,}864}$ 236r40

29. $73\overline{)19{,}710}$ 270 **30.** $846\overline{)6768}$ 8 **31.** $379\overline{)8768}$ 23r51 **32.** $842\overline{)42{,}100}$ 50

Divide. Refer to pages 108–117.

33. $12.32 \div 7$ 1.76 **34.** $2.96 \div 8$ 0.37 **35.** $243.2 \div 64$ 3.8 **36.** $381.80 \div 92$ 4.15

37. $7\overline{)0.14}$ 0.02 **38.** $9\overline{)2.7}$ 0.3 **39.** $8\overline{)2}$ 0.25 **40.** $3.1\overline{)105.4}$ 34

41. $7.3\overline{)70.08}$ 9.6 **42.** $0.3\overline{)6}$ 20 **43.** $0.08\overline{)64}$ 800 **44.** $0.12\overline{)6}$ 50

Find each quotient to the nearest tenth. Refer to page 118.

45. $4 \div 18$ 0.2 **46.** $59 \div 9$ 6.6 **47.** $2.5 \div 6$ 0.4 **48.** $11.8 \div 7$ 1.7

49. $5.6 \div 3.4$ 1.6 **50.** $4 \div 0.7$ 5.7 **51.** $6 \div 1.4$ 4.3 **52.** $1.5 \div 0.91$ 1.6

53. $8\overline{)14}$ 1.8 **54.** $9\overline{)6}$ 0.7 **55.** $0.6\overline{)0.5}$ 0.8 **56.** $0.09\overline{)1.7}$ 18.9

57. $2.6\overline{)4.5}$ 1.7 **58.** $0.23\overline{)10}$ 43.5 **59.** $0.3\overline{)2}$ 6.7 **60.** $4.8\overline{)3}$ 0.6

61. $8\overline{)4.6}$ 0.6 **62.** $0.03\overline{)0.47}$ 15.7 **63.** $4.2\overline{)9.36}$ 2.2 **64.** $15\overline{)13}$ 0.9

What part is shaded? Refer to page 136.

1. 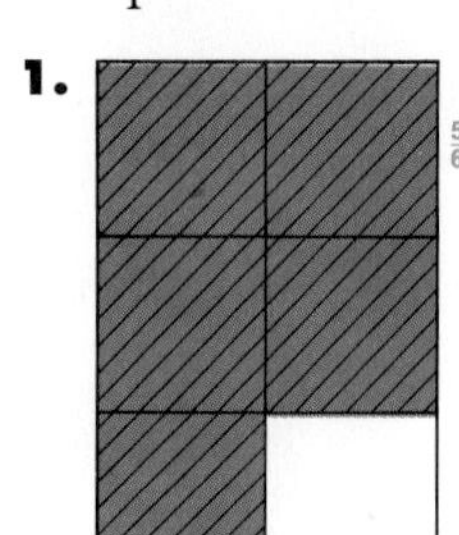 $\frac{5}{6}$

2. $\frac{3}{8}$

3. 1

4. 0

Compare. Use $>$ or $<$. Refer to page 142.

5. $\frac{2}{3}$ and $\frac{3}{4}$ $<$ **6.** $\frac{5}{6}$ and $\frac{4}{5}$ $>$ **7.** $\frac{7}{8}$ and $\frac{9}{10}$ $<$ **8.** $\frac{3}{2}$ and $\frac{4}{3}$ $>$

9. $\frac{5}{8}$ and $\frac{6}{8}$ $<$ **10.** $\frac{9}{10}$ and $\frac{8}{10}$ $>$ **11.** $\frac{7}{4}$ and $\frac{5}{4}$ $>$ **12.** $\frac{9}{8}$ and $\frac{12}{8}$ $<$

Convert to a fraction. Refer to page 150.

13. $2\frac{2}{3}$ $\frac{8}{3}$ **14.** $3\frac{3}{4}$ $\frac{15}{4}$ **15.** $1\frac{7}{8}$ $\frac{15}{8}$ **16.** $2\frac{1}{2}$ $\frac{5}{2}$

17. $7\frac{5}{8}$ $\frac{61}{8}$ **18.** $6\frac{1}{3}$ $\frac{19}{3}$ **19.** $3\frac{2}{5}$ $\frac{17}{5}$ **20.** $3\frac{7}{10}$ $\frac{37}{10}$

Convert to a mixed number. Refer to page 152.

21. $\frac{7}{4}$ $1\frac{3}{4}$ **22.** $\frac{9}{8}$ $1\frac{1}{8}$ **23.** $\frac{13}{2}$ $6\frac{1}{2}$ **24.** $\frac{17}{4}$ $4\frac{1}{4}$

25. $\frac{10}{3}$ $3\frac{1}{3}$ **26.** $\frac{16}{5}$ $3\frac{1}{5}$ **27.** $\frac{23}{10}$ $2\frac{3}{10}$ **28.** $\frac{17}{12}$ $1\frac{5}{12}$

Convert to a decimal. Refer to page 154.

29. $6\frac{3}{10}$ 6.3 **30.** $7\frac{25}{100}$ 7.25 **31.** $6\frac{327}{1000}$ 6.327 **32.** $5\frac{3}{100}$ 5.03

33. $9\frac{7}{8}$ 9.875 **34.** $4\frac{1}{4}$ 4.25 **35.** $87\frac{2}{5}$ 87.4 **36.** $36\frac{9}{20}$ 36.45

Multiply and simplify. Refer to page 174.

37. $\frac{3}{8} \cdot \frac{2}{3}$ $\frac{1}{4}$ **38.** $\frac{1}{2} \cdot \frac{8}{9}$ $\frac{4}{9}$ **39.** $\frac{7}{8} \cdot \frac{2}{3}$ $\frac{7}{12}$ **40.** $\frac{4}{12} \cdot \frac{3}{4}$ $\frac{1}{4}$

41. $\frac{1}{2} \cdot 8$ 4 **42.** $9 \cdot \frac{2}{3}$ 6 **43.** $3 \cdot \frac{1}{9}$ $\frac{1}{3}$ **44.** $\frac{3}{4} \cdot 12$ 9

45. $\frac{11}{9} \cdot \frac{3}{5}$ $\frac{11}{15}$ **46.** $\frac{5}{8} \cdot \frac{4}{3}$ $\frac{5}{6}$ **47.** $\frac{12}{5} \times \frac{7}{3}$ $5\frac{3}{5}$ **48.** $\frac{8}{3} \times \frac{7}{4}$ $4\frac{2}{3}$

Multiply and simplify. Refer to page 176.

1. $6 \times 2\frac{1}{2}$ 15 **2.** $9 \times 3\frac{1}{3}$ 30 **3.** $3\frac{3}{4} \times 8$ 30 **4.** $1\frac{2}{5} \times 10$ 14

5. $3\frac{1}{6} \times 1\frac{1}{2}$ $4\frac{3}{4}$ **6.** $4\frac{3}{8} \times 3\frac{2}{5}$ $14\frac{7}{8}$ **7.** $6\frac{3}{10} \times 1\frac{2}{3}$ $10\frac{1}{2}$ **8.** $2\frac{1}{2} \times 2\frac{1}{2}$ $6\frac{1}{4}$

Divide and simplify. Refer to page 180.

9. $12 \div \frac{1}{2}$ 24 **10.** $10 \div \frac{3}{5}$ $16\frac{2}{3}$ **11.** $\frac{3}{4} \div 2$ $\frac{3}{8}$ **12.** $\frac{5}{8} \div 5$ $\frac{1}{8}$

13. $\frac{2}{3} \div \frac{2}{3}$ 1 **14.** $\frac{3}{4} \div \frac{1}{2}$ $1\frac{1}{2}$ **15.** $\frac{1}{8} \div \frac{2}{3}$ $\frac{3}{16}$ **16.** $\frac{7}{8} \div \frac{3}{10}$ $2\frac{11}{12}$

Divide and simplify. Refer to page 182.

17. $10 \div 2\frac{1}{2}$ 4 **18.** $9 \div 3\frac{2}{3}$ $2\frac{5}{11}$ **19.** $7\frac{1}{2} \div 2$ $3\frac{3}{4}$ **20.** $4\frac{3}{4} \div 19$ $\frac{1}{4}$

21. $6\frac{2}{3} \div 3\frac{1}{3}$ 2 **22.** $7\frac{5}{6} \div 2\frac{1}{2}$ $3\frac{2}{15}$ **23.** $9\frac{3}{4} \div \frac{1}{4}$ 39 **24.** $8\frac{3}{10} \div \frac{3}{10}$ $27\frac{2}{3}$

Add and simplify. Refer to pages 196–202.

25. $\frac{3}{4} + \frac{3}{4}$ $1\frac{1}{2}$ **26.** $\frac{7}{8} + \frac{1}{8}$ 1 **27.** $\frac{5}{6} + \frac{5}{6}$ $1\frac{2}{3}$ **28.** $\frac{7}{10} + \frac{1}{10}$ $\frac{4}{5}$

29. $\frac{3}{12} + \frac{3}{12}$ $\frac{1}{2}$ **30.** $\frac{1}{2} + \frac{1}{2}$ 1 **31.** $\frac{7}{10} + \frac{5}{10}$ $1\frac{1}{5}$ **32.** $\frac{7}{3} + \frac{6}{3}$ $4\frac{1}{3}$

33. $\frac{1}{4} + \frac{1}{2}$ $\frac{3}{4}$ **34.** $\frac{3}{4} + \frac{1}{2}$ $1\frac{1}{4}$ **35.** $\frac{7}{8} + \frac{3}{4}$ $1\frac{5}{8}$ **36.** $\frac{7}{10} + \frac{3}{5}$ $1\frac{3}{10}$

37. $\frac{1}{3} + \frac{1}{2} + \frac{1}{4}$ $1\frac{1}{12}$ **38.** $\frac{3}{8} + \frac{7}{8} + \frac{1}{2}$ $1\frac{3}{4}$ **39.** $\frac{9}{10} + \frac{1}{5} + \frac{5}{6}$ $1\frac{14}{15}$ **40.** $\frac{3}{4} + \frac{3}{5} + \frac{1}{2}$ $1\frac{17}{20}$

41. $5\frac{1}{2} + 5\frac{1}{4}$ $10\frac{3}{4}$ **42.** $7\frac{2}{3} + 1\frac{2}{3}$ $9\frac{1}{3}$ **43.** $3\frac{1}{3} + 1\frac{5}{6}$ $5\frac{1}{6}$ **44.** $8\frac{1}{2} + 6\frac{6}{12}$ 15

45. $9\frac{4}{5} + 8$ $17\frac{4}{5}$ **46.** $8 + 12\frac{3}{8}$ $20\frac{3}{8}$ **47.** $2\frac{1}{2} + 2\frac{1}{4} + 2\frac{1}{3}$ $7\frac{1}{12}$ **48.** $7\frac{3}{4} + 2\frac{1}{6} + 3\frac{5}{12}$ $13\frac{1}{3}$

Subtract and simplify. Refer to page 206.

49. $\frac{9}{12} - \frac{3}{12}$ $\frac{1}{2}$ **50.** $\frac{7}{8} - \frac{1}{8}$ $\frac{3}{4}$ **51.** $\frac{6}{10} - \frac{1}{10}$ $\frac{1}{2}$ **52.** $\frac{5}{6} - \frac{1}{6}$ $\frac{2}{3}$

53. $\frac{7}{12} - \frac{1}{3}$ $\frac{1}{4}$ **54.** $\frac{8}{9} - \frac{2}{3}$ $\frac{2}{9}$ **55.** $\frac{1}{2} - \frac{3}{8}$ $\frac{1}{8}$ **56.** $\frac{4}{5} - \frac{7}{10}$ $\frac{1}{10}$

57. $\frac{7}{8} - \frac{5}{6}$ $\frac{1}{24}$ **58.** $\frac{3}{5} - \frac{4}{9}$ $\frac{7}{45}$ **59.** $\frac{4}{15} - \frac{1}{6}$ $\frac{1}{10}$ **60.** $\frac{2}{3} - \frac{9}{16}$ $\frac{5}{48}$

61. $\frac{3}{10} - \frac{1}{7}$ $\frac{11}{70}$ **62.** $\frac{11}{12} - \frac{3}{8}$ $\frac{13}{24}$ **63.** $\frac{5}{6} - \frac{2}{9}$ $\frac{11}{18}$ **64.** $\frac{1}{2} - \frac{1}{5}$ $\frac{3}{10}$

Subtract and simplify. Refer to page 208.

1. $9\frac{7}{12} - 6\frac{1}{12}$ $3\frac{1}{2}$ **2.** $8\frac{7}{8} - 1\frac{6}{8}$ $7\frac{1}{8}$ **3.** $7\frac{2}{3} - 3$ $4\frac{2}{3}$ **4.** $12\frac{7}{10} - 7$ $5\frac{7}{10}$

5. $4\frac{4}{5} - 3\frac{1}{10}$ $1\frac{7}{10}$ **6.** $8\frac{5}{6} - 1\frac{2}{3}$ $7\frac{1}{6}$ **7.** $13\frac{7}{8} - 4\frac{1}{2}$ $9\frac{3}{8}$ **8.** $17\frac{1}{3} - 8\frac{1}{4}$ $9\frac{1}{12}$

Subtract and simplify. Refer to page 210.

9. $9 - 2\frac{1}{2}$ $6\frac{1}{2}$ **10.** $6 - 3\frac{3}{4}$ $2\frac{1}{4}$ **11.** $7 - 5\frac{3}{10}$ $1\frac{7}{10}$ **12.** $6 - 5\frac{2}{3}$ $\frac{1}{3}$

13. $10\frac{1}{6} - 3\frac{5}{6}$ $6\frac{1}{3}$ **14.** $4\frac{3}{8} - 1\frac{7}{8}$ $2\frac{1}{2}$ **15.** $5\frac{1}{2} - 3\frac{3}{4}$ $1\frac{3}{4}$ **16.** $6\frac{1}{10} - 3\frac{1}{5}$ $2\frac{9}{10}$

Write the ratio. Simplify, if possible. Refer to page 224.

17. 8 pencils for 4 people $\frac{2}{1}$

18. 6 problems in 18 minutes $\frac{1}{3}$

19. 5 servings per 5 people $\frac{1}{1}$

20. 8 dimes for 16 nickels $\frac{1}{2}$

Find the unit rate. Refer to page 226.

21. 5 items cost 45¢. What is the cost per item? 9¢

22. $ 16.60 is received for 4 hours of work. What is the amount per hour? $4.15

23. A bus driver travels 3200 miles in 6 hours. How many miles per hour is this? Round to tenths. 533.3 mi/hr

24. A car travels 186 kilometers on 8 liters of gas. How many kilometers per liter is this? Round to tenths. 23.3 km/L

Solve. Refer to page 230.

25. $\frac{2}{3} = \frac{x}{15}$ 10 **26.** $\frac{3}{4} = \frac{15}{x}$ 20 **27.** $\frac{3}{4} = \frac{x}{9}$ $6\frac{3}{4}$ **28.** $\frac{5}{6} = \frac{12}{x}$ 14.4

29. $\frac{7}{x} = \frac{5}{8}$ 11.2 **30.** $\frac{x}{3} = \frac{9}{4}$ 6.75 **31.** $\frac{9}{x} = \frac{2}{7}$ $31\frac{1}{2}$ **32.** $\frac{x}{10} = \frac{5}{6}$ $8\frac{1}{3}$

Solve. Refer to page 232.

33. 4 pens are needed for every 2 students. How many are needed for 10 students? 20 pens

34. $ 9.75 is paid for 3 hours of work. How much is paid for 2 hours of work? $6.50

35. If 12 pounds of fruit cost 96¢, how much do 10 pounds cost? 80¢

36. If it takes 15 minutes to answer 5 questions, how long does it take to answer 7 questions? 21 minutes

Convert to a decimal. Refer to page 248.

1. 73% 0.73 **2.** 140% 1.40 **3.** 6% 0.06 **4.** 7.3% 0.073

5. 17% 0.17 **6.** 25% 0.25 **7.** 150% 1.50 **8.** 13.7% 0.137

Convert to a percent. Refer to page 248.

9. 0.37 37% **10.** 0.19 19% **11.** 0.06 6% **12.** 0.60 60%

13. 1.35 135% **14.** 0.003 0.3% **15.** 0.99 99% **16.** 1.78 178%

Convert to a fraction, mixed number, or whole number. Refer to page 250.

17. 27% $\frac{27}{100}$ **18.** 79% $\frac{79}{100}$ **19.** 35% $\frac{7}{20}$ **20.** 80% $\frac{4}{5}$

21. $66\frac{2}{3}\%$ $\frac{2}{3}$ **22.** $12\frac{1}{2}\%$ $\frac{1}{8}$ **23.** 175% $1\frac{3}{4}$ **24.** 300% 3

Convert to a percent. Refer to page 252.

25. $\frac{4}{5}$ 80% **26.** $\frac{7}{25}$ 28% **27.** $3\frac{3}{4}$ 375% **28.** 4 400%

29. $\frac{7}{8}$ $87\frac{1}{2}\%$ **30.** $1\frac{1}{3}$ $133\frac{1}{3}\%$ **31.** $3\frac{3}{7}$ 342.9% **32.** $\frac{5}{6}$ 83.3%

Solve. Refer to pages 256–263.

33. 75% of 10 is what number? 7.5

34. 90% of 20 is what number? 18

35. What is 83% of 100? 83

36. What is 25% of 40? 10

37. What percent of 14 is 7? 50%

38. What percent of 40 is 4? 10%

39. 9 is what percent of 27? $33\frac{1}{3}\%$

40. 30 is what percent of 40? 75%

41. 50 is 50% of what number? 100

42. 12 is 25% of what number? 48

43. 6 is $33\frac{1}{3}\%$ of what number? 18

44. 7 is $16\frac{2}{3}\%$ of what number? 42

Find the percent of increase or decrease. Refer to page 264.

45. From 2 to 3 50%

46. From 3 to 2 $33\frac{1}{3}\%$

47. From 15 to 20 $33\frac{1}{3}\%$

48. From 20 to 15 25%

Objective 196

SOLVING PROBLEMS

This activity could be completed orally with students providing additional examples.

For extra practice, use Making Practice Fun 61 with this lesson.

When solving a problem, we sometimes need to decide whether to do a mental estimate, a paper and pencil calculation, or use a calculator. For each of the following problem situations, tell whether you would use a calculator, do a paper and pencil calculation, or make a mental estimate.

SITUATION 1 You are at the ballgame. Do you have enough money to buy two hot dogs and a bag of peanuts? mental estimate

SITUATION 2 You are completing your income tax return. calculator

SITUATION 3 You are deciding about how much it will cost to buy furniture for a bedroom. paper and pencil calculation

SITUATION 4 You are planning a vacation. From a map you want to estimate the approximate number of kilometers you will drive. paper and pencil calculation

SITUATION 5 You are balancing your checkbook. calculator

SITUATION 6 You are in a pastry shop with \$ 6. You are to purchase a dozen pastries for a party. Pastries sell for \$ 1.20/dozen, but there is a sign telling that there is a 10% discount for the purchase of 5 dozen or more. Should you take advantage of the discount and buy 5 dozen? mental estimate

CHAPTER 10 PRE-TEST

Which is the likely measure? pp. 284–296

1. The length of a swimming pool:
20 mm, 20 cm, 20 m, 20 km **20 m**

2. The height of a chair:
40 mm, 40 cm, 40 m, 40 km **40 cm**

3. A glass of water:
300 mL or 300 L **300 mL**

4. A storage tank:
500 mL or 500 L **500 L**

5. An apple:
100 mg, 100 g, 100 kg, 100 t **100 g**

6. A box of apples:
5 mg, 5 g, 5 kg, 5 t **5 kg**

7. A chilly day:
40°C or 40°F **40°F**

8. A hot bath:
40°C or 40°F **40°C**

Make these changes. pp. 286–298

9. 19 m = __1900__ cm

10. 3284 m = __3.284__ km

11. 72 cm = __0.72__ m

12. 8.7 cm = __87__ mm

13. 6.2 L = __6200__ mL

14. 9.87 g = __9870__ mg

15. 8240 g = __8.24__ kg

16. 3.2 t = __3200__ kg

17. 9 yd = __27__ ft

18. 4 qt = __8__ pt

19. 80 oz = __5__ lb

20. 60 in. = __5__ ft

Find the measure of each angle. p. 300

21. 35°

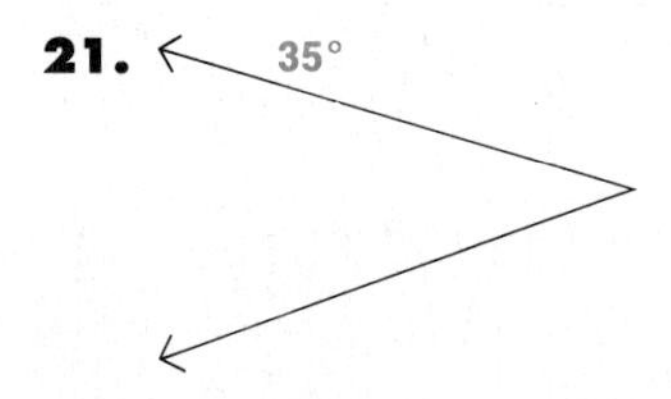

22. 150°

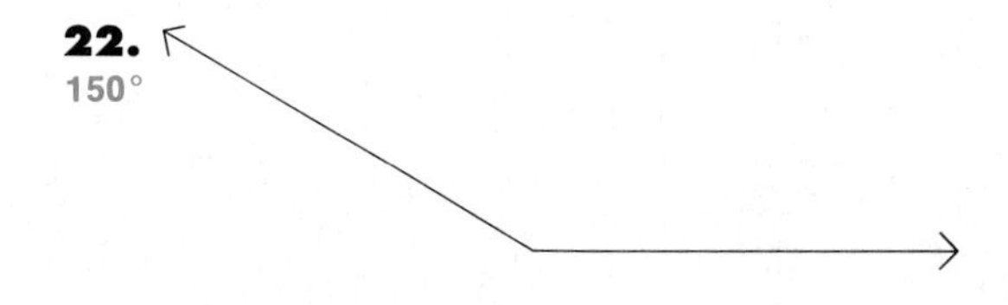

Application pp. 288, 294

23. A bus leaves New River, Arizona at 4:35 AM and arrives in Flagstaff, Arizona at 7:07 AM. How long does the trip take? **2:32**

24. Sales tax is $6\frac{1}{2}$%. How much is the sales tax for a calculator which costs $ 29? **$1.89**

10 MEASUREMENT

GETTING STARTED

Have your students compare their estimates. Which student estimated the closest? Have your students suggest other objects to measure. The student whose estimate is the closest to the actual measurement is the winner.

METRIC GUESTIMATION

A meter (m) is about the distance from the tip of your nose to the end of your outstretched hand, or 1 long step.

1. Estimate the length of your classroom's longest wall in meters.
2. Measure the wall. Was your estimate close?

A centimeter (cm) is about the width of your little finger.

3. Estimate the length of the long side of this book in centimeters.
4. Measure the side. Was your estimate close? 24 cm

A millimeter (mm) is about the thickness of a dime.

5. Estimate the thickness of this book from cover to cover in millimeters.
6. Measure the thickness. Was your estimate close? 25 mm

A kilometer (km) is about 1000 long steps.

7. Think of something one kilometer away. Estimate how long it would take to walk there.
8. It takes about 1 minute to walk 100 steps. Approximately how long would it take to walk 1 kilometer? Was your estimate close? 10 min

Objectives 100 and 101

METRIC LENGTH

Students should realize that there are other units of metric length. We have included only the most widely used units.
In *Systeme Internationale* (SI), mm, cm, m, and km are referred to as symbols, not abbreviations. Therefore a period is not required after any of them.

The **meter** (m) is the basic unit of length.

Unit	Size	Example
millimeter (mm)	0.001 m	Thickness of a dime
centimeter (cm)	0.01 m	Width of a pen
meter (m)	1 m	1 long step
kilometer (km)	1000 m	1000 long steps

EXAMPLES. To measure would you likely use mm, cm, m, or km?

1. The thickness of your book mm
2. The length of your pencil cm
3. The distance to the principal's office m
4. The distance to the North Pole km

Have students suggest other items and the appropriate units used to measure them.

TRY THIS

To measure would you likely use mm, cm, m, or km?

1. The distance from New York to Chicago km
2. The length of a pin mm
3. The length of a sheet of paper cm
4. The length of an athletic field m

EXAMPLES. Which is the likely measure?

5. The height of a 3-year-old child: 1 mm, 1 cm, 1 m, 1 km 1 m
6. The length of a river: 350 mm, 350 cm, 350 m, 350 km 350 km

TRY THIS

Which is the likely measure?

5. The length of a desk top: 30 mm, 30 cm, 30 m, 30 km 30 cm
6. The distance around a room: 40 mm, 40 cm, 40 m, 40 km 40 m

We can use a metric ruler to measure length.

EXAMPLE 7. Give the length in mm and cm.

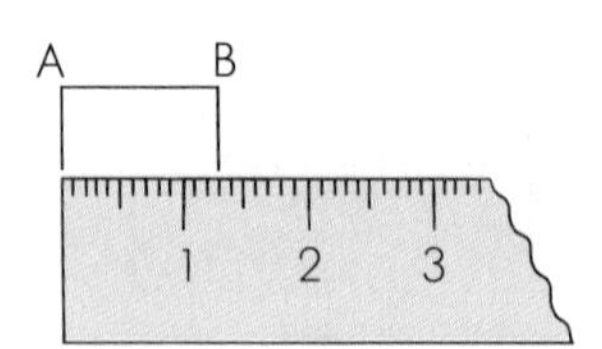

13 mm Count the mm from A to B.
1.3 cm 1 cm + 3 mm = 1.3 cm

Discuss the unit markings with your students. They should understand that the smaller units are millimeters and the larger units are centimeters.

TRY THIS

7. Give the length in mm and cm. 53 mm; 5.3 cm

PRACTICE

For extra practice, use Making Practice Fun 62 with this lesson.

To measure would you likely use mm, cm, m, or km?

1. The height of a building m

2. The length of a picture cm

3. A running dash m

4. A marathon run km

5. The thickness of a picture frame mm

6. The height of a chair cm

Which is the likely measure?

7. The height of a person:
180 mm, 180 cm, 180 m, 180 km 180 cm

8. The length of a baseball bat:
85 mm, 85 cm, 85 m, 85 km 85 cm

9. The length of a tennis court:
20 mm, 20 cm, 20 m, 20 km 20 m

10. The length of a key:
20 mm, 20 cm, 20 m, 20 km 20 mm

11. The distance per liter of gas:
10 mm, 10 cm, 10 m, 10 km 10 km

12. The length of a swimming pool:
50 mm, 50 cm, 50 m, 50 km 50 m

Give the length in mm and cm.

13. 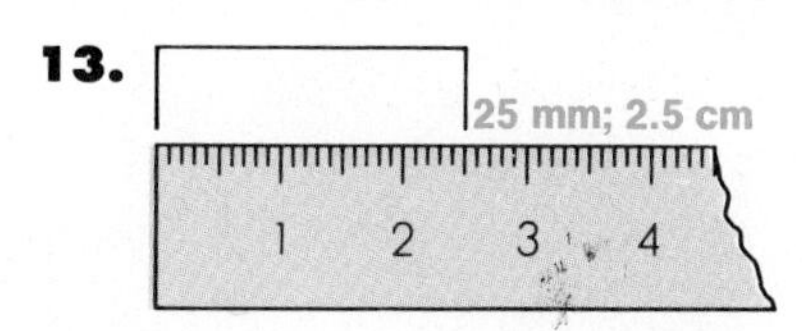

14. 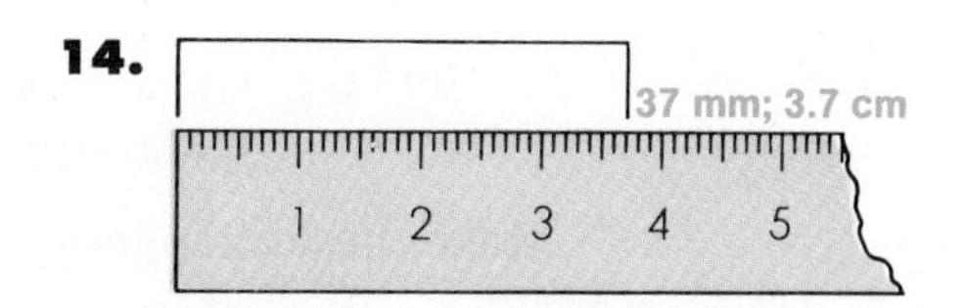

15. 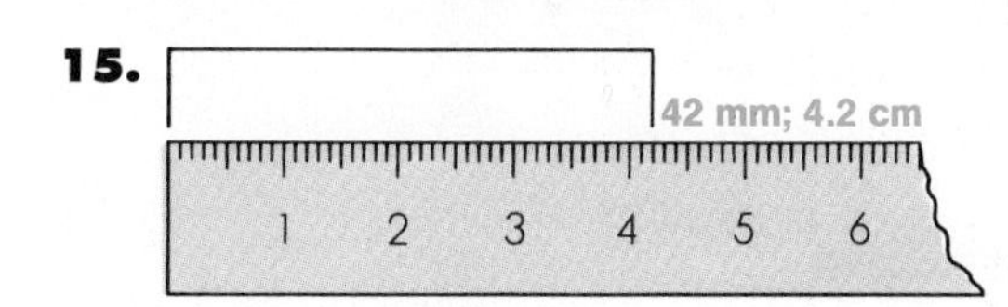

16. 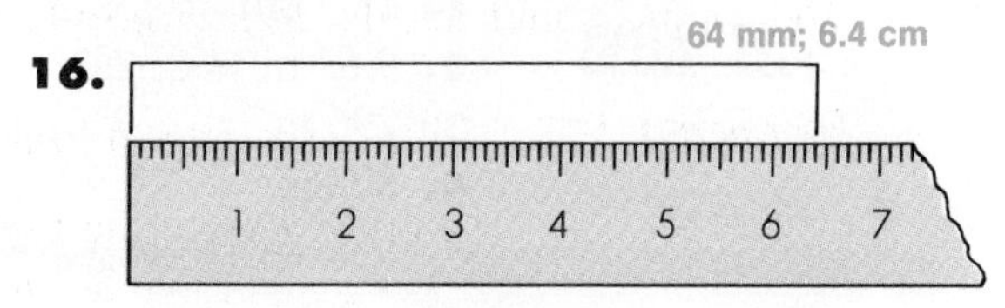

Objective 102

CHANGING METRIC UNITS OF LENGTH

The goal is to have students think in metric terms. Therefore we do not include lessons in converting between metric and customary measures.

1 km = 1000 m To change from a larger unit to a smaller unit, we multiply.
1 m = 100 cm We move the decimal point to the right.
1 cm = 10 mm To change from a smaller unit to a larger unit, we divide.
We move the decimal point to the left.

Make this change. 5 km = ______ m

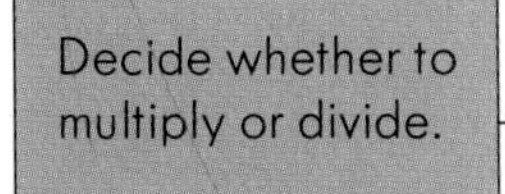

→ To multiply or divide by 10, 100, or 1000, move the decimal point. → Write the length.

5 km = ______ m	1 km = 1000 m	5 km = 5000 m
Larger to smaller:	Multiply by 1000.	
Multiply.	5.0 0 0	

EXAMPLES. Make these changes.

1. 7.5 m = ______ cm
7.5 0
7.5 m = 7 5 0 cm

Think: Larger to smaller. Multiply. 1 m = 100 cm Multiply by 100. Move the decimal point 2 places to the right.

2. 4 mm = ______ cm
4.
4 mm = 0.4 cm

Think: Smaller to larger. Divide. 1 cm = 10 mm Divide by 10. Move the decimal point 1 place to the left.

3. 5 2 7 8 m = ______ km
5 2 7 8.
5 2 7 8 m = 5.2 7 8 km

Think: Smaller to larger. Divide. 1 m = 1000 km Divide by 1000. Move the decimal point 3 places to the left.

TRY THIS

Make these changes.

1. 18.7 cm = __0.187__ m
2. 9.62 m = __962__ cm
3. 296 m = __0.296__ km
4. 87 cm = __870__ mm

PRACTICE

Make these changes.

1. 2 km = 2000 m
2. 12 km = 12,000 m
3. 6.4 km = 6400 m
4. 32.7 km = 32,700 m
5. 8.76 km = 8760 m
6. 42.58 km = 42,580 m
7. 9 m = 900 cm
8. 15 m = 1500 cm
9. 8.6 m = 860 cm
10. 24.7 m = 2470 cm
11. 9 mm = 0.9 cm
12. 3 mm = 0.3 cm
13. 38 mm = 3.8 cm
14. 97 mm = 9.7 cm
15. 498 mm = 49.8 cm
16. 1284 mm = 128.4 cm
17. 876 cm = 8.76 m
18. 378 cm = 3.78 m
19. 78 cm = 0.78 m
20. 42 cm = 0.42 m
21. 9.86 m = 986 cm
22. 32.04 m = 3204 cm
23. 8 cm = 0.08 m
24. 5 cm = 50 mm
25. 4 cm = 0.04 m
26. 5892 m = 5.892 km
27. 12,476 m = 12.476 km
28. 24 cm = 240 mm
29. 947 m = 0.947 km
30. 8.7 cm = 87 mm

PROBLEM CORNER

1. A track race is $1\frac{1}{2}$ kilometers. How many meters is this? 1500 m
2. A swimming race is 300 meters long. What part of a kilometer is this? 0.3 km

More Practice, page 306

Objective 180

APPLICATION

You can obtain local schedules for classroom use from airlines, bus depots, and train stations.

TIME SCHEDULES

AIRLINES SCHEDULE

SAN FRANCISCO/(S) OAKLAND/(K) SAN JOSE(J)

	Leave	Arrive	Flight	Service	Via
To Birmingham, Ala. (CST)					
	7:00a (S)	2:35p	344	🍸✕	1
	1:00p (S)	9:05p	782 / 896	🍸✕	Lax
From Birmingham, Ala. (CST)					
67	8:45a	12:38p (S)	887 / 1431	🍸✕	Lax
	8:45a	1:08p (S)	887	🍸✕	1
	7:40p	12:06a (S)	637 / 546	🍸✕	Lax
To Honolulu, Hi. (HST)					
	8:45a (K)	12:10p (H)	105	🎥🍸✕	0
	9:00a (S)	**12:20p (H)**	**181**	🎥🍸✕★	**0**
	10:00a (J)	1:25p (H)	69	🎥🍸✕	0
	1:30p (S)	**4:45p (H)**	**35**	🎥🍸✕★	**0**
	4:35p (S)	**7:40p (I)**	**121**	🎥🍸✕●	**0**
	4:35p (S)	**9:25p (H)**	**121**	🎥🍸✕●	**1**
	7:10p (S)	**10:25p (H)**	**189**	🎥🍸✕★	**0**
From Honolulu, Hi. (HST)					
	9:30a (H)	**4:20p (S)**	**180**	🎥🍸✕★	**0**
	12:45p (H)	**7:35p (S)**	**96**	🎥🍸✕★	**0**
	1:15p (H)	8:10p (J)	68	🎥🍸✕	0
	1:25p (H)	**9:55p (S)**	**186**	🎥🍸✕●	**1**
	2:30p (H)	9:25p (K)	106	🎥🍸✕	0
	3:20p (I)	**9:55p (S)**	**186**	🎥🍸✕●	**0**
	11:50p (H)	**6:40a (S)**	**22**	♫ 🍸▬★	**0**
To Providence, R.I. (EST)					
	7:05a (J)	**8:29p**	630 / **240** 562	§ 🍸✕●	**Den** /Chi
	10:00a (S)	8:29p	**126** / 562	§ 🍸✕●	**Chi**
From Providence, R.I. (EST)					
	7:45a	12:20p (J)	445 / 943	🍸✕	Chi
	7:45a	**12:40p (S)**	445 / **123**	§ 🍸✕●	**Chi**
	2:20p	**9:00p (J)**	937 / **157** 921	🍸✕●	**Cle** /Chi
X6	3:55p	9:00p (J)	809 / 921	🍸✕	Chi
X6	3:55p	**9:05p (S)**	809 / **135**	§ 🍸✕●	**Chi**
X6	3:55p	9:15p (K)	809 / 841	🍸✕	Chi
To Vancouver, B.C. (PST)					
	11:30a (S)	3:14p	696 / 308	🍸✕	Sea
		Above Eff. Oct. 29			
	6:35p (K)	10:29p	468 / 342	🍸✕	Sea
	7:30p (S)	10:29p	246 / 342	🍸✕	Sea
From Vancouver, B.C. (PST)					
	7:10a	10:45a (S)	479 / 603	🍸✕	Sea
	4:00p	7:45p (S)	641 / 339	🍸✕	Sea
	4:00p	9:11p (J)	641 / 807	🍸✕	Sea
	4:00p	9:59p (K)	641 / 807	🍸✕	Sea
		Above Eff. Feb. 8			

✕ **Meal** ▬ **Snacks** 🍸 **Cocktails**
🎥 **Movies** ♫ **Audio** § **Short-Subject**

(I) – Hilo/Hawaii (S) – San Francisco
(K) – Oakland (H) – Honolulu
(J) – San Jose

1–Monday 2–Tuesday 3–Wednesday
4–Thursday 5–Friday 6–Saturday
7–Sunday X–Except

★ indicates 747 ▲ indicates DC-10 to 747
● indicates DC-10 jet ⊕ indicates Night Coach may apply
♦ indicates 747 to DC-10

Joins Flight #896 Los Angeles to Birmingham

Indicates the airport when there is more than one major airport in an area.

Which flight or flights can you take for each?

1. to arrive in Birmingham after 6 PM 782/896
2. to return to San Francisco from Vancouver in the morning 479/603
3. to go to Honolulu on a 747 181, 35, 189

On flights with a stop, flight time includes time of layover. How long is each?

4. Flight #937/157 from Providence 9:40
5. Flight #22 from Honolulu? 4:50
6. Which flight to Vancouver takes the least time? 246/342
7. Which flight from Birmingham takes the most time? 637/546
8. Which flight from Providence does not operate on Saturday? 809

Montana Highway Bus Service						
Billings to Butte Read down			Billings—Butte Service	Butte to Billings Read up		
Daily	Mon–Fri	Mon–Fri		Daily	Mon–Fri	Mon–Fri
am + pm 47/67	am + pm 22/32	91	← Bus Number →	am + pm 147/167	am + pm 122/132	191
6:00	4:35	9:50a	dep BILLINGS arr	12:17	10:25	3:15p
6:35	↓	↓	↓ LAUREL ↑	11:43	↑	↑
7:25	↓	↓	↓ COLUMBUS ↑	10:53	↑	↑
7:59	↓	↓	↓ BIG TIMBER ↑	10:18	↑	↑
8:45	7:00	12:15a	↓ LIVINGSTON ↑	9:32	8:02	12:48p
9:15	7:30	↓	↓ BOZEMAN ↑	9:00	7:30	↑
10:06	↓	↓	↓ THREE FORKS ↑	8:05	↑	↑
10:35	↓	↓	↓ WHITE HALL ↑	7:24	↑	↑
11:30	9:30	2:40p	arr BUTTE dep	6:45	5:30	10:30a
See local schedules for town to town times. Bus remains 5 minutes at each scheduled stop before departure.						

Which bus can you take for each trip?

9. from Columbus to Butte? 47/67
10. from Boseman to Livingston 147/167

How long does it take for each trip?

11. from Billings to Big Timber on #47 1:59
12. from Columbus to Billings 1:24
13. from White Hall to Laurel 4:19
14. from Columbus to Three Forks 2:41

How much longer does it take for each trip?

15. from Billings to Livingston on #47 than on #91 20 min
16. from Butte to Billings on #147 than on #122 37 min
17. Suzie Plumhoff lives in Livingston and works in Bozeman. She rides the #47 bus to work. She rides a local bus home. The local takes 35% longer than the non-stop bus. How much longer does the local bus take? $11\frac{1}{5}$ min
18. Suzie buys a commuter ticket book of 20 rides for $ 36. The regular fare is $ 2.30 per ride. What percent does she save on each ride? 22%

Objectives 103 and 104

METRIC CAPACITY

There are other units of metric capacity. We have included only the most widely used units.

The amount a container will hold is called capacity.
The **liter** is the basic unit of capacity.

Unit	Size	Example
milliliter (mL)	0.001 L	An eyedropper will hold 1 mL
liter (L)	1000 mL	About 4 glasses of water

EXAMPLES. Which is the likely measure? Discuss other items and their likely measures with students.

1. A glass of milk: 200 mL or 200 L 200 mL
2. Gas for a car: 20 mL or 20 L 20 L

TRY THIS

Which is the likely measure?

1. A doctor's prescription: 10 mL or 10 L 10 mL
2. A barrel of oil: 400 mL or 400 L 400 L

EXAMPLES. Make these changes:

3. 7 L = ______ mL

7.0 0 0

7 L = 7 0 0 0 mL

1 L = 1000 mL
Multiply by 1000.
Move the decimal point 3 places to the right.

4. 3 7 5 mL = ______ L

3 7 5.

3 7 5 mL = 0.3 7 5 L

1 L = 1000 mL
Divide by 1000.
Move the decimal point 3 places to the left.

5. 0.7 8 L = ______ mL

0.7 8 0

0.7 8 L = 7 8 0 mL

Move the decimal point 3 places to the right.

TRY THIS

Make these changes.

3. 3.2 L = 3200 mL
4. 0.5 L = 500 mL
5. 250 mL = 0.25 L

PRACTICE

For extra practice, use Making Practice Fun 63 with this lesson.

Which is the likely measure?

1. A can of soda:
   360 mL or 360 L 360 mL
2. A carton of cream:
   200 mL or 200 L 200 mL
3. Water in a swimming pool:
   10,000 mL or 10,000 L 10,000 L
4. A cup of soup:
   150 mL or 150 L 150 mL
5. A bathtub:
   300 mL or 300 L 300 L
6. A bottle of medicine:
   150 mL or 150 L 150 mL
7. A bottle of perfume:
   100 mL or 100 L 100 mL
8. A bucket:
   10 mL or 10 L 10 L

Make these changes.

9. 8 L = 8000 mL
10. 12 L = 12,000 mL
11. 3.7 L = 3700 mL
12. 4.9 L = 4900 mL
13. 2598 mL = 2.598 L
14. 4782 mL = 4.782 L
15. 8000 mL = 8 L
16. 14,000 mL = 14 L
17. 0.49 L = 490 mL
18. 478 mL = 0.478 L
19. 750 mL = 0.75 L
20. 0.98 L = 980 mL
21. 0.3 L = 300 mL
22. 87 mL = 0.087 L

PROBLEM CORNER

1. Some containers are labeled $\frac{1}{2}$ liter. How many milliliters is this? 500 mL
2. Jonnie bought $\frac{1}{4}$ of a liter of juice. How many milliliters is this? 250 mL
3. A container held 750 milliliters. What part of a liter is this? 0.75 L

Objectives 105 and 106

METRIC MASS

There are other units of metric mass. We have included only the most widely used units.

The basic unit of mass is the **kilogram.**

Unit	Size	Example
milligram (mg)	0.001 g	An aspirin is about 500 mg
gram (g)	0.001 kg	An average paper clip
kilogram (kg)	1 kg	The mass of this book
metric ton (t)	1000 kg	A truckload of coal

Have your students suggest other items and their likely measures.

EXAMPLES. Which is the likely measure?

1. A potato chip: 400 mg, 400 g, 400 kg, 400 t 400 mg
2. A potato: 50 mg, 50 g, 50 kg, 50 t 50 g
3. A sack of potatoes: 5 mg, 5 g, 5 kg, 5 t 5 kg
4. A truckload of potatoes: 2 mg, 2 g, 2 kg, 2 t 2 t

TRY THIS

Which is the likely measure?

1. A mouse: 50 mg, 50 g, 50 kg, 50 t 50 g
2. A man: 70 mg, 70 g, 70 kg, 70 t 70 kg
3. An elephant: 2 mg, 2 g, 2 kg, 2 t 2 t
4. A fly: 250 mg, 250 g, 250 kg, 250 t 250 mg

EXAMPLES. Make these changes.

5. 4 2 8 6 mg = ______ g

4 2 8 6.

4 2 8 6 mg = 4.2 8 6 g

1 g = 1000 mg
Move the decimal point 3 places to the left.

6. 7.8 5 kg = ______ g

7.8 5 0

7.8 5 kg = 7 8 5 0 g

1 kg = 1000 g
Move the decimal point 3 places to the right.

TRY THIS

Make these changes.

5. 3786 g = 3.786 kg
6. 7.86 g = 7860 mg
7. 1286 kg = 1.286 t
8. 730 mg = 0.73 g

PRACTICE

For extra practice, use Making Practice Fun 64 with this lesson.
Use Quiz 21 after this Practice.

Which is the likely measure?

1. A pin:
   250 mg, 250 g, 250 kg, 250 t 250 mg
2. A pencil:
   6 mg, 6 g, 6 kg, 6 t 6 g
3. A ping pong ball:
   750 mg, 750 g, 750 kg, 750 t 750 mg
4. A bowling ball:
   7 mg, 7 g, 7 kg, 7t 7 kg
5. A truckload of bricks:
   2 mg, 2 g, 2 kg, 2 t 2t
6. A wheelbarrow of bricks:
   60 mg, 60 g, 60 kg, 60 t 60 kg

Make these changes.

7. 8479 mg = 8.479 g
8. 250 mg = 0.25 g
9. 5482 g = 5.482 kg
10. 100 g = 0.1 kg
11. 9 t = 9000 kg
12. 7.47 t = 7470 kg
13. 6 kg = 6000 g
14. 3.12 kg = 3120 g
15. 8249 kg = 8.249 t
16. 912 kg = 0.912 t
17. 3 g = 3000 mg
18. 4.75 g = 4750 mg
19. 768 mg = 0.768 g
20. 7 t = 7000 kg
21. 7.4 kg = 7400 g
22. 7.9 g = 7900 mg

PROBLEM CORNER

1. A large animal looked as if it weighed half a metric ton. How many kilograms is this? 500 kg
2. A tablet was labeled 250 milligrams. What part of a gram is this? 0.25 g
3. A prescription called for $\frac{3}{4}$ grams. How many milligrams is this? 750 mg

More Practice, page 306

Objectives 181 and 182

APPLICATION

RATE SCHEDULES

STATE DEPARTMENT OF TAXATION											
3 1/2% BRACKET CARD											
Transaction			Tax	Transaction			Tax	Transaction			Tax
$0.01	to	$0.14	$0.00	$16.72	to	$16.99	$0.59	$33.58	to	$33.85	$1.18
0.15	to	0.38	0.01	17.00	to	17.28	0.60	33.86	to	34.14	1.19
0.39	to	0.64	0.02	17.29	to	17.57	0.61	34.15	to	34.42	1.20
0.65	to	0.88	0.03	17.58	to	17.85	0.62	34.43	to	34.71	1.21
0.89	to	1.18	0.04	17.86	to	18.14	0.63	34.72	to	34.99	1.22
1.19	to	1.57	0.05	18.15	to	18.42	0.64	35.00	to	35.28	1.23
1.58	to	1.85	0.06	18.43	to	18.71	0.65	35.29	to	35.57	1.24
1.86	to	2.14	0.07	18.72	to	18.99	0.66	35.58	to	35.85	1.25
2.15	to	2.42	0.08	19.00	to	19.28	0.67	35.86	to	36.14	1.26
2.43	to	2.71	0.09	19.29	to	19.57	0.68	36.15	to	36.42	1.27
2.72	to	2.99	0.10	19.58	to	19.85	0.69	36.43	to	36.71	1.28
3.00	to	3.28	0.11	19.86	to	20.14	0.70	36.72	to	36.99	1.29

In most states, customers pay sales tax on their purchases. Sales clerks use a chart to find sales tax.

Use the table. How much tax is there on these sales?

1. $ 17.56 $0.61 2. $ 2.18 $0.08 3. 45¢ $0.02 4. $ 36.28 $1.27

Nevada charges $ 1.18 for all sales between $ 33.58 and $ 33.85.

5. For what exact sales total is the $3\frac{1}{2}\%$ tax closest to $ 1.18? $33.71

$3\frac{1}{2}\%$ of ? = $1.18

Compute the sales tax for these sales.

6. Connecticut–7%; sale–$ 29.85 $2.09 7. Missouri–$3\frac{1}{8}\%$; sale–$ 158.90 $4.97
8. Maryland–5%; sale–$ 253.87 $12.69 9. California–$4\frac{3}{4}\%$; sale–$ 18.78 $0.89

Allan bought a fish tank for $ 12.95. He paid a state tax of $3\frac{1}{2}\%$.

10. What was the total charge? $13.40
11. He gave the clerk a twenty-dollar bill. How much change did he get back? $6.60

The Act of 1789 created the U.S. Post Office. The first postage rate law was enacted in 1792. It established these rates for single-sheet letters.

Rates In 1792

Distance in miles	0-30	30-60	60-100	100-150	150-200	200-250	250-350	350-450	over 450
Rate	6¢	8¢	10¢	$12\frac{1}{2}$¢	15¢	17¢	20¢	22¢	25¢

Double sheets cost double, and so forth, to 1 ounce. Items in excess of 1 ounce cost 4 times the rate for each ounce or fraction.

How much would it have cost to mail these items in 1792?

12. a 1-page letter mailed 220 miles 17¢
13. a 1-ounce letter mailed 650 miles 25¢
14. 2 pages mailed 30 miles 16¢
15. a $4\frac{1}{2}$-ounce package not enough information
16. Why might this rate schedule have caused some disputes? Answers will vary.

In 1879 the rates were simplified. The 4-class mail system was invented and is still used today. Postage rates depend on the weight and type of item being mailed.

Rates In 1879

1st class	3¢ per $\frac{1}{2}$ ounce (letters); 1¢ (post cards)
2nd class	special rate for publishers
3rd class	1¢ per 2 ounces (circulars, books, newspapers, etc.)
4th class	1¢ per ounce (merchandise)

How much would it have cost to mail these items in 1879?

17. a 2-ounce letter 12¢
18. books weighing $4\frac{1}{2}$ pounds 36¢
19. a $\frac{3}{4}$-pound hat 12¢
20. an 8-ounce newspaper 4¢
21. What is the current postage rate for each of the four classes? Answers will change.
22. How much would it cost to mail each of the items today? Answers will change.

Objective 107

TEMPERATURE

Temperatures can be measured on the Celsius (C) scale or the Fahrenheit (F) scale. Both use the degree symbol (°) to represent their unit of measure.

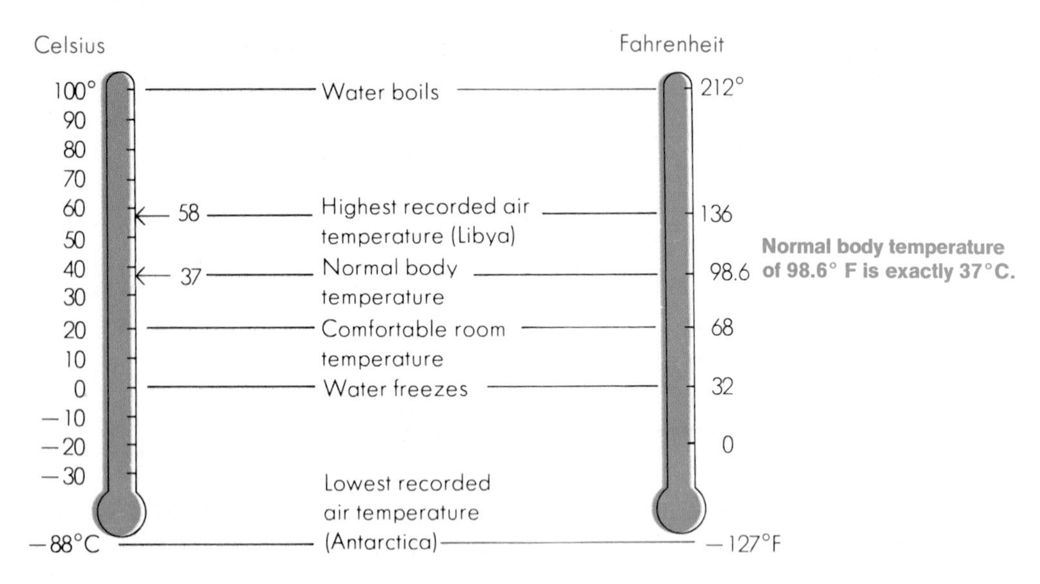

Normal body temperature of 98.6° F is exactly 37°C.

EXAMPLES. Is the temperature likely to be C or F?

1. A pleasant summer day: 23° C

2. A cold winter day: 15° F

23°C is read 23 degrees Celsius.

TRY THIS

Is the temperature likely to be C or F?

1. Baking bread: 350° F

2. Freezing ice cream: −10° C

EXAMPLES. Which is the likely temperature?

3. Warm soup: 40° C or 40° F 40° C

4. A chilled glass of milk: 38° C or 38° F 38° F

TRY THIS

Which is the likely temperature?

3. A high fever: 100° C or 100° F 100° F

4. A hot day: 40° C or 40° F 40° C

PRACTICE

Is the temperature likely to be C or F?

1. A hot summer day: 42° C

2. Bathtub water: 85° F

3. Cold enough to have the heater on: 40° F

4. Warm enough to have the air conditioner on: 40° C

5. A snowball fight: 28° F

6. Sunbathing: 35° C

7. Water freezes: 32° F

8. Water boils: 100° C

9. Swimming pool water: 75° F

10. A teapot steaming: 90° C

Which is the likely temperature?

11. A warm stove:
50° C or 50° F 50° C

12. A hot stove:
75° C or 75° F 75° C

13. Cold enough for a jacket:
35° C or 35° F 35° F

14. Warm enough to go without a jacket:
30° C or 30° F 30° C

15. A summer day in the Sahara Desert: 55° C or 55° F 55° C

16. A winter day at the North Pole:
−20° C or −20° F −20° C

17. A rainstorm:
20° C or 20° F 20° C

18. A snowstorm:
20° C or 20° F 20° F

19. Snow skiing:
40° C or 40° F 40° F

20. Water skiing:
30° C or 30° F 30° C

PROBLEM CORNER

1. During a storm the temperature dropped from 87° F to 73° F. How many degrees did it drop? 14° F

2. The temperature was 15° C in the morning. It was 32° C at noon. How many degrees did it rise? 17° C

Objective 108

CUSTOMARY MEASURES

Length
12 inches (in.) = 1 foot (ft)
3 feet = 1 yard (yd)
36 inches = 1 yard
5280 feet = 1 mile (mi)
1760 yards = 1 mile

Capacity
8 fluid ounces (fl oz) = 1 cup (c)
2 cups = 1 pint (pt)
2 pints = 1 quart (qt)
4 quarts = 1 gallon (gal)

Weight
16 ounces (oz) = 1 pound (lb)
2000 pounds = 1 ton

EXAMPLES. Make these changes.

1. 8 ft = ______ in.
8 ft = 96 in.

Think: Larger to smaller. Multiply.
1 ft = 12 in. $8 \times 12 = 96$

2. $4\frac{1}{2}$ gal = ______ qt
$4\frac{1}{2}$ gal = 18 qt

Think: Larger to smaller. Multiply.
1 gal = 4 qt $4\frac{1}{2} \times 4 = 18$

3. 20 oz = ______ lbs
20 oz = $1\frac{1}{4}$ lbs

Think: Smaller to larger. Divide.
16 oz = 1 lb $20 \div 16 = 1\frac{1}{4}$

4. 13,200 ft = ______ mi
13,200 ft = $2\frac{1}{2}$ mi

Think: Smaller to larger. Divide.
5280 ft = 1 mi
$13{,}200 \div 5280 = 2\frac{1}{2}$

TRY THIS

Make these changes.

1. 5 yd = __15__ ft

2. $2\frac{1}{4}$ tons = __2250__ lbs

3. 20 pt = __10__ qts

4. 40 in. = __$1\frac{1}{9}$__ yds

PRACTICE

For extra practice, use Making Practice Fun 65 with this lesson.

Make these changes.

1. 5 ft = 60 in.
2. 3 yd = 108 in.
3. 2 mi = 3520 yd
4. 4 pt = 8 cups
5. 5 qt = 10 pt
6. 4 ton = 8000 lb
7. $5\frac{2}{3}$ yd = 17 ft
8. $1\frac{1}{4}$ c = 10 fl oz
9. $7\frac{3}{4}$ gal = 31 qt
10. $2\frac{3}{4}$ lb = 44 oz
11. 12 ft = 4 yd
12. 64 fl oz = 8 c
13. 56 qt = 14 gal
14. 18 in. = $1\frac{1}{2}$ ft
15. 54 in. = $1\frac{1}{2}$ yd
16. 5 cups = $2\frac{1}{2}$ pt
17. 10 qt = $2\frac{1}{2}$ gal
18. 40 oz = $2\frac{1}{2}$ lb
19. 7 yd = 21 ft
20. 6 gal = 24 qt
21. 18 pt = 9 qt
22. $2\frac{1}{2}$ yd = 90 in.
23. 3000 lb = $1\frac{1}{2}$ ton
24. 24 fl oz = 3 c
25. 3 mi = 15,840 ft
26. 108 in. = 3 yd
27. 3 lb = 48 oz
28. 24 in. = 2 ft

PROBLEM CORNER

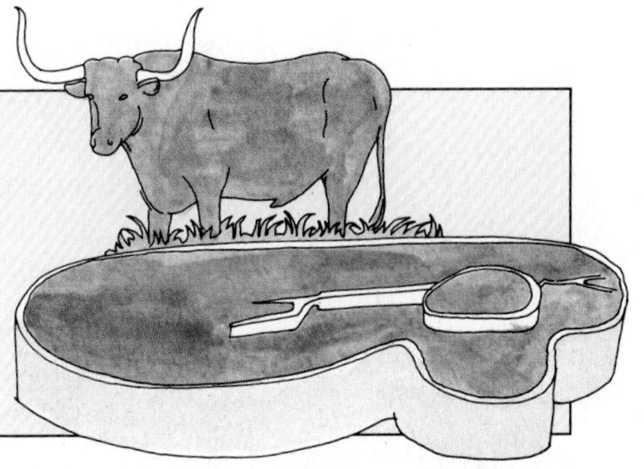

1. Jerry jumped 16 feet. George jumped $5\frac{2}{3}$ yards. Who jumped farther? George
2. Mrs. Harold bought a 20-ounce steak for $ 4.35. Mr. Harold bought a $1\frac{1}{4}$ pound steak for $ 4.35. Which was the more economical buy? They were the same.

More Practice, page 307

Objectives 109 and 110

MEASURING ANGLES

Students will need protractors for this lesson.

Discuss with your students angles and their measures in the physical world and nature.

A protractor measures angles in degrees (°).

Find the measure of this angle.

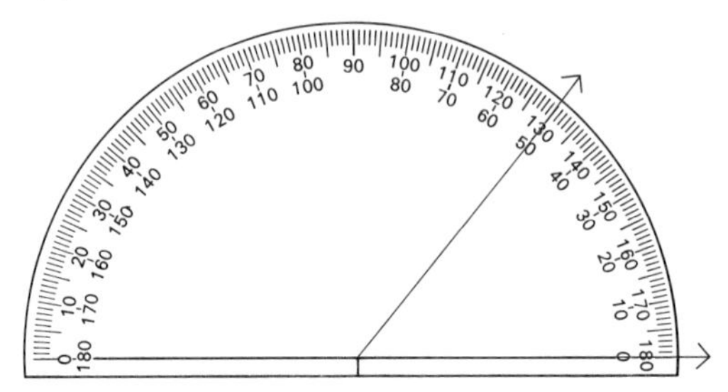

1. Place the center of the protractor where the sides of the angle meet.
2. Lay the line of the protractor along the side of the angle.
3. Read the degrees: 50°

Notice the angle that makes a square corner.
This angle measures 90°. It is called a right angle.

EXAMPLES. Tell whether each angle is a right angle, larger than a right angle, or smaller than a right angle.

1. smaller **2.** right **3.** larger

TRY THIS

Tell whether each angle is a right angle, larger than a right angle, or smaller than a right angle.

1. right **2.** larger **3.** smaller

EXAMPLES. Find the measure of each angle.

4. The angle is smaller than a right angle, so the measure is less than 90°. Read the numbers less than 90°. The measure is 45°.

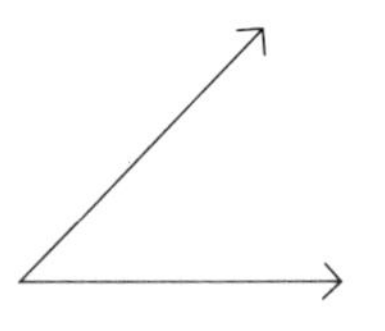

5. The angle is larger than a right angle, so the measure is greater than 90°. Read the numbers greater than 90°. The measure is 130°.

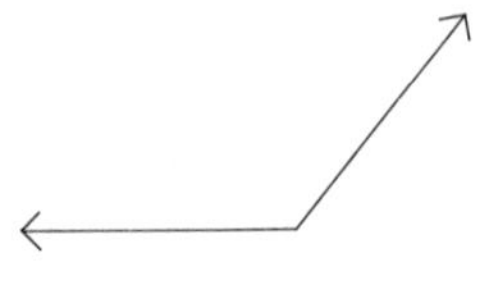

TRY THIS

Find the measure of each angle.

4. 31°

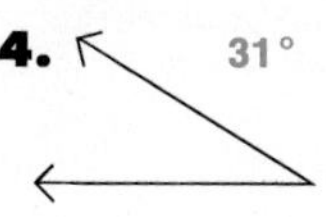

5. 145°

PRACTICE

Use Quiz 22 after this Practice.

Tell whether each is a right angle, larger than a right angle, or smaller than a right angle.

1. right

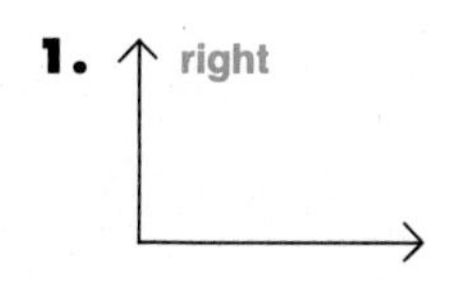

2. larger

3. smaller

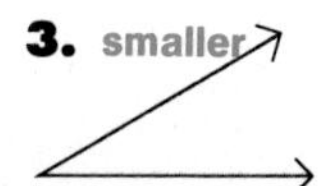

4. smaller

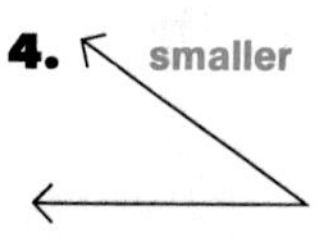

5. larger

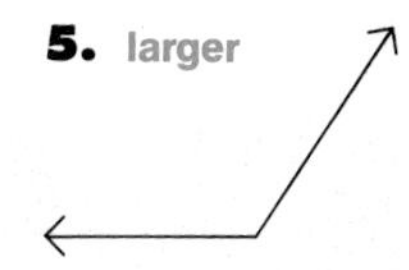

6. right

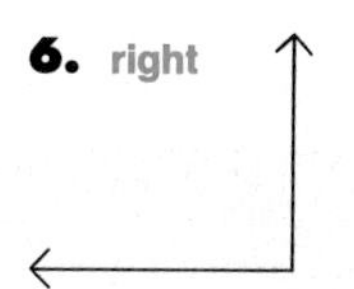

7. right

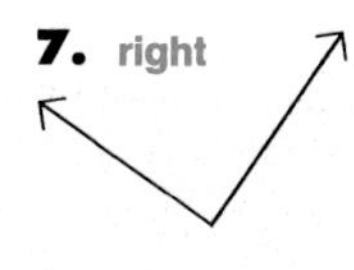

8. smaller

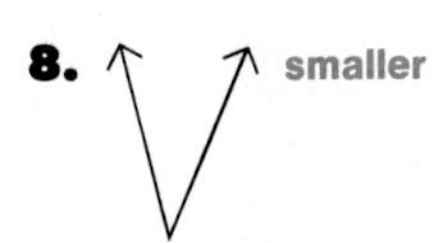

Find the measure of each angle.

9. 25°

10. 28°

11. 90°

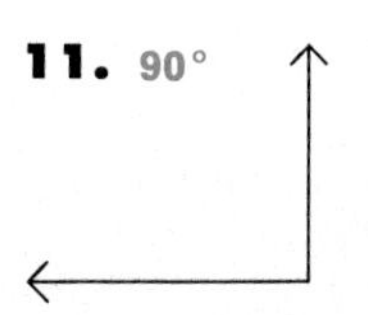

12. 90°

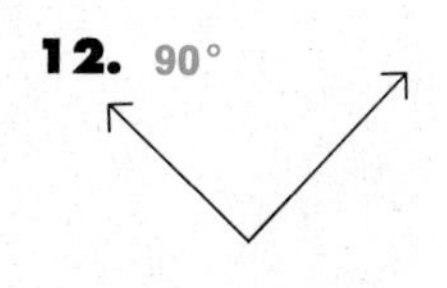

13. 130°

14. 120°

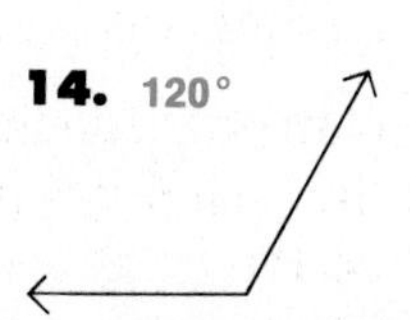

15. 55°

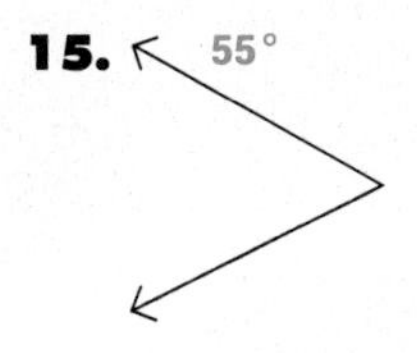

16. 25°

17. 110°

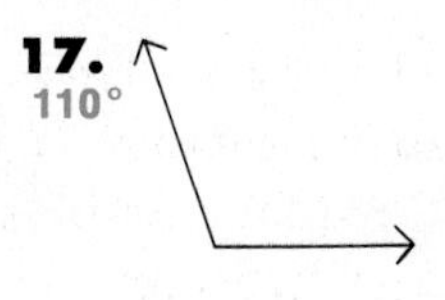

More Practice, page 307

CAREER CAREER CAREER CAREER CAREER

Copies of civil service examinations may be obtained from libraries and bookstores.

CIVIL SERVICE

The U.S. Government employs about one million people in Civil Service jobs. Applicants must take a Civil Service Exam which contains problems similar to the following. Which letter indicates the correct answer?

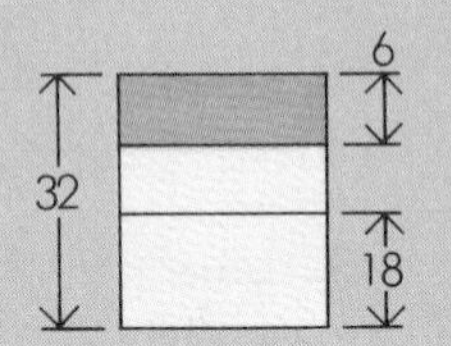

1. What part of the whole piece of wood is shaded? B

   A. $\frac{1}{3}$ B. $\frac{3}{16}$ C. $\frac{1}{4}$ D. $\frac{3}{4}$ E. None of these

2. What is $\frac{1}{4}$ of $\frac{1}{8}$? C

   A. $\frac{1}{2}$ B. $\frac{1}{16}$ C. $\frac{1}{32}$ D. $\frac{1}{4}$ E. 2

3. Which is the smallest amount? D

   A. $\frac{1}{4}$ B. 0.4×0.4 C. 0.4 D. $\frac{0.4}{4}$ E. $\frac{0.4}{0.4}$

4. $5413.68 \div 36$ is: B

   A. 15.038 B. 150.38 C. 150.382 D. 1503.8 E. None of these

5. A reamer cuts 3 holes per minute. There are 1125 holes to be cut. How many hours will it take? B

   A. $6\frac{5}{12}$ B. $6\frac{1}{4}$ C. 37.5 D. $3\frac{3}{4}$ E. $18\frac{3}{4}$

6. Two bottles, X and Y, are the same size. Bottle X is $\frac{1}{4}$ full. Bottle Y is $\frac{1}{3}$ full. Pour the contents of X into Y. How full is Y? D

   A. $\frac{1}{12}$ B. $\frac{2}{7}$ C. $\frac{3}{4}$ D. $\frac{7}{12}$ E. None of these

7. Nails cost \$0.76 per kilogram. How much do 4.75 kg of nails cost? B

   A. \$6.17 B. \$3.61 C. \$36.10 D. \$3.55 E. None of these

8. A sheet of metal is 1 m^2. About how many strips 2.5 cm by 1 m can be cut from the sheet of metal? B

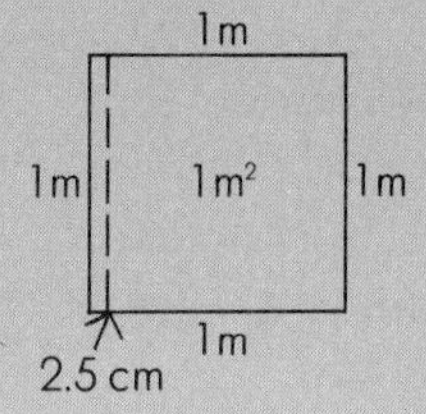

   A. 4 B. 40 C. 400
   D. 4000 E. None of these

9. A sheet of plywood is 4′ × 8′. How many 2′ × 2′ pieces can be cut from the sheet? A

   A. 8 B. 16 C. 4
   D. 10 E. None of these

10. A rod is 15.3 cm long. A piece 58 mm long is cut off. How long is the piece that is left? B

    A. 9.5 mm B. 9.5 cm C. 95 cm
    D. 42.7 mm E. None of these

11. Between which two speeds does the rate of gas consumption decrease the fastest? D

    A. 30 to 40 B. 40 to 50
    C. 50 to 60 D. 60 to 70
    E. None of these

Miles Per Gallon (mpg)
22 20 18 16 14 12 10 8
20 30 40 50 60 70
Miles Per Hour (mph)

12. How many degrees must this pole be tilted to stand erect (perpendicular to the base)? E

    A. 44° B. 0.4° C. 40°
    D. 8° E. None of these

CHAPTER 10 REVIEW

For extra practice, use Making Practice Fun 66 with this lesson.

Which is the likely measure? pp. 284–296

1. The height of a roof:
8 mm, 8 cm, 8 m, 8 km 8 m

2. The thickness of a pancake:
250 mm, 250 cm, 250 m, 250 km 250 mm

3. A bottle of pickles:
365 mL or 365 L 365 mL

4. Water to take a shower:
60 mL or 60 L 60 L

5. A lion:
300 mg, 300 g, 300 kg, 300 t 300 kg

6. A canary:
300 mg, 300 g, 300 kg, 300 t 300 g

7. Baseball season:
75° C or 75° F 75° F

8. Football season:
40° C or 40° F 40° F

Make these changes. pp. 286-298

9. 4349 m = 4.349 km

10. 9.6 cm = 96 mm

11. 85 cm = 0.85 m

12. 36 m = 3600 cm

13. 8.62 L = 8620 mL

14. 7.24 g = 7240 mg

15. 5147 kg = 5.147 t

16. 9215 g = 9.215 kg

17. 8 ft = 96 in.

18. 6 gal = 24 qt

19. 4000 lb = 2 ton

20. 27 ft = 9 yd

Find the measure of each angle. p. 300

134°
21.

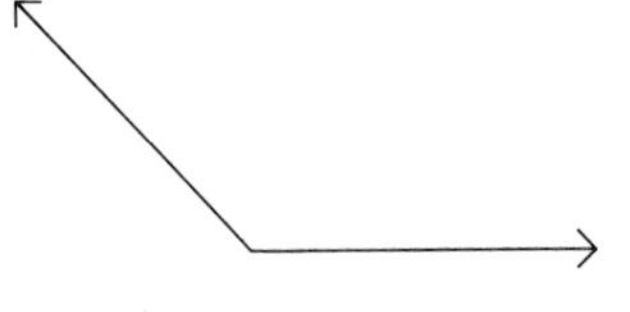

22. 31°

APPLICATION pp. 288, 294

23. A commuter train takes 20% less time than the regular train. If a trip takes 50 minutes on the regular train, how long does it take on the commuter train? 40 minutes

24. Sales tax is 5%. What is the total price for a stereo which costs $ 235? $246.75

CHAPTER 10 TEST

Which is the likely measure?

1. The length of a hall:
25 mm, 25 cm, 25 m, 25 km 25 m

2. The length of a river:
30 mm, 30 cm, 30 m, 30 km 30 km

3. A truck's gas tank:
200 mL or 200 L 200 L

4. A glass of juice:
400 mL or 400 L 400 mL

5. A box of cereal:
425 mg, 425 g, 425 kg, 425 t 425 g

6. A truck:
10 mg, 10 g, 10 kg, 10 t 10 t

7. Ice skating:
25° C or 25° F 25° F

8. Water skiing:
28° C or 28° F 28° C

Make these changes.

9. 95 cm = __0.95__ m

10. 6280 m = __6.28__ km

11. 9.6 cm = __96__ mm

12. 87 m = __8700__ cm

13. 398 mL = __0.398__ L

14. 2387 kg = __2.387__ t

15. 9.31 g = __9310__ mg

16. 3984 g = __3.984__ kg

17. 36 yd = __108__ ft

18. 7 qt = __14__ pt

19. 108 in. = __9__ ft

20. 160 oz = __10__ lb

Find the measure of each angle.

20°
21.

150°
22.

APPLICATION

23. A plane leaves Honolulu, Hawaii at 10:18 PM and arrives in Hilo, Hawaii at 11:03 PM. How long does the trip take? 45 min

24. Sales tax is $5\frac{1}{2}$%. How much is the sales tax for a bicycle which costs $98? $5.39

MORE PRACTICE

Give the length in mm and cm. Use after page 285.

1.

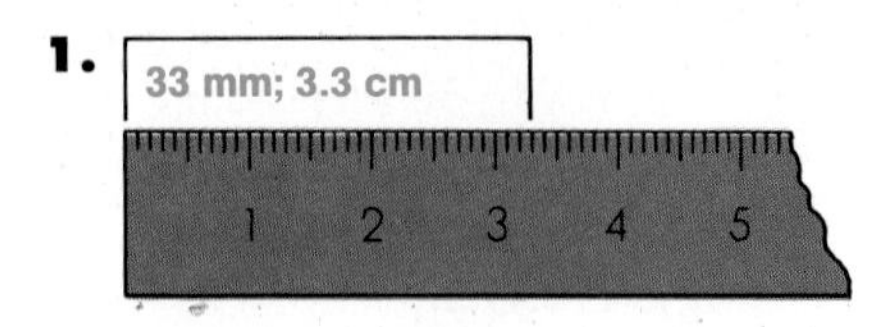

2.

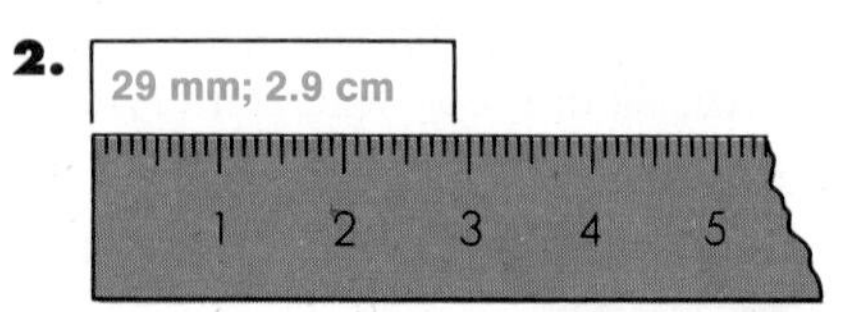

3.

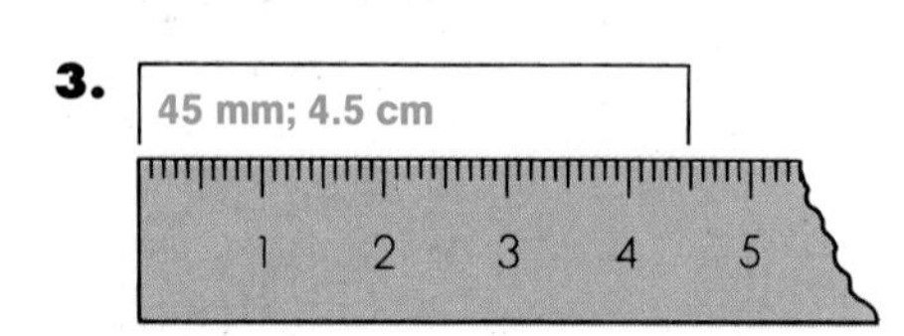

4.

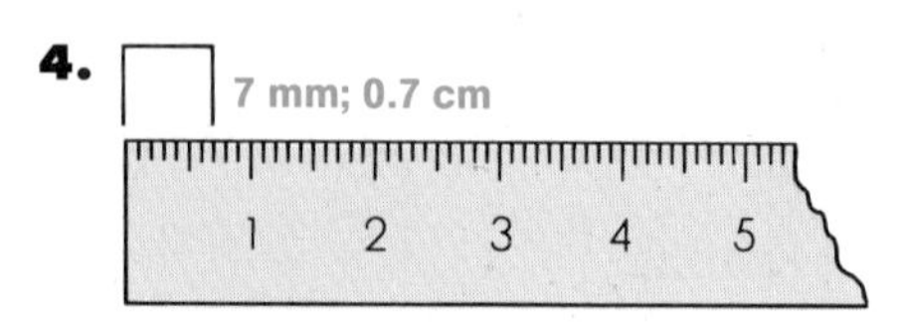

Make these changes. Use after page 287.

5. 3 km = 3000 m

6. 96 cm = 0.96 m

7. 48 mm = 4.8 cm

8. 7 m = 700 cm

9. 9.6 km = 9600 m

10. 6284 m = 6.284 km

11. 764 mm = 76.4 cm

12. 6.7 cm = 67 mm

Make these changes. Use after page 291.

13. 9 L = 9000 mL

14. 747 mL = 0.747 L

15. 3462 mL = 3.462 L

16. 0.7 L = 700 mL

17. 4.9 L = 4900 mL

18. 64 mL = 0.064 L

19. 7000 mL = 7 L

20. 450 mL = 0.450 L

Make these changes. Use after page 293.

21. 6 t = 6000 kg

22. 478 g = 0.478 kg

23. 492 mg = 0.492 g

24. 6.3 kg = 6300 g

25. 4.7 t = 4700 kg

26. 6123 kg = 6.123 t

27. 7305 g = 7.305 kg

28. 6 g = 6000 mg

Make these changes. Use after page 299.

1. 6 ft = 72 in.

2. $4\frac{1}{2}$ ft = 54 in.

3. 10 yd = 30 ft

4. $6\frac{1}{3}$ yd = 19 ft

5. 4 yd = 144 in.

6. $3\frac{1}{2}$ yd = 126 in.

7. 5 mi = 26,400 ft

8. 5 mi = 8800 yd

9. 4 c = 32 fl oz

10. $7\frac{1}{2}$ c = 60 fl oz

11. 7 pt = 14 cups

12. $2\frac{1}{2}$ qt = 5 pt

13. $3\frac{1}{2}$ gal = 14 qt

14. 6 lb = 96 oz

15. 2 ton = 4000 lb

16. $1\frac{1}{2}$ ton = 3000 lb

17. 60 in. = 5 ft

18. 72 in. = 2 yd

19. 20 cups = 10 pt

20. 40 qt = 10 gal

21. 30 in. = $2\frac{1}{2}$ ft

22. 14 qt = $3\frac{1}{2}$ gal

23. 44 in. = $1\frac{2}{9}$ yd

24. 7000 lb = $3\frac{1}{2}$ ton

Find the measure of each angle. Use after page 301.

25. 30°

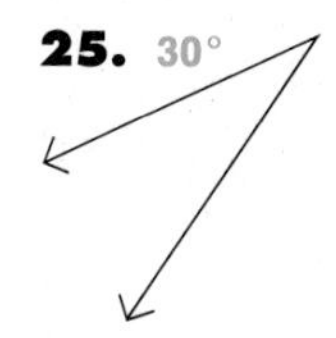

26. 75°

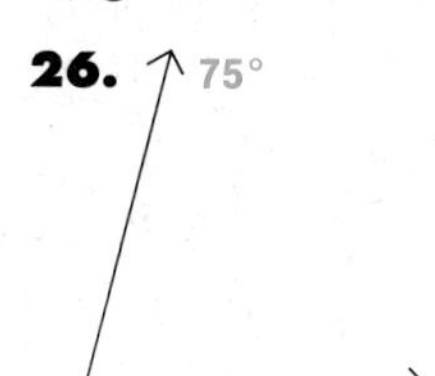

27. 90°

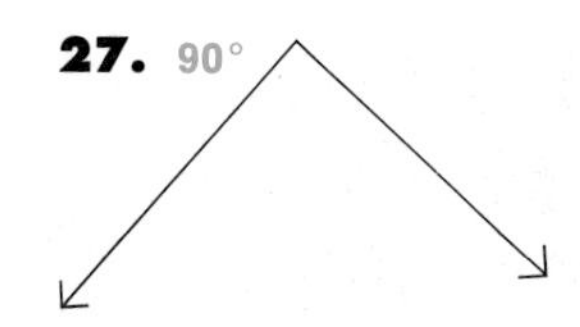

28. 125°

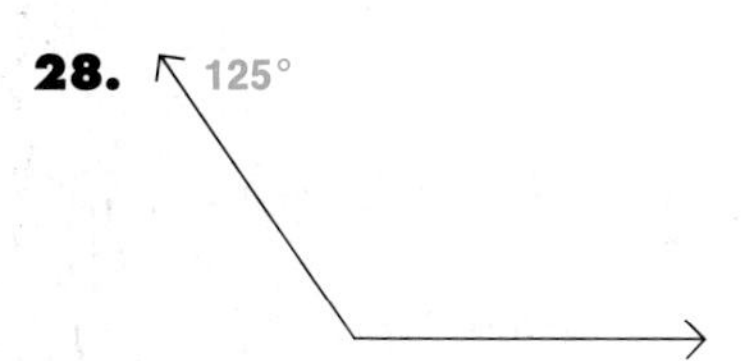

29. 170°

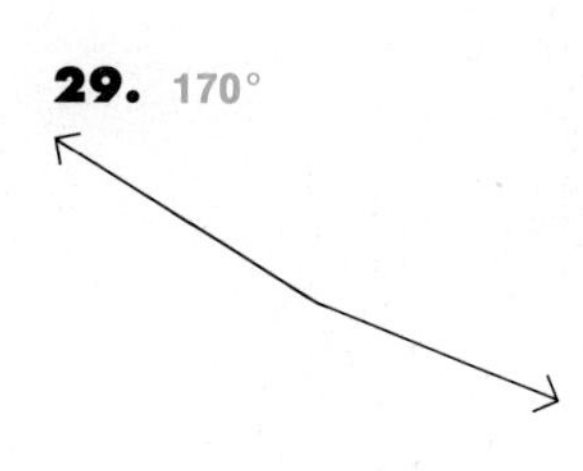

30. 115°

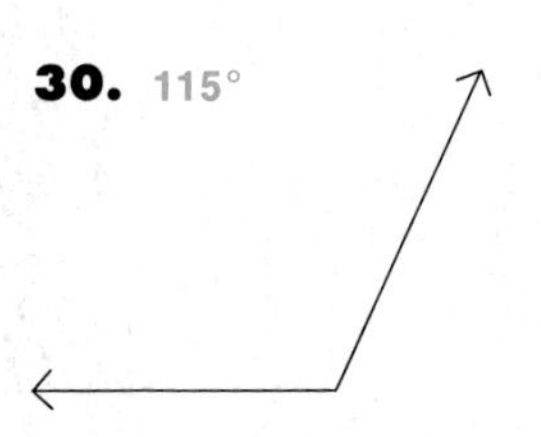

CALCULATOR CORNER

PUZZLING PATTERNS

Pattern A Make a pattern according to these rules.

Rule 1 If the number is even, divide by 2.
Rule 2 If the number is odd, multiply by 3 and add 1.

Try this with your calculator.
Start with 9.
9 is odd, so multiply by 3 and add 1. Write the result: 28
28 is even, so divide by 2. Write the result: 14
14 is even, so divide by 2. Write the result: 7
Continue until you make a pattern.

Result: 9, 28, 14, 7, 22, 11, 34, 17, 52, 26,
13, 40, 20, 10, 5, 16, 8, 4, 2, 1, . . .

The pattern that occurs is that you will always reach the number 1. The length of the sequence varies with the starting number.

Use the following as starting numbers. Complete the pattern.

1. 10 2. 18 3. 15 4. 13 5. 84

Pattern B In each of the following do the first four calculations using your calculator. Write down your results. Look for a pattern. Then do the last calculation without your calculator.

6.
24 — 4×6
264 — 44×6
2664 — 444×6
26664 — 4444×6
266664 — 44444×6

7.
9×9 — 81
99×89 — 8811
999×889 — 888111
9999×8889 — 88881111
99999×88889 — 8888811111

8.
6006 — 77×78
60606 — 777×78
606606 — 7777×78
6066606 — 77777×78
60666606 — 777777×78

9.
3×37 — 111
6×37 — 222
9×37 — 333
12×37 — 444
15×37 — 555
18×37 — 666
21×37 — 777

CHAPTER 11 PRE-TEST

Find the perimeter. p. 312

1. 28 m

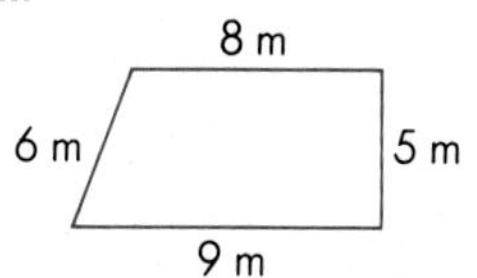

2. 26 cm

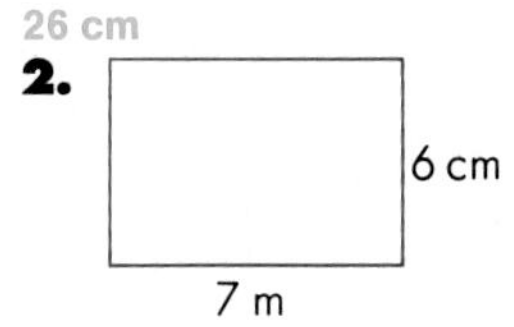

3. 28 ft

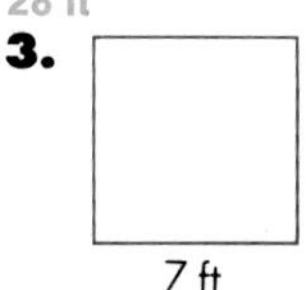

Find the circumference. p. 314

4. 28.26 m

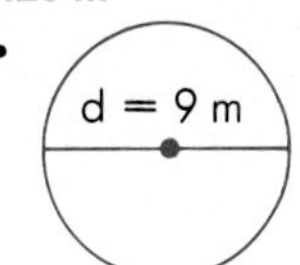

5. 27.318 cm

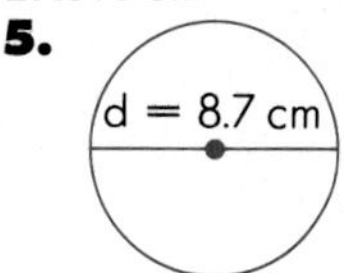

6. 25.12 yd

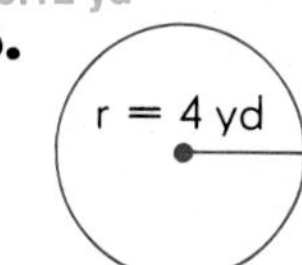

Find the area. pp. 318–326

7. 48 m^2

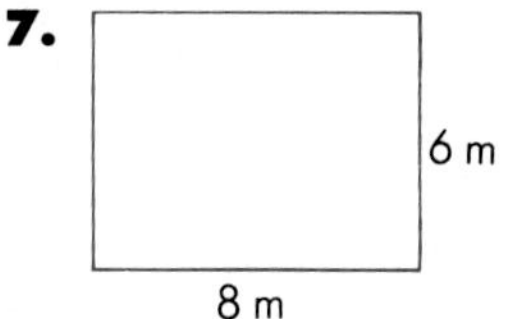

8. 60.9 m^2

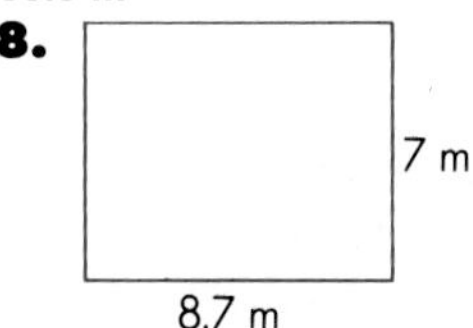

9. 25 sq in.

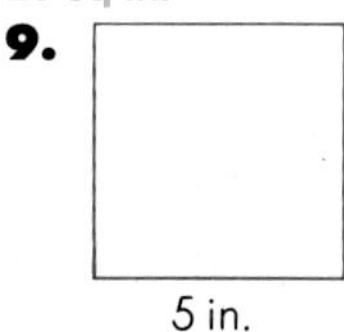

10. Parallelogram
$b = 12$ cm
$h = 8$ cm 96 cm^2

11. Triangle
$b = 9$ m
$h = 7$ m 31.5 m^2

12. Triangle
$b = 12$ ft
$h = 8$ ft 48 sq ft

13. Trapezoid
$h = 4$ m
$b_1 = 13$ m
$b_2 = 9$ m 44 m^2

14. Trapezoid
$h = 3$ yd
$b_1 = 4$ yd
$b_2 = 8$ yd 18 sq yd

15. Trapezoid
$h = 5$ cm
$b_1 = 20$ cm
$b_2 = 10$ cm 75 cm^2

16. Circle
$d = 12$ cm
113.04 cm^2

17. Circle
$r = 6$ m
113.04 m^2

18. Circle
$r = 20$ in.
1256 sq in.

Application pp. 316, 324

19. Darlene budgets 22% of her net income for housing. Her net income is \$ 550 per month. How much does Darlene budget for housing? \$121

20. A house is 50 feet long and 25 feet wide. Construction costs \$ 30 per square foot. What is the cost of constructing the house? \$37,500

11
PERIMETER AND AREA

GETTING STARTED

MATH PI

The **circumference** of a circle is the distance around it. The **diameter** is the distance across the circle through the center.

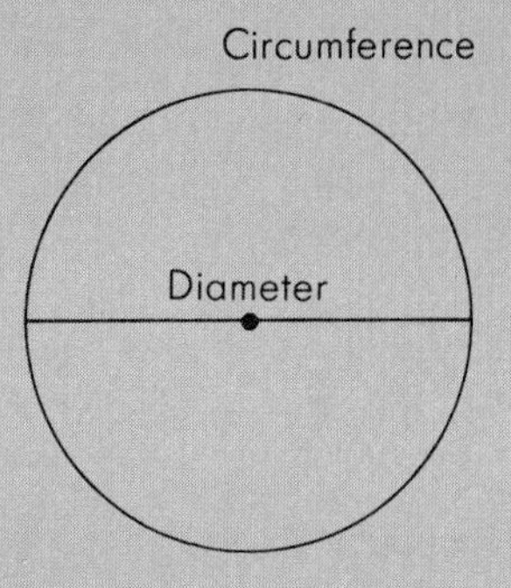

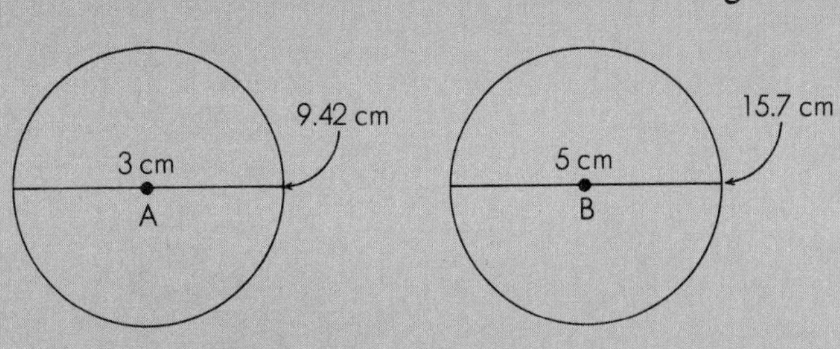

1. What is the circumference of circle A?
2. What is the diameter?
3. Divide the circumference by the diameter. $9.42 \div 3$
   What is the quotient? 3.14
4. Divide the circumference of circle B by the diameter. Divide to hundredths.
5. Compare the quotient to problem 3. What do you find? The quotients are the same.

For all circles, the circumference divided by the diameter gives the same number. We use the Greek letter π (pronounced pi) for this number. It is approximately 3.14. $\frac{C}{d} = \pi$.

6. Find a circular object. Estimate its circumference. Wrap a string around it to find the circumference. Measure the string on a meter stick to the nearest millimeter.
7. Measure the diameter with a meter stick.
8. Divide the circumference by the diameter. Is the quotient close to π? Have students measure several circular objects to verify the relationship.

Objective 111

PERIMETER

It is important that students realize that the perimeter of a figure is a number and not part of the figure.

The **perimeter** (P) of a polygon is the distance around the figure. To find the perimeter we add the lengths of the sides.

EXAMPLE 1. Find the perimeter.

$P = 3 + 4 + 4 + 3 + 5$
$P = 19$

The perimeter is 19 cm.

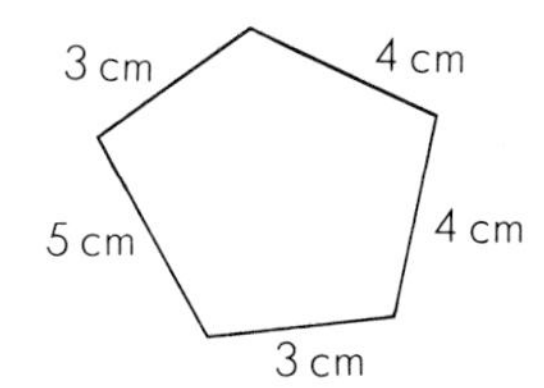

Formulas can be used to find perimeters of rectangles and squares.

EXAMPLE 2. Find the perimeter of the rectangle.

You may need to remind students that $2l + 2w$ means $(2 \times l) + (2 \times w)$.

Perimeter $= 2 \times$ length $+ 2 \times$ width
$P = 2l + 2w$
$P = (2 \times 8) + (2 \times 4)$
$P = 16 + 8$
$P = 24$

Do the calculations inside the parentheses first.

The perimeter is 24 cm.

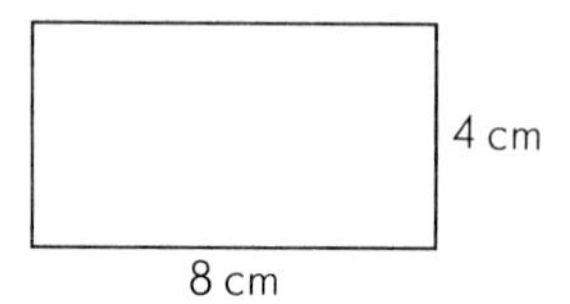

EXAMPLE 3. Find the perimeter of the square.

Perimeter $= 4 \times$ length of a side
$P = 4s$
$P = 4 \times 5$
$P = 20$

The perimeter is 20 m.

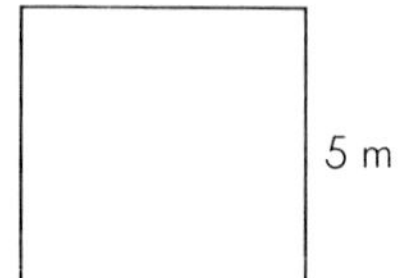

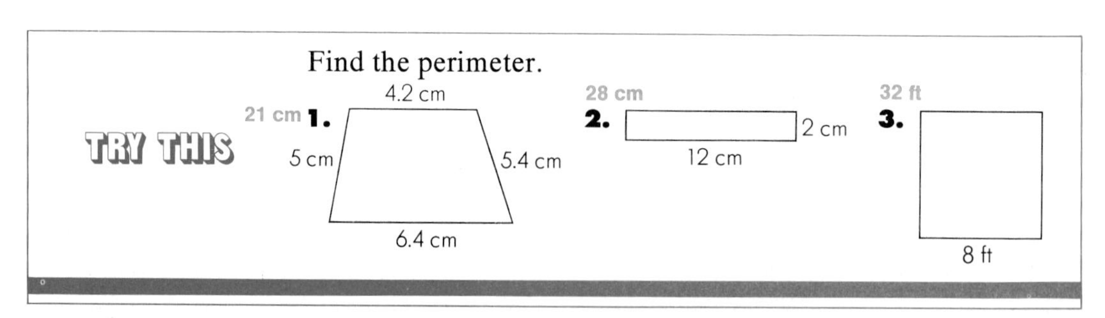

PRACTICE

For extra practice, use Making Practice Fun 67 with this lesson.

Find the perimeter.

12 cm
1.

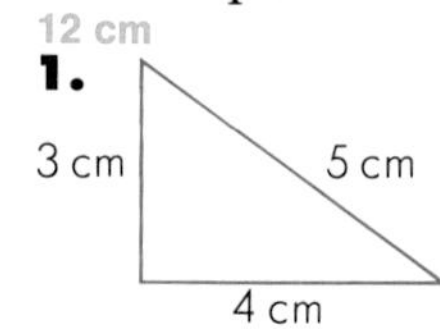

23 m
2.

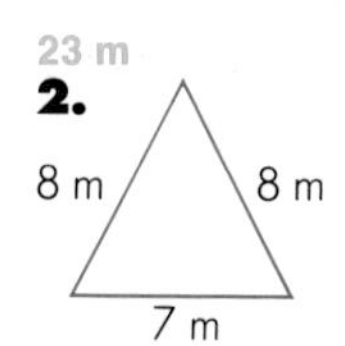

22.5 mm
3.

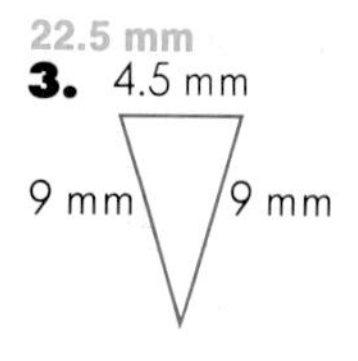

23 in.
4.

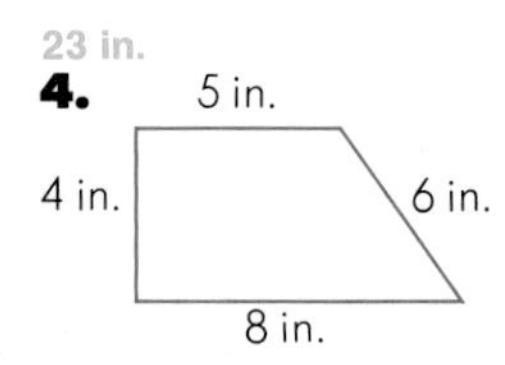

34 cm
5.

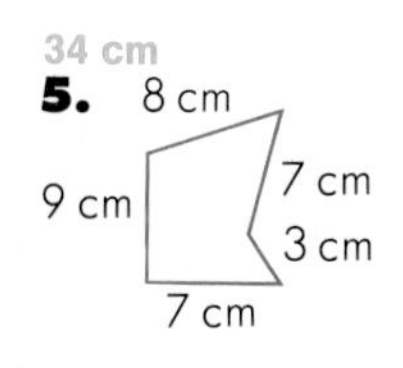

16 yd
6.

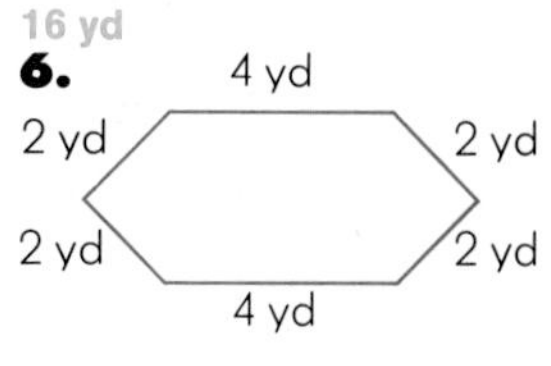

16 m
7.

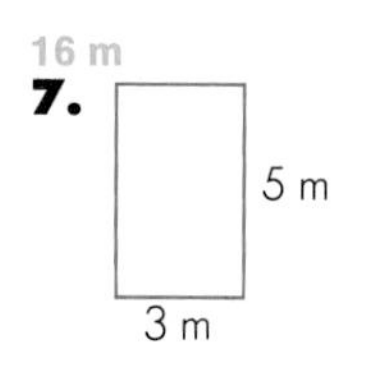

34 cm
8.

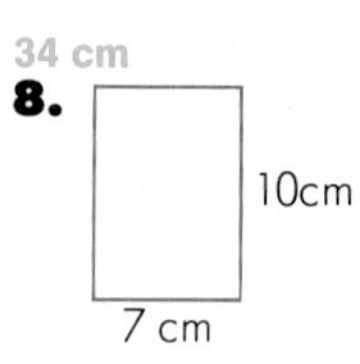

28 mm
9.

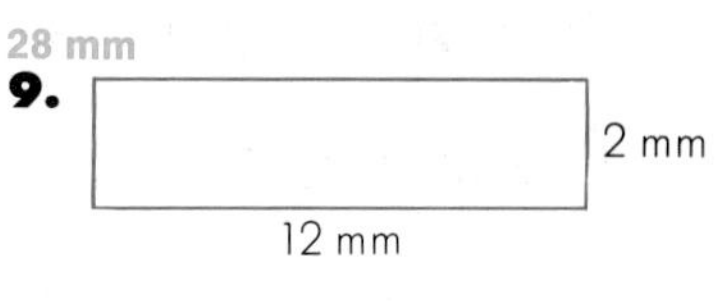

10. Rectangle
$l = 9.1$ cm
$w = 3.4$ cm 25 cm

11. Rectangle
$l = 8$ ft
$w = 7$ ft 30 ft

12. Rectangle
$l = 3\frac{1}{2}$ yd
$w = 1\frac{1}{2}$ yd 10 yd

13. Square
$s = 3$ m 12 m

14. Square
$s = 8$ cm 32 cm

15. Square
$s = 6$ mm 24 mm

16. Square
$s = 2.5$ m 10 m

17. Square
$s = 1$ cm 4 cm

18. Square
$s = 2\frac{1}{2}$ ft 10 ft

PROBLEM CORNER

1. Find the perimeter of rectangle *A*. 10 cm

2. In rectangle *B* each side is double that of rectangle *A*. Find the perimeter of rectangle *B*. 20 cm

3. Compare the perimeters. Is the perimeter of rectangle *B* double that of rectangle *A*? yes

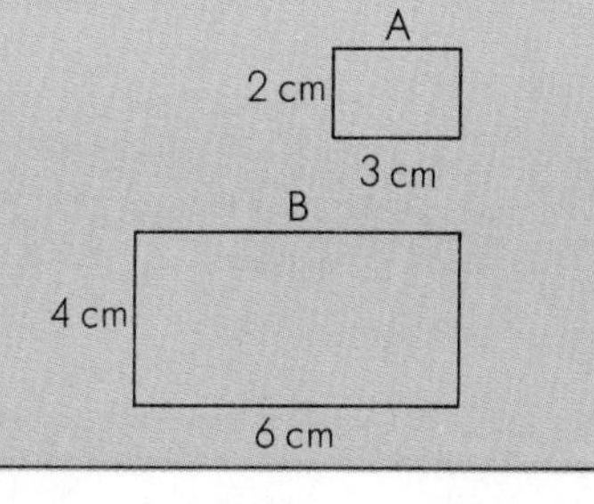

More Practice, page 334

Objective 112

CIRCUMFERENCE

Students may be proud to know π to 11 digits. The number of letters in each word of "How I wish I could recollect in simple terms its value" gives 3.1415926535.

Students should realize that the circumference is a number and not part of the circle.

The distance around a circle is the **circumference** (C). The distance across a circle through the center is the **diameter** (d). The **radius** (r) is the distance from the center to the outer edge of the circle.

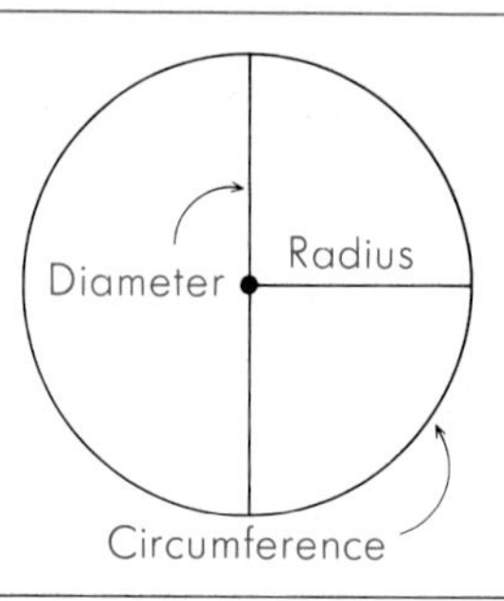

The ratio $\frac{C}{d}$ is the same for all circles. We use pi (π) for this number: $\frac{C}{d} = \pi$, or $C = \pi d$. Since the diameter is twice the radius, we also have the formula $C = 2\pi r$.

EXAMPLE 1. Find the circumference.

Since 3.14 is an approximation for π, the answer is also only an approximation.

$C = \pi d$
$C = 3.14 \times 4$
$C = 12.56$

The circumference is approximately 12.56 cm.

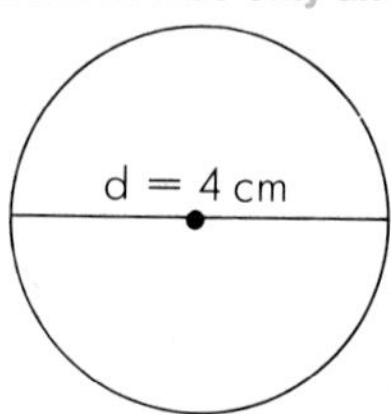

TRY THIS

Find the circumference.

28.26 m

1. 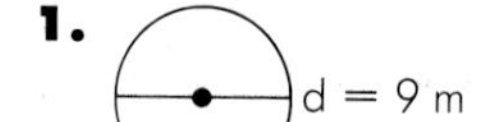

35.482 cm

2. 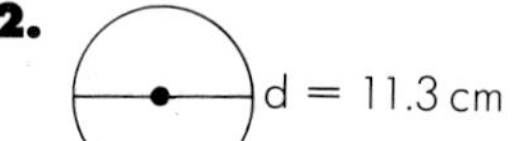

EXAMPLE 2. Find the circumference.

$C = 2\pi r$
$C = 2 \times 3.14 \times 5$
$C = 31.4$

The circumference is approximately 31.4 cm.

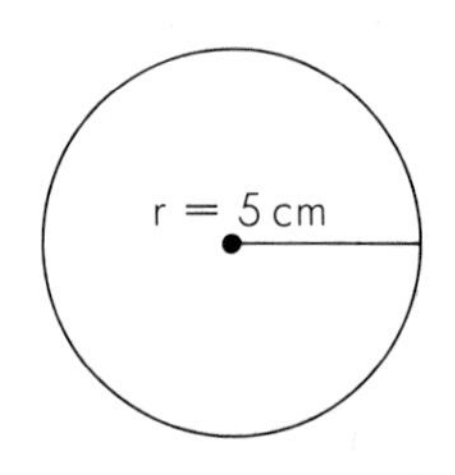

TRY THIS

Find the circumference.

18.84 cm

3. 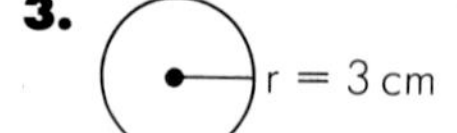

38.308 mm

4. 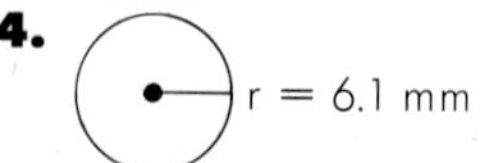

PRACTICE

For extra practice, use Making Practice Fun 68 with this lesson.

Find the circumference.

25.12 m

1.

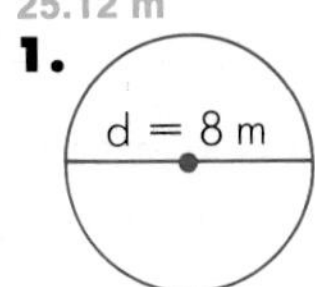

9.42 cm

2.

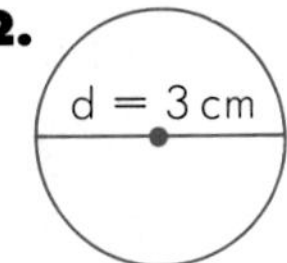

15.7 cm

3.

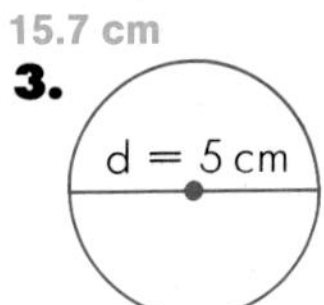

10.048 cm

4.

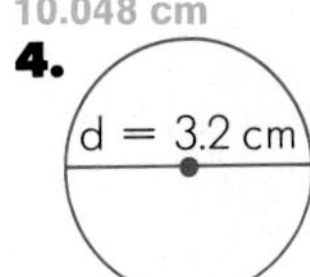

27.318 cm

5.

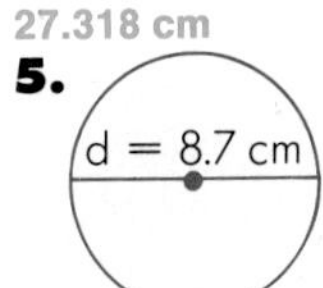

10.99 cm

6.

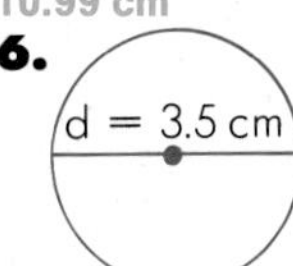

7. Circle
$d = 12$ cm 37.68 cm

8. Circle
$d = 6$ in. 18.84 in.

9. Circle
$d = 4$ ft 12.56 ft

50.24 cm

10.

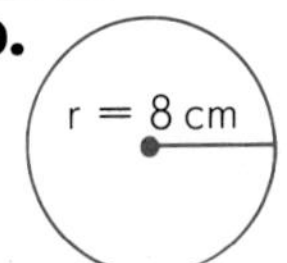

12.56 m

11.

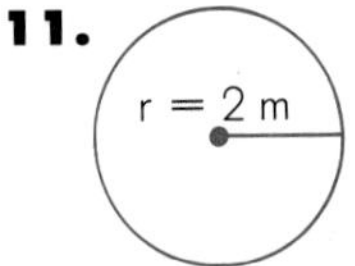

37.68 m

12.

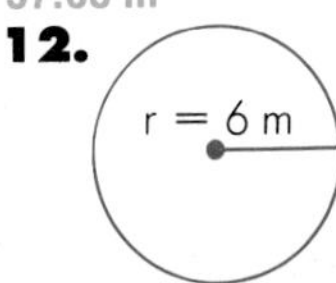

10.676 cm

13.

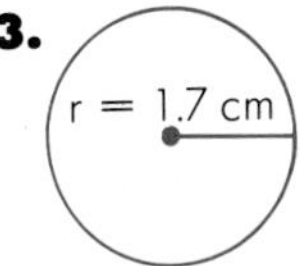

24.492 m

14.

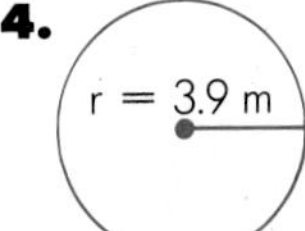

20.096 m

15.

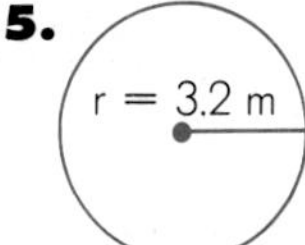

16. Circle
$r = 16$ mm 100.48 mm

17. Circle
$r = 12$ in. 75.36 in.

18. Circle
$r = 16$ ft 100.48 ft

PROBLEM CORNER

1. Find the circumference of circle A. 18.84 cm
2. In circle B the radius is double that of circle A. Find the circumference of circle B. 37.68 cm
3. Compare the circumferences. Is the circumference of circle B double that of circle A? yes

A

r = 3 cm

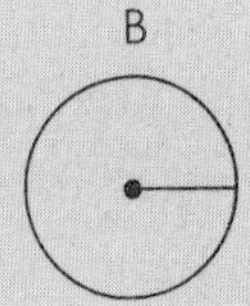

Objectives 183 and 184

APPLICATION

Have students compare local apartments, rent, and salaries by looking in a newspaper.

RENTING AN APARTMENT

Money spent for rent is generally between 20% and 25% of the net income.

EXAMPLE Net monthly income: $ 602.58
The rent budget is between 20% and 25% of about $ 600.
20% of $ 600 = $ 120 25% of 600 = $ 150
The budget for rent is about $ 120 to $ 150 per month.

For each of the following estimate amount that should be budgeted for rent as 20% to 25% of net income.

1. Net monthly income: $ 847.95 $170 to $213
2. Net monthly income: $ 514.32 $100 to $125
3. Net weekly income: $ 318.96 $260 to $325
4. Net weekly income: $ 163.25 $130 to $163
5. Vince Valdez is a heavy equipment operator earning $ 62.25 take-home pay per day for a five day week. $250 to $313
6. Jane Childs takes home $ 9640 per year as a teacher. $160 to $200
7. Kris Karman manages The Pizza Place. He takes home $ 8175 yearly in salary plus 10% in tips. $150 to $188
8. Bonnie Baker earns $ 13,119 working in a U.S. Post Office. About 20% is taken out in deductions. $175 to $219

Utility bills must also be considered when renting an apartment.

The manager tells prospective tenants that electricity costs are about 20% of the monthly rent. About how much is the full cost of these apartments?

9. studio $156
10. 2 bedroom $282
11. 1 bedroom, full bath $192
12. 3 bedroom $336

HAINES MANOR
Furnished Apartments

Pool—Tennis
Covered Parking $15 per mo

Close to schools, shopping and transportation

Studio Efficiency	$130 mo
1 Bdrm, full bath	$160 mo
1 Bdrm, 1½ bath	$195 mo
2 Bdrm, 1½ bath	$235 mo
3 Bdrm, 2 bath	$280 mo

Utilities included except electricity
Laundry Facilities—Health Club
Day Care Center available

Gloria Wren earns $975 net per month. She budgeted 22% of her salary for rent.

13. Can Gloria rent the studio apartment with covered parking? yes
14. Diane and Chris Wild earn $1875 net monthly. They budget 19% for housing. For how much under their budget can they rent a 2-bedroom apartment at Alberti Apartments? $81.25

FURNISHED APTS
All Utilities Included
2 bedroom, 1 bath apts $275
800 sq ft

ALBERTI APARTMENTS
Centerville 453-2146

How much will it cost per month to rent each of the following?

15. a 1-bedroom apartment, 400 sq ft $260
16. a 3-bedroom apartment, 650 sq ft $422.50
17. a 2-bedroom apartment, 540 sq ft $351.00

How large are these apartments?

18. $240.50 monthly rent 370 sq ft
19. $377.00 monthly rent 580 sq ft

5 STAR APARTMENTS
1 BDRM—from $165 up
5 sizes to pick from
Pay by the size you want
All Utilities included
65¢ per square ft/mo
All Electric Cooking-Solar Heat
All Luxury Apartments
Pools-Tennis Courts
Health Clubs-Laundry

ALL INCLUDED IN ONE LOW PRICE

Objectives 113 and 114

AREA OF RECTANGLES AND SQUARES

Area is measured with square units. This is a square centimeter. It is 1 cm on each side. We write: 1 cm². We say: 1 square centimeter or 1 centimeter squared.

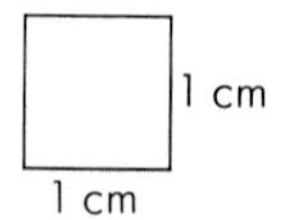

Find the area of the rectangle.
Area = length × width

$A = lw$
$A = 3 \times 2$
$A = 6$

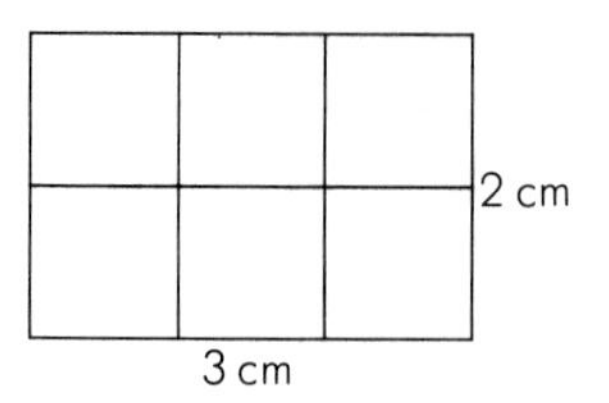

The area is 6 cm². Students can count the squares to verify the formula.

EXAMPLE 1. Find the area.

$A = lw$
$A = 8 \times 4$
$A = 32$

The area is 32 m².

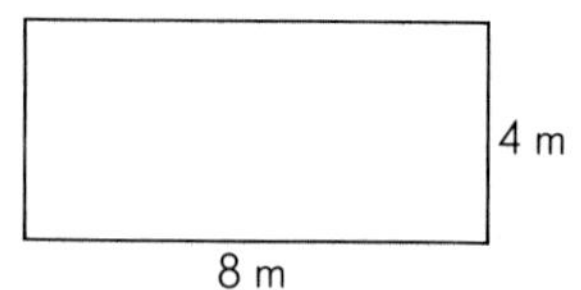

A square is a special rectangle. All sides have the same length.
Area = side × side

$A = s \times s$ or s^2 (s^2 is read: side squared)

EXAMPLE 2. Find the area.

$A = s \times s$
$A = 5 \times 5$
$A = 25$

The area is 25 cm².

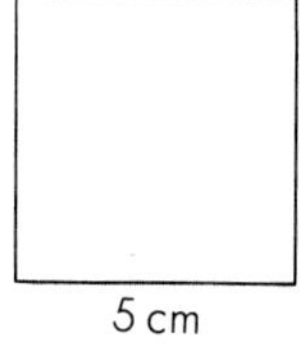

TRY THIS

Find the area.

1. 18 km² — 9 km, 2 km
2. 5 sq ft — 2 ft, $2\frac{1}{2}$ ft
3. 6.25 cm² — 2.5 cm
4. 12.25 sq yd — $3\frac{1}{2}$ yd

PRACTICE

For extra practice, use Making Practice Fun 69 with this lesson.
Use Quiz 23 after this Practice.

Find the area.

21 m²
1.
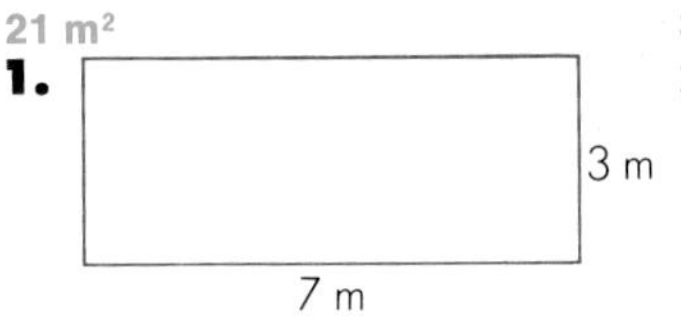

32 cm²
2.
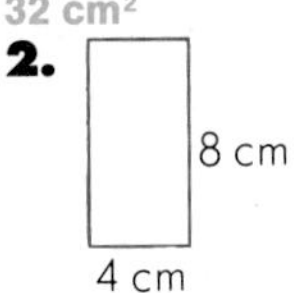

108 mm²
3.
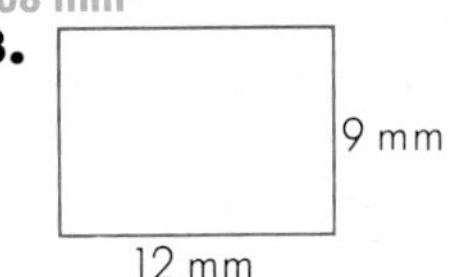

32 sq ft
4.
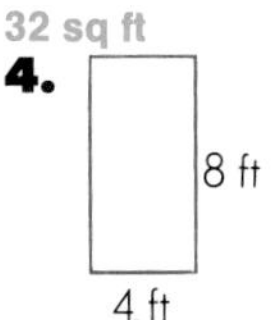

54 sq yd
5.
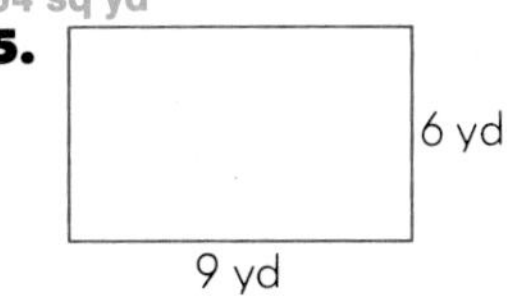

15 sq in.
6.
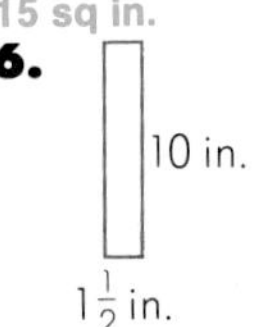

7. Rectangle
$l = 8.4$ cm 50.4 cm²
$w = 6$ cm

8. Rectangle
$l = 12$ m 55.2 m²
$w = 4.6$ m

9. Rectangle
$l = 8.7$ m 32.19 m²
$w = 3.7$ m

49 m²
10.
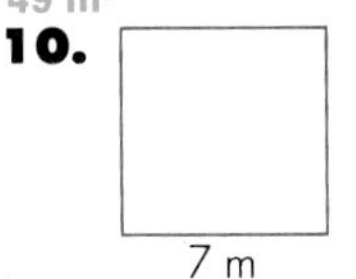

81 cm²
11.
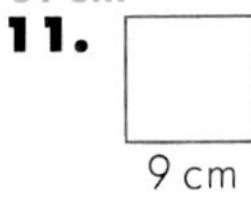

64 mm²
12.
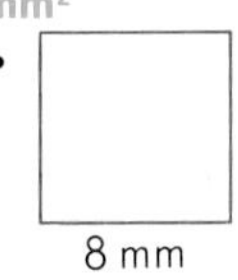

25 sq ft
13.
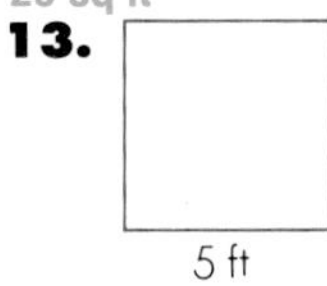

$42\frac{1}{4}$ sq yd
14.
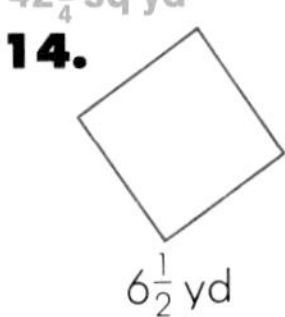

$33\frac{1}{16}$ sq in.
15.
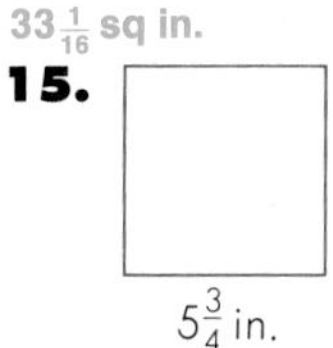

16. Square
$s = 4.6$ m 21.16 m²

17. Square
$s = 9.7$ m 94.09 m²

18. Square
$s = 3.7$ cm 13.69 cm²

PROBLEM CORNER

1. Find the perimeters of rectangle *A* and square *B*. Compare the perimeters. A: 12 cm; B: 12 cm

2. Find the areas of rectangle *A* and square *B*. Compare the areas. A: 8 cm; B: 9 cm

3. Which type of figure has more area for a given perimeter? the square

A
2 cm
4 cm
B
3 cm

More Practice, page 334

Objectives 115 and 116

AREA OF PARALLELOGRAMS, TRIANGLES

Parallelogram

Think of sliding the left triangle over to make a rectangle.

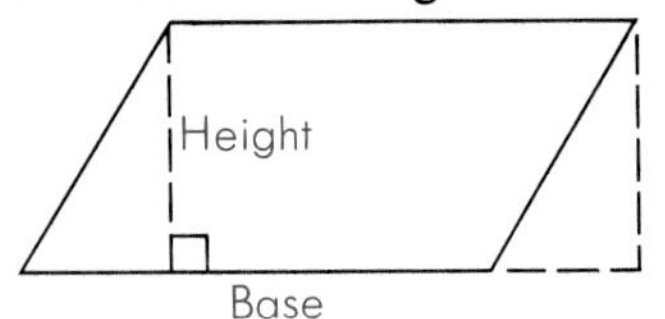

Area = base × height

$A = bh$

Triangle

Think of a triangle as half of a parallelogram.

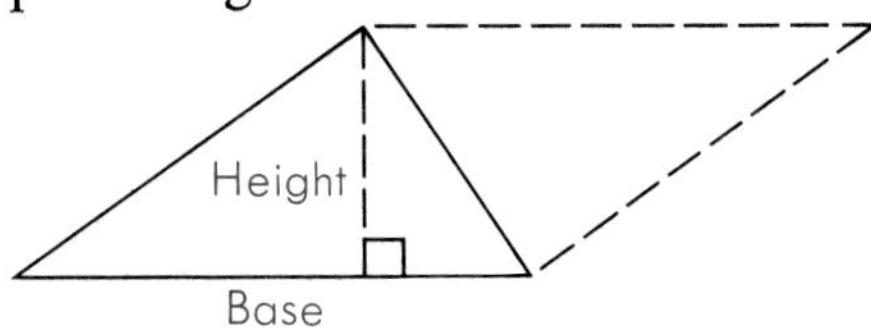

Area = $\frac{1}{2}$ × base × height

$A = \frac{1}{2}bh$

EXAMPLE 1. Find the area.

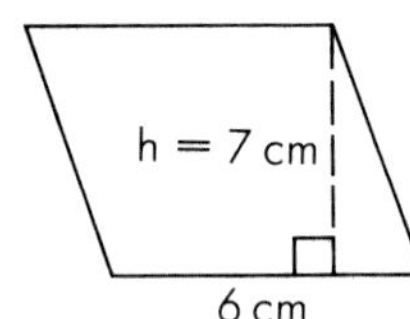

In finding the numbers to use for b and h, be sure students use one base and the height measured *from that base.*

$A = bh$

$A = 6 \times 7$

$A = 42$

The area is 42 cm².

TRY THIS

Find the area.

72 cm²

1.

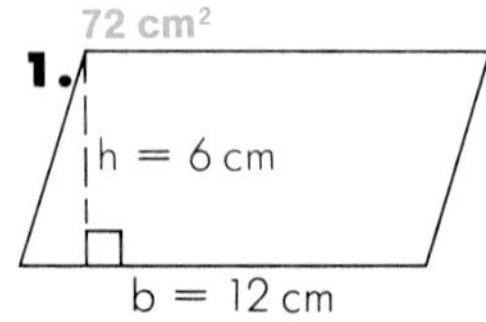

32.1 mm²

2.

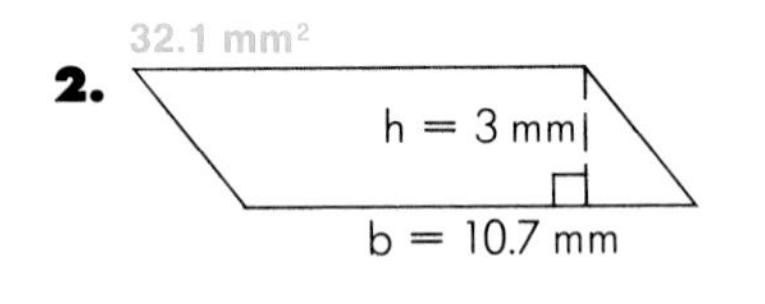

EXAMPLES. Find the area.

2.

h = 4 cm

b = 5 cm

$A = \frac{1}{2}bh$

$A = \frac{1}{2} \times 5 \times 4$

$A = \frac{1}{2} \times 20 = 10$

The area is 10 cm².

3.

h = 8 m

b = 9.4 m

$A = \frac{1}{2}bh$

$A = \frac{1}{2} \times 9.4 \times 8$

$A = \frac{1}{2} \times 75.2 = \frac{75.2}{2} = 37.6$

The area is 37.6 m².

TRY THIS

Find the area.

3. $7\frac{1}{2}$ sq ft

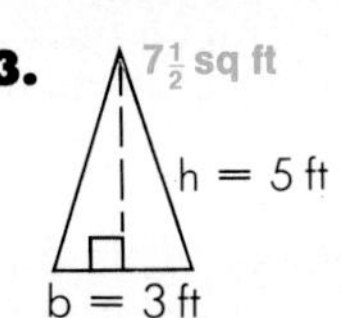

4. 21.5 m²

h = 8.6 m

b = 5 m

5. 7 km²

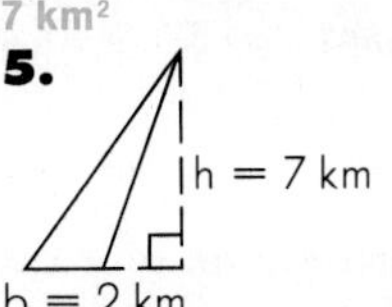

PRACTICE

For extra practice, use Making Practice Fun 70 with this lesson.

Find the area.

1. 108 m²

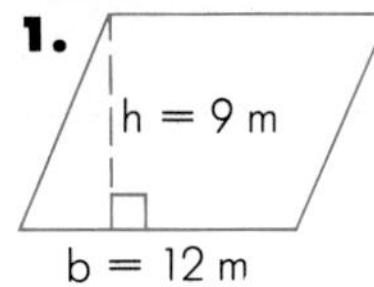

2. 120 sq yd

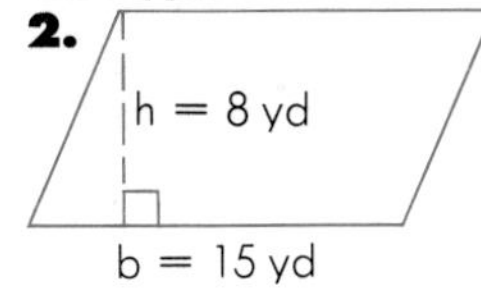

3. 125 sq in.

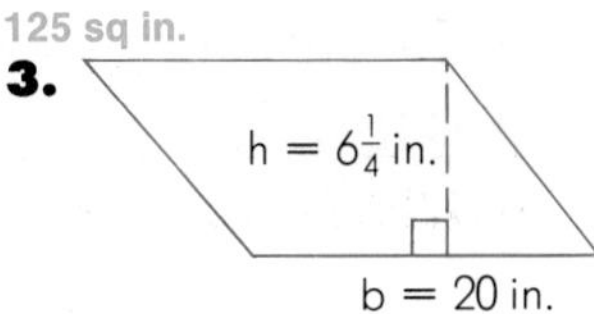

4. 65.8 cm²

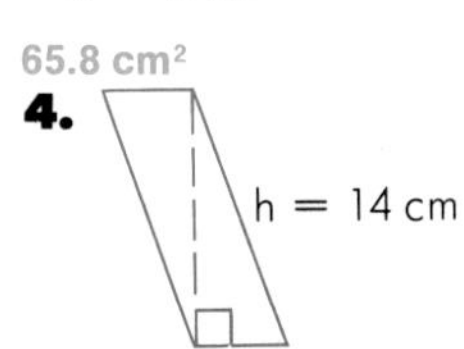

5. 43.2 m²

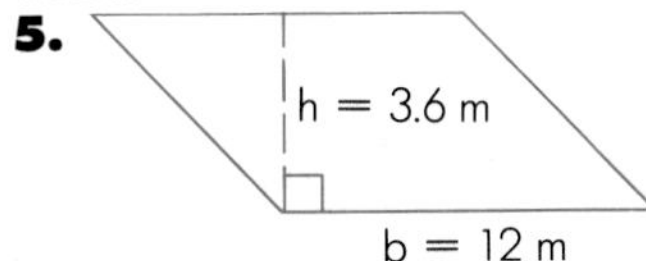

6. 54.6 mm²

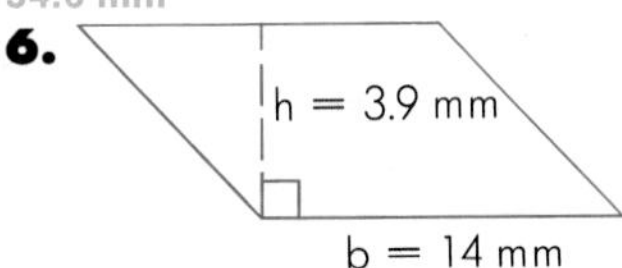

7. Parallelogram
$b = 20$ ft 170 sq ft
$h = 8\frac{1}{2}$ ft

8. Parallelogram
$b = 11$ cm 88 cm²
$h = 8$ cm

9. Parallelogram
$b = 30$ mm 300 mm²
$h = 10$ mm

10. 18 m²

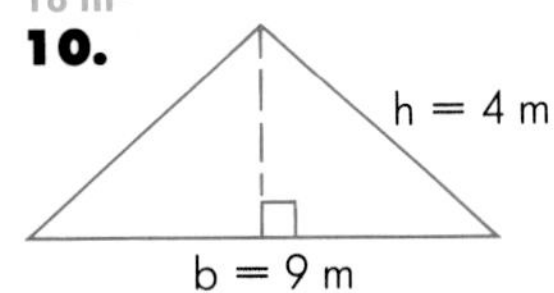

11. 96 sq in.

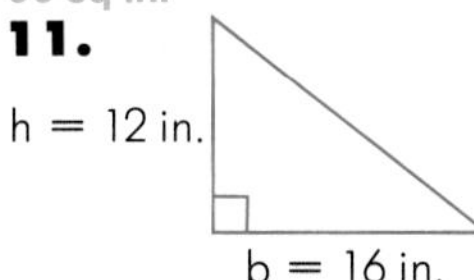
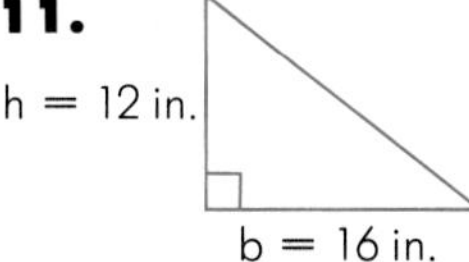

12. 150 sq yd

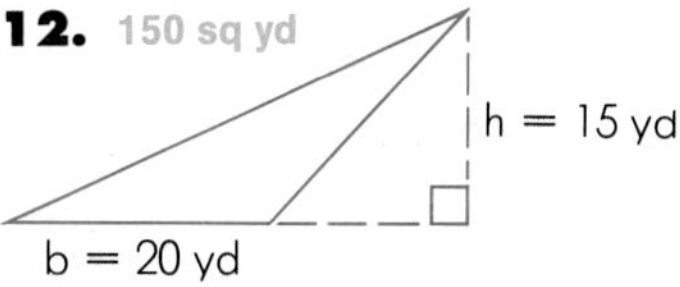

13. 14.4 cm²

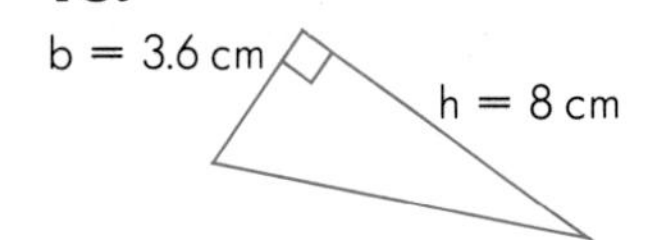

14. 36 m²

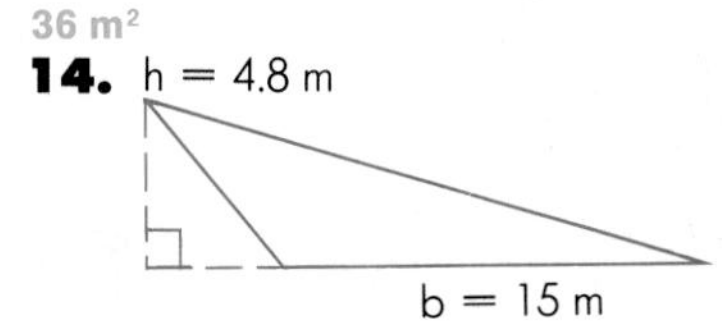

15. 24.79 m²

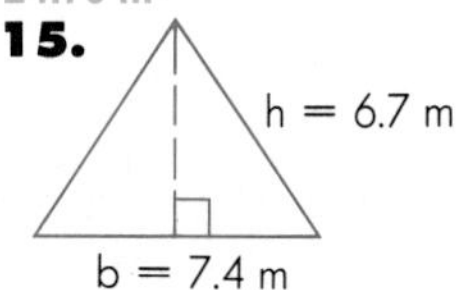

16. Triangle
$b = 7$ ft 10.5 sq ft
$h = 3$ ft

17. Triangle
$b = 21$ cm 294 cm²
$h = 28$ cm

18. Triangle
$b = 40$ mm 620 mm²
$h = 31$ mm

More Practice, page 335

Objective 117

AREA OF TRAPEZOIDS

The trapezoid area formula comes from combining the areas of two triangles. Another method is to divide the trapezoid into two triangles and a rectangle and add the areas of the three figures.

To find the area of a trapezoid we multiply the average of its bases ($\frac{1}{2}$ of base 1 + $\frac{1}{2}$ of base 2) times the height.

Area $= \frac{1}{2} \times$ (base 1 + base 2) $\times$ height

$A = \frac{1}{2} \times (b_1 + b_2) \times h$

$A = \frac{1}{2}(b_1 + b_2)h$

The formula is usually written $A = \frac{1}{2}h(b_1 + b_2)$.

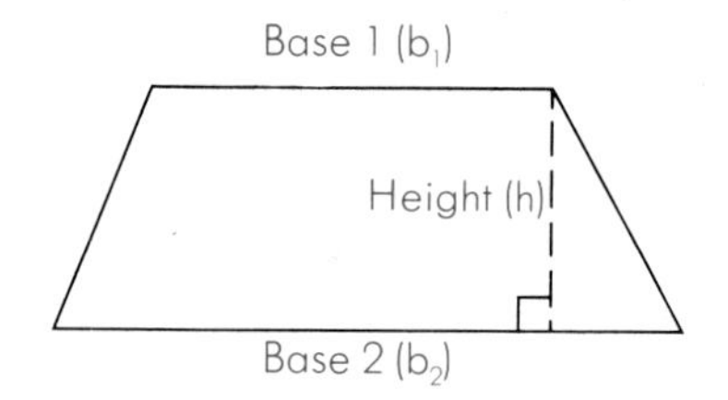

EXAMPLE 1. Find the area.

8 cm
h = 5 cm
10 cm

$A = \frac{1}{2}h(b_1 + b_2)$

$A = \frac{1}{2} \times 5 \times (8 + 10)$

$A = \frac{1}{2} \times 5 \times 18$ *Add.*

$A = 45$ *Then multiply.*

The area is 45 cm².

EXAMPLE 2. Find the area.

The height is one of the sides.

$A = \frac{1}{2}h(b_1 + b_2)$

$A = \frac{1}{2} \times 6.8 \times (8 + 12)$

$A = \frac{1}{2} \times 6.8 \times 20$

$A = \frac{1}{2} \times 136 = 68$

The area is 68 mm².

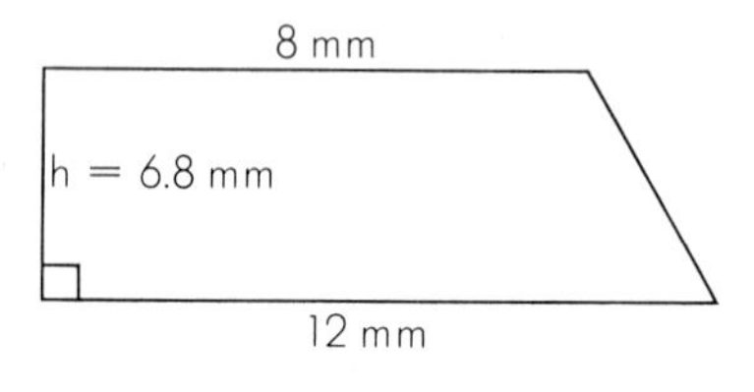

TRY THIS

Find the area.

24.4 m²

1. 3.2 m, h = 4 m, 9 m

8 km²

2.

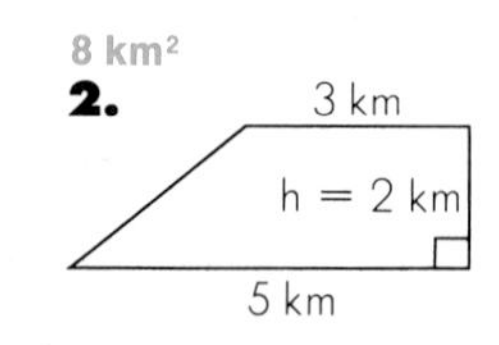

PRACTICE

Find the area.

1. 16 m² 3 m
h = 4 m
5 m

2. 41 cm² 7.4 cm
h = 5 cm
9 cm

3. 337.5 mm² 25 mm
h = 15 mm
20 mm

4. Trapezoid
$h = 5$ in. $36\frac{1}{4}$ sq in.
$b_1 = 6$ in.
$b_2 = 8\frac{1}{2}$ in.

5. Trapezoid
$h = 8$ ft 128 sq ft
$b_1 = 20$ ft
$b_2 = 12$ ft

6. Trapezoid
$h = 10$ yd 225 sq yd
$b_1 = 30$ yd
$b_2 = 15$ yd

7. Trapezoid
$h = 8$ cm 84 cm²
$b_1 = 9$ cm
$b_2 = 12$ cm

8. Trapezoid
$h = 6$ mm 36 mm²
$b_1 = 8$ mm
$b_2 = 4$ mm

9. Trapezoid
$h = 8$ m 88.4 m²
$b_1 = 10$ m
$b_2 = 12.1$ m

59.5 m²
10. 8 m
h = 7 m
9 m

52.5 cm²
11.

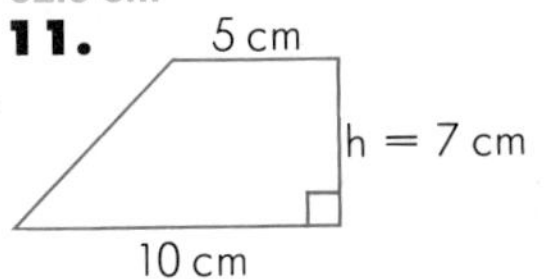

265 mm²
12.

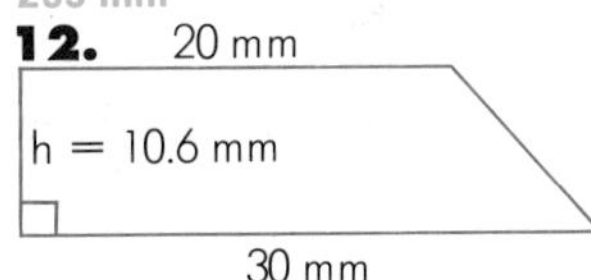

13. Trapezoid
$h = 4$ in. 46 sq in.
$b_1 = 15$ in.
$b_2 = 8$ in.

14. Trapezoid
$h = 7\frac{1}{4}$ ft $32\frac{5}{8}$ sq ft
$b_1 = 6$ ft
$b_2 = 3$ ft

15. Trapezoid
$h = 8$ yd 104 sq yd
$b_1 = 16$ yd
$b_2 = 10$ yd

PROBLEM CORNER

1. Find the area of trapezoid A. 9 cm²
2. In trapezoid B the lengths are tripled. Find the area of trapezoid B. 81 cm²
3. Compare the areas. Is the area of trapezoid B triple that of trapezoid A? What is the ratio of the area of trapezoid B to the area of trapezoid A? No; $\frac{9}{1}$

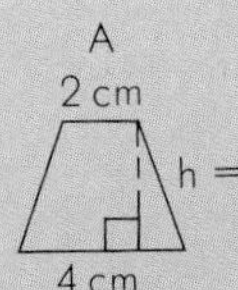
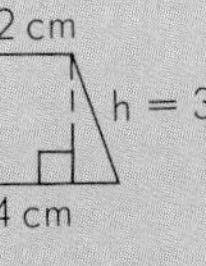

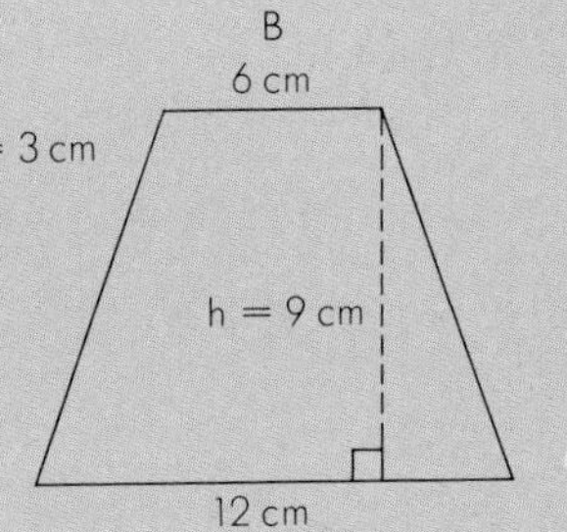

More Practice, page 335

Objective 185

APPLICATION

BUYING A HOUSE

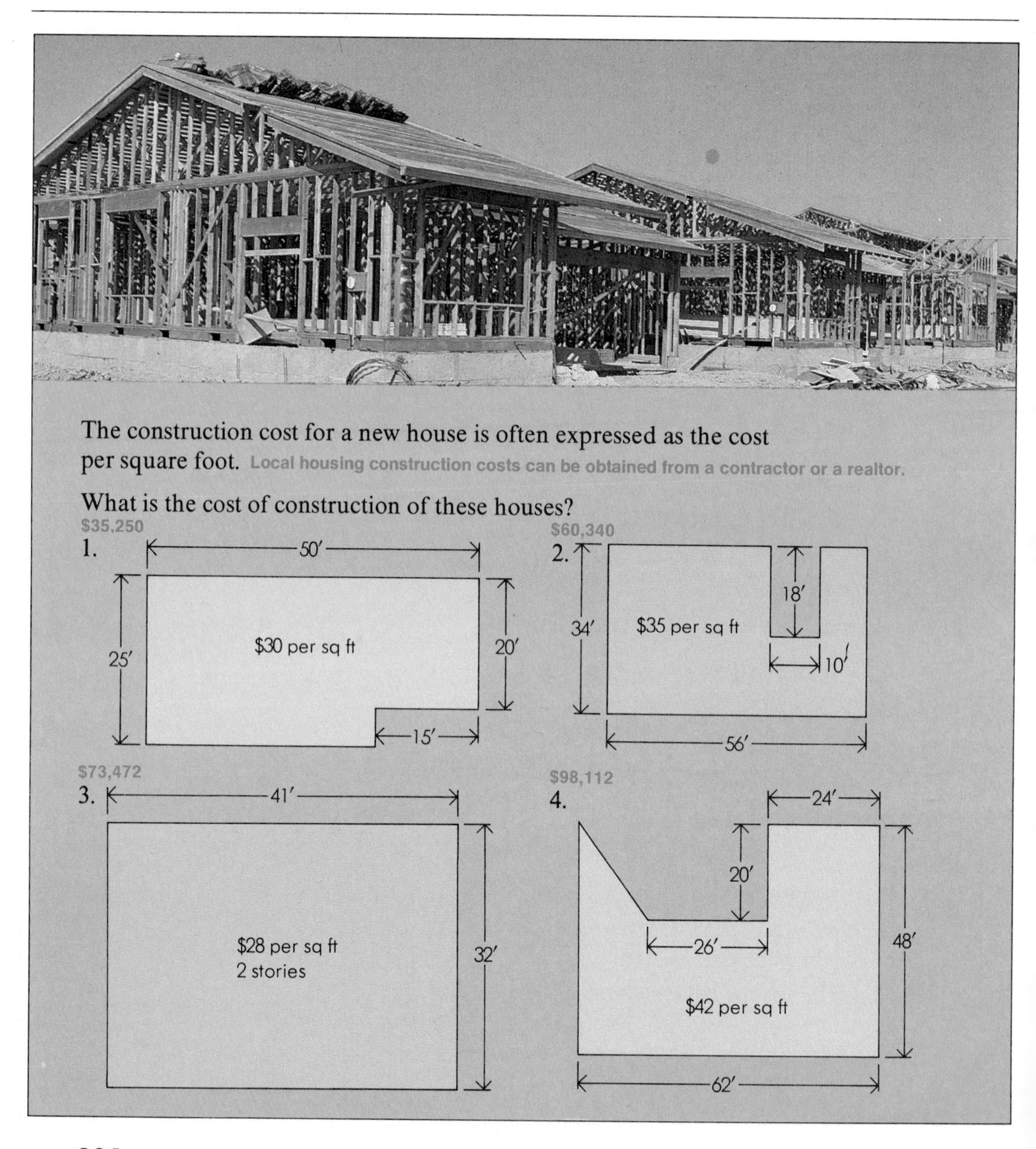

The construction cost for a new house is often expressed as the cost per square foot. Local housing construction costs can be obtained from a contractor or a realtor.

What is the cost of construction of these houses?

Many people consider buying mobile homes.

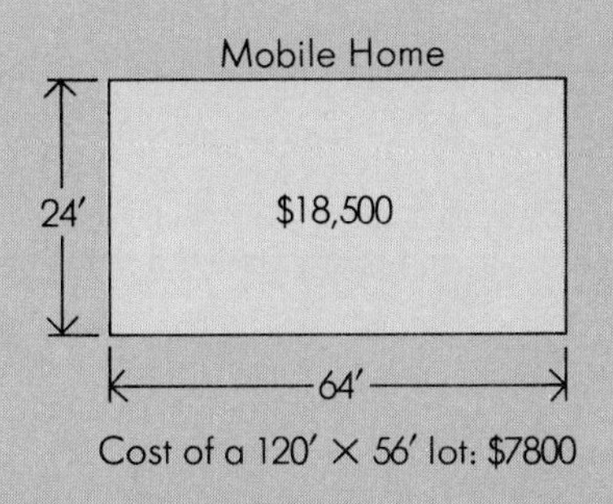

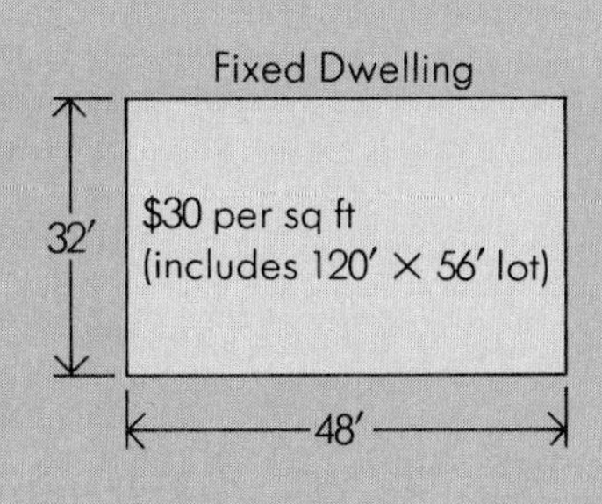

5. What is the total cost of the mobile home? $26,300
6. What is the total cost of the fixed dwelling? $53,880
7. Compare the total costs. What percent of the total cost of the fixed dwelling is the cost of the mobile home? 48.8%
8. List some advantages and disadvantages of owning each type of dwelling. Answers will vary.

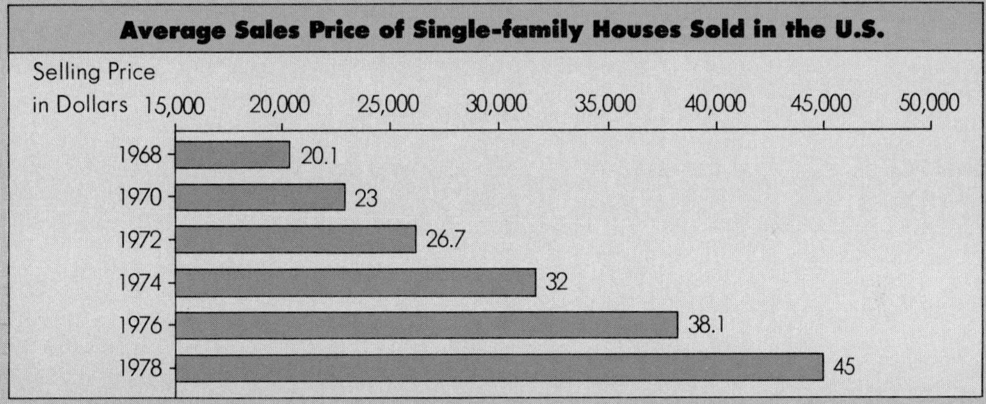

Find the percent of increase between the following years?

9. 1968 and 1970 14.4%
10. 1970 and 1972 16.1%
11. 1972 and 1974 19.9%
12. 1974 and 1976 19.1%
13. 1976 and 1978 18.1%

Predict the average selling price in these years.

14. 1980
15. 1990 Answers will vary.

Objective 118

AREA OF CIRCLES

Make the area formula reasonable by showing that the area of three squares with edge r covers a circle of radius r.

To find the area of a circle, we multiply $\pi(3.14)$ by the radius squared. ($r^2 = r \times r$). Area $= \pi \times r \times r$

$$A = \pi r^2$$

EXAMPLE 1. Find the area.

$A = \pi r^2$

$A = 3.14 \times 3^2$ $\quad (3^2 = 3 \times 3 = 9)$

$A = 3.14 \times 9$

$A = 28.26$

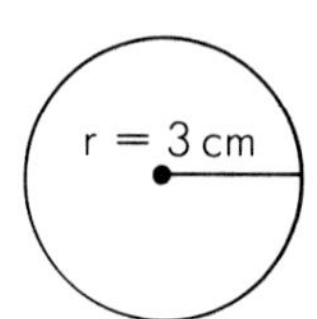

The area is approximately 28.26 cm^2.

TRY THIS

Find the area.

50.24 m^2

1. r = 4 m

452.16 km^2

2. r = 12 km

To find the area when the diameter is given we first find the radius.

EXAMPLE 2. Find the area.

$d = 14$

$r = \frac{1}{2} \cdot 14 = 7$

$A = \pi r^2$

$A = 3.14 \times 7^2$

$A = 3.14 \times 49$

$A = 153.86$

d = 14 cm

The area is approximately 153.86 cm^2.

TRY THIS

Find the area.

28.26 m^2

3.

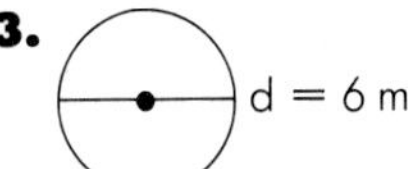

78.5 sq ft

4.

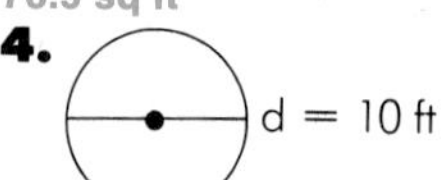

PRACTICE

Use Quiz 24 after this Practice.

Find the area.

78.5 cm^2

1.

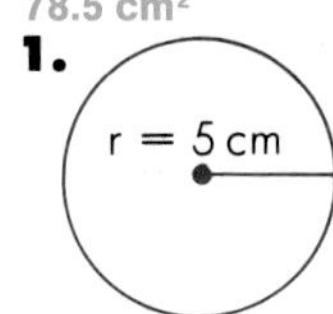

153.86 m^2

2.

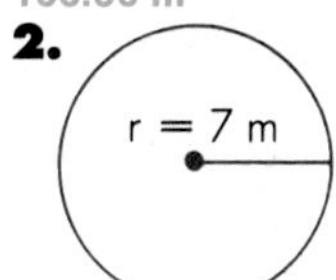

200.96 mm^2

3.

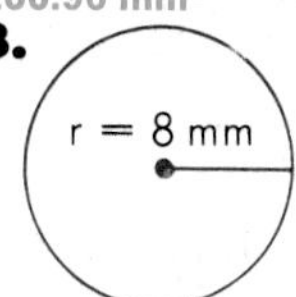

314 sq in.

4.

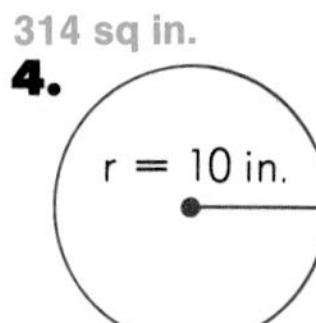

706.5 sq ft

5.

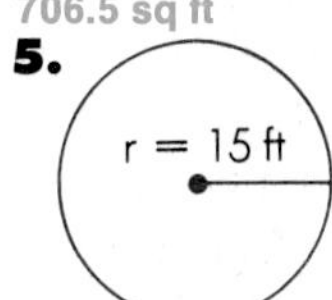

452.16 sq yd

6.

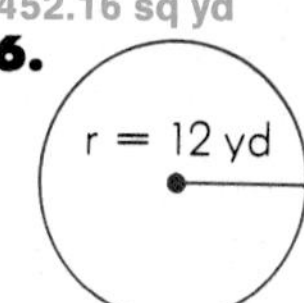

7. Circle
$r = 4$ m 50.24 m^2

8. Circle
$r = 8$ cm 200.96 cm^2

9. Circle
$r = 3$ mm 28.26 mm^2

113.04 m^2

10.

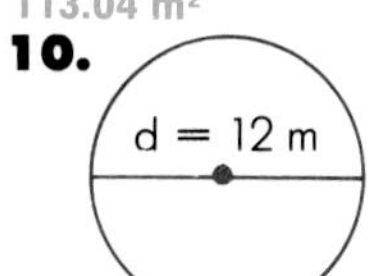

50.24 cm^2

11.

d = 8 cm

254.34 mm^2

12.

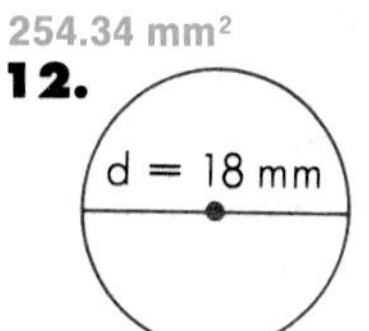

452.16 sq ft

13.

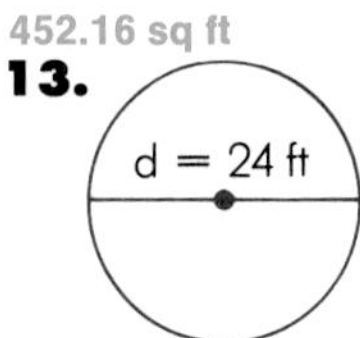

706.5 sq in.

14.

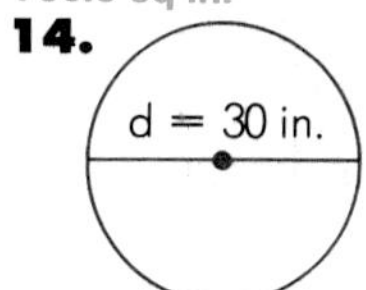

1133.54 sq yd

15.

d = 38 yd

16. Circle
$d = 6$ m 28.26 m^2

17. Circle
$d = 18$ mm 254.34 mm^2

18. Circle
$d = 10$ cm 78.5 cm^2

PROBLEM CORNER

1. Find the area of circle A. 12.56 cm^2

2. The radius of circle B is triple that of circle A. Find the area of circle B. 113.04 cm^2

3. What is the ratio of area B to area A? $\frac{9}{1}$

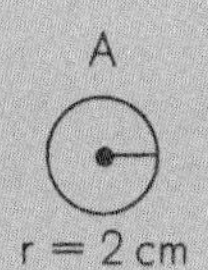

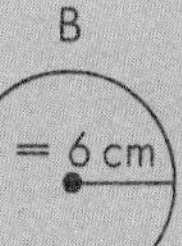

Objective 196

SOLVING PROBLEMS

Drawing a picture can sometimes help with the solution of a problem.

EXAMPLE. The Yingling family needs to build a wood fence around their new swimming pool. The fence will have five sides and a gate. Mr. Yingling drew this picture.

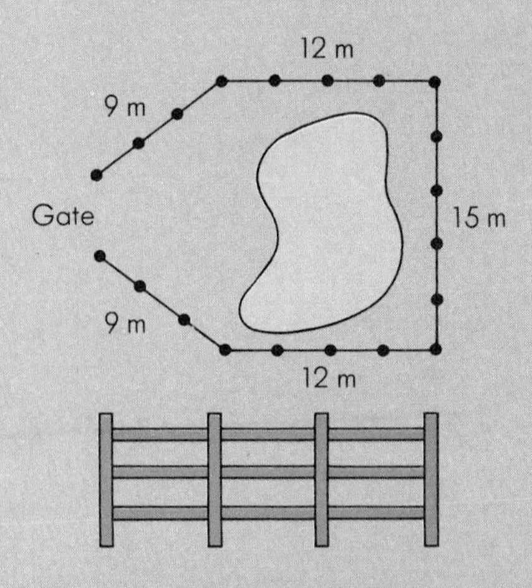

1. What is the approximate distance between fence posts? 3 m
2. The cost of posts is $ 4.25 each. How much must be spent for posts? $85
3. The fence will consist of three rails. What is the total length of rails needed? 171 m
4. Rails cost $ 3.25 per linear meter. What will be the cost of the rails? $555.75
5. A gate will cost $ 12.50. Find the total cost of materials for the fence. $653.25

TRY THIS

Peninsula Garden Service planted rose bushes around two gardens. One garden was rectangular, measuring 12 m by 32 m. The other garden was square, measuring 20 m on a side. The rose bushes were planted 4 meters apart. How many more rosebushes were needed for the rectangular garden than the square garden? Make drawings to help solve the problem. 2

PRACTICE For extra practice, use Making Practice Fun 71 with this lesson.

1. Joe Gori plans to fence in his garden. The shape is a quadrilateral which has sides of 3 m, 5 m, 5 m, and 6 m. The posts will be 1 meter apart and cost $ 3.25 each. Wire garden fence will cost $ 4.95 per meter. There will be no gate. Find the total cost of the fence. $155.80

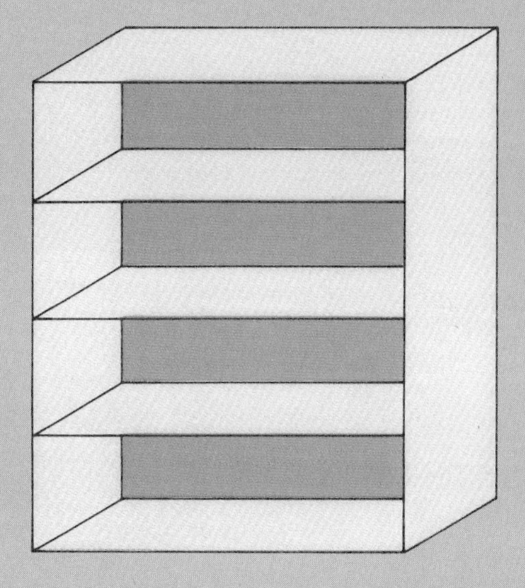

2. A bookcase is 4′ high by 5′ long by 1′ deep. There are four shelves 1′ apart. The backing is one solid piece 4′ × 5′. A quart of paint will cover 44 square feet. How many quarts of paint are needed to paint the bookcase completely? $2\frac{9}{22}$ or 3 qt
3. A round pool has a radius of 15 feet. A trapezoidal pool has a width of 16 feet and opposite base sides of 14 feet and 20 feet. Which of these swimming pools has the greater surface area? How much greater is it? Make a drawing to prove (show) your answer. the round pool; 434.5 sq ft
4. Carpet costs $ 14.25 per square yard. Follow these steps to find how much it costs to carpet a square room whose perimeter is 48 feet. Draw a square whose perimeter is 48 feet. Use a scale where $\frac{1}{2}$ inch represents 1 foot.

   How long is each side on the scale drawing? 6 in.
   How many square feet are there in the room? 144 sq ft
   How much does the carpet cost? (9 square feet = 1 square yard) $228

ARCHITECT

Maria Constantine is an architect. Architects design and draw plans for all types of structures.

In a home she is designing, Maria plans to put decorative brick 5 feet high around the base of the house. There is an 8-foot door in the back and a 6-foot door in the front.

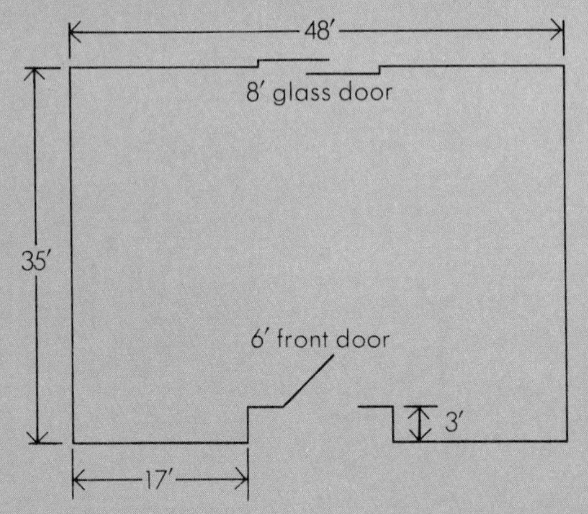

1. How many linear feet of brick will be needed? 158′
2. How many square feet of brick will be needed? 790 sq ft
3. Brick costs $ 12 per square foot. Find the total cost of the brick. $9480

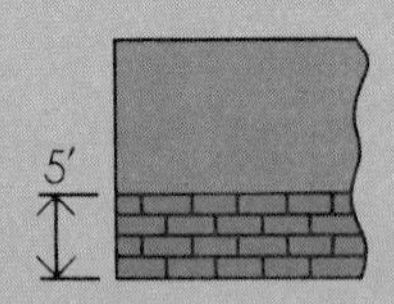

Architect's fees are often based on a percentage of the final construction cost of a house.

4. Maria designs a house with an estimated construction cost of $ 55,800. She is to be paid 6% for preparing the working drawings. About how much can she expect to be paid? $3350
5. For preparing working drawings and supervising the construction, Maria was paid $8\frac{1}{2}$% of the construction cost of an $ 83,650 home. What was her fee? $7110.25

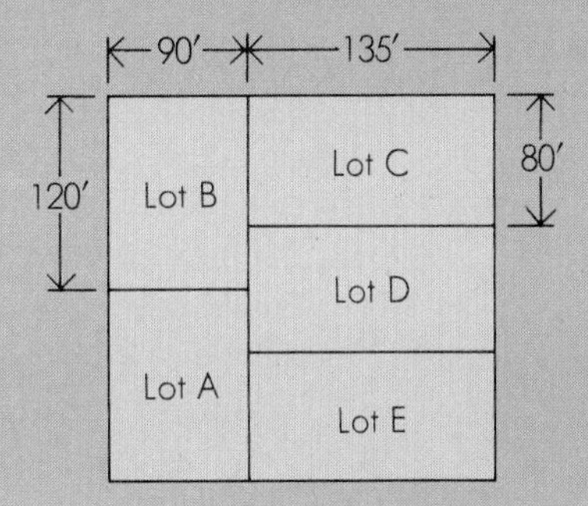

Architects frequently work with developers to plan communities. Maria has a tract to divide into five lots. She made two lots 120′ × 90′ and three lots 135′ × 80′.

6. What is the area of lot A? 10,800 sq ft
7. What is the area of lot D? 10,800 sq ft
8. What is the perimeter of lot B? 420′
9. What is the perimeter of lot E? 430′
10. How much fencing will be needed to enclose all five lots completely? (Remember, some parts of the fence will be shared by two lots.) 1530′

Fencing costs $ 6 per running foot. Lots sharing a fence are assessed an equal part of the cost.

How much is each of the lots assessed for fencing?

11. lot A $1890 12. lot B $1890 13. lot C $1935 14. lot D $1530 15. lot E $1935

CHAPTER 11 REVIEW

For extra practice, use Making Practice Fun 72 with this lesson.

Find the perimeter. p. 312

31 cm
1.
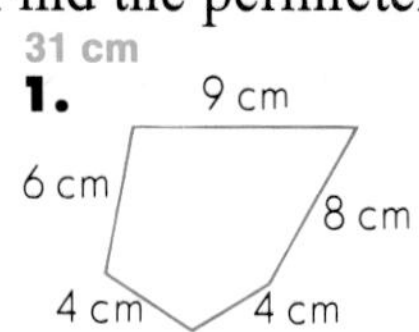

40 m
2.
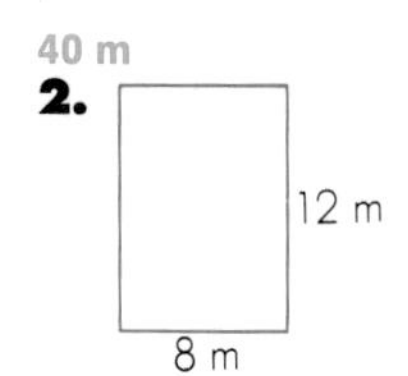

20 in.
3.
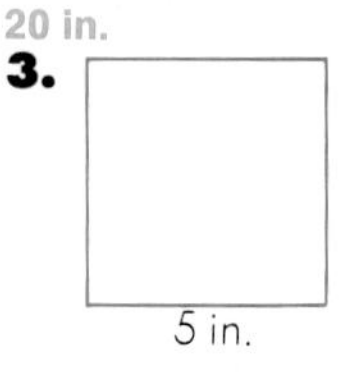

Find the circumference. p. 314

25.12 cm
4.
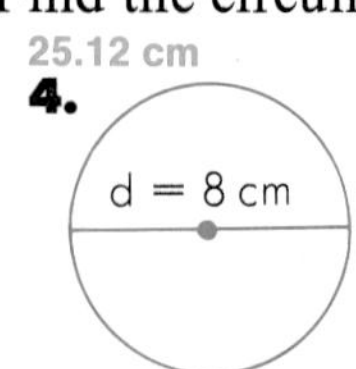

24.492 m
5. d = 7.8 m

37.68 in.
6.

Find the area. pp. 318-326

63 cm²
7.
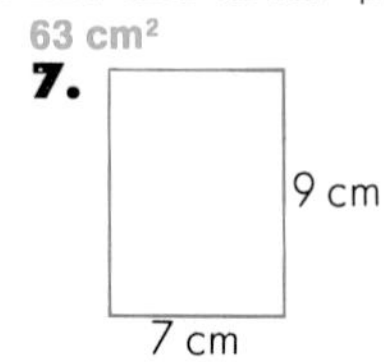

28.8 m²
8.
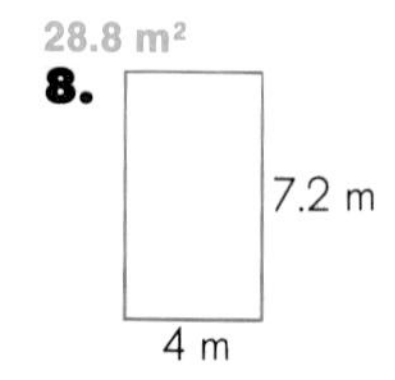

144 sq ft
9.
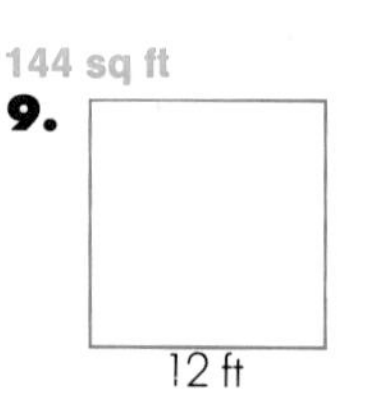

10. Parallelogram
$b = 12$ m
$h = 9$ m 108 m²

11. Triangle
$b = 10$ cm
$h = 8$ cm 40 cm²

12. Triangle
$b = 8$ in.
$h = 6$ in. 24 sq in.

13. Trapezoid
$h = 4$ mm
$b_1 = 15$ mm
$b_2 = 7$ mm 44 mm²

14. Trapezoid
$h = 6$ m 78 m²
$b_1 = 18$ m
$b_2 = 8$ m

15. Trapezoid
$h = 3$ yd 22.5 sq yd
$b_1 = 10$ yd
$b_2 = 5$ yd

16. Circle
$d = 18$ m 254.34 m²

17. Circle
$r = 3$ cm 28.26 cm²

18. Circle
$r = 7$ ft 153.86 sq ft

Application pp. 316, 324

19. Randy budgets 23% of his net income for housing. His net income is \$ 650 per month. How much does Randy budget for housing? \$149.50

20. In 1970 the average sales price of a house was \$ 23,000. In 1974, the price was \$ 32,000. What was the percent of increase in the sales price? 39%

CHAPTER 11 TEST

Find the perimeter.

22 m
1.

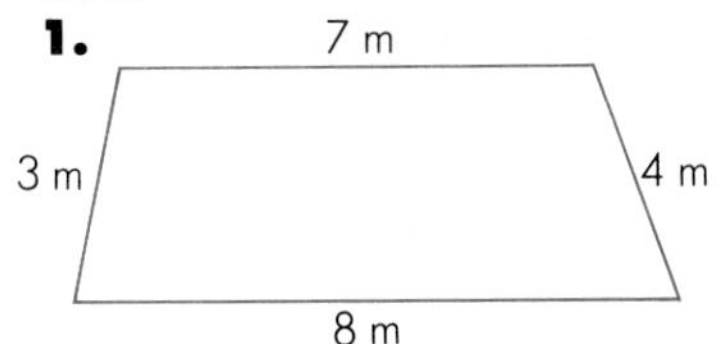

30 cm
2.

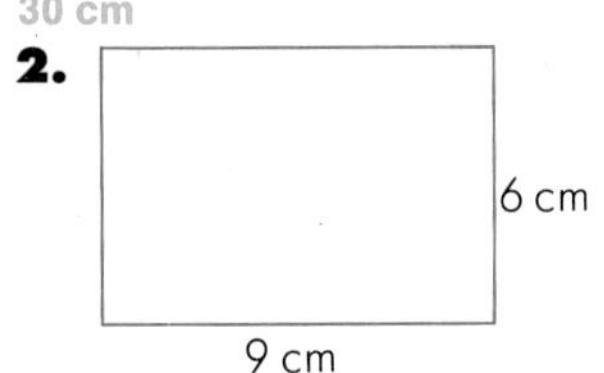

44 ft
3.

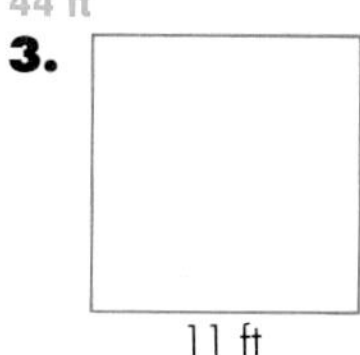

Find the circumference.

47.1 in.
4.

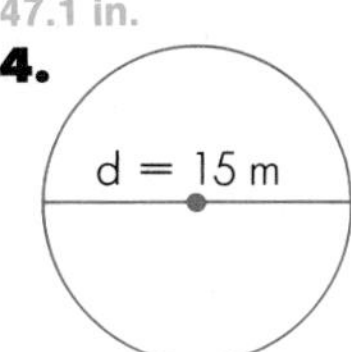

15.7 cm
5.

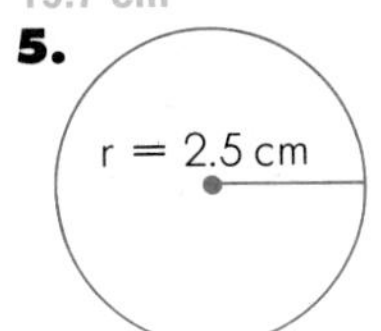

37.68 in.
6.

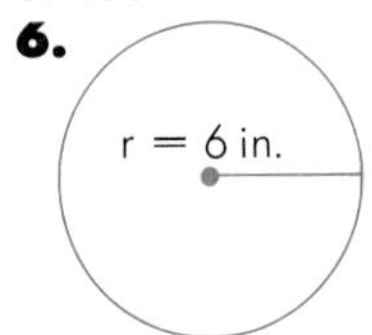

Find the area.

104 m²
7.

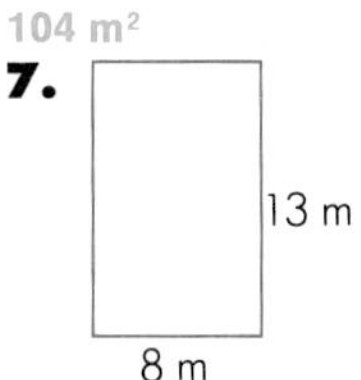

7.2 cm²
8.

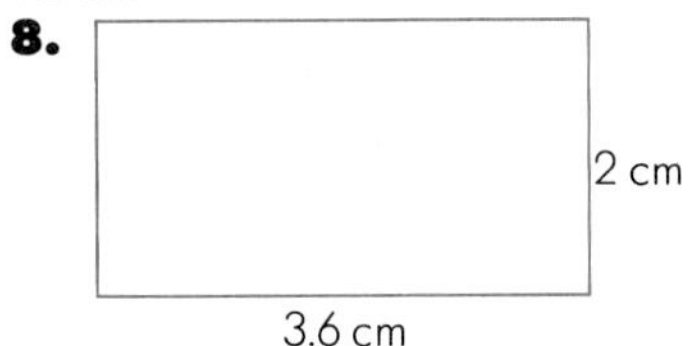

169 sq in.
9.

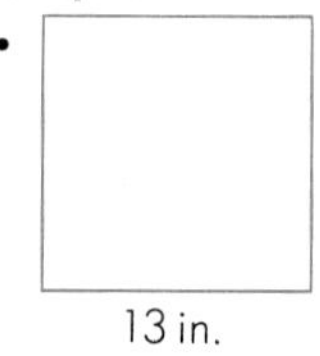

10. Parallelogram
$b = 14$ cm
$h = 8$ cm 112 cm²

11. Triangle
$b = 13$ m
$h = 5$ m 32.5 m²

12. Triangle
$b = 24$ yd
$h = 18$ yd 216 sq yd

13. Trapezoid
$h = 8$ m
$b_1 = 13$ m
$b_2 = 9$ m 88 m²

14. Trapezoid
$h = 5$ cm
$b_1 = 10$ cm
$b_2 = 5$ cm 37.5 cm²

15. Trapezoid
$h = 8$ ft
$b_1 = 12$ ft
$b_2 = 12$ ft 96 sq ft

16. Circle
$d = 24$ cm 452.16 cm²

17. Circle
$r = 9$ m 254.34 m²

18. Circle
$r = 8$ in. 200.96 sq in.

Application

19. Jane budgets 22% of her net income for housing. Her net income is \$550 per month. How much does Jane budget for housing? \$121

20. A 2-story house is 40 feet long and 35 feet wide. Construction costs \$35 per square foot. What is the cost of constructing the house? \$98,000

MORE PRACTICE

Find the perimeter. Use after page 313.

23 ft
1.
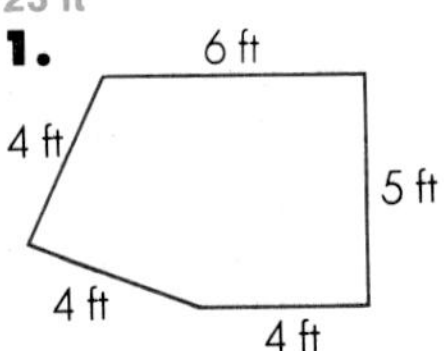

18 in.
2.
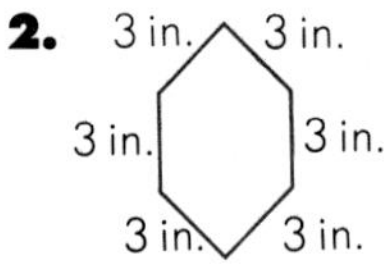

44 m
3.
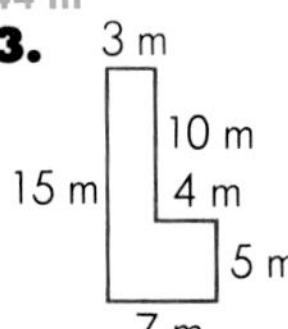

44 cm
4.
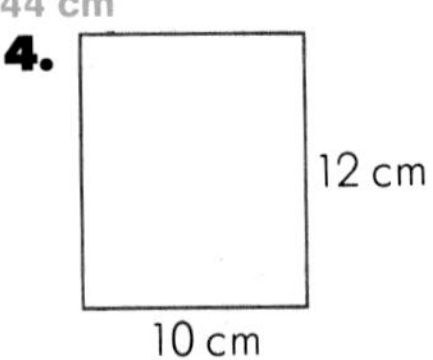

36 mm
5.
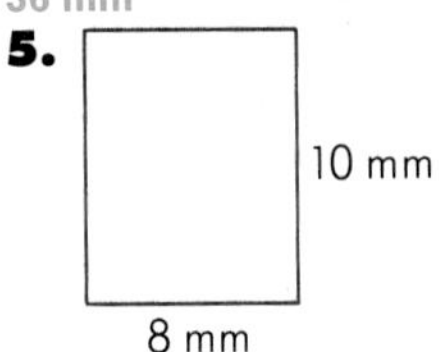

24 m
6.
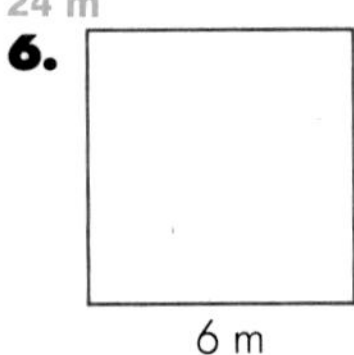

7. Rectangle
$l = 20$ yd 50 yd
$w = 5$ yd

8. Square
$s = 5$ cm 20 cm

9. Square
$s = 4\frac{1}{2}$ ft 18 ft

Find the circumference. Use after page 315.

75.36 in.
10.
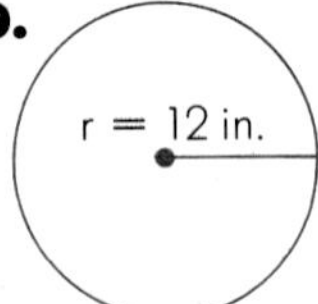

16.642 m
11.
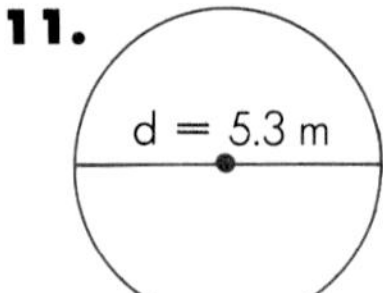

21.98 cm
12.
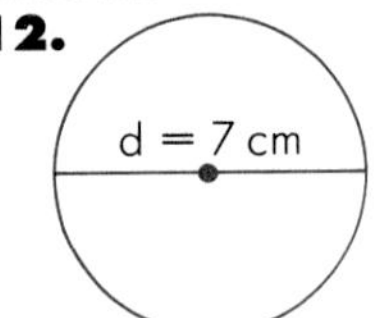

13. Circle
$d = 4$ mm 12.56 mm

14. Circle
$r = 10$ ft 62.8 ft

15. Circle
$r = 9.3$ cm 58.404 cm

Find the area. Use after page 319.

100 sq in.
16.
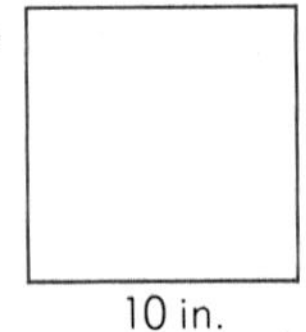

48 m^2
17.
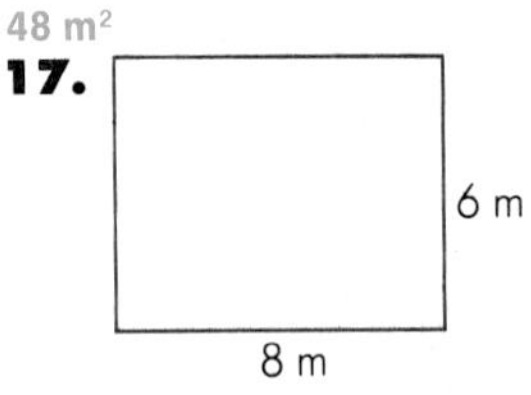

28 sq ft
18.
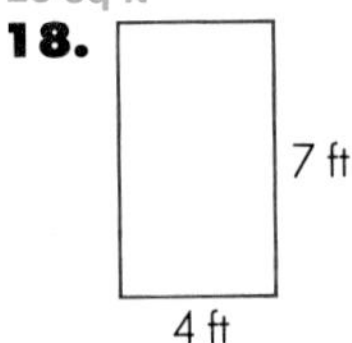

19. Rectangle
$l = 9.8$ cm 39.2 cm^2
$w = 4$ cm

20. Square
$s = 6.3$ m 39.69 m^2

21. Square
$s = 4$ mm 16 mm^2

Find the area. Use after page 321.

104 m²

1.

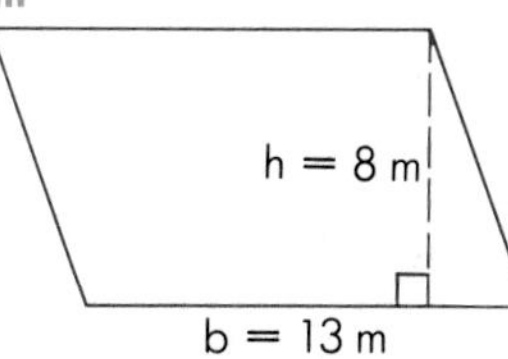

72 sq ft

2.

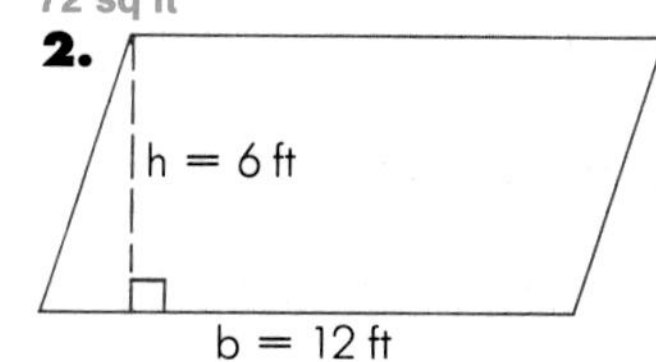

242 cm²

3.

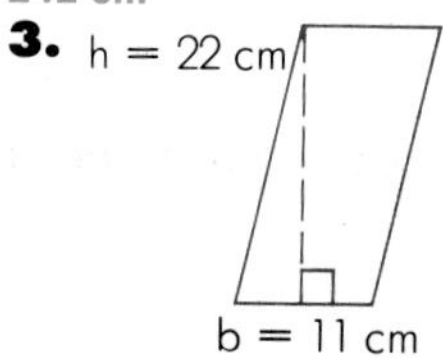

100 mm²

4.

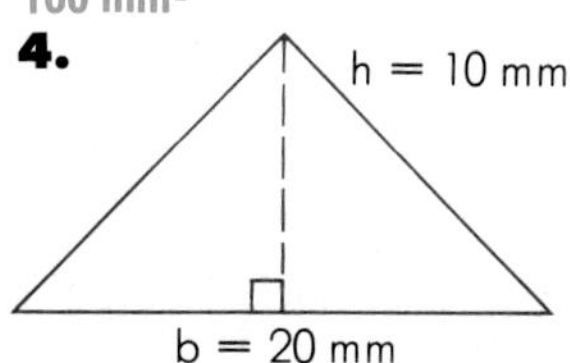

3 m²

5.

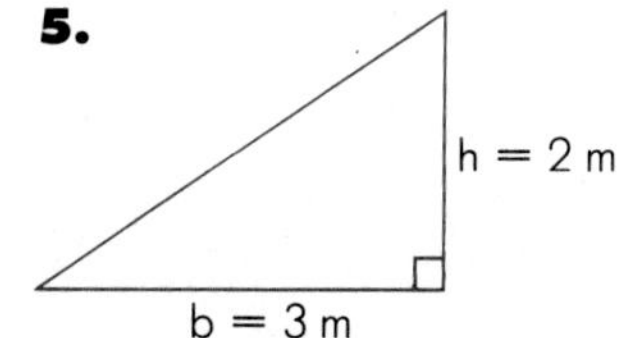

4 sq yd

6.

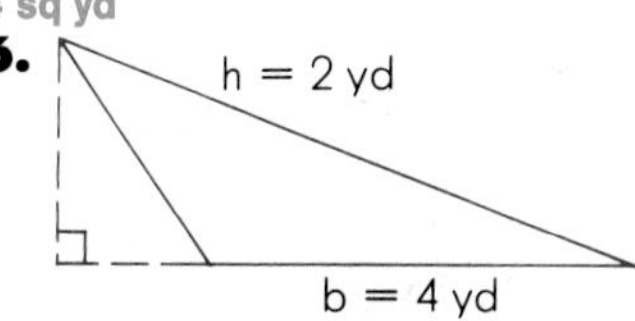

7. Triangle
$b = 12$ cm 54 cm²
$h = 9$ cm

8. Triangle
$b = 8$ in. 20 sq in.
$h = 5$ in.

9. Triangle
$b = 4.4$ m 11.66 m²
$h = 5.3$ m

Find the area. Use after page 323.

36 m²

10.

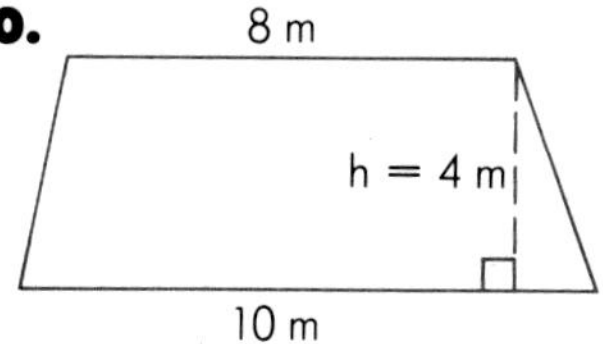

66 cm²

11.

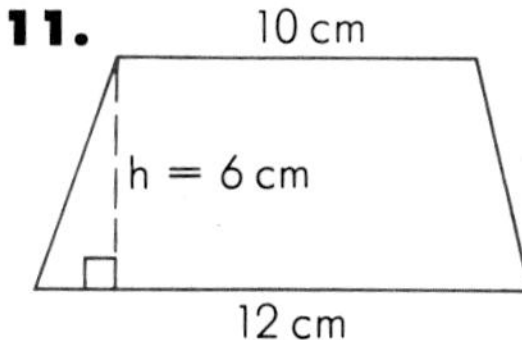

45 sq in.

12.

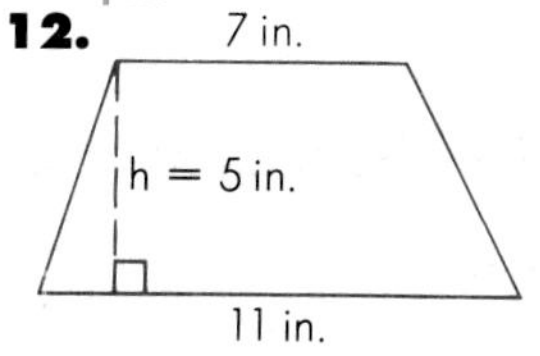

13. Trapezoid
$h = 5$ ft 25 sq ft
$b_1 = 4$ ft
$b_2 = 6$ ft

14. Trapezoid
$h = 7$ m 91 m²
$b_1 = 11$ m
$b_2 = 15$ m

15. Trapezoid
$h = 15$ mm 180 mm²
$b_1 = 10$ mm
$b_2 = 14$ mm

Find the area. Use after page 327.

78.5 m²

16.

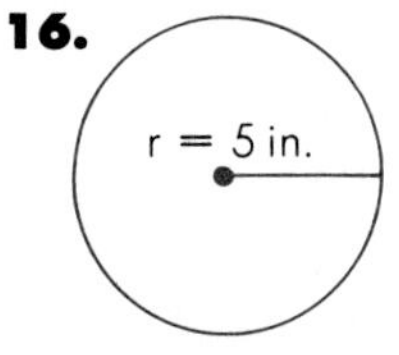

50.24 sq ft

17.

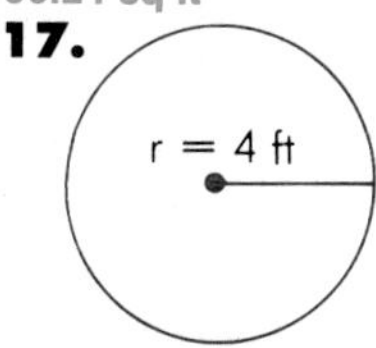

200.96 cm²

18.

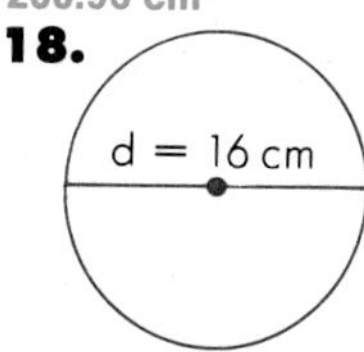

19. Circle
$r = 9$ mm 254.34 mm²

20. Circle
$d = 12$ cm 113.04 cm²

21. Circle
$d = 6$ m 28.26 m²

CALCULATOR CORNER

This activity emphasizes number relationships and place value.

The activity could be varied by making it into TWO OF A KIND or FOUR OF A KIND, depending on the ability level.

THREE OF A KIND

Players one or more
Materials one calculator, paper and pencil

Suppose a calculator has 345.6 on the display.

How many ways can you get a number ▢▢▢. 8 on the display where the three boxes contain the same digit?

Try these with your calculator.

1. Add 321.2 to 345.6
   Result: 666.8
2. Subtract 122.8 from 345.6
   Result: 222.8

Find 4 more ways to get such a number on the display. Write down your steps.

EXERCISES Write down your steps for each of the following. Answers will vary.

1. Start with 236.5 on the display. Find 5 ways to get a number ▢▢▢.7 on the display, where the three boxes contain the same digit.
2. Start with 342.16 on the display. Find 5 ways to get a number ▢▢▢.48 on the display, where the three boxes contain the same digit.
3. Start with 423.152 on the display. Find 5 ways to get number 6▢▢.▢66 on the display, where the three boxes contain the same digit.

CHAPTER 12 PRE-TEST

Find the volume. pp. 340–348

108 cm³
1.

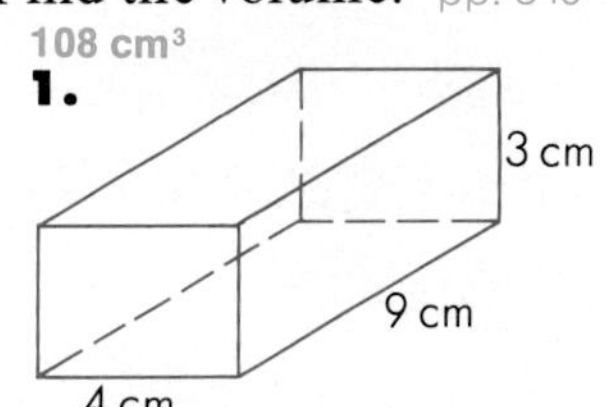

96 cu in.
2.

8 in.
4 in.
3 in.

343 m³
3.

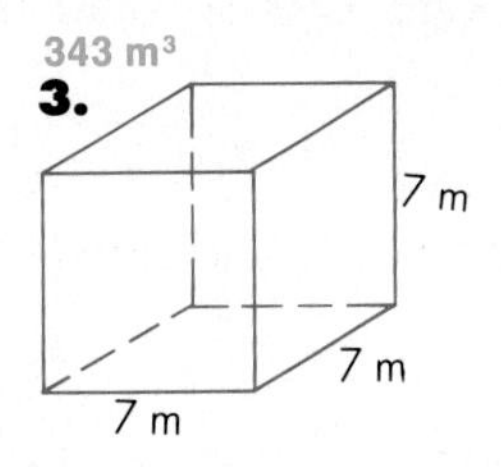

4. Rectangular prism
$l = 12$ m
$w = 6$ m
$h = 5$ m 360 m³

5. Cube
$s = 3$ cm 27 cm³

6. Cube
$s = 6$ ft 216 cu ft

401.92 cm³
7.

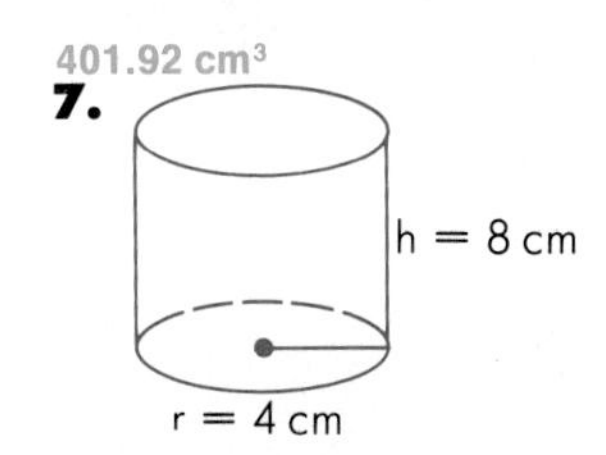

678.24 cu in.
8.

h = 6 in.
r = 6 in.

1130.4 m³
9.

h = 10 m
d = 12 m

10. Cylinder
$r = 3$ m
$h = 7$ m
197.82 m³

11. Cylinder
$d = 18$ cm
$h = 12$ cm
3052.08 cm³

12. Cylinder
$d = 3$ ft
$h = 2$ ft
14.13 cu ft

504 m³
13.

h = 12 m
l = 9 m
w = 14 m

6782.4 cu in.
14.

h = 20 in.
r = 18 in.

5572.45 cm³
15.

r = 11 cm

Application pp. 342, 350

16. A bill for electricity is $ 98 for 2000 kilowatt-hours of electricity. What is the cost per kilowatt-hour? $0.05

17. A barrel of oil generates 500 kilowatt-hours of electricity. How many kilowatt-hours are generated by 100,000 barrels? 50,000,000

12
VOLUME

GETTING STARTED

CUBE COST

If 1 cube costs 5¢, how much would each stack of cubes cost?

60¢

1.

This activity could be varied by changing the unit cost.

$2

2.

50¢

3.

$1.15

4.

30¢

5.

45¢

6.

70¢

7.

$1.05

8.

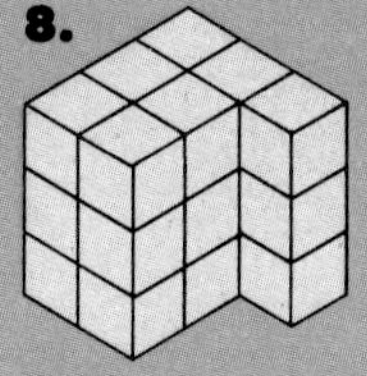

Objectives 119 and 120

RECTANGULAR PRISMS AND CUBES

Volume is measured with cubic units.
This is a cubic centimeter. It is 1 cm on each side.
We write: 1 cm^3.
We say: 1 cubic centimeter or 1 centimeter cubed.

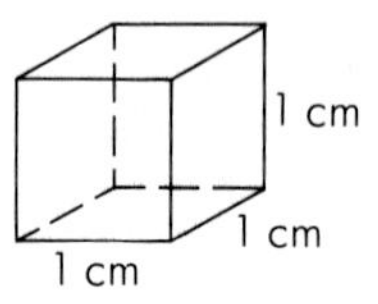

Find the volume of the rectangular prism.

Volume = length × width × height

$V = lwh$
$V = 4 \times 3 \times 2$
$V = 24 \text{ m}^3$

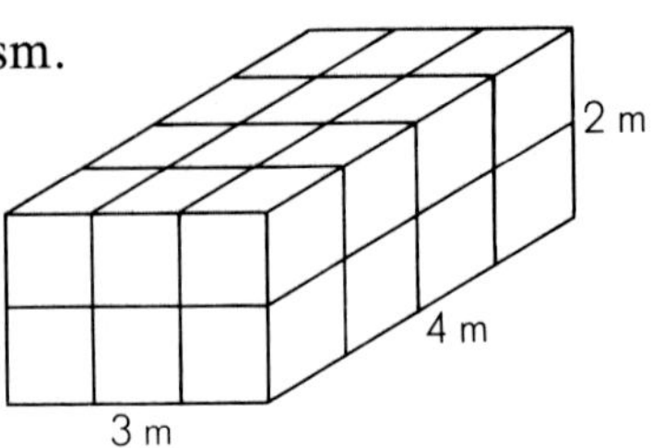

Students can count the cubic units to verify the formula.

EXAMPLE 1. Find the volume.

$V = lwh$
$V = 3 \times 2 \times 5$
$V = 30 \text{ cm}^3$

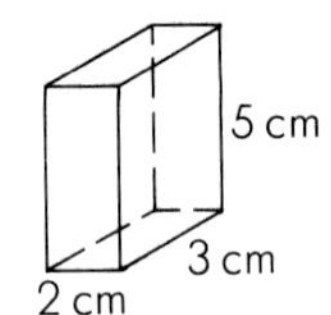

A cube is a special rectangular prism. Its length, width, and height are the same.

EXAMPLE 2. Find the volume.

Volume = side × side × side

$V = s \times s \times s$
$V = 3 \times 3 \times 3$
$V = 27 \text{ cm}^3$

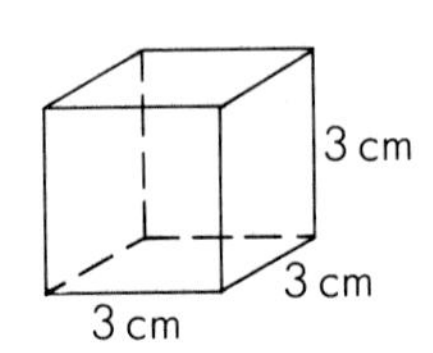

TRY THIS

Find the volume.

24 cm³
1.

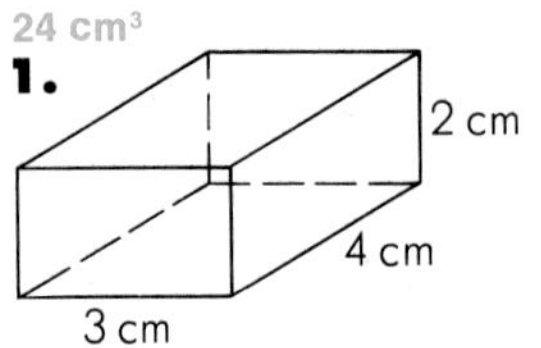

1000 m³
2.

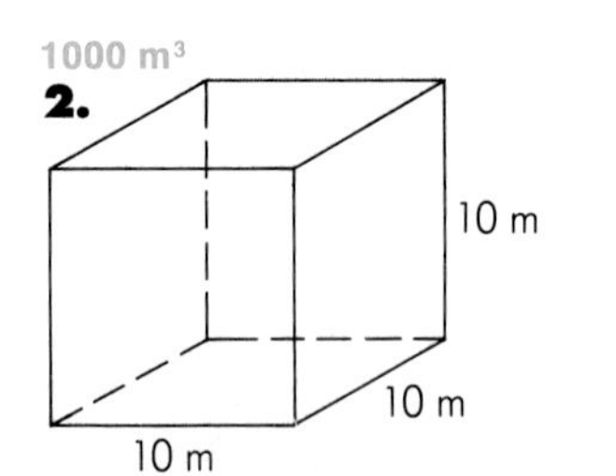

PRACTICE

For extra practice, use Making Practice Fun 73 with this lesson.

Find the volume.

30 m³
1.

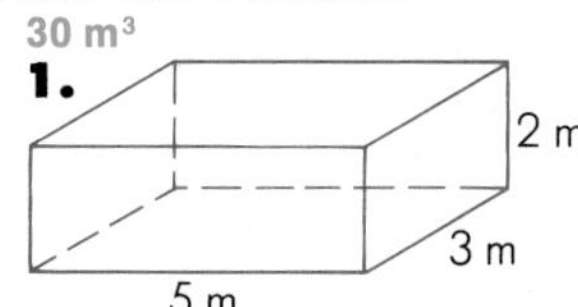

37.8 cm³
2.

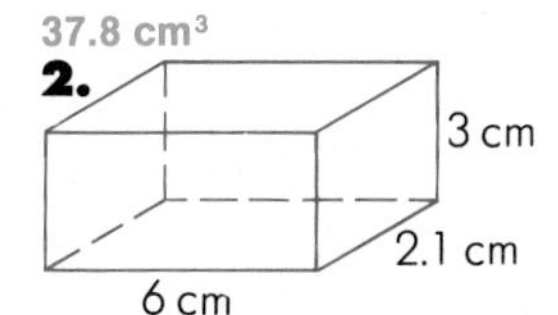

84 mm³
3.

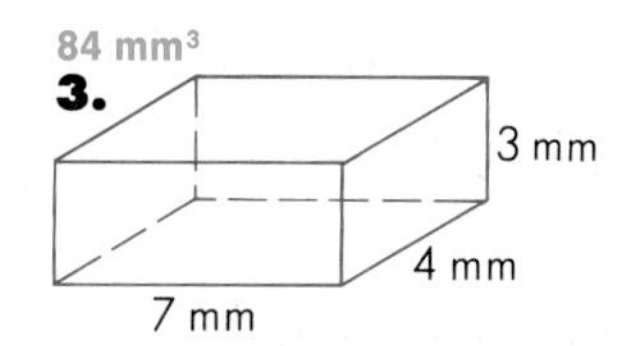

4. Rectangular prism
$l = 3\frac{1}{2}$ in.
$w = 3$ in.
$h = 7$ in. $73\frac{1}{2}$ cu in.

5. Rectangular prism
$l = 3$ ft
$w = 7$ ft
$h = 5$ ft 105 cu ft

6. Rectangular prism
$l = 6$ yd
$w = 2$ yd
$h = 7$ yd 84 cu yd

7. Rectangular prism
$l = 8$ mm
$w = 6$ mm
$h = 4$ mm 192 mm³

8. Rectangular prism
$l = 4$ m
$w = 3$ m
$h = 5$ m 60 m³

9. Rectangular prism
$l = 6$ cm
$w = 2$ cm
$h = 5.6$ cm 67.2 cm³

343 mm³
10.

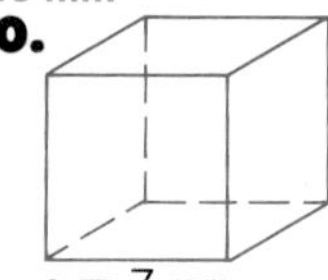

216 cm³
11.

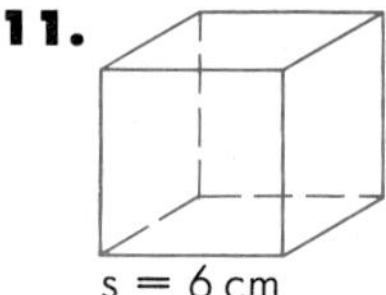

1000 m³
12.

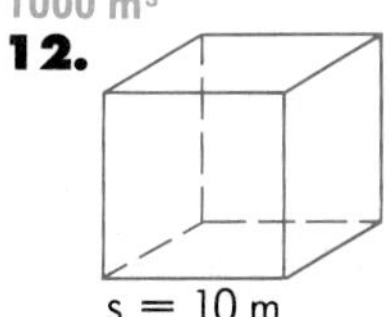

13. Cube
$s = 4\frac{1}{2}$ yd $91\frac{1}{8}$ cu yd

14. Cube
$s = 8$ ft 512 cu ft

15. Cube
$s = 9$ in. 729 cu in.

16. Cube
$s = 3$ m 27 m³

17. Cube
$s = 5$ cm 125 cm³

18. Cube
$s = 12$ mm 1728 mm³

PROBLEM CORNER

1. Find the volume of cube *A*. 8 cm³

2. In cube *B* the sides are doubled. Find the volume of cube *B*. 64 cm³

3. Is the volume of cube *B* double that of cube *A*? What is the ratio of the volume of cube *B* to the volume of cube *A*? No; $\frac{8}{1}$

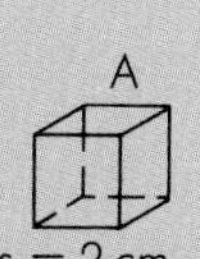

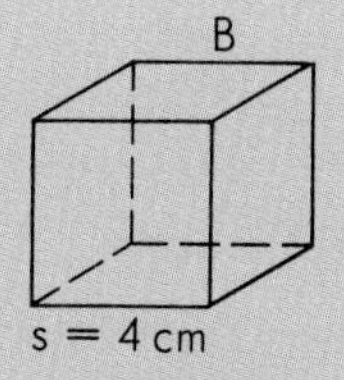

More Practice, page 356

Objectives 186 and 187

Try to have some samples of electric, gas, and water bills available. Have students find out the local unit rates.

ENERGY BILLS

The cost of electricity, gas, and water add to the cost of housing. Their cost depends on the number of units used.

Utility	Unit
Electricity	Kilowatt-hour (kWh)
Gas	Cubic Foot
Water	Gallon or Cubic Foot

$$\text{UNIT RATE} = \frac{\text{TOTAL COST}}{\text{NUMBER OF UNITS}}$$

What is the cost per kWh for these electric bills?

	kWh	Cost	
1. San Francisco, California	2160	$ 114.48	$0.05
2. New York City, New York	1800	156.60	$0.09
3. Dallas, Texas	1995	95.76	$0.05
4. Pittsburgh, Pennsylvania	1460	75.92	$0.05
5. Atlanta, Georgia	978	34.23	$0.04

6. Find an electric bill from your city or town. What is the cost per kWh? Answers will vary.

ONE KILOWATT = 1000 WATTS

ONE KILOWATT × HOURS USED = kWh

Each electric appliance has a watt rating. The watt rating can be used to compute the cost of energy.

$$\frac{\text{WATTAGE}}{1000} \times \text{COST PER KILOWATT-HOUR} = \begin{array}{l}\text{COST OF ELECTRICITY FOR}\\ \text{1 HOUR OF OPERATION}\end{array}$$

5¢ per kilowatt-hour is commonly used to calculate energy costs.

EXAMPLE. An old color TV has a watt rating of 350.

$$\frac{350}{1000} \times \$0.05 = \$0.0175 \text{ cents per hour}$$

If you watch TV from 5 PM to 10 PM every day, the cost would be:

5 hours × 365 days × $0.0175 = $31.94 per year *Round to the nearest cent.*

Find the cost to operate these electric appliances for 1 year.

7. a 1000 W hair dryer for 1 hour per week $2.60
8. a 650 W microwave oven for 3 hours per week $5.07
9. a 1400 W toaster for 1 hour per week $3.64
10. a 230 W stereo for 3 hours per day $12.59
11. a 2 W clock continuously $0.88
12. a 700 W vacuum cleaner for 15 minutes per day $3.19
13. a 100 W bulb for 6 hours every day $10.95
14. a 40 W bulb for 6 hours every day $4.38

How much money will you save in a year in each situation?

15. replacing ten 100 W bulbs by 60 W bulbs, used 6 hours per day, each day of the year $43.80
16. using a $7\frac{1}{2}$ W night light instead of a 60 W bulb for 8 hours each night of the year $7.67

Objective 121

CYLINDERS

The base of a cylinder is a circle.

Volume = area of the base × height
Volume = $\pi r^2 \times h$
$V = \pi r^2 h$

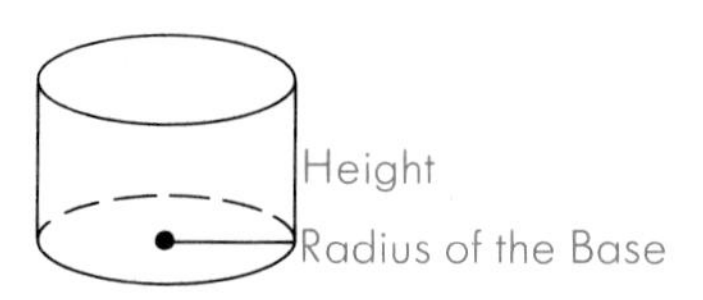

EXAMPLE 1. Find the volume.

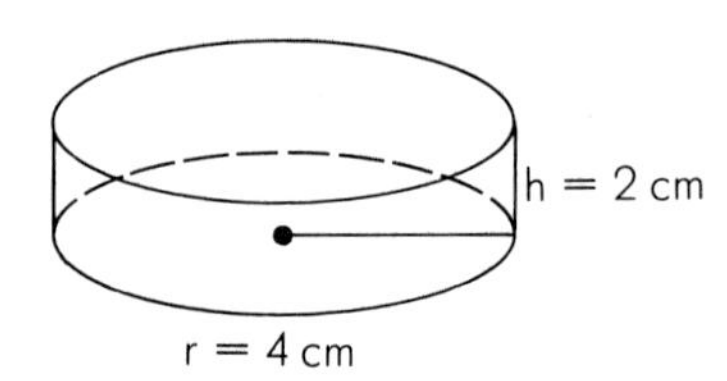

$V = \pi r^2 h$
$V = 3.14 \times 4^2 \times 2$ *Use 3.14 for π.*
$V = 3.14 \times 16 \times 2$ $4^2 = 4 \times 4 = 16$
$V = 100.48 \text{ cm}^3$

To find the volume when the diameter is given, we first find the radius.

EXAMPLE 2. Find the volume.

$d = 6$
$r = \frac{1}{2} \cdot 6 = 3$
$V = \pi r^2 h$
$V = 3.14 \times 3^2 \times 6$
$V = 3.14 \times 9 \times 6$
$V = 169.56 \text{ cm}^3$

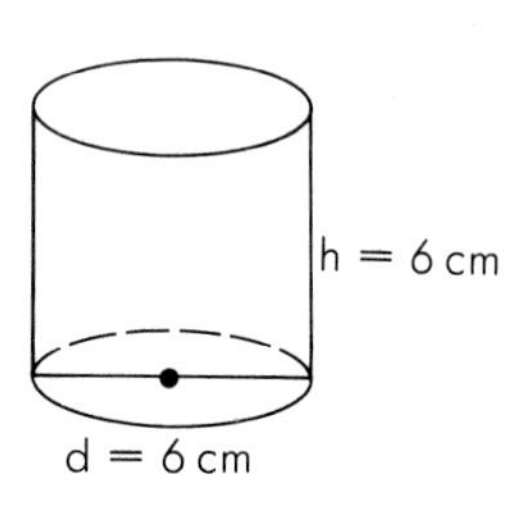

TRY THIS

Find the volume.

62.8 m^3
1.

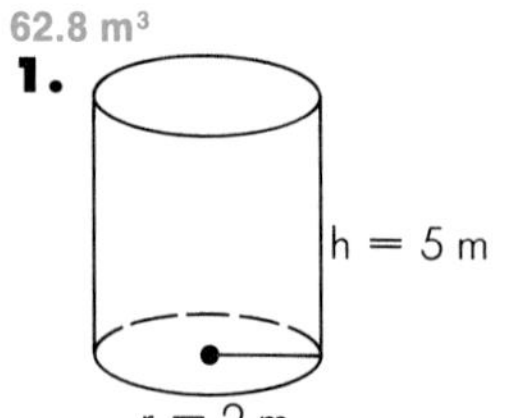

138.16 mm^3
2.

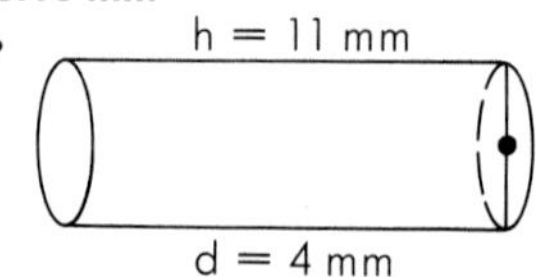

PRACTICE

Find the volume.

197.82 cm³

1.

h = 7 cm

r = 3 cm

2034.72 mm³

2.

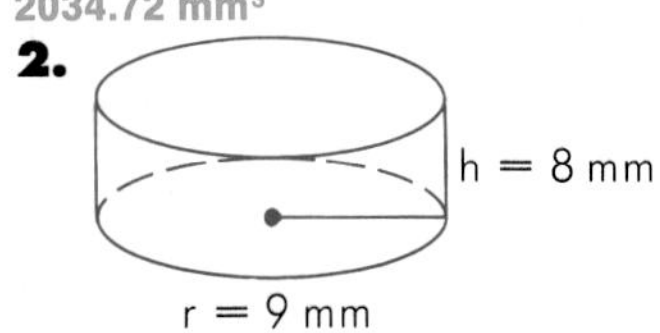

339.12 m³

3.

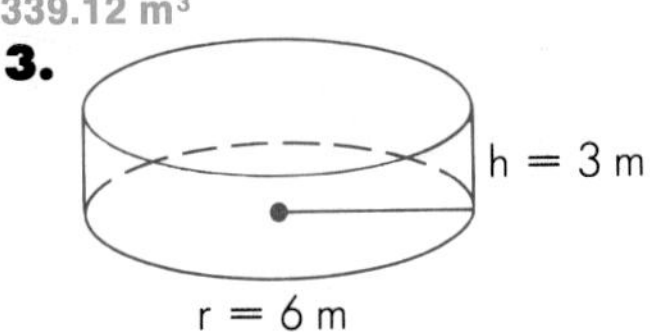

4. Cylinder
$r = 9$ m; $h = 10$ m
2543.4 m³

5. Cylinder
$r = 8$ cm; $h = 6$ cm
1205.76 cm³

6. Cylinder
$r = 20$ mm; $h = 20$ mm
25,120 mm³

7. Cylinder
$r = 6$ in.; $h = 5$ in.
565.2 cu in.

8. Cylinder
$r = 3$ ft; $h = 2$ ft
56.52 cu ft

9. Cylinder
$r = 4$ yd; $h = 3$ yd
150.72 cu yd

628 cm³

10.

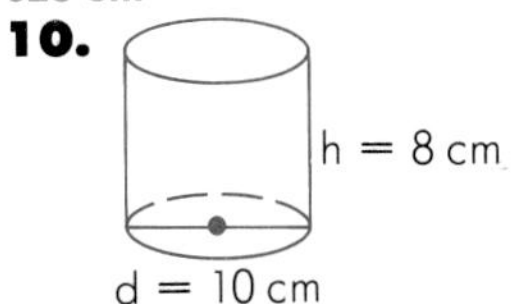

2260.8 mm³

11.

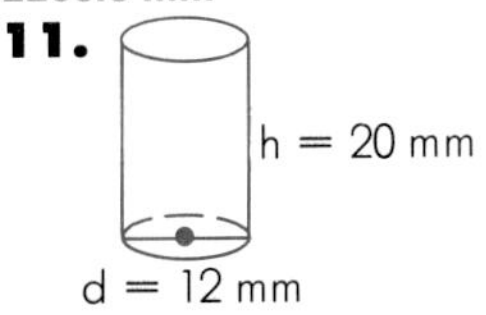

1017.36 m³

12.

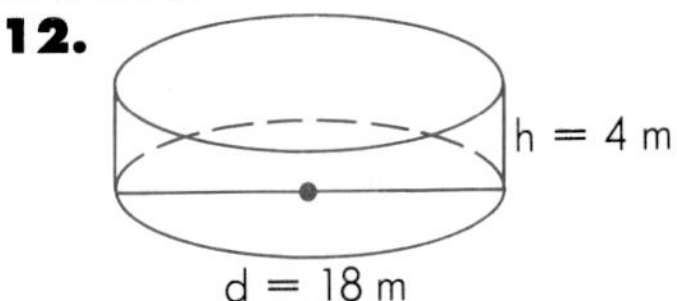

13. Cylinder
$d = 16$ m; $h = 30$ m
6028.8 m³

14. Cylinder
$d = 14$ cm; $h = 10$ cm
1538.6 cm³

15. Cylinder
$d = 12$ mm; $h = 40$ mm
4521.6 mm³

16. Cylinder
$d = 2$ yd; $h = 3$ yd
9.42 cu yd

17. Cylinder
$d = 8$ ft; $h = 5$ ft
251.2 cu ft

18. Cylinder
$d = 6$ in.; $h = 6$ in.
169.56 cu in.

PROBLEM CORNER

1. Find the volume of cylinder *A*. 50.24 cm³

2. In cylinder *B* the height is double that of cylinder *A*. The radius is the same. Find the volume of cylinder *B*. What is the ratio of the volume of cylinder *B* to the volume of cylinder *A*? 100.48 cm³; $\frac{2}{1}$

3. In cylinder *C* the radius is double that of cylinder *A*. The height is the same. Find the volume of cylinder *C*. What is the approximate ratio of the volume of cylinder *C* to the volume of cylinder *A*? 200.96 cm³; $\frac{4}{1}$

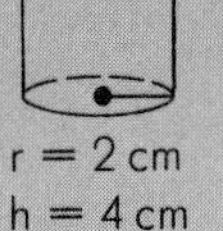

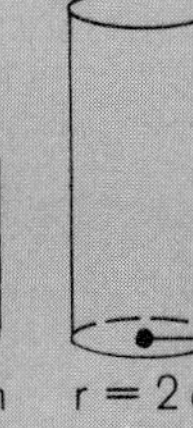

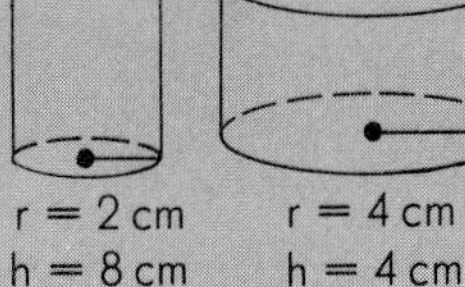

Objectives 122 and 123

PYRAMIDS AND CONES

To draw a pyramid, draw the base as a parallelogram. Connect opposite corners with dotted lines. The peak of the pyramid is directly above the intersection point.

Imagine a pyramid full of sand emptied into a prism which has the same base and height as the pyramid.

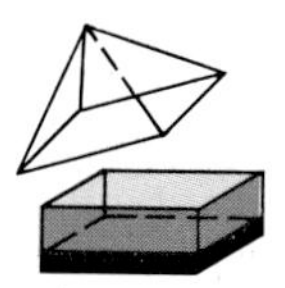

The volume of the pyramid is $\frac{1}{3}$ the volume of the prism.

Prism: $V = lwh$

Pyramid: $V = \frac{1}{3}lwh$

Imagine a cone full of sand emptied into a cylinder which has the same base and height as the cone.

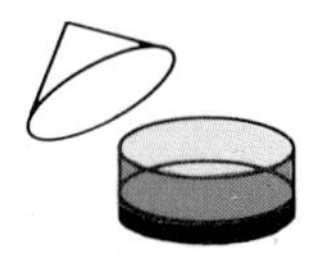

The volume of the cone is $\frac{1}{3}$ the volume of the cylinder.

Cylinder: $V = \pi r^2 h$

Cone: $V = \frac{1}{3}\pi r^2 h$

EXAMPLE 1. Find the volume.

$V = \frac{1}{3}lwh$

$V = \frac{1}{3} \times 5 \times 4 \times 3$

$V = 20 \text{ cm}^3$

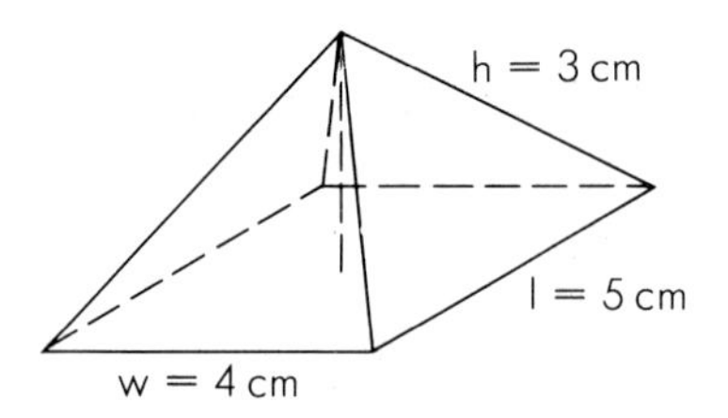

EXAMPLE 2. Find the volume.

$V = \frac{1}{3}\pi r^2 h$

$V = \frac{1}{3} \times 3.14 \times 3^2 \times 5$

$V = \frac{1}{3} \times 3.14 \times 9 \times 5$

$V = \frac{1}{3} \times \frac{3.14}{1} \times \frac{9}{1} \times \frac{5}{1}$

$V = \frac{141.30}{3} = 47.1 \text{ cm}^3$

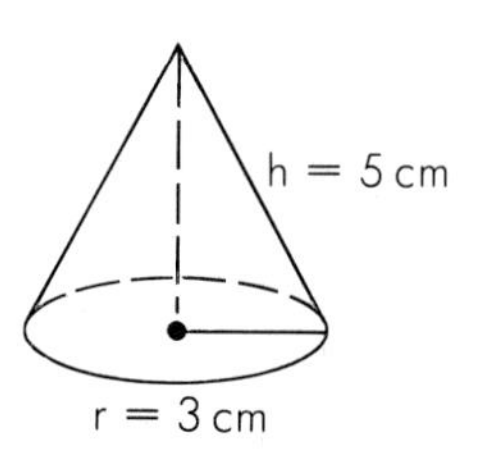

TRY THIS

Round to the nearest tenth or hundredth when necessary.

Find the volume.

8 m³

1.

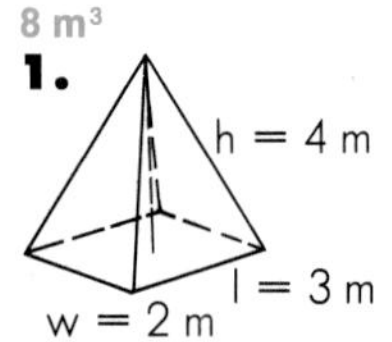

847.8 mm³

2.

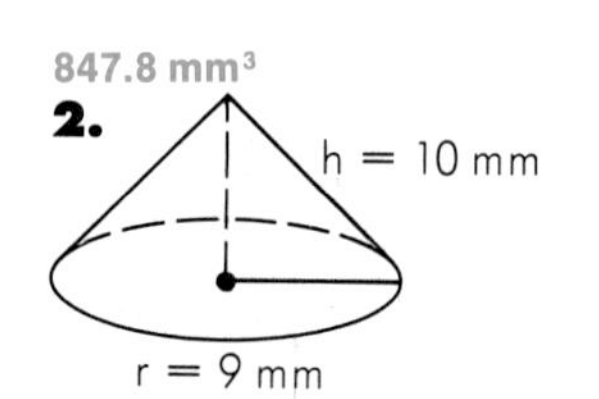

PRACTICE

For extra practice, use Making Practice Fun 74 with this lesson.

Find the volume.

8 m³
1.

105 cm³
2.

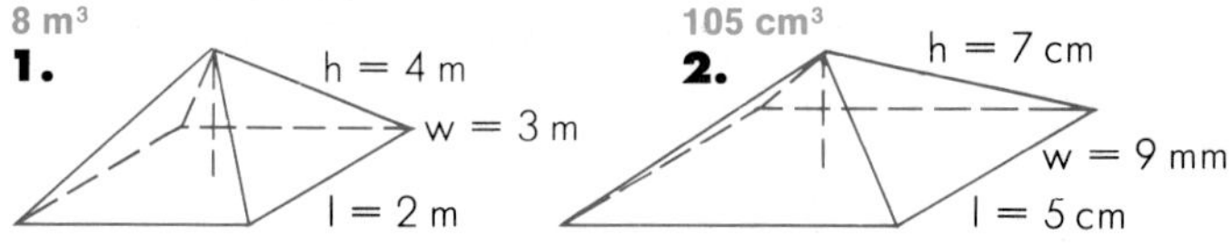

36 mm³
3.

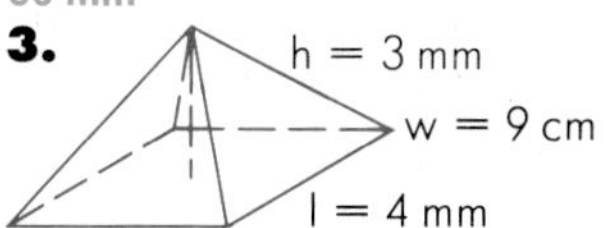

4. Pyramid
$l = 9$ mm
$w = 12$ mm
$h = 18$ mm 648 mm³

5. Pyramid
$l = 10$ cm
$w = 10$ cm
$h = 10$ cm 333.3 cm³

6. Pyramid
$l = 4$ m
$w = 7$ m
$h = 20$ m 186.67 m³

7. Pyramid
$l = 7$ in.
$w = 8$ in.
$h = 4\frac{1}{2}$ in. 84 cu in.

8. Pyramid
$l = 9$ ft
$w = 4$ ft
$h = 6$ ft 72 cu ft

9. Pyramid
$l = 6\frac{1}{2}$ yd
$w = 10$ yd
$h = 20$ yd $433\frac{1}{3}$ cu yd

117.23 cm³
10. h = 7 cm, r = 4 cm

113.04 m³
11.

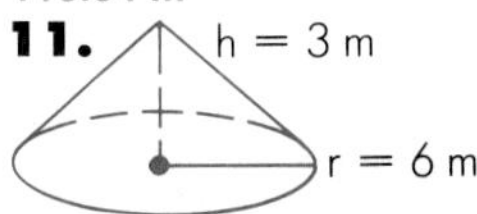

847.8 mm³
12.

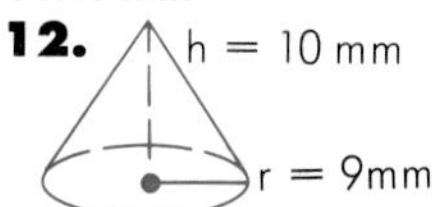

13. Cone
$r = 12$ mm
$h = 15$ mm 2260.8 mm³

14. Cone
$r = 7$ cm
$h = 3$ cm 153.86 cm³

15. Cone
$r = 3$ m
$h = 10$ m 94.2 m³

16. Cone
$r = 4$ in.
$h = 6$ in. 100.48 cu in.

17. Cone
$r = 2\frac{1}{2}$ ft
$h = 3$ ft 19.625 cu ft

18. Cone
$r = 7$ yd
$h = 10$ yd 512.87 cu yd

PROBLEM CORNER

1. Find the volume of cone A. 100.48 cm³

2. In cone B the lengths are doubled. Find the volume of cone B. 803.84 cm³

3. Is the volume of cone B double that of cone A? What is the ratio of the volume of cone B to the volume of cone A? No; $\frac{8}{1}$

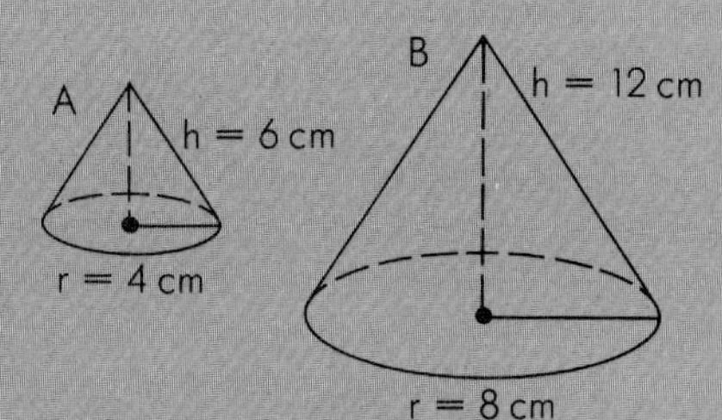

Objective 124

SPHERES

The **radius** of a sphere is the distance from the center to any point on the sphere.

Volume $= \frac{4}{3} \times \pi \times$ radius cubed ($r^3 = r \times r \times r$)

$V = \frac{4}{3}\pi r^3$

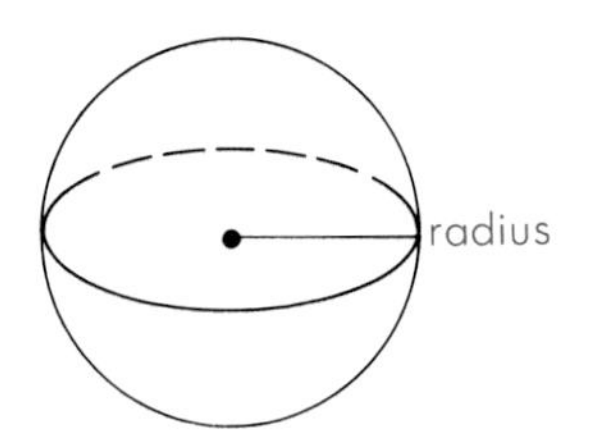

EXAMPLE 1. Find the volume. Round to the nearest tenth or hundredth when necessary.

$V = \frac{4}{3}\pi r^3$

$V = \frac{4}{3} \times 3.14 \times 2^3$

$V = \frac{4}{3} \times 3.14 \times 8$

$V = \frac{4}{3} \times \frac{3.14}{1} \times \frac{8}{1}$

$V = \frac{100.48}{3} = 33.49 \text{ cm}^3$

r = 2 cm

To find the volume when the diameter is given, we first find the radius.

EXAMPLE 2. Find the volume.

$d = 10$

$r = \frac{1}{2} \cdot 10 = 5$

$V = \frac{4}{3}\pi r^3$

$V = \frac{4}{3} \times 3.14 \times 5^3$

$V = \frac{4}{3} \times \frac{3.14}{1} \times \frac{125}{1}$

$V = \frac{1570}{3} = 523.33 \text{ m}^3$

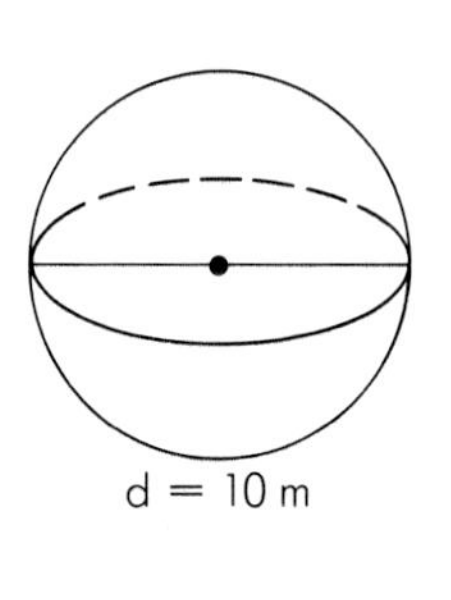

TRY THIS

Find the volume.

1. 267.95 cm³

r = 4 cm

2. 113.04 m³

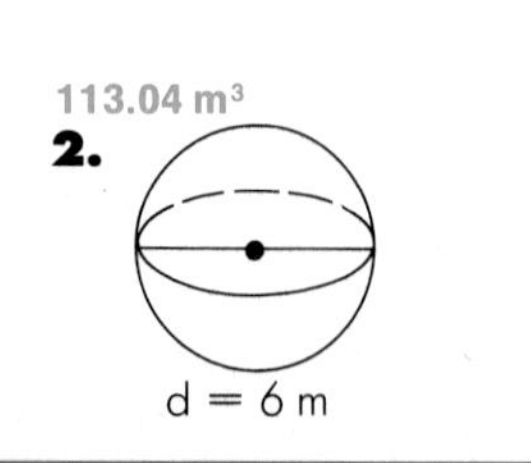

PRACTICE Use Quiz 25 after this Practice.

267.95 cm³
1.

r = 4 cm

2143.57 m³
2.

r = 8 m

33.49 mm³
3.

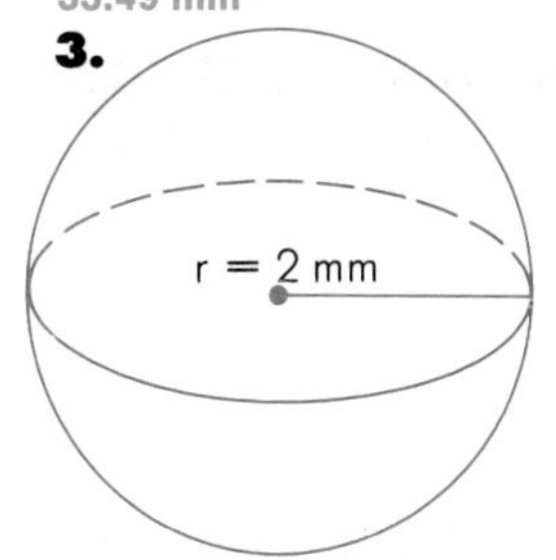

4. Sphere
$r = 6$ in. 904.32 cu in.

5. Sphere
$r = 3$ yd 113.04 cu yd

6. Sphere
$r = 5$ ft 523.33 cu ft

7. Sphere
$r = 9$ mm 3052.08 mm³

8. Sphere
$r = 1$ m 4.19 m³

9. Sphere
$r = 7$ cm 1436.03 cm³

1436.03 cm³
10.

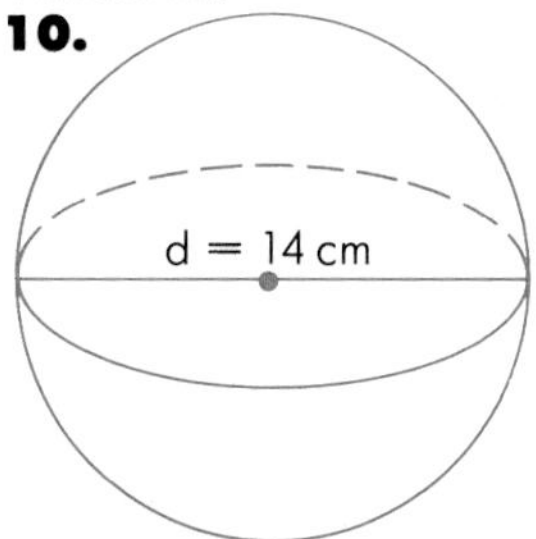

3052.08 m³
11.

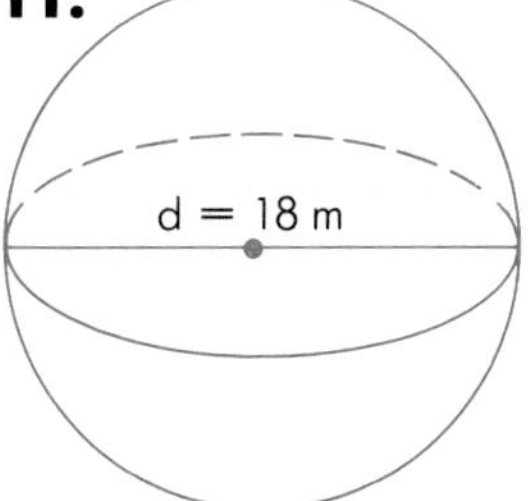

904.32 mm³
12.

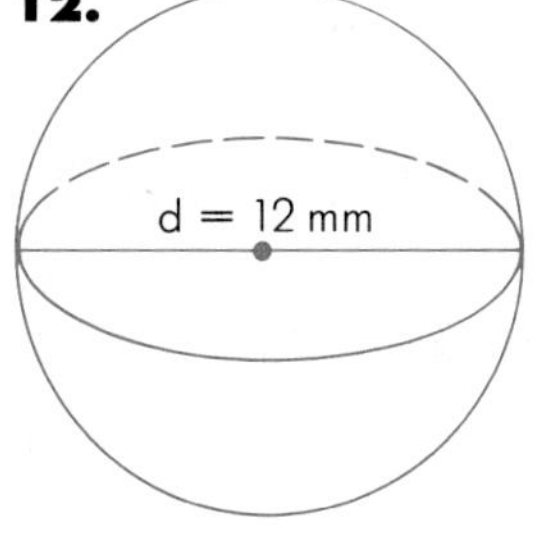

13. Sphere
$d = 10$ in. 523.33 cu in.

14. Sphere
$d = 8$ ft 267.95 cu ft

15. Sphere
$d = 2$ yd 4.19 cu yd

16. Sphere
$d = 100$ mm 523,333.33 mm³

17. Sphere
$d = 8$ cm 267.95 cm³

18. Sphere
$d = 24$ cm 7234.56 cm³

PROBLEM CORNER

1. Find the volume of the hemisphere. 452.16 cm³
(Hint: the prefix "hemi" means half.)

2. A hemisphere dome on the top of a silo has a diameter of 40 feet. Find the volume of the dome. 16,746$\frac{2}{3}$ cu ft

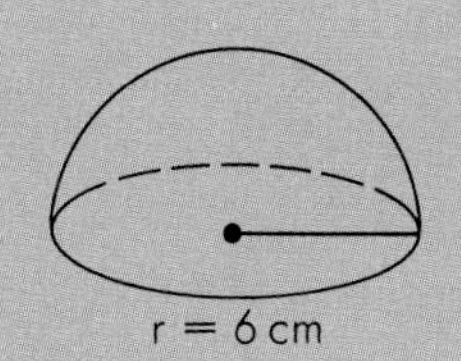

Objective 188

APPLICATION

ENERGY CONSERVATION

There are 42 gallons of oil in a barrel. A barrel of oil generates 500 kWh of electricity.

Suppose thermostats in all homes in the U.S. are lowered an average of 6° in winter. The country would save 570,000 barrels of oil each day.

1. How many gallons of oil would be saved daily? 23,940,000 gal
2. How many gallons would be saved in a 90 day winter? 2,154,600,000 gal
3. How many kWh would be saved in a 90 day winter? 25,650,000,000 kWh
4. At an average rate of $ 0.05 per kWh, how much money would be saved? $1,282,500,000
5. The present consumption is about 35,000,000 barrels of oil per day in the U.S. What percent could be saved per day by lowering thermostats 6° in winter? 1.6%

A large amount of oil can be saved daily by not wasting electricity. Suppose you leave two 100 W lights on when you leave for school each day. How much oil would be wasted?

Remember: $\frac{\text{WATTS}}{1000} \times \text{HOURS} = \text{kWh}$ $\quad$ $\text{kWh} \times \text{RATE} = \text{COST}$

$\frac{200 \text{ watts}}{1000} \times$ 8 hours of unneeded lights each day $\times$ 180 days of school $=$ 288 wasted kWh

It takes about one gallon of oil to generate 11.9 kWh of electricity.

$288 \div 11.9 = 24.2$ gallons of oil wasted *(Rounded to tenths)*

For each situation tell:

A. how much money is spent in one year (365 days) at the rate of $0.05 per kWh
B. how much oil is used yearly (1 gallon for 11.9 kWh)
C. how much oil is saved by the conservation way

Have students suggest other ways to conserve energy.

Wasteful Way

6. Leave two 75 W lights burning at night in bathrooms from 7 PM to 7 AM. A. $32.85 B. 55.2

Conservation Way

7. Leave a $7\frac{1}{2}$ W night light on from 7 PM to 7 AM. A. $1.64 B. 2.8 C. 52.4 gallons

8. An electric clothes dryer uses 4.8 kilowatts per hour, full or empty. Use the electric dryer for 10 hours weekly with partial loads. A. $124.80 B. 210

9. Use the dryer at the manufacturer's recommended full load for 6 hours weekly. A. $74.88 B. 125.8 C. 84.2 gallons

10. Hang most clothing outdoors. Use the dryer one hour per week for emergency loads or special clothes. A. $12.48 B. 21.0 C. 189 gallons

11. Uninsulated hot water tanks and pipes can use 600 kWh monthly for an average family's hot water needs. A. $360.00 B. 605.0

12. Properly insulated tanks and pipes save 30% of the electricity that is used to produce hot water. A. $252 B. 423.5 C. 181.5 gallons

CAREER CAREER CAREER CAREER CAREER

OCEANOGRAPHER

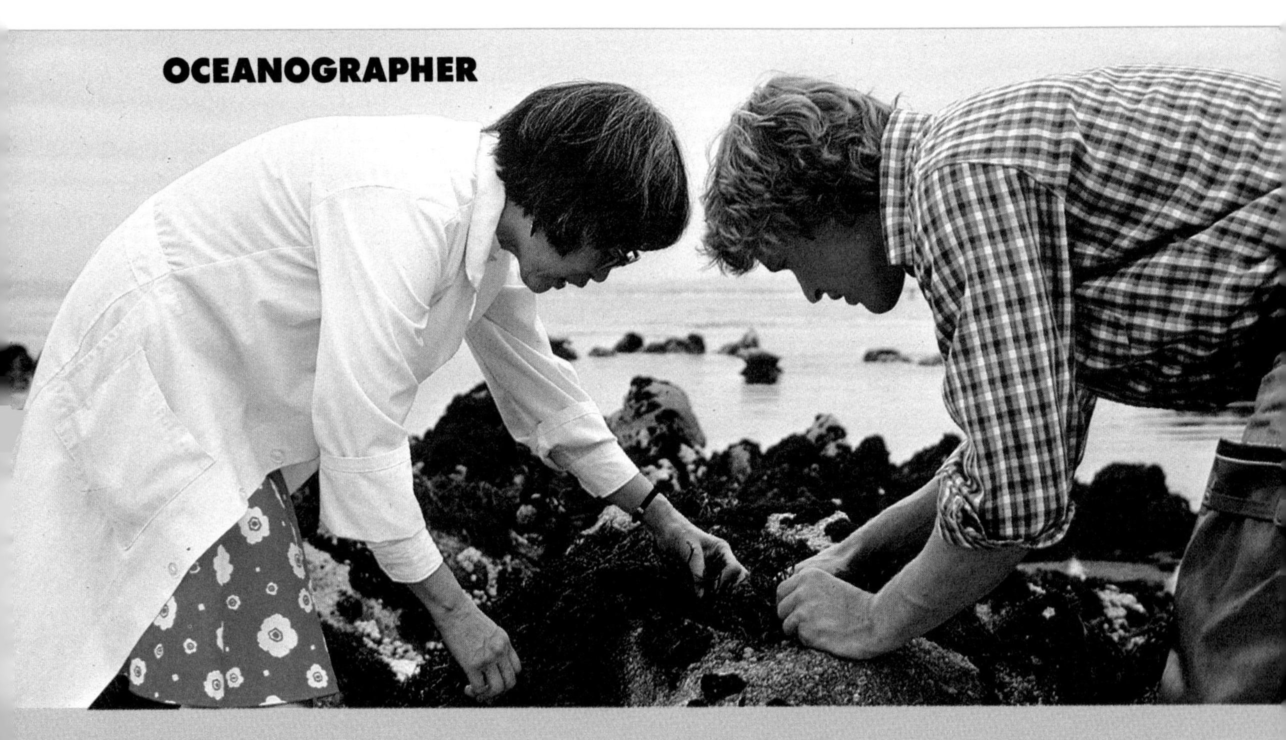

Chris Tanaka is an oceanographer. She is an expert on the way oceans are formed. She studies the things that live in the ocean.

To go deep in the ocean, Chris uses a bathysphere. A bathysphere is made to withstand great water pressures. Only rounded objects have been found to be strong enough. Water pressure at one mile in depth is 2280 pounds per square inch. The pressure increases at the rate of 2280 pounds per square inch for each mile of depth.

What is the water pressure per square inch at the following depths?

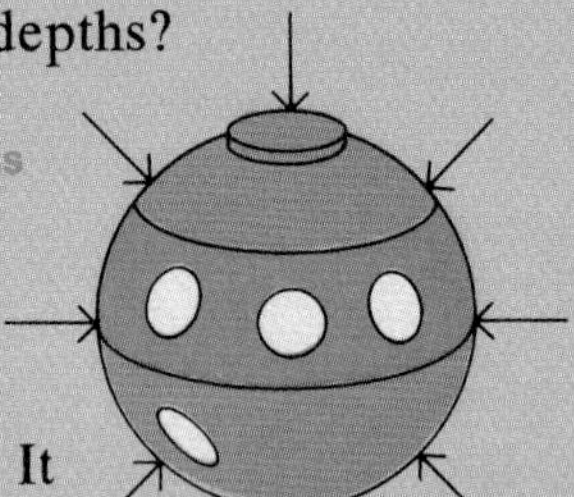

1. 4 miles 9120 pounds 2. 6.5 miles 14,820 pounds 3. 15,840 feet 6840 pounds

SURFACE AREA OF A SPHERE = $4\pi r^2$

4. A bathysphere is 12 feet in diameter. What is the surface area of the bathysphere? 452.16 square feet
5. A bathysphere has a surface area of 45,216 square inches. It descends into the Marianas Trench which is about 6.5 miles deep. What is the total weight of water on the bathysphere at that depth? 670,101,120 pounds

CAREER CAREER CAREER CAREER CAREER

Chris often takes samples of the ocean floor. To do so, she drills a hole that is the shape of a cylinder. What is the volume of each of the following?

6. a core sample 46 cm in diameter and 56.4 meters in length retrieved from a depth of 3 miles 9.37 m^3
7. a core 28 cm in diameter, 120 meters long 7.39 m^3
8. a core 37 cm in diameter, 184 meters long 19.77 m^3

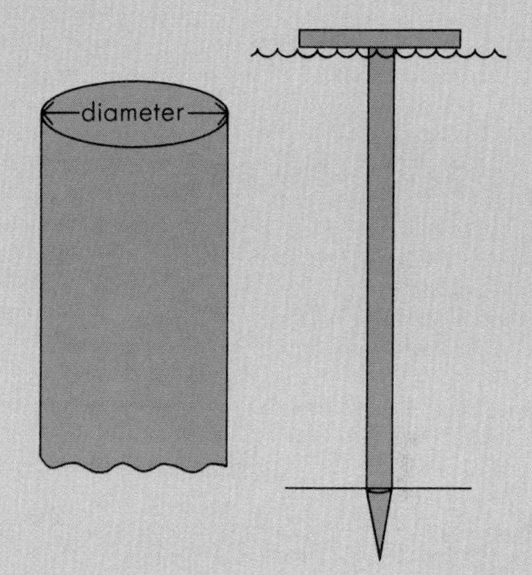

Some oceanographers specialize in the chemistry of the sea. Almost every element known is found in the ocean. Each cubic meter of sea water contains 0.004 mg of gold. How much gold is there in these average seawater samples?

9. 8.14 mg

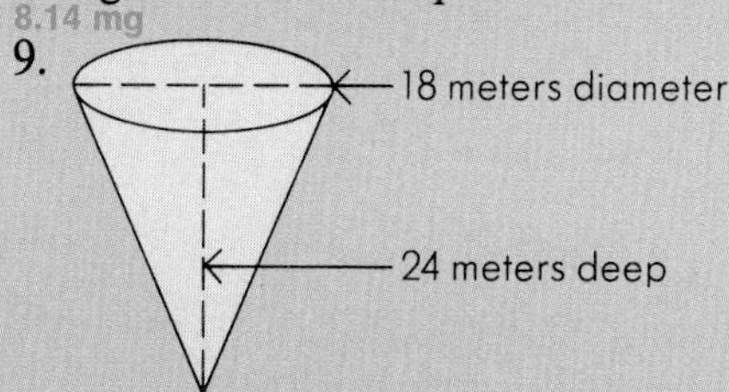

10. 154.34 mg

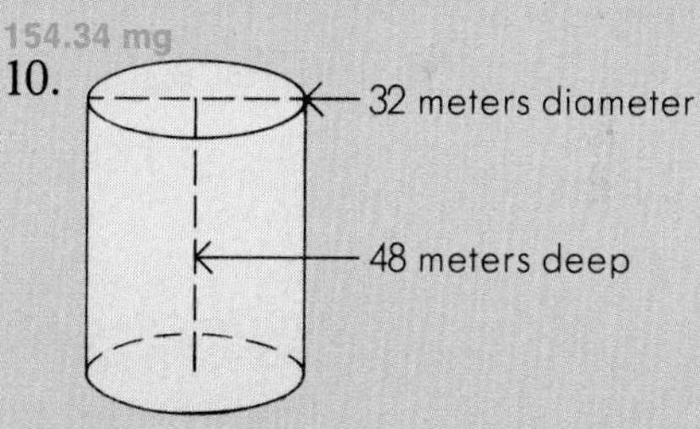

Icebergs are studied by oceanographers from the Arctic to the Antarctic. Some icebergs are so large that early explorers thought they were islands.

11. One iceberg measured 30 kilometers wide by 100 kilometers long and 800 meters thick. What was the volume of this iceberg? 2400 km^3
12. Another iceberg measured 25 kilometers wide by 75 kilometers long and 600 meters thick. What was the volume of this iceberg? 1125 km^3

CHAPTER 12 REVIEW

For extra practice, use Making Practice Fun 75 with this lesson.

Find the volume. pp. 340–348

160 cm³
1.

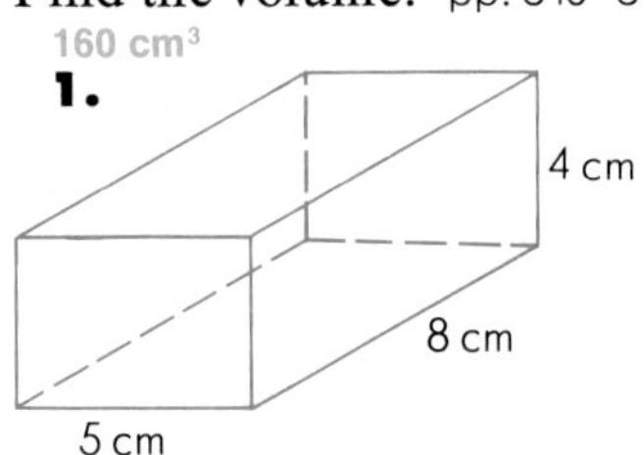

168 cu ft
2.

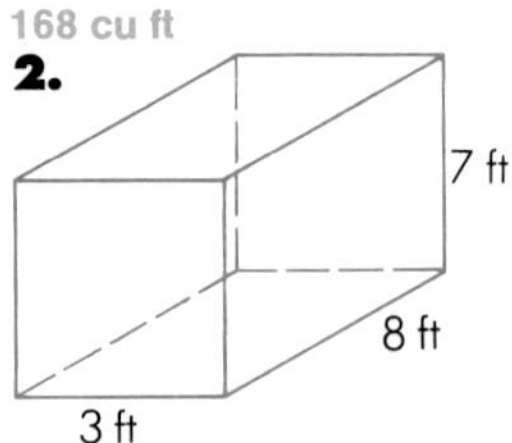

512 m³
3.

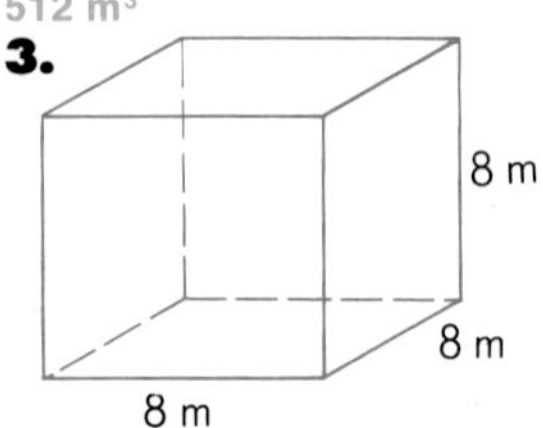

4. Rectangular prism
$l = 8$ m
$w = 5$ m
$h = 6$ m 240 m³

5. Cube
$s = 4$ cm 64 cm³

6. Cube
$s = 5$ in. 125 cu in.

1538.6 cm³
7.

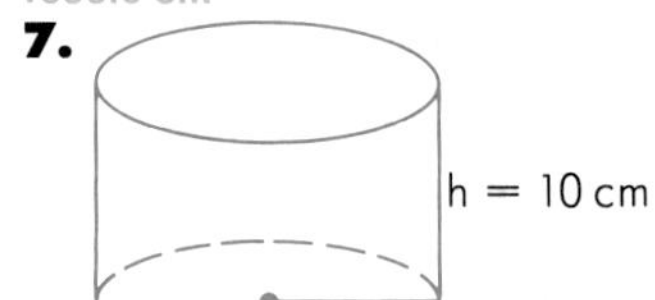

84.78 cu ft
8.

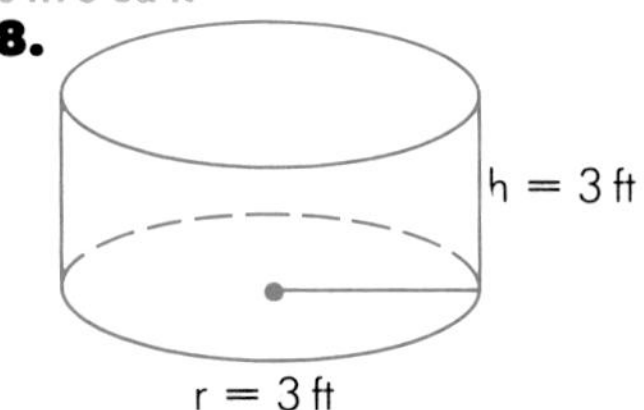

141.3 m³
9.

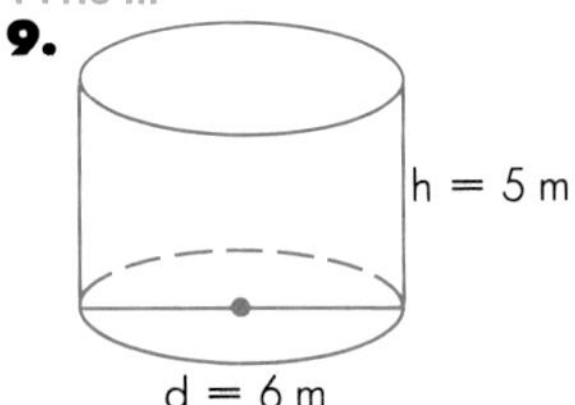

10. Cylinder
$r = 6$ m
$h = 14$ m 1582.56 m³

11. Cylinder
$d = 9$ cm
$h = 6$ cm 381.51 cm³

12. Cylinder
$d = 6$ in.
$h = 2$ in. 56.52 cu in.

56 m³
13.

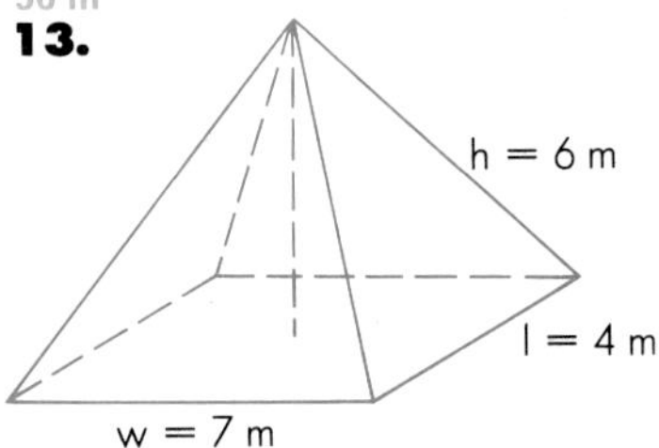

847.8 cu in.
14.

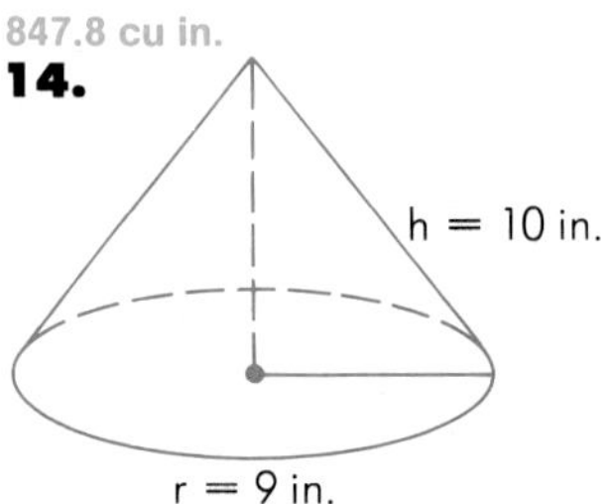

523.33 cm³
15.

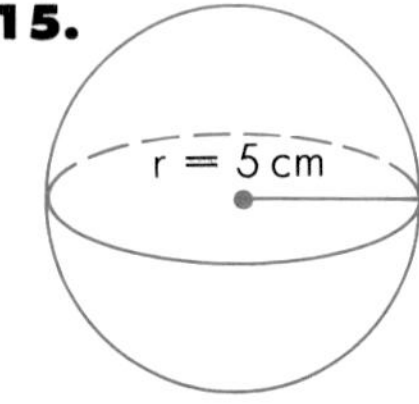

Application pp. 342, 350

16. A bill for electricity was $ 112.50. At an average rate of $ 0.05 per kilowatt-hour, how many kilowatt-hours were used? 2250 kWh

17. During the month of January the Duncans left extra lights on in their house. They used 350 more kilowatt-hours than usual. At $ 0.06 per kWh, how much higher was their electricity bill? $21

CHAPTER 12 TEST

Find the volume.

270 m³

1.

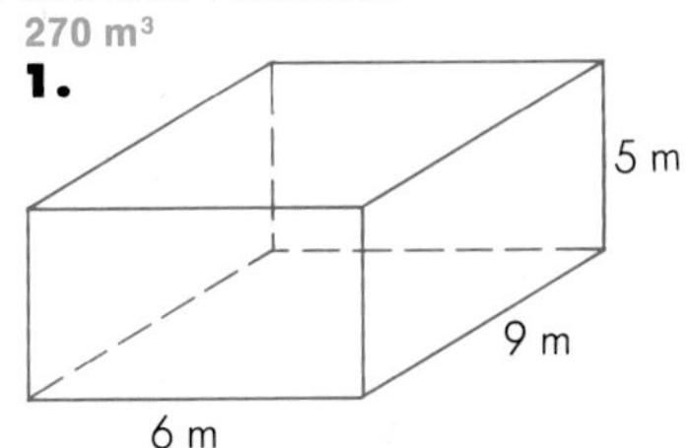

128 cu in.

2.

4 in.

8 in.

4 in.

729 cm³

3.

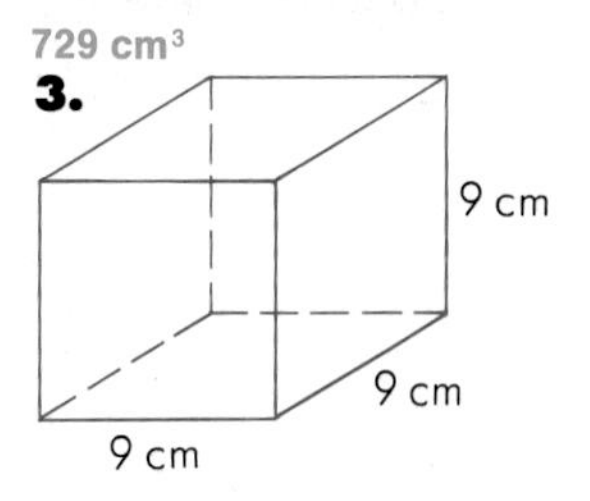

4. Rectangular prism
$l = 9$ cm
$w = 6$ cm
$h = 7$ cm 378 cm³

5. Cube
$s = 2$ m 8 m³

6. Cube
$s = 6$ ft 216 cu ft

2210.56 m³

7.

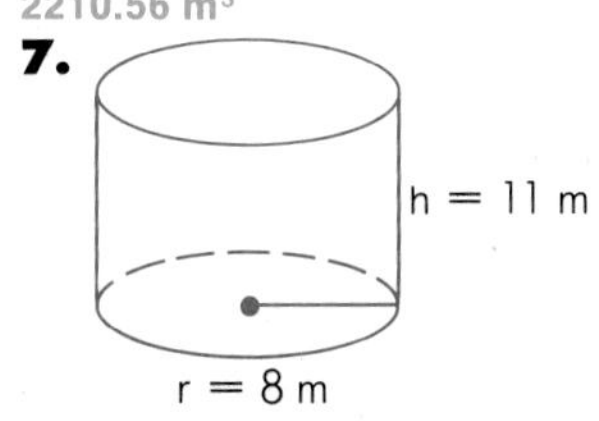

251.2 cu yd

8.

h = 5 yd

r = 4 yd

1570 cm³

9.

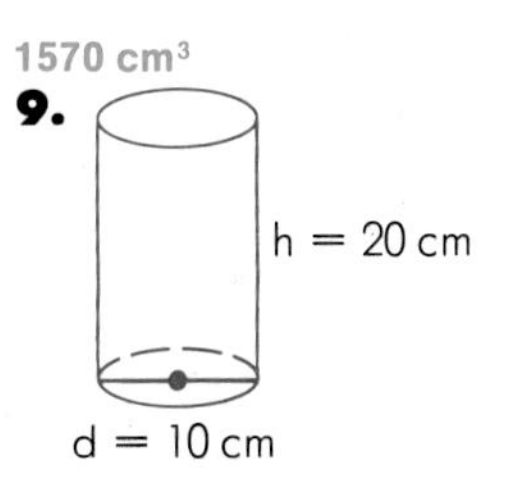

10. Cylinder
$r = 7$ cm
$h = 15$ cm 2307.9 cm³

11. Cylinder
$d = 8$ m
$h = 6$ m 301.44 m³

12. Cylinder
$d = 3$ ft
$h = 4$ ft 28.26 cu ft

144 cm³

13.

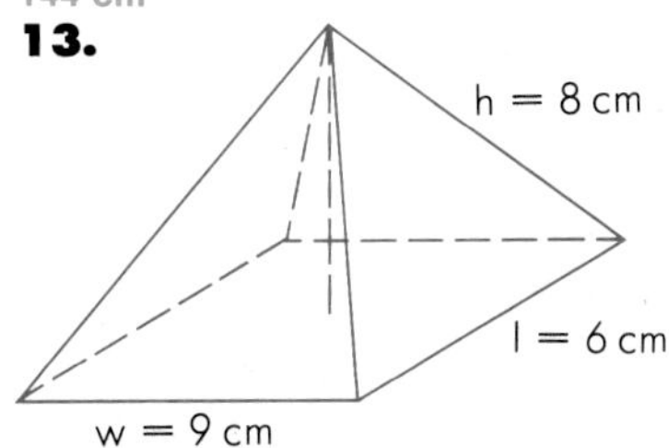

267.95 cu ft

14.

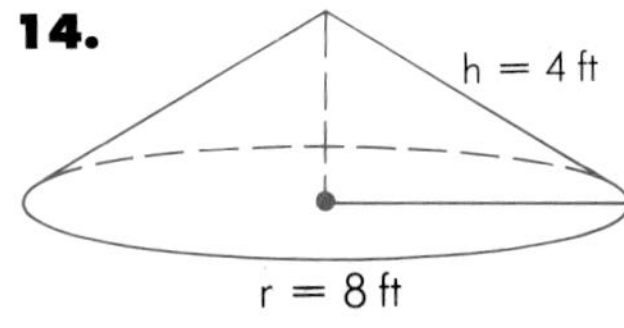

7234.56 cm³

15.

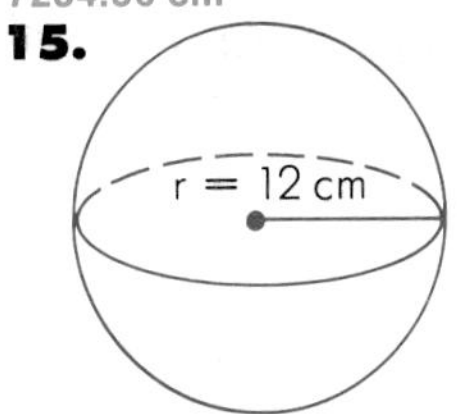

Application

16. A bill for electricity is \$ 73.47 for 1800 kilowatt-hours of electricity. What is the cost per kilowatt-hour? \$0.04

17. The Perrys saved 1542 kilowatt-hours of electricity in a year. At an average rate of \$ 0.05 per kWh, how much money was saved? \$77.10

MORE PRACTICE

Find the volume. Use after page 341.

84 cu ft
1.

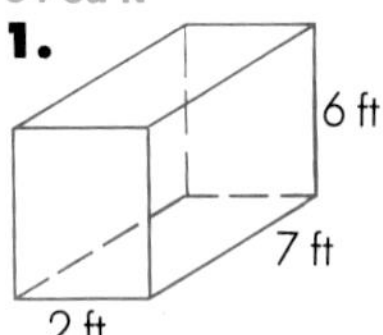

360 cm^3
2.

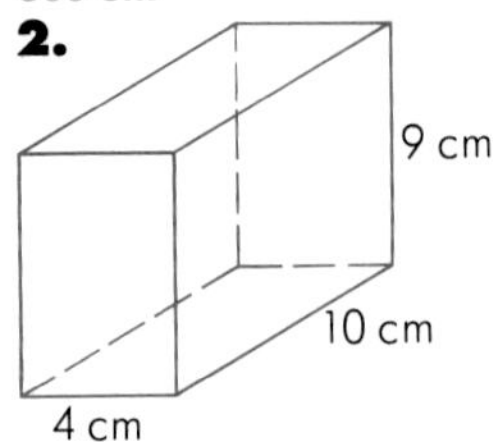

960 m^3
3.

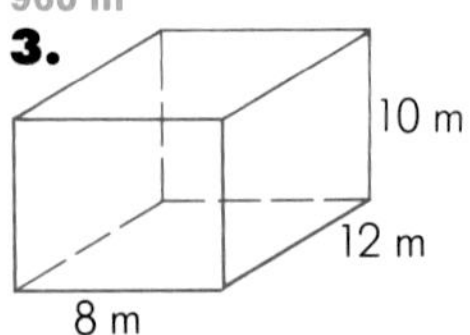

4. Rectangular prism
$l = 7$ in.
$w = 7$ in.
$h = 3$ in. 147 cu in.

5. Rectangular prism
$l = 9$ m
$w = 10$ m
$h = 5$ m 450 m^3

6. Rectangular prism
$l = 8$ cm
$w = 6$ cm
$h = 4$ cm 192 cm^3

7. Cube
$s = 6$ in. 216 cu in.

8. Cube
$s = 10$ m 1000 m^3

9. Cube
$s = 11$ cm 1331 cm^3

Find the volume. Use after page 345.

461.58 m^3
10.

h = 3 m
r = 7 m

803.84 cm^3
11.

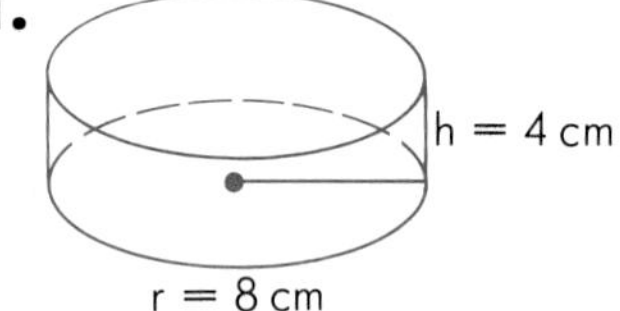

1271.7 cu in.
12.

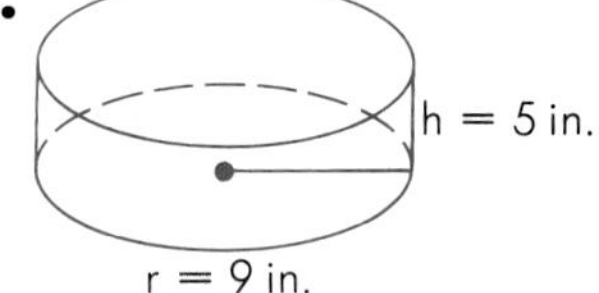

13. Cylinder
$r = 2$ ft
$h = 3$ ft 37.68 cu ft

14. Cylinder
$r = 10$ mm
$h = 6$ mm 1884 mm^3

15. Cylinder
$r = 5$ cm
$h = 5$ cm 392.5 cm^3

16. Cylinder
$d = 20$ cm
$h = 4$ cm 1256 cm^3

17. Cylinder
$d = 4$ in.
$h = 20$ in. 251.2 cu in.

18. Cylinder
$d = 6$ cm
$h = 6$ cm 169.56 cm^3

19. Cylinder
$d = 8$ cm
$h = 10$ cm 502.4 cm^3

20. Cylinder
$d = 4$ ft
$h = 2$ ft 25.12 cu ft

21. Cylinder
$d = 6$ m
$h = 5$ m 141.3 m^3

Find the volume. Use after page 347.

16 cm³
1.

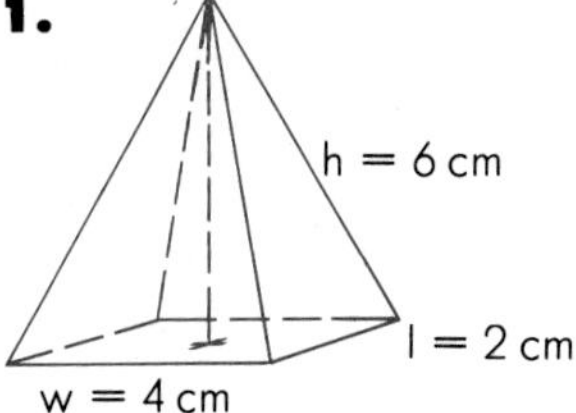

40 mm³
2.

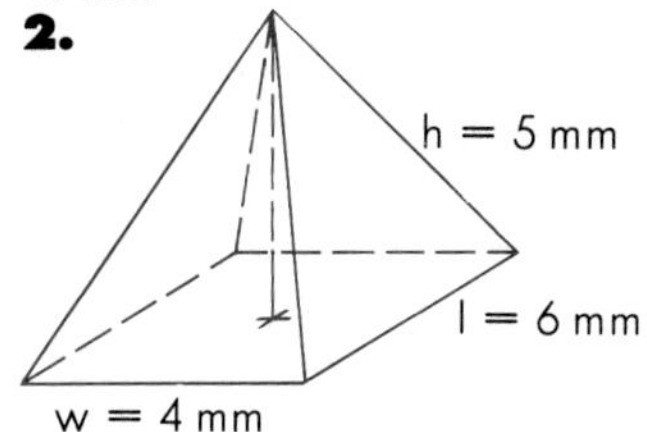

168 cm³
3.

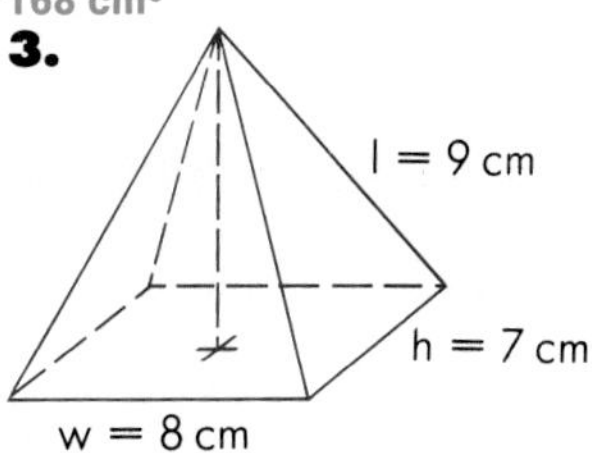

4. Pyramid
$l = 10$ in.
$w = 2$ in.
$h = 6$ in. 40 cu in.

5. Pyramid
$l = 4$ cm
$w = 9$ cm
$h = 4$ cm 48 cm³

6. Pyramid
$l = 5$ m
$w = 5$ m
$h = 5$ m 41.67 m³

65.94 cm³
7.

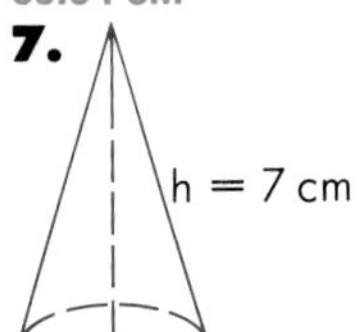

169.56 mm³
8.

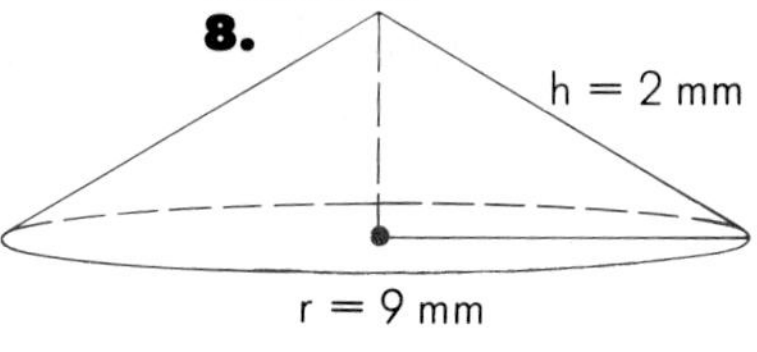

37.68 cu in.
9.

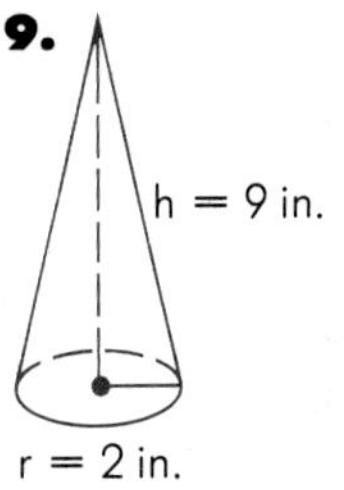

10. Cone
$r = 7$ cm
$h = 6$ cm 307.72 cm³

11. Cone
$r = 8$ m
$h = 10$ m 669.87 m³

12. Cone
$r = 4$ cm
$h = 8$ cm 133.97 cm³

Find the volume. Use after page 349.

4186.67 cm³
13.

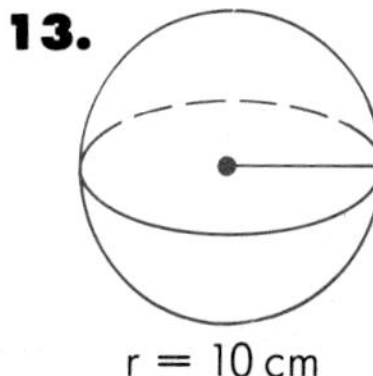

2143.57 m³
14.

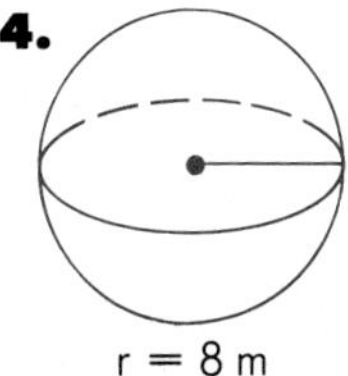

267.95 cu in.
15.

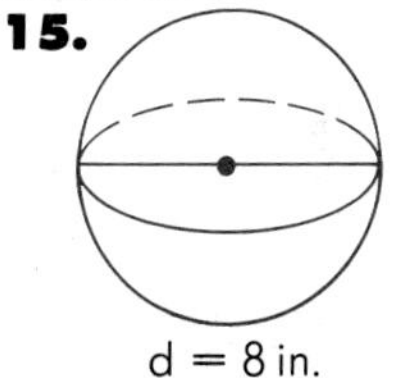

16. Sphere
$d = 12$ cm 904.32 cm³

17. Sphere
$d = 10$ m 523.33 m³

18. Sphere
$d = 6$ ft 113.04 cu ft

19. Sphere
$r = 4$ ft 267.95 cu ft

20. Sphere
$r = 9$ m 3052.08 m³

21. Sphere
$r = 6$ cm 904.32 cm³

SKILLS REVIEW

Multiply and simplify. Refer to pages 168–177.

1. $\frac{3}{4} \cdot \frac{3}{4}$ $\frac{9}{16}$

2. $\frac{7}{8} \times \frac{4}{5}$ $\frac{7}{10}$

3. $\frac{3}{8} \times \frac{4}{9}$ $\frac{1}{6}$

4. $\frac{4}{5} \cdot \frac{5}{6} \cdot \frac{2}{5}$ $\frac{4}{15}$

5. $\frac{1}{2} \times 12$ 6

6. $9 \cdot \frac{2}{3}$ 6

7. $5 \times \frac{4}{5}$ 4

8. $\frac{2}{3} \times 8$ $5\frac{1}{3}$

9. $8 \cdot 2\frac{2}{3}$ $21\frac{1}{3}$

10. $1\frac{2}{3} \times \frac{4}{9}$ $\frac{20}{27}$

11. $7\frac{1}{2} \cdot 3\frac{2}{3}$ $27\frac{1}{2}$

12. $6\frac{1}{4} \cdot 2\frac{2}{5}$ 15

Divide and simplify. Refer to pages 180–183.

13. $6 \div \frac{1}{3}$ 18

14. $7 \div \frac{4}{7}$ $12\frac{1}{4}$

15. $\frac{4}{5} \div 4$ $\frac{1}{5}$

16. $10 \div \frac{1}{2}$ 20

17. $\frac{2}{3} \div \frac{7}{3}$ $\frac{2}{7}$

18. $\frac{3}{4} \div \frac{1}{2}$ $1\frac{1}{2}$

19. $\frac{7}{10} \div \frac{3}{5}$ $1\frac{1}{6}$

20. $\frac{4}{5} \div \frac{4}{10}$ 2

21. $1\frac{1}{2} \div 1\frac{1}{2}$ 1

22. $6\frac{2}{3} \div 1\frac{1}{4}$ $5\frac{1}{3}$

23. $7\frac{3}{4} \div 2\frac{1}{2}$ $3\frac{1}{10}$

24. $6\frac{4}{5} \div 5\frac{3}{10}$ $1\frac{15}{53}$

Add and simplify. Refer to pages 196–203.

25. $\frac{2}{3} + \frac{2}{3}$ $1\frac{1}{3}$

26. $\frac{7}{10} + \frac{3}{10}$ 1

27. $\frac{9}{12} + \frac{6}{12}$ $1\frac{1}{4}$

28. $\frac{4}{5} + \frac{2}{5}$ $1\frac{1}{5}$

29. $\frac{3}{4} + \frac{1}{2}$ $1\frac{1}{4}$

30. $\frac{2}{3} + \frac{3}{4}$ $1\frac{5}{12}$

31. $\frac{7}{8} + \frac{1}{4} + \frac{1}{2}$ $1\frac{5}{8}$

32. $\frac{3}{5} + \frac{7}{10} + \frac{1}{2}$ $1\frac{4}{5}$

33. $2\frac{1}{2} + 3\frac{1}{4}$ $5\frac{3}{4}$

34. $3\frac{5}{8} + 2\frac{3}{4}$ $6\frac{3}{8}$

35. $4\frac{3}{10} + 1\frac{4}{5}$ $6\frac{1}{10}$

36. $3\frac{1}{2} + 2\frac{3}{4} + 3\frac{7}{8}$ $10\frac{1}{8}$

Subtract and simplify. Refer to pages 206–211.

37. $\frac{9}{10} - \frac{4}{10}$ $\frac{1}{2}$

38. $\frac{7}{8} - \frac{1}{8}$ $\frac{3}{4}$

39. $\frac{11}{12} - \frac{3}{4}$ $\frac{1}{6}$

40. $\frac{7}{9} - \frac{2}{3}$ $\frac{1}{9}$

41. $8\frac{7}{10} - 4\frac{3}{10}$ $4\frac{2}{5}$

42. $9\frac{3}{4} - 6\frac{1}{4}$ $3\frac{1}{2}$

43. $4\frac{9}{10} - 1\frac{2}{5}$ $3\frac{1}{2}$

44. $8\frac{3}{4} - 7\frac{1}{2}$ $1\frac{1}{4}$

45. $6 - 3\frac{3}{4}$ $2\frac{1}{4}$

46. $10 - 3\frac{1}{2}$ $6\frac{1}{2}$

47. $12\frac{1}{3} - 6\frac{2}{3}$ $5\frac{2}{3}$

48. $5\frac{1}{5} - 4\frac{7}{10}$ $\frac{1}{2}$

49. $17\frac{15}{16} - 9$ $8\frac{15}{16}$

50. $12\frac{2}{5} - 6\frac{2}{3}$ $5\frac{11}{15}$

51. $22\frac{1}{4} - 8\frac{4}{5}$ $13\frac{9}{20}$

52. $10\frac{3}{8} - 8\frac{5}{6}$ $1\frac{13}{24}$

Solve. Refer to page 230.

1. $\frac{3}{4} = \frac{x}{20}$ 15

2. $\frac{4}{5} = \frac{20}{x}$ 25

3. $\frac{2}{3} = \frac{7}{x}$ $10\frac{1}{2}$

4. $\frac{7}{10} = \frac{x}{3}$ 2.1

5. $\frac{x}{5} = \frac{11}{4}$ 13.75

6. $\frac{4}{x} = \frac{9}{8}$ $3\frac{5}{9}$

7. $\frac{x}{8} = \frac{7}{10}$ 5.6

8. $\frac{9}{x} = \frac{2}{5}$ 22.5

Solve. Refer to page 232.

9. If you drive 110 miles in 2 hours, how far will you drive in 3 hours? 165 miles

10. It takes Carie 2 hours to drive 110 miles. How many hours does it take her to drive 330 miles? 6 hours

11. If you receive $ 17.00 for 4 hours of work, how much will you receive for 6 hours of work? $25.50

12. Bo receives $ 17.00 for 4 hours of work. How many hours must he work to receive $ 25.50? 6 hours

13. If 3 tapes cost $ 9, how many tapes can you buy for $ 15? 5

14. If 5 pounds of mushrooms cost $ 7, how much do 2 pounds cost? $2.80

15. A team won 4 out of 7 games. At this rate, how many will it win in 21 games? 12 games

16. If you save $ 50 every 3 months, how much will you save in 12 months? $200

Solve. Refer to pages 256–263.

17. 60% of 10 is what number? 6

18. What is 70% of 100? 70

19. What percent of 10 is 6? 60%

20. 12 is what percent of 24? 50%

21. 75 is 50% of what number? 150

22. 9 is $12\frac{1}{2}$% of what number? 72

23. 75% of 20 is what number? 15

24. What is $33\frac{1}{3}$% of 27? 9

25. What percent of 50 is 5? 10%

26. 100 is what percent of 200? 50%

27. 12 is 50% of what number? 24

28. 10 is $66\frac{2}{3}$% of what number? 15

29. 90% of 50 is what number? 45

30. What is $87\frac{1}{2}$% of 8? 7

31. What percent of 300 is 200? $66\frac{2}{3}$%

32. 13 is what percent of 39? $33\frac{1}{3}$%

Make these changes. Refer to page 284.

1. 6 km = 6000 m
2. 4378 m = 4.378 km
3. 76 cm = 0.76 m
4. 814 mm = 81.4 cm
5. 79 mm = 7.9 cm
6. 4.1 cm = 41 mm
7. 8 m = 800 cm
8. 9.21 km = 9210 m
9. 6.2 km = 6200 m
10. 371 m = 0.371 km

Make these changes. Refer to page 290.

11. 4 L = 4000 mL
12. 7.4 L = 7400 mL
13. 6247 mL = 6.247 L
14. 9000 mL = 9 L
15. 0.91 L = 910 mL
16. 438 mL = 0.438 L
17. 0.6L = 600 mL
18. 73 mL = 0.073 L
19. 17,000 mL = 17 L
20. 750 mL = 0.75 L

Make these changes. Refer to page 292.

21. 8605 g = 8.605 kg
22. 4 g = 4000 mg
23. 7.3 t = 7300 kg
24. 2476 kg = 2.476 t
25. 814 mg = 0.814 g
26. 8.8 kg = 8800 g
27. 9 t = 9000 kg
28. 704 g = 0.704 kg
29. 7141 mg = 7.141 g
30. 12 kg = 12,000 g

Make these changes. Refer to page 298.

31. 7 ft = 84 in.
32. 8 pt = 16 cups
33. 5 ft = $1\frac{2}{3}$ yd
34. $3\frac{1}{2}$ qt = 7 pt
35. $2\frac{1}{2}$ yd = 90 in.
36. 96 oz = 6 lb
37. 3 mi = 15,840 ft
38. 48 in. = 4 ft
39. 40 fl oz = 5 cups
40. 3000 lb = $1\frac{1}{2}$ ton

Find the perimeter. Refer to page 312.

23 cm
1.

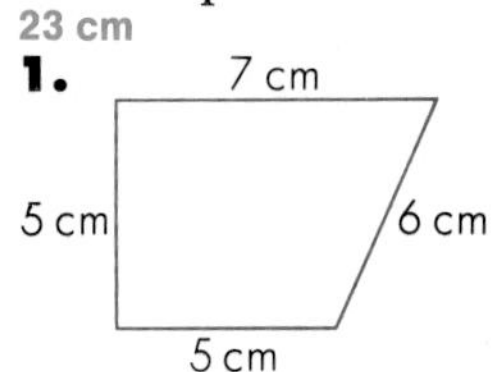

10 m
2.

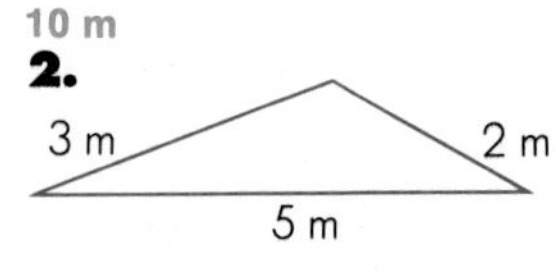

9 ft
3.

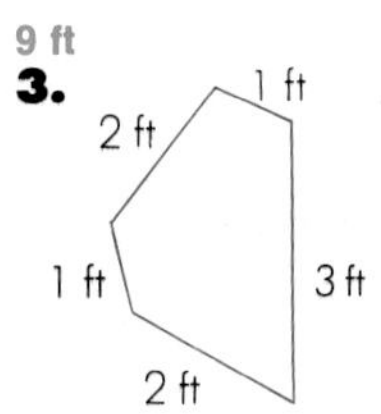

18 ft
4.

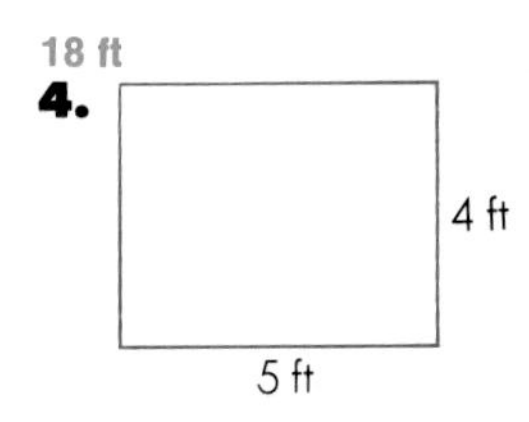

48 in.
5.

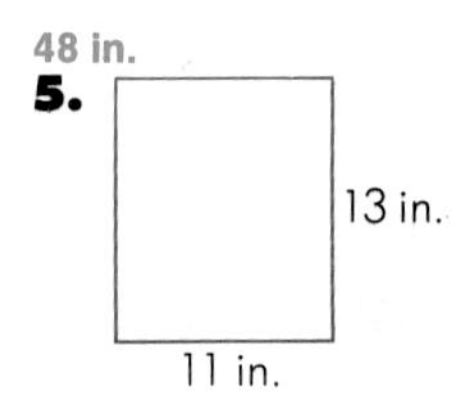

18.8 m
6.

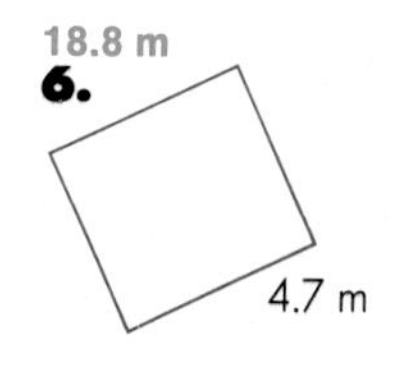

30 in.
7.

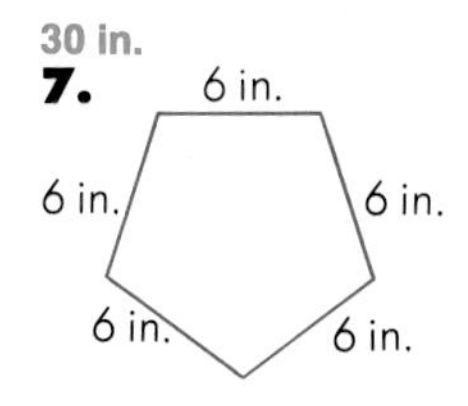

42 cm
8.

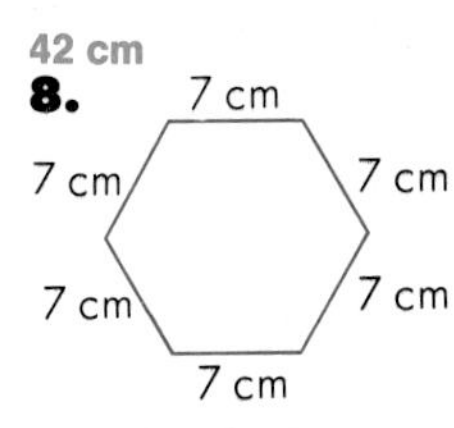

23.8 m
9.

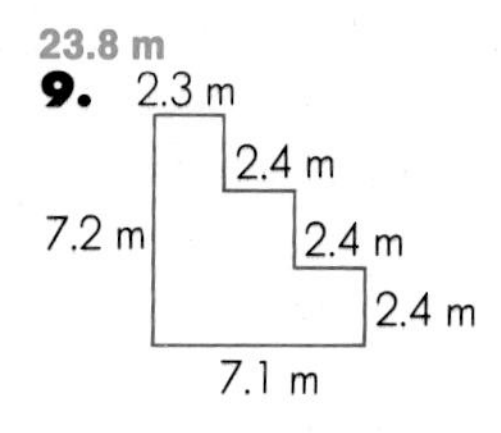

10. Square
$s = 5$ cm 20 cm

11. Square
$s = 12$ mm 48 mm

12. Rectangle
$l = 4\frac{1}{2}$ yd
$w = 2\frac{1}{2}$ yd 14 yd

Find the circumference. Refer to page 314.

28.26 cm
13.

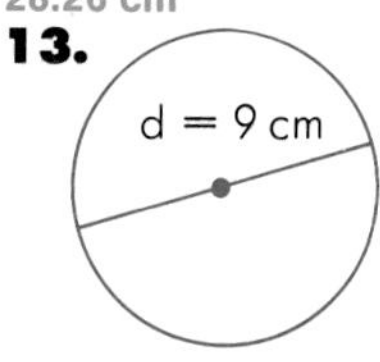

25.12 m
14.

d = 8 m

31.4 in.
15.

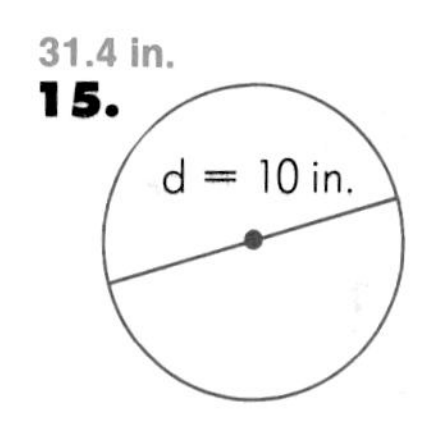

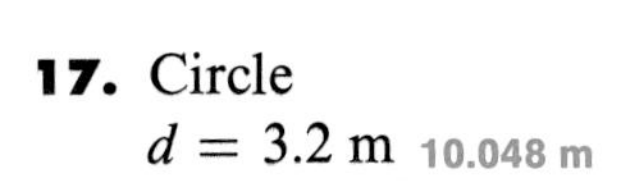

16. Circle
$d = 12$ mm 37.68 mm

17. Circle
$d = 3.2$ m 10.048 m

18. Circle
$d = 10.3$ cm 32.342 cm

12.56 ft
19.

r = 2 ft

25.12 in.
20.

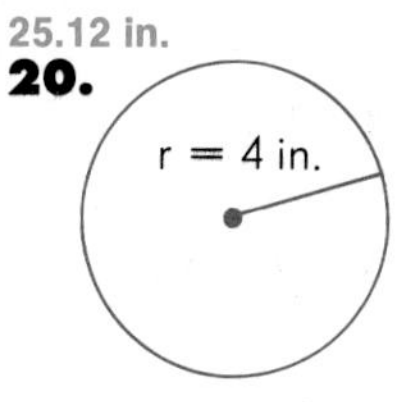
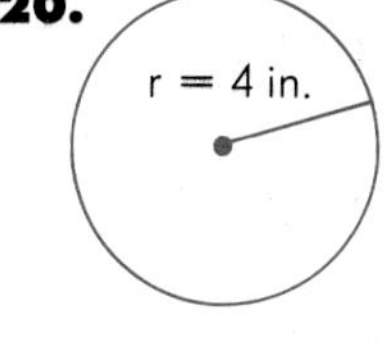

43.96 m
21.

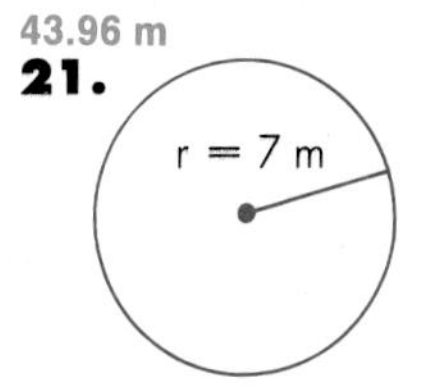

22. Circle
$r = 6$ cm 37.68 cm

23. Circle
$r = 1.3$ m 8.164 m

24. Circle
$r = 2.2$ cm 13.816 cm

Find the area. Refer to page 318.

35 cm²
1.
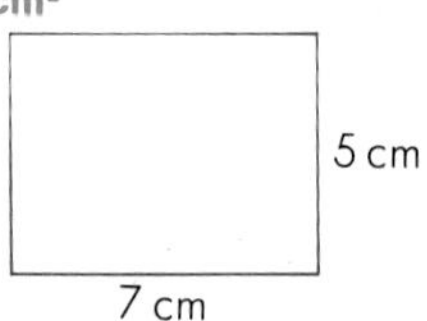

18 sq ft
2.
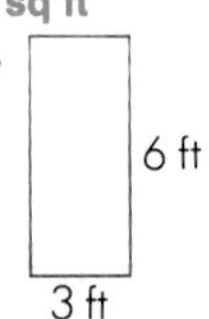

3.61 mm²
3.
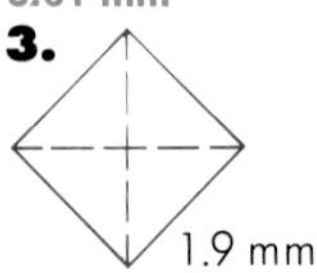

4. Square
$s = 7$ in. 49 sq in.

5. Square
$s = 3.7$ m 13.69 m²

6. Rectangle
$l = 8.1$ m
$w = 2.7$ m 21.87 m²

Find the area. Refer to page 320.

54 m²
7.
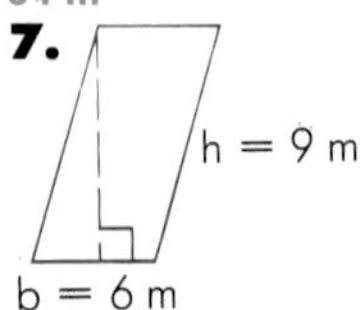

120 cm²
8.
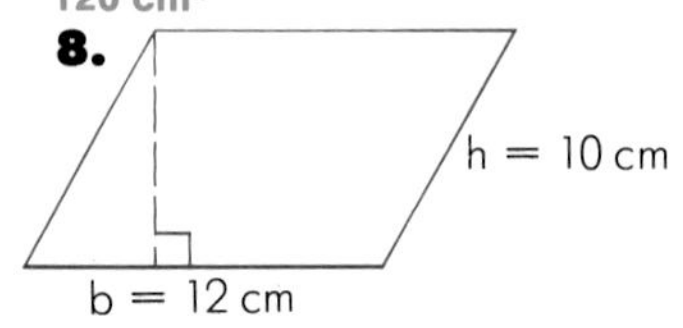

54 sq in.
9.
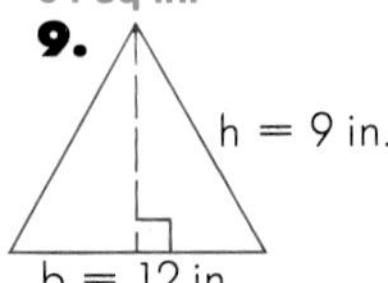

10. Triangle
$b = 10$ cm
$h = 6$ cm 30 cm²

11. Triangle
$b = 4$ m
$h = 3.6$ m 7.2 m²

12. Triangle
$b = 5.2$ m
$h = 4.4$ m 11.44 m²

Find the area. Refer to page 322.

30 m²
13.
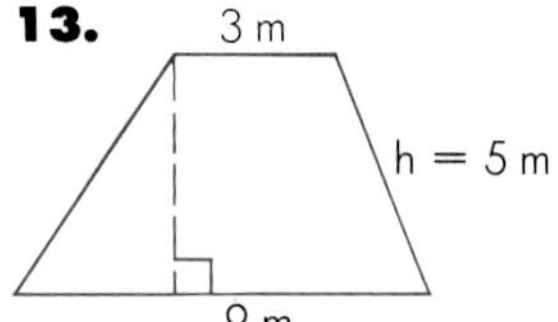

63 cm²
14.
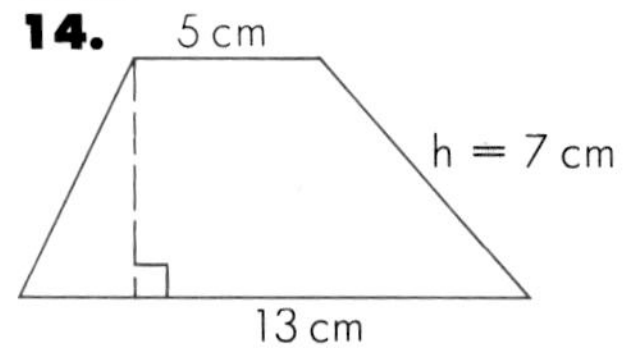

22 sq ft
15.
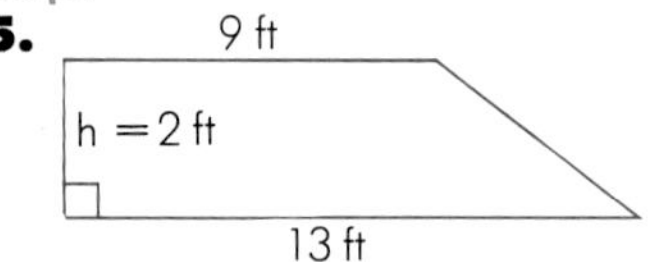

16. Trapezoid
$h = 20$ mm
$b_1 = 12$ mm
$b_2 = 16$ mm 280 mm²

17. Trapezoid
$h = 10$ m
$b_1 = 10$ m
$b_2 = 12$ m 110 m²

18. Trapezoid
$h = 4$ in.
$b_1 = 8$ in.
$b_2 = 10$ in. 36 sq in.

Find the area. Refer to page 326.

28.26 mm²
19.
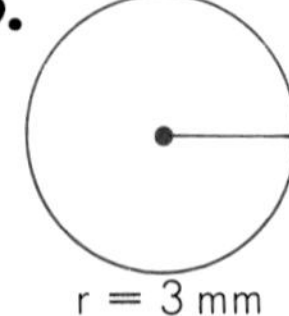

113.04 sq ft
20.
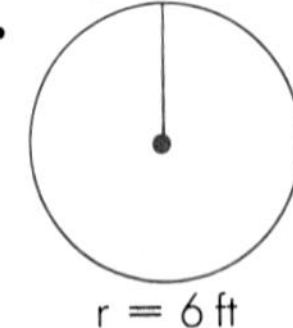

153.86 cm²
21.
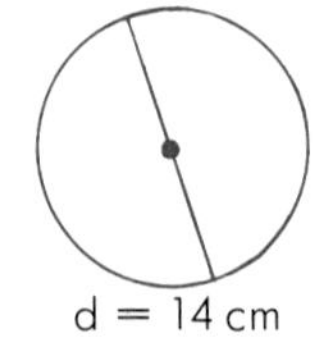

22. Circle
$d = 10$ m 78.5 m²

23. Circle
$d = 8$ cm 50.24 cm²

24. Circle
$r = 4$ in. 50.24 sq in.

Find the volume. Refer to page 340.

84 cm³
1.
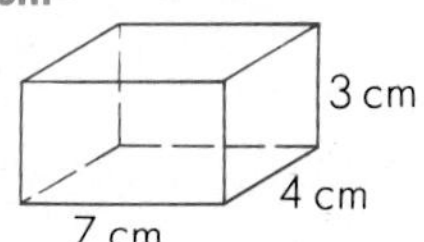

240 cu in.
2.
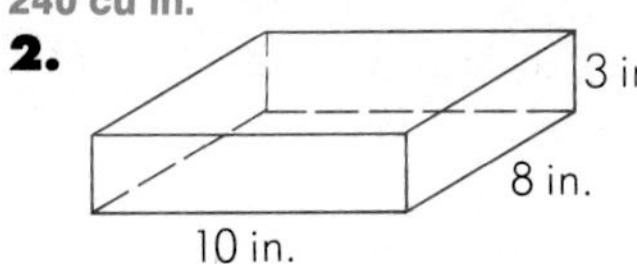

729 cu in.
3.
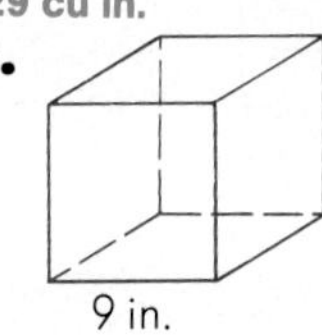

4. Cube
$s = 3$ m 27 m³

5. Cube
$s = 8$ cm 512 cm³

6. Rectangular prism
$l = 3$ ft
$w = 2$ ft
$h = 4$ ft 24 cu ft

Find the volume. Refer to page 344.

1526.04 cm³
7.
h = 6 cm
r = 9 cm

1570 cu in.
8.
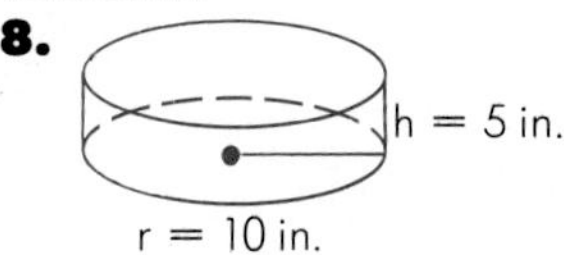

169.56 cu ft
9.
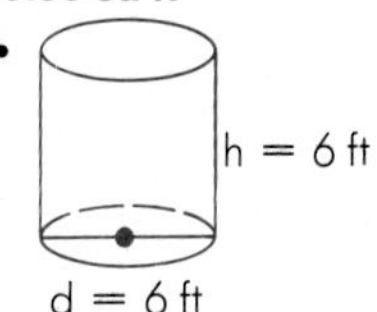

10. Cylinder
$d = 12$ mm
$h = 12$ mm 1356.48 mm³

11. Cylinder
$d = 20$ cm
$h = 5$ cm 1570 cm³

12. Cylinder
$r = 5$ ft
$h = 10$ ft 785 cu ft

Find the volume. Refer to page 346.

168 cm³
13.
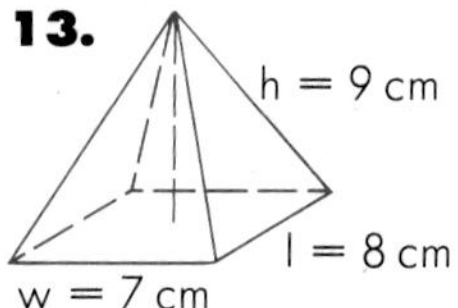

133.33 m³
14.
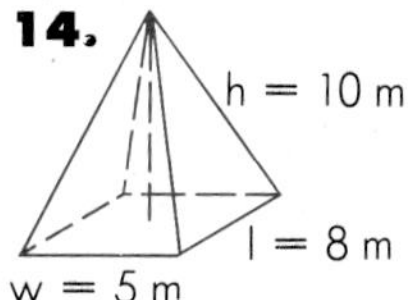

535.89 m³
15.
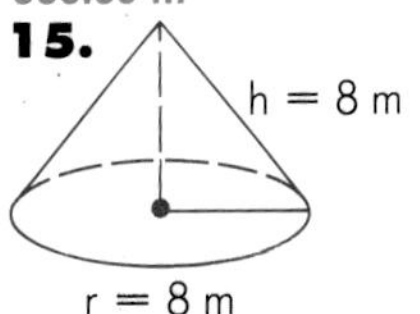

16. Cone
$r = 6$ in.
$h = 4$ in. 150.72 cu in.

17. Cone
$r = 4$ cm
$h = 6$ cm 100.48 cm³

18. Pyramid
$l = 12$ cm
$w = 3$ cm
$h = 10$ cm 120 cm³

Find the volume. Refer to page 348.

267.95 m³
19.
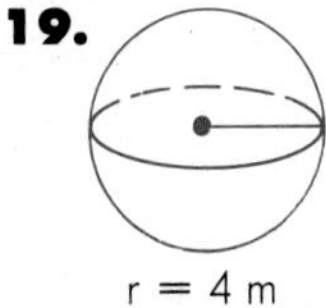

7234.56 mm³
20.
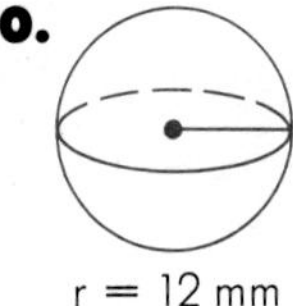

4186.67 cu in.
21.
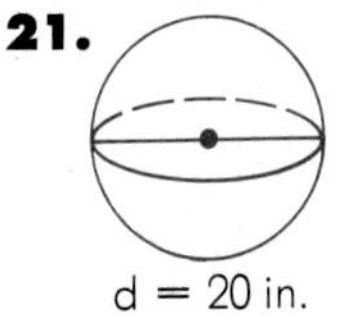

22. Sphere
$d = 16$ ft
2143.57 cu ft

23. Sphere
$d = 10$ m
523.33 m³

24. Sphere
$r = 9$ in.
3052.08 cu in.

Objective 196

SOLVING PROBLEMS

Mental calculation is an important skill. Other mental calculations which might be stressed are multiplying and dividing by powers of 10, dividing by 8, dividing by 25, and multiplying by 99, 999, etc.

Sometimes problems can be solved best by doing a quick mental calculation.

SITUATION 1 You are out to lunch with 3 friends. The bill is \$ 16.56. You decide to divide the bill four ways.

To mentally divide by 4, take half and then take half again. Divide the bill by 4 mentally. How much should each person pay? \$4.14

SITUATION 2 You are making a fertilizer mixture which calls for $1\frac{1}{2}$ tablespoons of fertilizer per gallon of water. You wish to make 8 gallons.

To mentally multiply by 8, double the number, double again and finally double a third time. Multiply $1\frac{1}{2}$ by 8 mentally. How much fertilizer should you use? 12 tablespoons

SITUATION 3 Your dinner bill is \$ 14.80. You wish to leave a 15% tip.

To mentally take 15%, take 10% and then add half of your result. Mentally take 15% of \$ 14.80. How much tip should you leave? \$2.22

SITUATION 4 You are chairman of a committee to plan a party for 25 people. The caterer says that the cost will be \$ 8.50 per person.

To mentally multiply by 25, multiply by 100 and then divide by 4. Mentally multiply \$ 8.50 by 25. How much will the party cost? \$212.50

CHAPTER 13 PRE-TEST

1. Graph the ordered pair (4,5) on a grid. p. 368 See answer section.

2. Find an ordered pair for the point A on the grid. p. 368 (2,3)

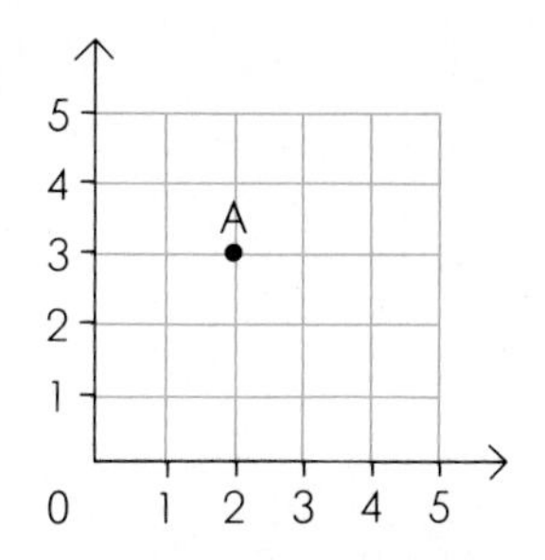

3. Make a line graph for the data in the table. p. 370 See answer section.

Height (in thousands of feet)	View (in miles)
5	87
10	122
15	149
20	172

4. Make a bar graph for the data in the table. p. 372 See answer section.

Causes of Death	Number of Deaths (in thousands)
Heart Trouble	230
Cancer	175
Accidents	80

5. Find the angle for each percent and make a circle graph. p. 374 See answer section.

House Color	Percent
White	48%
Gray	34%
Yellow	16%
Other	2%

Find these statistics. pp. 378–381

6. the mean: 14, 28, 16, 10, 11 15.8

7. the median: 78, 75, 73, 77, 80 77

8. the mode: 96, 89, 96, 97, 99, 103 96

Application pp. 376, 382

9. Rona went on a weekend ski trip and spent $ 38.60 for transportation, $ 18.25 for food, $ 14 for lift tickets, and $ 41.63 for a hotel room. How much did she spend in all on her trip? $112.48

10. A football team won 13 out of 16 games. Find the percentage of games won. 0.813 or 81.3%

13 GRAPHING AND STATISTICS

GETTING STARTED

COORDINATE MATCH

WHAT'S NEEDED A 5 × 5 grid drawn on the blackboard

THE RULES

1. On a sheet of paper each player makes a scorecard. Then each player writes down five ordered pairs.
2. The leader marks ten points on the grid.
3. The players compare their ordered pairs with the points on the grid and check the ones that match. Each player records on the scorecard the number of points that match.
4. Play three rounds and total the scores. The player with the highest total wins.

SAMPLE

Round 1

(3,3)
(0,4)
(1,2)
(2,1)
(5,0)

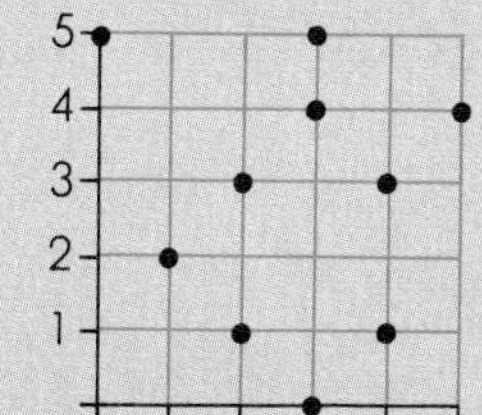

Round 2

(4,1)
(5,5)
(3,2)
(0,4)
(1,5)

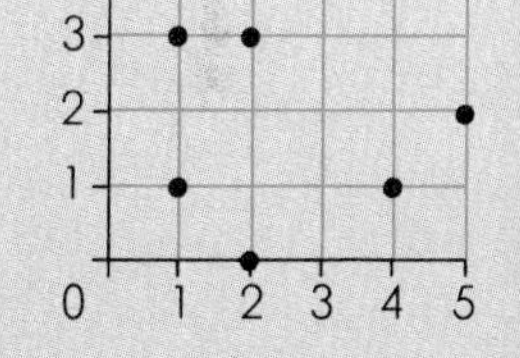

Round 3

(0,0)
(2,4)
(3,5)
(4,0)
(1,1)

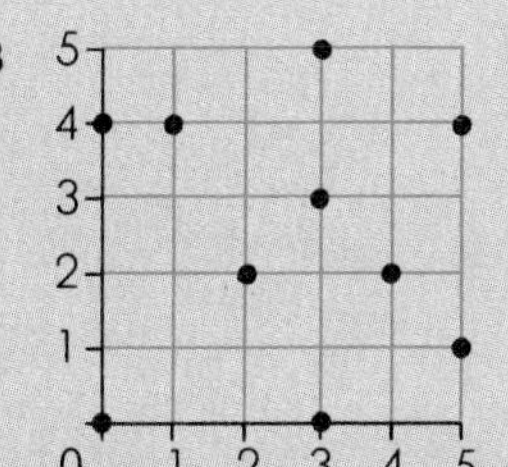

Scorecard

1. 2

2. 3

3. 2

Total 7

Objectives 125 and 126

GRAPHING ORDERED PAIRS

The first number in an ordered pair is the number of units to the right of (0,0). The second number is the number of units up.

EXAMPLE 1. Graph this ordered pair on a grid. (3,4).

(3,4)
Go 3 units to the right of (0,0).
Go 4 units up.
Mark and label the point.

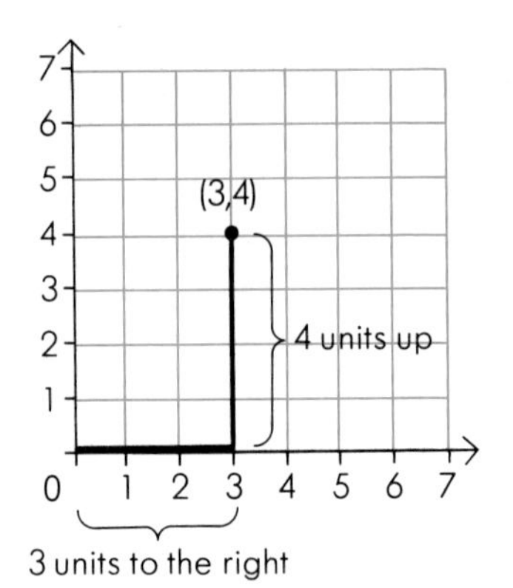

TRY THIS Graph each ordered pair on a grid. See answer section.

1. (4,5) **2.** (1,2) **3.** (0,3) **4.** (2,0)

EXAMPLE 2. Find an ordered pair for point A on the grid.

Point A is 7 units to the right of (0,0). → (,)

Point A is 5 units up. →

The ordered pair is (7,5).

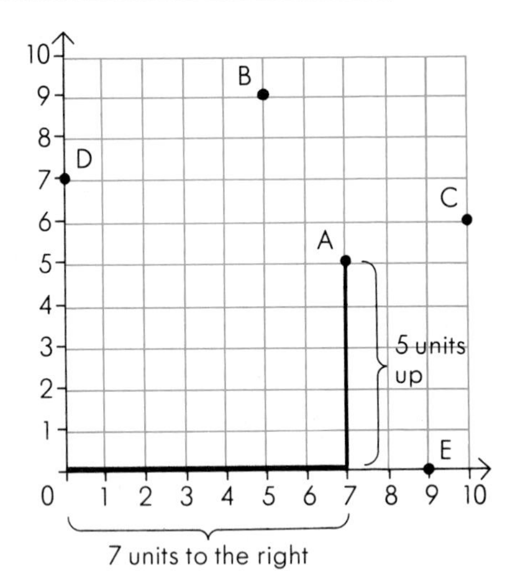

TRY THIS Find an ordered pair for each point on the grid in Example 2.

5. B (5,9) **6.** C (10,6) **7.** D (0,7) **8.** E (9,0)

PRACTICE

Graph each ordered pair on a grid. See answer section.

1. (2,1) **2.** (3.7) **3.** (2,5) **4.** (8,4)

5. (0,6) **6.** (0,8) **7.** (0,5) **8.** (0,10)

9. (5,0) **10.** (3,0) **11.** (7,0) **12.** (1,0)

13. (4,4) **14.** (9,9) **15.** (10,10) **16.** (2,2)

Find an ordered pair for each point on the grid.

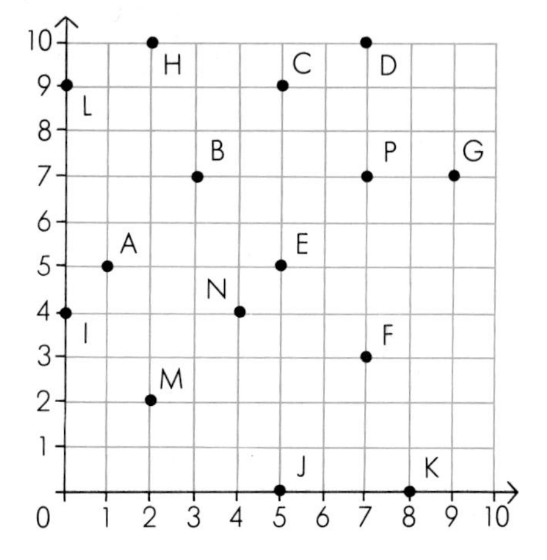

17. A (1,5) **18.** B (3,7) **19.** C (5,9)

20. D (7,10) **21.** E (5,5) **22.** F (7,3)

23. G (9,7) **24.** H (2,10) **25.** I (0,4)

26. J (5,0) **27.** K (8,0) **28.** L (0,9)

29. M (2,2) **30.** N (4,4) **31.** P (7,7)

PROBLEM CORNER

A bottle washed up on shore contains the following message.

Clues to Buried Treasures

Treasure A: (2,3)
Treasure B: (7,9)
Treasure C: (10,5)
Treasure D: (5,0)

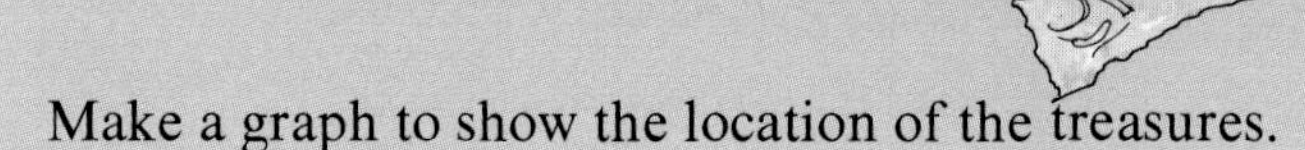

Make a graph to show the location of the treasures.
See answer section.

Objectives 127 and 128

MAKING LINE GRAPHS

It is important that students learn to label the axes before they start to draw the graph.

A line graph shows the change in data.

EXAMPLE 1. Make a line graph for the data in the table.

The table shows the decreasing temperature in a glass of water as it cools.

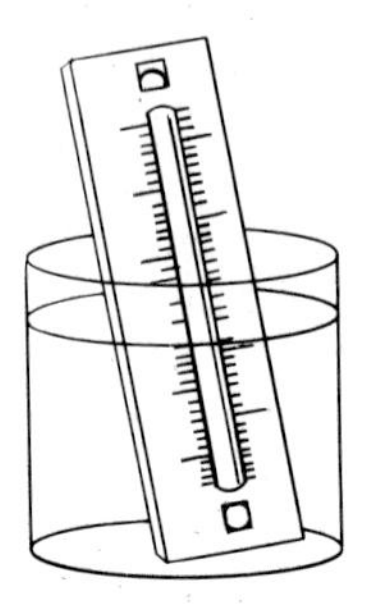

Time (in minutes)	Temperature (in Fahrenheit)
0	160°
30	120°
60	90°
90	70°
120	70°

To make a line graph follow these steps.

1. Draw and label a vertical axis. It does not have to start at 0. We start at some number less than the smallest number in the table. The numbers increase along the axis by the same amount.
2. Draw and label a horizontal axis. This time it is convenient to start with 0.
3. Graph ordered pairs from the table. For example, the table gives us the ordered pair (30,120°).
4. Connect each pair of points with a line.

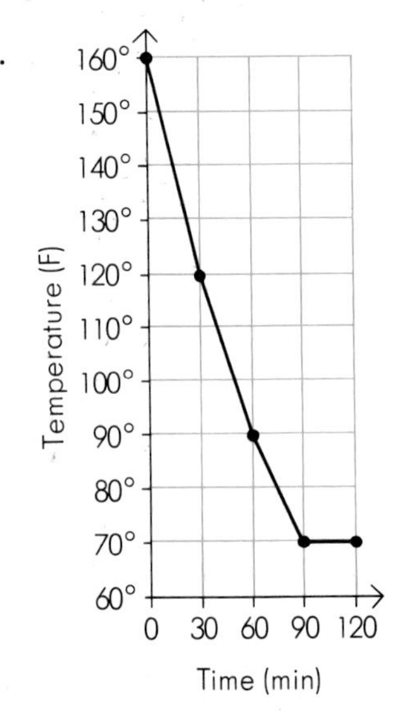

TRY THIS

Make a line graph for the data in the table.

See answer section.

Year	Value of a One-Carat Diamond
1940	$900
1950	$1500
1960	$2500
1970	$6300

PRACTICE

For extra practice, use Making Practice Fun 77 with this lesson.

Make a line graph. See answer section.

1.

Time	Temperature (C)
9:00	18°
10:00	20°
11:00	21°
12:00	23°
1:00	22°

2.

Year	Earnings
1940	$ 5500
1950	$ 6000
1960	$ 8000
1970	$14,000
1980	$12,000

3.

Year	Rainfall (Inches)
1940	28
1950	25
1960	30
1970	27
1980	30

4.

Speed (mph)	Gas Mileage (miles per gallon)
20	17
30	18
40	19
50	18
60	16

PROBLEM CORNER

How many TV's were sold each year?

1. 1978 200,000

2. 1979 400,000

3. 1980 550,000

4. 1981 475,000

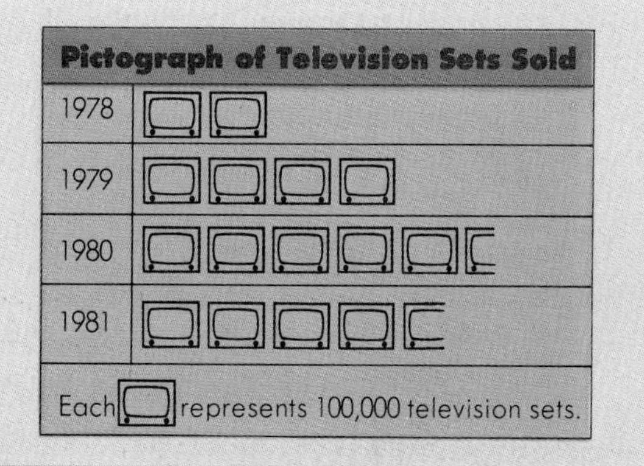

Objectives 129 and 130

MAKING BAR GRAPHS

Students should be reminded that the numbers increase along the axes by the same amount, even though they need not start at 0.

A bar graph shows the relationship of data.

EXAMPLE 1. Make a bar graph for the data in the table.

To make a bar graph follow these steps.

1. Draw and label a vertical axis. The last number should be large enough so that every number in the table can be graphed.
2. Draw and label a horizontal axis. Space the names of the nations evenly.
3. Draw vertical bars to show the percents.

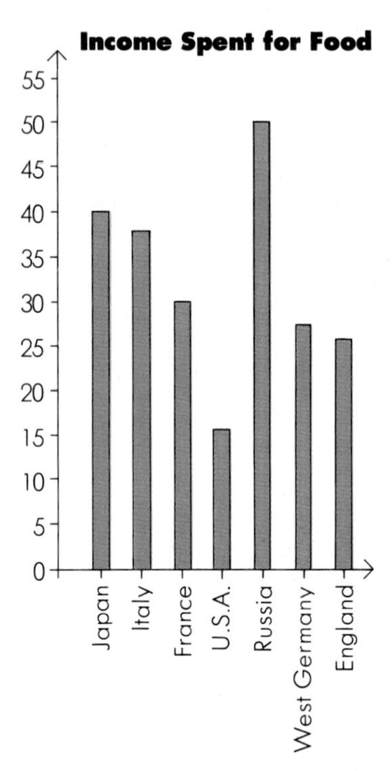

Nation	Percent of Income Spent for Food
Japan	40
Italy	38
France	31
U.S.A.	17
Russia	50
West Germany	28
England	26

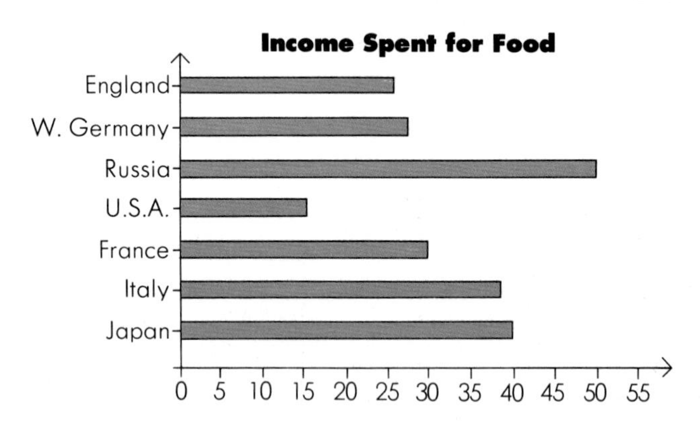

TRY THIS

1. Make a table and a bar graph for the following data. The number of calories in a one-scoop ice cream cone: Chocolate Fudge, 230; French Vanilla, 220; Strawberry, 170; Peach, 170; Butter Pecan, 200; Chocolate Mint, 190.

See answer section.

PRACTICE

Make a bar graph. See answer section.

1.

Activity	Calories Burned Per Hour by Person Weighing 152 lbs
Tennis	420
Jogging	650
Hiking	590
Office Work	180
Sleeping	70

2.

Speed (mph)	Gas Mileage (miles per gallon)
20	17
30	18
40	19
50	18
60	16

3.

Education (years)	Income
8	$ 8000
12	$12,000
16	$18,000

4.
Make a table and a bar graph for the following data. The number of passengers in one year (in millions) to fly out of a given airport: Chicago (O'Hare), 41; Atlanta, 27; Los Angeles, 26; London (Heathrow), 24; New York (Kennedy), 21; San Francisco, 18.

PROBLEM CORNER

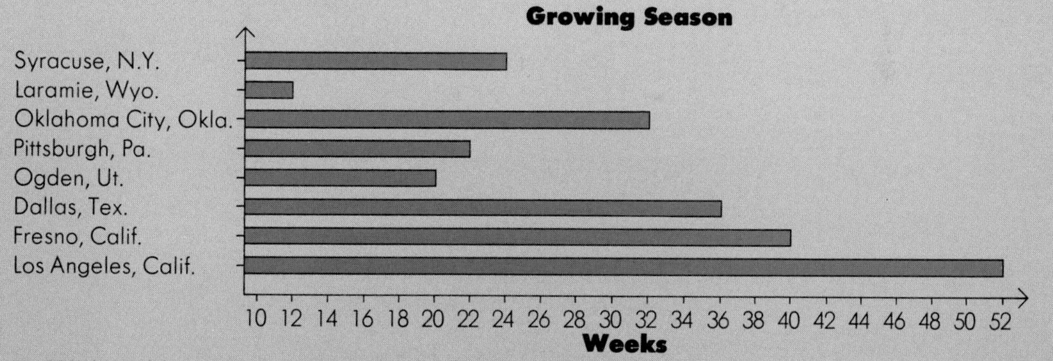

1. Which city has the longest growing season? Los Angeles, California
2. Which city has the shortest growing season? Laramie, Wyoming
3. How many times longer is the growing season in Fresno, California than in Ogden, Utah? in Dallas, Texas than in Laramie, Wyoming? 2 times; 3 times

Objectives 131 and 132

MAKING CIRCLE GRAPHS

Students will need protractors for this lesson.

A circle graph shows data as percents.

EXAMPLE 1. Find the angle for each percent in the table and make a circle graph.

It is important that students realize the percents should add up to 100% and the angle measures should add up to 360°.

1. To find the angle convert each percent to a decimal and multiply by 360, the number of degrees in a circle.

Hair Color	Percent
Brown	36%
Black	28%
Blonde	24%
Red	12%

$0.36 \times 360° \approx 130°$
$0.28 \times 360° \approx 101°$
$0.24 \times 360° \approx 86°$
$0.12 \times 360° \approx 43°$

Some students may need extra instruction to draw the angles given by the degree measures.

2. Draw a circle and a radius. Draw the angles given by the degree measures from Step 1. Label each angle as it is drawn.

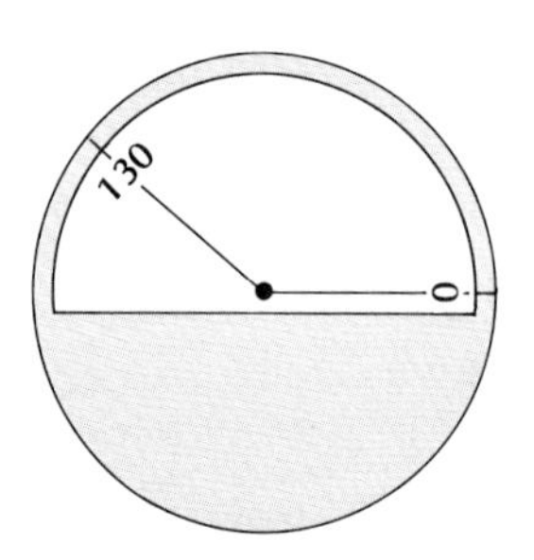

Hair Color

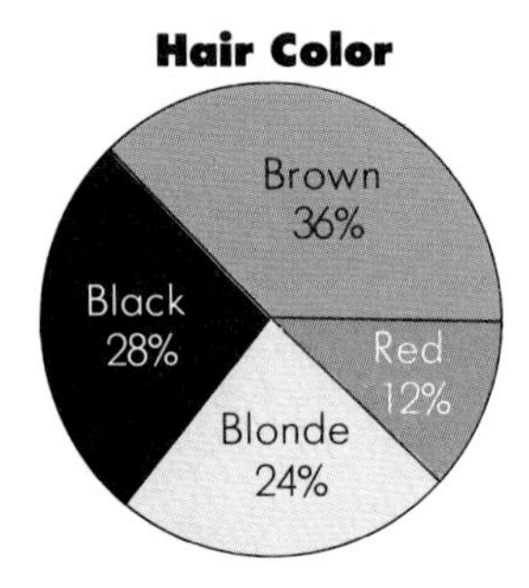

TRY THIS

Find the angle for each percent and make a circle graph.
See answer section.

1.

Family Expenses	Percent
Food	20%
Auto	14%
Clothing	10%
Housing	26%
Other	30%

PRACTICE

For extra practice, use Making Practice Fun 78 with this lesson.
Use Quiz 26 after this Practice.

Find the angle for each percent in the table and make a circle graph. See answer section.

1.

Study Time	Percent
Math	24%
English	38%
Social Studies	20%
Science	18%

2.

Sales	Percent
Shoes	10%
Suits	50%
Shirts	15%
Ties	5%
Pants	20%

3.

Favorite TV Shows	Percent
Movies	12%
Sports	20%
News	4%
Drama	16%
Comedy	20%
Music	28%

4.

Personal Income in 1980	Percent
$25,000 and over	13%
$15,000–$25,000	33%
$10,000–$15,000	23%
$7000–$10,000	12%
Below $7000	19%

PROBLEM CORNER

Compare the graphs. Does the percent increase or decrease from 1965 to 1975 for each situation?

1. Households with no TV sets decrease
2. Households with 1 TV set decrease
3. Households with 2 or more TV sets increase

What percent of U.S. households had at least 1 TV set in the following years?

4. 1965 92% 5. 1975 97%

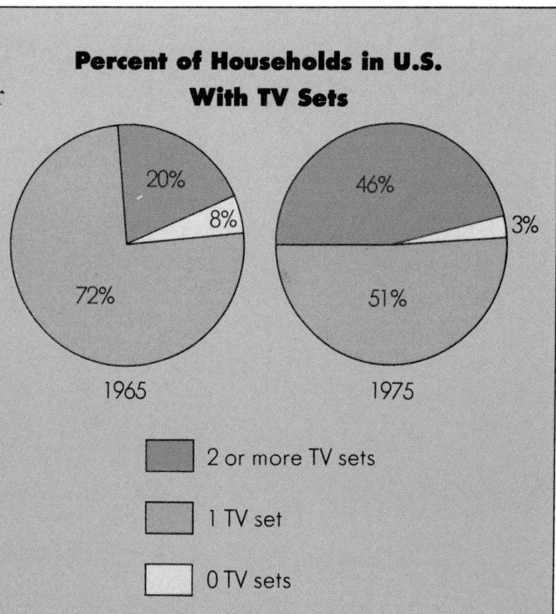

Objective 189

Discuss with your students some of the reasons why travel costs fluctuate. Have them suggest ways to save money in traveling.

TRAVEL COSTS

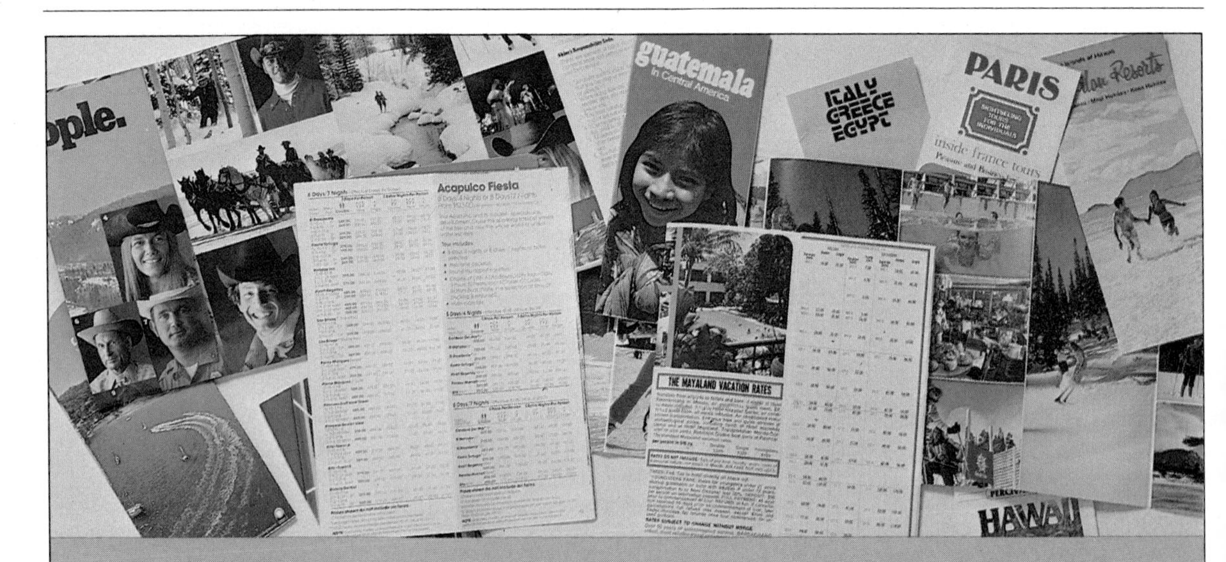

Each year many more people travel for business and for pleasure. The amount of money they spend is also increasing.

Find how much the average American traveler spent abroad in each of the following years.

1. 1960 $1605.26
2. 1965 $1436.52
3. 1970 $1174.90
4. 1975 $1596.32
5. 1980 $1507.04
6. In which year was the most amount of money spent per traveler? the least amount? 1960; 1970
7. Are travelers to the U.S. spending more or less money each year? more
8. Draw a graph to show the comparison of dollars spent by travelers to and from the U.S. See answer section.

Overseas Travel from the U.S.

Year	Number of Travelers	Millions of Dollars Spent
1960	1,634,000	2623
1965	2,623,000	3768
1970	5,260,000	6180
1975	6,354,000	10,143
1980	7,100,000	10,700

Travel to the U.S.

Year	Millions of Dollars Spent
1960	1025
1965	1545
1970	2708
1975	5606
1980	7200

Business travelers must often report their travel expenses. Tax deductible expenses must be shown with receipts. It is necessary to keep accurate records of travel expenses.

Betty Riggs is a buyer for a department store. On a business trip last month she kept a list of her expenses.

Travel Expenses	Amount	Paid With
Airfare	$ 115.98	Credit card
Hotel	62.30	Credit card
Meals	22.89	Cash
Tips	4.00	Cash
Cab fare	18.25	Cash
Convention	25.00	Check
Phone	8.50	Cash

9. How much did Betty charge on her credit card? $178.28
10. What was the total amount of cash spent? $53.64
11. How much did Betty spend in all? $256.92

Betty could have saved 20% on the airfare by traveling at night.

12. How much money could she have saved? $23.20
13. What was the price of the night airfare? $92.78
14. James went skiing last weekend. He spent $ 29.75 for transportation, $ 18.34 for food, $ 12 for lift tickets, and $ 35 for a hotel room. How much did he spend in all? $95.09
15. On her vacation Tanja spent $ 153.50 for airfare, $ 75.80 for hotel rooms, $ 42.37 for food, and $ 23.40 for souvenirs. How much did she spend in all? $ 295.07

The Clark twins compared their vacation expenses. Jerry spent $ 83 for transportation. Jill spent $ 28 for food.

16. Jerry spent 25% more than Jill for food. How much did he spend for food? $35
17. Jill spent 30% more than Jerry for transportation. How much did she spend for transportation? $107.90
18. Make a chart to show their individual and total expenses. See answer section.

Objective 133

FINDING THE MEAN

The terms *mean* and *average* are used interchangeably. Some texts, however, consider mean, median, and mode to be averages.

The mean of a set of numbers is the average of the numbers.

Find the mean. 78, 81, 82, 79

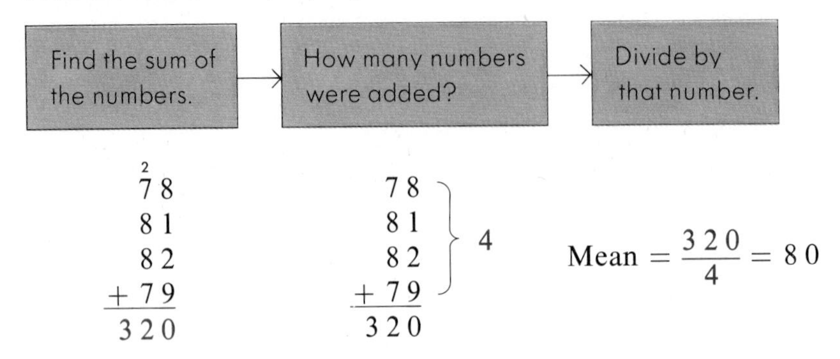

EXAMPLE 1. Find the mean. 75, 36.8, 95.7, 12.1, 47.6

Find the sum of the numbers.	*How many numbers were added?*	*Divide by that number.*
$75 + 36.8 + 95.7 + 12.1 + 47.6 = 267.2$	75, 36.8, 95.7, 12.1, 47.6 → 5	$\frac{267.2}{5} = 53.44$
		Mean = 53.44

TRY THIS

Find the mean.

1. 86, 91, 68, 76 80.25

2. 105.2, 108.9, 110, 200, 240 152.82

Sometimes the sum is given and all we need to do is divide.

EXAMPLE 2. A car went 432 miles on 24 gallons of gasoline. What was the average number of miles per gallon?

$$\frac{432 \text{ miles}}{24 \text{ gallons}} = 18 \text{ miles per gallon}$$

We divide 432 miles by 24 gallons.

TRY THIS 3. A car went 522 miles on 18 gallons of gasoline. What was the average number of miles per gallon? 29 miles per gallon

PRACTICE

For extra practice, use Making Practice Fun 79 with this lesson.

Find the mean.

1. 8, 7, 15, 15, 15, 12 12
2. 72, 83, 85, 88, 92 84 88 80
3. 5, 10, 15, 20, 25, 30, 35 20
4. 13, 13, 25, 27, 32 22
5. 1.2, 4.3, 5.7, 7.4, 7.4 5.2
6. 13.4, 13.4, 12.6, 42.9 20.575
7. Sandra Palmer scored 78, 74, 71, and 72 to win the U.S. Open Golf Tournament. What was her average score? 73.75
8. The following temperatures were recorded for seven days: 43°, 40°, 23°, 38°, 54°, 35°, 47° What was the average temperature? 40°
9. The following prices per pound of steak were found at five supermarkets: $ 3.79, $ 3.59, $ 3.69, $ 3.79, $ 3.89. What was the average price per pound? $3.75
10. The following prices per pound of hamburger were found at five supermarkets: $ 1.39, $ 1.29, $ 1.49, $ 1.09, $ 0.99. What was the average price per pound? $1.25
11. The attendance at five basketball games was 30,050, 28,740, 35,170, 29,080, and 32,610. What was the average attendance? 31,130
12. O. J. Simpson set an NFL rushing record gaining 2003 yards in a 14-game season. What was the average number of yards gained per game? 143.1

PROBLEM CORNER

Sally, 124; Randy, 178; Chip, 136; Leanne, 167
Total: 617, 618, 578 Average: 154, 155, 145

1. What was each person's average score?
2. What was the team total and average in Game 1? Game 2? Game 3?
3. What was the team average for the 3 games? 604.3

The Oakwood Bowling Team Scores

	Sally	Randy	Chip	Leanne
Game 1	115	196	144	162
Game 2	132	182	138	166
Game 3	124	156	126	172

Objectives 134 and 135

FINDING THE MEDIAN AND MODE

In studying data it is often required to find the median and mode of a set of numbers.

Find the median. 78, 81, 82, 76, 84

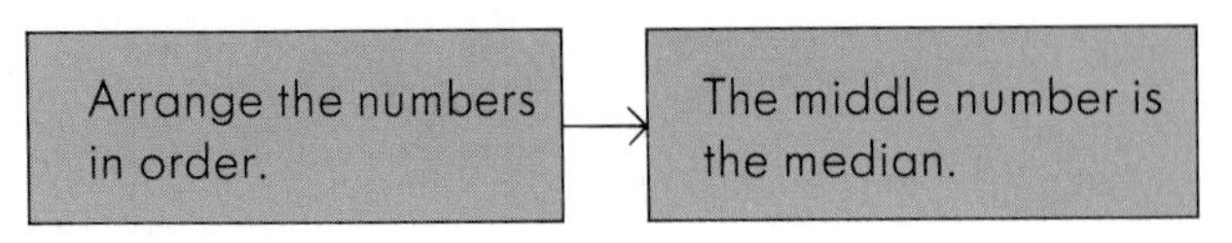

76, 78, 81, 82, 84 Median = 81

If there are two middle numbers, the median is the number halfway between.

EXAMPLE 1. Find the median. 23, 24, 25, 26

The median is the average of 24 and 25.

23, 24, 25, 26 *The median is halfway between 24 and 25.*

Median = 24.5 $\frac{24 + 25}{2} = \frac{49}{2} = 24.5$

TRY THIS

Find the median.

1. 20, 13, 17, 14, 18 17
2. 37, 57, 27, 47, 27 37
3. 18, 14, 16, 13, 19, 20 17
4. 8, 7, 8, 6, 9, 10 8

Find the mode. 14, 23, 25, 23, 27, 27, 29, 27

If all numbers occur an equal number of times, then there is no mode.

Arrange the numbers in order. → The number which appears the greatest number of times is the mode.

14, 23, 23, 25, 27, 27, 27, 29 Mode = 27

EXAMPLE 2. Find the mode. 69, 60, 60, 61, 62, 63, 63, 59

59, 60, 60, 61, 62, 63, 63, 69

Mode (60, 60) Mode (63, 63)

Both 60 and 63 appear twice. This set of numbers has 2 modes, 60 and 63.

TRY THIS

Find the mode.

5. 78, 66, 78, 81, 81, 78, 78, 81, 102 78
6. 103, 101, 101, 101, 102, 103, 103, 100, 104 101, 103

PRACTICE

For extra practice, use Making Practice Fun 80 with this lesson.
Use Quiz 27 after this Practice.

Find the median.

1. 72, 83, 85, 92, 88 85
2. 80, 81, 79, 78, 76 79
3. 8, 7, 15, 15, 12, 13, 14 13
4. 11, 9, 17, 13, 27, 29, 28 17
5. 36, 37, 39, 40 38
6. 41, 42, 44, 45 43
7. 36, 37, 38, 39 37.5
8. 41, 42, 43, 44 42.5
9. 8, 7, 15, 15, 15, 12 13.5
10. 13, 32, 25, 27, 13 25

Find the mode.

11. 1, 3, 3, 3, 4 3
12. 5, 5, 6, 7, 8 5
13. 14, 14, 15, 15, 15, 12, 14 14, 15
14. 19, 17, 19, 17, 19, 20, 20, 17 17, 19
15. 9, 9, 9, 9, 9, 9, 9, 7 9
16. 20, 20, 20, 20, 20, 18, 20 20
17. 34, 34, 35, 35, 36, 34, 34 34
18. 102, 101, 102, 103, 102, 102, 101 102
19. 38, 38, 37, 38, 39, 39, 40, 39 38, 39
20. 97, 98, 99, 97, 98, 99, 97, 98, 100 97, 98

PROBLEM CORNER

Marcy received the following scores on five tests:

1. What was the average? 90
2. What was the median? 91
3. What was the mode? 91

Objective 190

APPLICATION

SPORTS STATISTICS

In many different sports an important figure is the percentage of games won.

$$\text{PERCENTAGE OF GAMES WON} = \frac{\text{NUMBER OF GAMES WON}}{\text{NUMBER OF GAMES PLAYED}}$$

The Boston Celtics won the Eastern Conference Championship of the NBA 14 times from 1957 to 1976. They won each year from 1957 to 1965 and each year from 1972 to 1976. Here are their records from 1957 to 1965.

1. How many games were played each year?
2. What was the percentage of games won each year?
3. In 1976, the Celtics played 82 games. Their percentage of games won was 0.659. How many games did they win? 54 lose? 28
4. In 1975, the Chicago Bulls played 82 games. Their percentage of games won was 0.573. How many games did they win? 47 lose? 35

Year	Games Won	Games Lost		%
1957	44	28	72	0.611
1958	49	23	72	0.681
1959	52	20	72	0.722
1960	59	16	75	0.787
1961	57	22	79	0.722
1962	60	20	80	0.750
1963	58	22	80	0.725
1964	57	21	78	0.731
1965	62	18	80	0.775

A basketball game can be won or lost from the free throw line.

$$\text{FREE THROW PERCENTAGE} = \frac{\text{NUMBER OF COMPLETIONS}}{\text{NUMBER OF ATTEMPTS}}$$

What is the free throw percentage of each player?

5. Manny 39%
6. Philip 33%
7. Sam 39%
8. Ron 45%
9. Bob 38%
10. What is the team free throw percentage? 39%

Player	Attempts	Completions
Manny	62	24
Philip	57	19
Sam	41	16
Ron	71	32
Bob	45	17

We often use batting averages to compare baseball players' performances.

$$\text{BATTING AVERAGE} = \frac{\text{NUMBER OF HITS}}{\text{NUMBER OF TIMES AT BAT}}$$

In one season Melanie had 25 hits out of 58 times at bat. What was her batting average?

$$\text{BATTING AVERAGE} = \frac{25}{58} = 0.431$$

Player	Hits	Times at Bat
Jay	20	60
Rita	12	62
Ken	21	55
Connie	26	63
Paul	16	61
Julie	15	65

Find the batting average for each player.

11. Jay 0.333
12. Rita 0.194
13. Ken 0.382
14. Connie 0.413
15. Paul 0.262
16. Julie 0.231

The 100-meter women's butterfly stroke has been an event in the summer Olympics since 1956. The graph shows the winning times.

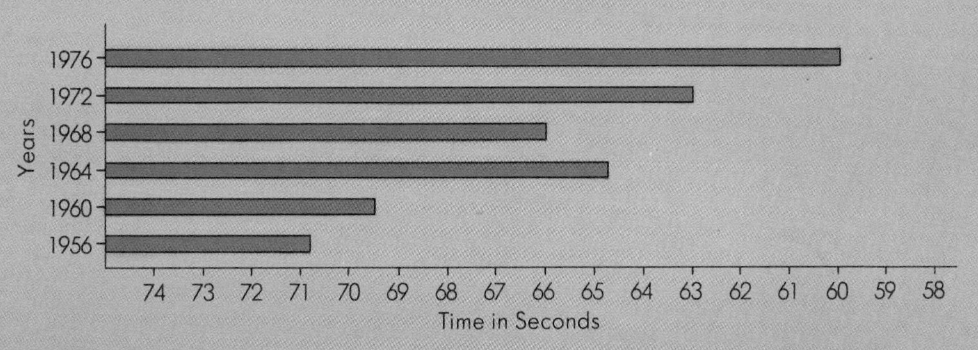

17. Did the winning times get faster or slower from 1956 to 1976? faster
18. Which year does not follow the pattern? 1968
19. How many seconds difference is there between the winning time in 1956 and 1976? 11 seconds

If this race is held in 1988, would you expect the winning time to be faster than each of these times?

20. 68 seconds yes
21. 63 seconds yes
22. 59 seconds maybe

CAREER CAREER CAREER CAREER CAREER

ADVERTISING LAYOUT ARTIST

Valerie Edwards is an advertising layout artist. She helps plan advertising campaigns. She works with the research staff. Her job is to present information about products in an attractive way. Simple graphs can be made more appealing with a creative design.

Valerie designed an ad for salad dressing. It shows the growth in sales from 1979 to 1982.

1. What type of graph is this? pictograph
2. Does it show exactly how many bottles were sold each year? no
3. In which year was the increase in growth the greatest? least? 1982; 1981
4. About how many bottles must be sold in 1983 to increase the 1982 sales by $\frac{1}{3}$? 93,000

Look How We've Grown in Just 4 Years

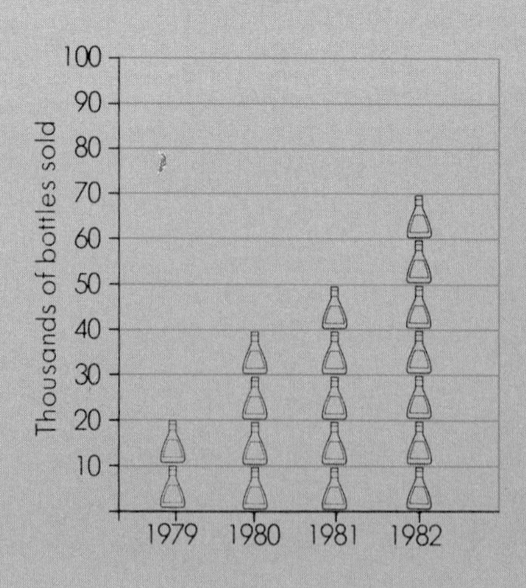

Valerie must be creative in her advertising designs. She needs to design several graphs and labels from the information in the chart. Can you help her out?

5. Design an attractive label to show the information in the table. The label is to be 3 in. by 3 in. Answers will vary.
6. Design a pictograph to advertise the vitamins. Answers will vary.

Vital Vitamins

One DNG capsule provides the adult RDA

Vitamin A – 125%
B_1 – 250%
B_2 – 225%
C – 125%
D – 100%

Valerie knows she can design graphs which give either a positive or negative impression. She designed Graphs A and B using the same data.

Graph A

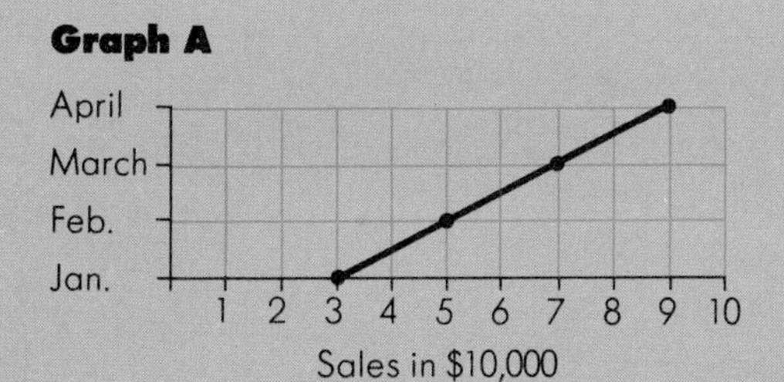

Graph B

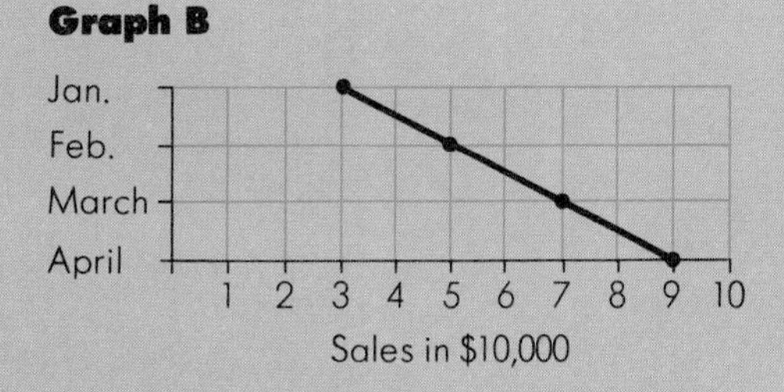

7. Which graph gives a positive impression of the company? Graph A
8. Use the information in Graph A. Draw a line graph which would give a more positive impression than Graph A. See answer section for 8, 9, 10.
9. Draw a bar graph to show the information in Graph A.
10. Draw a bar graph to show the information in Graph B.

For her monthly report, Valerie must keep a record of the time she spends each week on various projects.

Week of June 15	Time	12.
TV ads	10 hours	$\frac{1}{4}$
Magazine ads	12 hours	$\frac{3}{10}$
Company promotions	8 hours	$\frac{1}{5}$
Annual reports	5 hours	$\frac{1}{8}$
Client conferences	5 hours	$\frac{1}{8}$

11. How many hours in all did Valerie work during the week of June 15? 40 hours
12. What fractional part of the total time did she spend on each project?
13. What percent of the total time did she spend on each project?
14. Make a circle graph to show the information in the table. See answer section.

13. TV ads: 25%; Magazine ads: 30%; Company promotions: 20%; Annual reports: $12\frac{1}{2}$%; Conferences: $12\frac{1}{2}$%

CHAPTER 13 REVIEW

For extra practice, use Making Practice Fun 81 with this lesson.

1. Graph the ordered pair (5,4) on a grid. p. 368 See answer section.

2. Find an ordered pair for the point A on the grid. p. 368 (4,1)

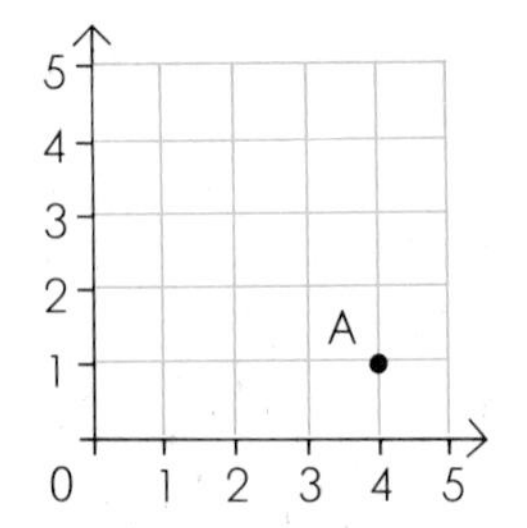

3. Make a line graph for the data in the table. p. 370 See answer section.

Distance (in miles)	Cost of 3 Min Evening Direct-Dial Phone Call
20	37¢
30	44¢
40	50¢
50	59¢

4. Make a bar graph for the data in the table. p. 372 See answer section.

Baseball Player	Lifetime Batting Average
Babe Ruth	0.342
Roger Maris	0.263
Rogers Hornsby	0.358

5. Find the angle for each percent and make a circle graph. p. 374 See answer section.

Sales	Percent
Bats	18%
Balls	21%
Rackets	13%
Shoes	14%
Other	34%

Find these statistics. pp. 378–381

6. the mean: 8, 10, 11, 7, 6, 5 7.83

7. the median: 8, 10, 11, 7, 6, 5 7.5

8. the mode: 24, 23, 27, 25, 27, 31, 27 27

Application pp. 376, 382

9. On vacation Jason spent $ 114 for plane fare, $ 53.75 for hotel rooms, $ 37.80 for food, and $ 18.27 for souvenirs. How much did he spend in all on his trip? $223.82

10. A basketball team won 23 out of 25 games. Find the percentage of games won. 0.920 or 92%

CHAPTER 13 TEST

1. Graph the ordered pair (1,4) on a grid. See answer section.

2. Find an ordered pair for the point A on the grid. (4,3)

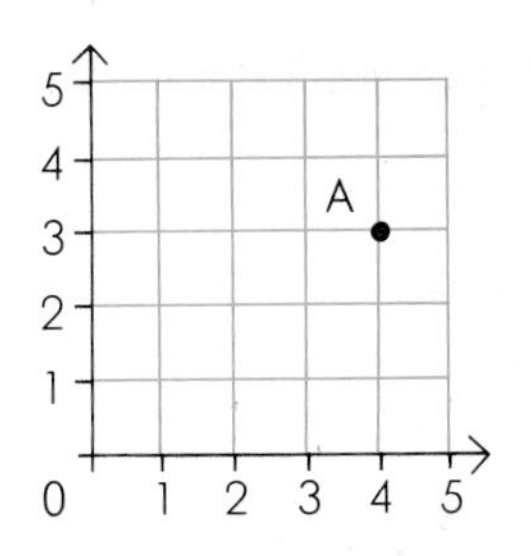

3. Make a line graph for the data in the table. See answer section.

Distance (in miles)	Cost of 3 Min Weekday Direct-Dial Phone Call
20	55¢
30	67¢
40	77¢
50	89¢

4. Make a bar graph for the data in the table. See answer section.

Player	Earnings in Women's Billiards Championship
Jean Balukas	$3000
Mieko Harada	$1500
Gerry Titcomb	$800
Becky Brown	$800
Palmer Byrd	$600

5. Find the angle for each percent and make a circle graph. See answer section.

Batters	Percent
Right-Handed	78%
Left-Handed	20%
Both	2%

Find these statistics.

6. the mean: 88, 86, 76, 20, 100, 20 65
7. the median: 88, 86, 76, 20, 100, 20 81
8. the mode: 88, 86, 76, 20, 100, 20 20

Application

9. Marguerite went to a convention and spent $ 89 for transportation, $ 3.50 for tips, $ 63.47 for a hotel room, and $ 22.75 for food. How much did she spend in all on her trip? $178.72

10. A football team won 12 out of 16 games. Find the percentage of games won. 0.750 or 75%

MORE PRACTICE

Graph each ordered pair on a grid. Use after page 369. See answer section.

1. (2,1) **2.** (3,8) **3.** (3,5)

4. (7,4) **5.** (0,5) **6.** (0,9)

7. (4,0) **8.** (0,1) **9.** (4,4)

10. (5,0) **11.** (1,9) **12.** (9,1)

Find an ordered pair for each point on the grid. Use after page 369.

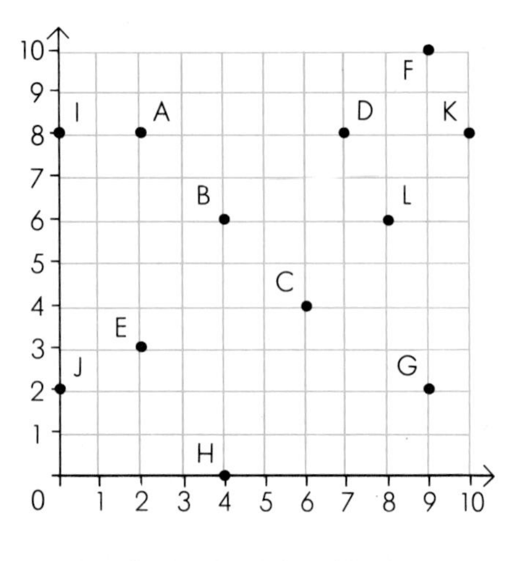

13. A (2,8) **14.** B (4,6) **15.** C (6,4)

16. D (7,8) **17.** E (2,3) **18.** F (9,10)

19. G (9,2) **20.** H (4,0) **21.** I (0,8)

22. J (0,2) **23.** K (10,8) **24.** L (8,6)

Make a line graph for the data in each table. Use after page 371. See answer section.

25.

Year	Demand for Natural Gas (in quadrillion BTU)
1950	19
1960	21
1970	22
1980	27

26.

Olympic Runner	Time in 100 Meter Dash (in seconds)
Betty Cuthbert	11.5
Wilma Rudolph	11.0
Wyomia Tyus	11.4
Renate Stecher	11.07

Make a bar graph for the data in the table. Use after page 373. See answer section.

27.

Cricket Chirps per Minute	Temperature (F°)
60	55°
76	59°
88	62°
100	65°

Make a table and a bar graph for the following data. Use after page 373. See answer section.

1. The population of the U.S. (in millions) for a given year:
1976, 216; 1977, 219; 1978, 222; 1979, 226; 1980, 229.

Find the angle for each percent and make a circle graph. Use after page 375. See answer section.

2.

Investment	Percent
Savings Account	48%
Stock	12%
Mutual Funds	16%
Retirement Fund	24%

3.

Family Expenses	Percent
Food	22%
Auto	18%
Clothing	11%
Housing	25%
Recreation	10%
Other	14%

Find the mean. Use after page 379.

4. 18, 17, 22, 48, 50 31

5. 102, 103, 110, 98, 120, 100 105.5

6. $ 1.79, $ 1.89, $ 1.99, $ 1.95 $1.91

7. 32°, 27°, 45°, 50°, 10°, 20°, 12° 28°

8. 87, 63, 72, 80 75.5

9. 59, 95, 83, 74, 69 76

Find the median. Use after page 381.

10. 18, 17, 48, 22, 50 22

11. 102, 103, 110, 98, 120, 100, 110 103

12. 102, 103, 110, 98, 120, 100 102.5

13. 78, 77, 79, 80 78.5

14. 78, 90, 62, 73, 89 78

15. 65, 45, 81, 57, 77, 63 64

Find the mode. Use after page 381.

16. 17, 16, 17, 19, 20, 14 17

17. 23, 44, 23, 45, 44, 22, 28 23,44

18. 96, 97, 95, 95, 96, 99, 100 95,96

19. 10, 8, 11, 8, 8, 7, 6, 8 8

20. 82, 59, 73, 59, 91 59

21. 79, 74, 62, 74, 88, 82, 62 62,74

CALCULATOR CORNER

TUG OF WAR

Players two
Materials two calculators, paper and pencils

More advanced students might be allowed to add or subtract decimals which would result in a longer game.

Sample Try this with your calculators.

Jay starts with a two-digit number, such as 15, on the display. Gina starts with a four-digit number, such as 1000, on the display.

Jay must always add a whole number to the number that shows on his calculator. His answer must always be larger than his preceding number.

Gina must always subtract a whole number from the number that shows on her calculator. Her answer must always be smaller than her preceding number.

If Jay allows his number to become equal to or greater than Gina's number, he loses.

If Gina allows her number to become less than or equal to Jay's number, she loses.

	Jay		Gina
	15		1000
Play #1	+ 10	**Play #2**	− 1
	25		999
Play #3	+ 500	**Play #4**	− 10
	525		989
Play #5	+ 200	**Play #6**	− 9
	725		980
Play #7	+ 100	**Play #8**	− 80
	825		900
Play #9	+ 74	**Play #10**	− 1
	899		899

Gina loses, since her number is equal to Jay's.

Play the game on your own.

CHAPTER 14 PRE-TEST

Give a positive or negative number for each situation. p. 394

1. a loss of \$ 148 −\$148 **2.** 14,254 ft above sea level 14,254 **3.** 4 hours ago −4

Compare. Use $<$ or $>$. p. 394

4. -3 and 7 $<$ **5.** -8 and -7 $<$ **6.** 0 and -2 $>$

Add. p. 398

7. $5 + (-3)$ 2 **8.** $-9 + 6$ −3 **9.** $-7 + (-4)$ −11

10. $-19 + (-16)$ −35 **11.** $24 + (-11)$ 13 **12.** $-18 + 10$ −8

Subtract. p. 404

13. $2 - 7$ −5 **14.** $-3 - 6$ −9 **15.** $-4 - (-5)$ 1

16. $8 - (-8)$ 16 **17.** $7 - 10$ −3 **18.** $-3 - (-9)$ 6

Multiply. p. 406

19. -4×5 −20 **20.** $-8 \times (-6)$ 48 **21.** $10 \times (-9)$ −90

Divide. p. 408

22. $54 \div (-6)$ −9 **23.** $-63 \div 7$ −9 **24.** $-75 \div (-25)$ 3

Application p. 410

25. Was it colder in Ft. Wayne or Indianapolis? Ft. Wayne

26. What was the difference between the high and low temperatures in each city? Indianapolis: 22°; Ft. Wayne: 25°

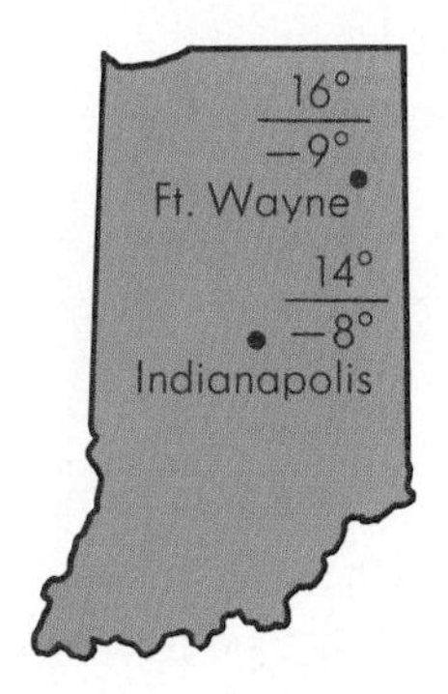

14 POSITIVE AND NEGATIVE NUMBERS

GETTING STARTED

THE GREATEST GREEN

WHAT'S NEEDED 20 squares of paper or cardboard about this size

Write the digits 0 through 9 in green on 10 squares, one digit on each square. Write the digits 0 through 9 in blue on 10 squares, one digit on each square. Put them in a box and mix well.

THE RULES

1. On a sheet of paper each player makes a scorecard. A leader draws two digits from the box. If both numbers are the same color, the players add. The sum is labeled the same color as the digits. If one number of each color is drawn, the players subtract. The difference is labeled the color of the larger number.
2. The leader continues to draw digits, one at a time. Each number is added or subtracted from the previous total, depending on the colors of the digits.
3. The game continues until five digits have been drawn. A player may stop after any number. The goal is to get the highest green number.
4. Play three games and find the total score. The player with the highest green total wins.

SAMPLE

Game 1

5 Green
9 Green
14 Green
7 Blue
7 Green
9 Blue
2 Blue
3 Blue
5 Blue

Scorecard

1.	5 Blue
2.	9 Green
Subtotal	4 Green
3.	5 Green
Total	9 Green

Objectives 136 and 137

POSITIVE AND NEGATIVE NUMBERS

We use positive and negative numbers to describe many situations.

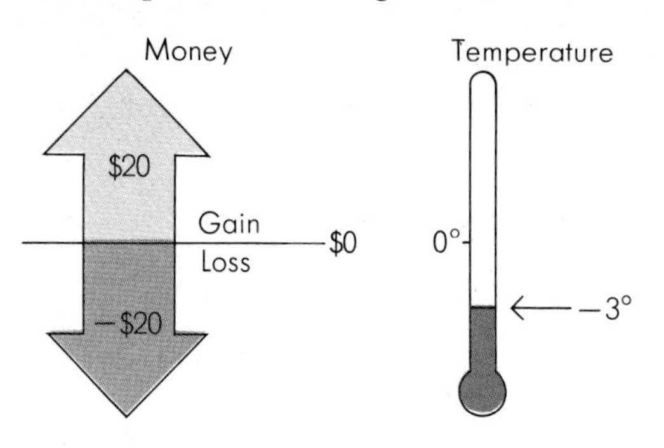

After students learn that positive and negative numbers are used to record *changes,* encourage them to find positive and negative numbers in their daily activities.

A loss of \$ 20 is −\$ 20. 3° below zero is −3°.
A gain of \$ 20 is \$ 20. 70° above zero is 70°.

EXAMPLES. Give a positive or negative number for each situation.

1. 280 ft below sea level −280
2. 4 hours from now 4
3. 2 hours before departure −2

TRY THIS

Give a positive or negative number for each situation.

1. a loss of 8 yards −8 **2.** 6 hours ago −6
3. a gain of 6 pounds 6 **4.** 10 seconds before liftoff −10

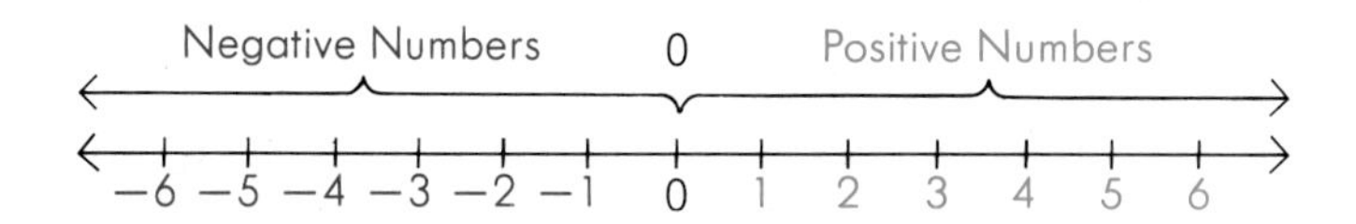

One number is less than another if it is to the left of the other on the number line. One number is greater than another if it is to the right of the other on the number line.

EXAMPLES. Compare. Use $<$ or $>$.

4. −3 and 4 $-3 < 4$ *−3 is to the left of 4*
5. −2 and −1 $-2 < -1$ *−2 is to the left of −1*
6. −4 and −6 $-4 > -6$ *−4 is to the right of −6*

TRY THIS

Compare. Use $>$ or $<$.

5. -5 and 2 $<$ **6.** 0 and -3 $>$ **7.** -5 and -2 $<$

8. 0 and 3 $<$ **9.** 5 and -5 $>$ **10.** -3 and 3 $<$

PRACTICE

Give a positive or negative number for each situation.

1. a loss of $65 −$65 **2.** a loss of $74 −$74 **3.** a loss of $51 −$51

4. a gain of $34 $34 **5.** a gain of $17 $17 **6.** a gain of $23 $23

7. 3 hours ago −3 **8.** 4 hours ago −4 **9.** 6 hours ago −6

10. 3 hours from now 3 **11.** 5 hours from now 5 **12.** 8 hours from now 8

13. gain 5 lbs 5 **14.** lose 4 lbs −4 **15.** lose 8 lbs −8

16. 230 ft below sea level −230 **17.** 240 ft above sea level 240 **18.** 30,000 ft above sea level 30,000

19. 15° below zero −15° **20.** 17° below zero −17° **21.** 1° below zero −1°

22. 10 seconds before rocket liftoff −10 **23.** 10 seconds after rocket liftoff 10 **24.** 30 minutes before departure −30

Compare. Use $>$ or $<$.

25. -2 and 4 $<$ **26.** -1 and 5 $<$ **27.** -4 and 8 $<$

28. 4 and -2 $>$ **29.** 5 and -1 $>$ **30.** 8 and -4 $>$

31. -4 and -5 $>$ **32.** -6 and -8 $>$ **33.** -7 and -9 $>$

34. -3 and -2 $<$ **35.** -4 and -1 $<$ **36.** -9 and -7 $<$

37. 5 and 0 $>$ **38.** -9 and 0 $<$ **39.** -9 and 5 $<$

40. -6 and 6 $<$ **41.** -7 and 7 $<$ **42.** -8 and -5 $<$

43. -5 and -11 $>$ **44.** -3 and -4 $>$ **45.** -6 and -5 $<$

More Practice, page 416

Objective 138

ADDING ON A NUMBER LINE

Add, using a number line. $2 + (-5)$

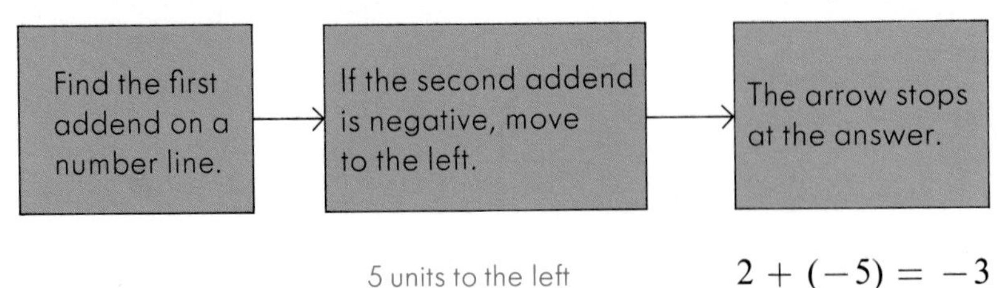

$2 + (-5) = -3$

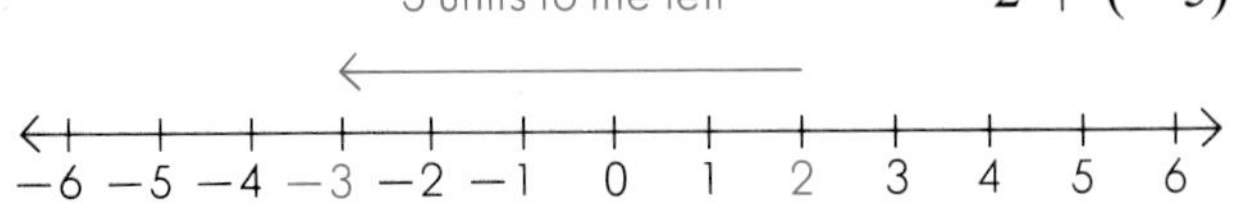

EXAMPLE 1. Add, using a number line. $-2 + (-4)$

$-2 + (-4) = -6$ *The sum of two negative numbers is always negative.*

4 units to the left

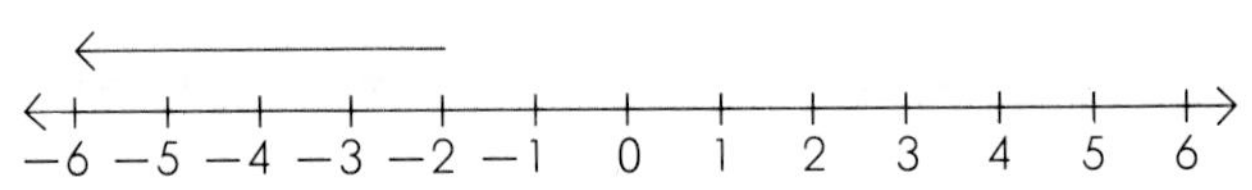

TRY THIS

Add, using a number line.

1. $3 + (-5)$ -2
2. $-1 + (-3)$ -4
3. $4 + (-4)$ 0

Add, using a number line. $-6 + 4$

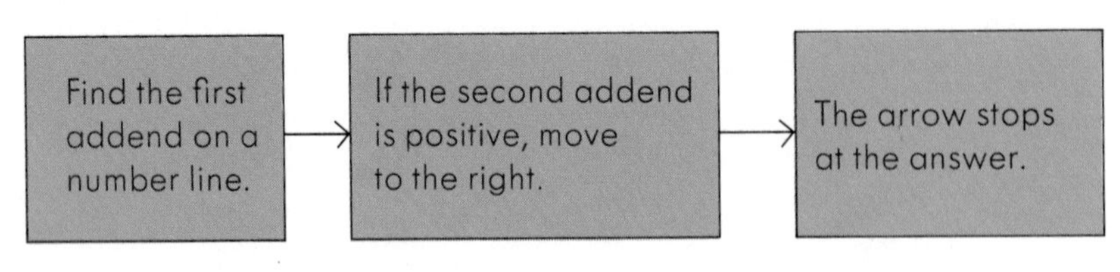

4 units to the right

$-6 + 4 = -2$

−6 −5 −4 −3 −2 −1 0 1 2 3 4 5 6

EXAMPLE 2. Add, using a number line. $-3 + 3$

3 units to the right $-3 + 3 = 0$

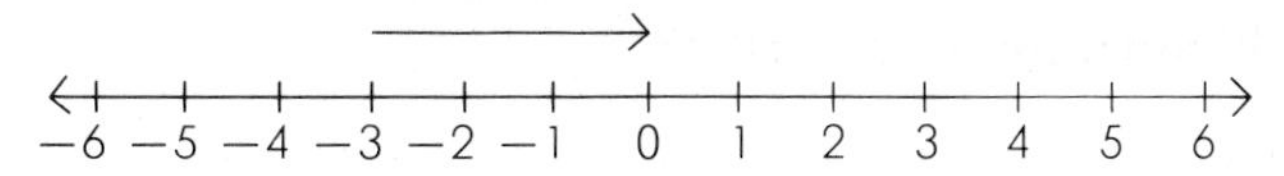

TRY THIS

Add, using a number line.

4. $-2 + 2$ 0 **5.** $-3 + 5$ 2 **6.** $-6 + 2$ −4

PRACTICE

Add, using a number line.

1. $8 + (-5)$ 3 **2.** $7 + (-5)$ 2 **3.** $9 + (-3)$ 6 **4.** $6 + (-5)$ 1

5. $7 + (-8)$ −1 **6.** $8 + (-10)$ −2 **7.** $6 + (-8)$ −2 **8.** $4 + (-7)$ −3

9. $-7 + (-5)$ −12 **10.** $-4 + (-2)$ −6 **11.** $-8 + (-7)$ −15 **12.** $-6 + (-3)$ −9

13. $-2 + 6$ 4 **14.** $-4 + 7$ 3 **15.** $-6 + 9$ 3 **16.** $-5 + 6$ 1

17. $-4 + 4$ 0 **18.** $-2 + 2$ 0 **19.** $-10 + 10$ 0 **20.** $-9 + 9$ 0

21. $5 + (-5)$ 0 **22.** $6 + (-6)$ 0 **23.** $8 + (-8)$ 0 **24.** $7 + (-7)$ 0

PROBLEM CORNER

1. The temperature one day was 5°C and it dropped 8°. What was the new temperature? −3°C

2. The temperature one day was −3°F and it rose 12°. What was the new temperature? 9°F

3. The temperature one day was −15°C and it dropped 9°. What was the new temperature? −24°C

Objective 139

ADDING

Students may feel uneasy about subtracting to find an addition answer. On page 404 they will add to find subtraction answers. The meanings of the operations expand as the math becomes more sophisticated.

The sum of two positive numbers is positive.

The sum of two negative numbers is negative.

Add. $-4 + (-5)$

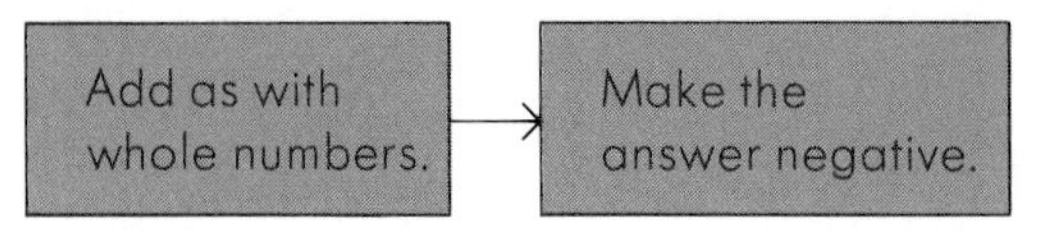

$4 + 5 = 9$ $\quad$ $-4 + (-5) = -9$

EXAMPLES. Add.

1. $-7 + (-8) = -15$ $\quad$ **2.** $-12 + (-14) = -26$

TRY THIS Add.

1. $-3 + (-8)$ -11 $\quad$ **2.** $-11 + (-18)$ -29

The sum of a positive number and a negative number has the same sign as the addend further from 0.

Add. $-14 + 2$

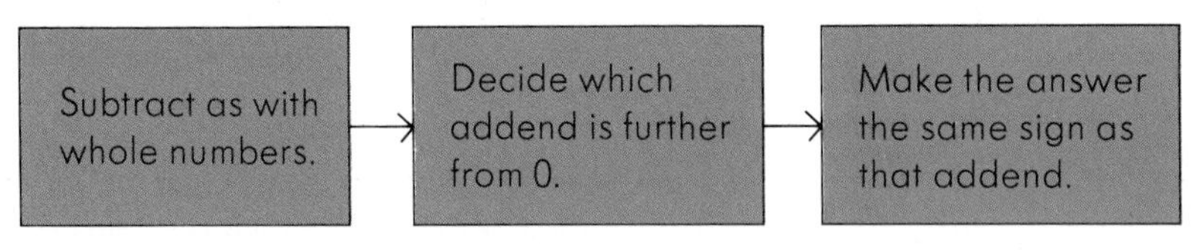

$14 - 2 = 12$ $\quad$ $-14 + 2$, -14 is further from 0. $\quad$ $-14 + 2 = -12$

EXAMPLES. Add.

3. $-5 + 12 = 7$ $\quad$ **4.** $-9 + 23 = 14$

5. $45 + (-18) = 27$ $\quad$ **6.** $-27 + 15 = -12$

TRY THIS Add.

3. $-5+8$ 3 **4.** $-9+4$ −5 **5.** $9+(-2)$ 7

PRACTICE

For extra practice, use Making Practice Fun 82 with this lesson.
Use Quiz 28 after this Practice.

Add.

1. $-5+6$ 1 **2.** $-3+9$ 6 **3.** $-5+7$ 2

4. $-8+7$ −1 **5.** $-10+8$ −2 **6.** $-8+2$ −6

7. $-3+(-6)$ −9 **8.** $-7+(-5)$ −12 **9.** $-2+(-4)$ −6

10. $6+(-2)$ 4 **11.** $9+(-7)$ 2 **12.** $7+(-3)$ 4

13. $18+(-7)$ 11 **14.** $29+(-8)$ 21 **15.** $17+(-12)$ 5

16. $-19+(-14)$ −33 **17.** $-18+(-17)$ −35 **18.** $-26+(-17)$ −43

19. $-32+17$ −15 **20.** $18+(-45)$ −27 **21.** $29+(-41)$ −12

22. $63+(-18)$ 45 **23.** $78+(-29)$ 49 **24.** $100+(-42)$ 58

25. $-65+47$ −18 **26.** $-36+19$ −17 **27.** $-28+(-53)$ −81

28. $-57+(-72)$ −129 **29.** $-103+(-75)$ −178 **30.** $-96+84$ −12

31. $-35+250$ 215 **32.** $-43+230$ 187 **33.** $-73+(-104)$ −177

PROBLEM CORNER

Use positive and negative numbers to solve each problem.

1. Sal spent $ 15 for a shirt and $ 7 for a record. Find how much he spent in all. −$22 (He spent $22.)

2. Lorraine owed $ 28 for a tennis racquet. She made a $ 15 payment. Find how much she still owes. −$13 (She owes $13.)

Objective 191

APPLICATION Students should understand the advantages of organizing information in charts and deriving information from charts.

READING CHARTS

Wind speed does not affect the actual temperature, but it can make a person feel colder. The wind chill temperature is what the temperature would have to be without wind to give the same chilling effect.

Wind Chill Temperature in °F

Wind Speed	Actual Thermometer Reading in °F								
Calm	50	40	30	20	10	0	−10	−20	−30
5 MPH	48	37	27	16	6	−5	−15	−26	−36
10 MPH	40	28	16	4	−9	−21	−33	−46	−58
15 MPH	36	22	9	−5	−18	−36	−45	−58	−72
20 MPH	32	18	4	−10	−25	−39	−53	−67	−82
25 MPH	30	16	0	−15	−29	−44	−59	−74	−88
30 MPH	28	13	−2	−18	−33	−48	−63	−79	−94
35 MPH	27	11	−4	−20	−35	−49	−67	−82	−98
40 MPH	26	10	−6	−21	−37	−53	−69	−85	−100

(Wind speeds greater than 40 mph have little additional effect.)

What is the wind chill temperature when the actual temperature is 20° F and the wind speed is 30 mph?

Locate 30 mph on the left side of the chart. Go across until you are under 20° F. The wind chill temperature is −18° F.

Find the wind chill temperatures for the following actual temperatures and wind speeds.

1. 20°; 35 mph −20°F
2. −10°; 20 mph −53°F
3. 0°; 25 mph −44°F
4. 10°; 10 mph −9°F
5. −30°; 40 mph −100°F
6. 30°; 25 mph 0°F
7. The actual temperature is 10° F and the wind speed is 20 mph. The actual temperature drops 20°. How much does the wind chill temperature drop? 28°F
8. The actual temperature is 20° F. The wind speed changes from 20 mph to 40 mph. How much does the wind chill temperature drop? 11°F

	Chicago, Ill	Denver, Colo.	Houston, Tex.	Las Vegas, Nev.	Mexico, D.F.	Miami, Fla.	Montreal, Que.	Nashville, Tenn.	New York, N.Y.	San Francisco, Calif.	Seattle, Wash.	Toronto, Ont.	Vancouver, B.C.	Winnipeg, Man.
Chicago, Ill.		996	1067	1772	2045	1329	828	446	802	2142	2013	465	2268	860
Denver, Colo.	996		1019	777	1746	2037	1815	1156	1771	1235	1307	1479	1453	1039
Houston, Tex.	1067	1019		1417	979	1190	1827	769	1608	1912	2274	1491	2420	1515
Las Vegas, Nev.	1772	777	1417		1834	2521	2591	1801	2548	564	1152	2255	1298	1704
Mexico, D.F.	2045	1746	979	1834		2169	2805	1747	2587	2291	2852	2469	2998	2421
Miami, Fla.	1329	2037	1190	2521	2169		1654	897	1308	3053	3273	1509	3419	2154
Montreal, Que.	828	1815	1827	2591	2805	1654		1074	378	2961	2685	341	2940	1450
Nashville, Tenn.	446	1156	769	1801	1747	897	1074		892	2333	2376	754	2522	1257
New York, N.Y.	802	1771	1608	2548	2587	1308	378	892		2934	2815	477	3070	1662
San Francisco, Calif.	2142	1235	1912	564	2291	3053	2961	2333	2934		808	2625	954	1919
Seattle, Wash.	2013	1307	2274	1152	2852	3273	2685	2376	2815	808		2600	146	1444
Toronto, Ont.	465	1479	1491	2255	2469	1509	341	754	477	2625	2600		2800	1310
Vancouver, B.C.	2268	1453	2420	1298	2998	3419	2940	2522	3070	954	146	2800		1490
Winnipeg, Man.	860	1039	1515	1704	2421	2154	1450	1257	1662	1919	1444	1310	1490	

MILEAGE CHART: CANADIAN, MEXICAN, AND U.S. CITIES

The time needed to drive between cities depends not only on distance, but on traffic conditions, types of roads, and speed limits.

To the nearest hour, find how long it takes to drive from city to city. Assume a maximum speed limit of 55 mph.

9. Miami to Montreal 30 hours
10. Winnipeg to Houston 28 hours
11. Seattle to Mexico 52 hours
12. Chicago to Toronto 8 hours
13. Vancouver to Denver 26 hours
14. New York to San Francisco 53 hours
15. Nashville to Las Vegas 33 hours
16. Mexico to Winnipeg 44 hours
17. Is Houston or Nashville further from Seattle? How much further? Nashville; 1002 miles
18. Is Winnipeg or Montreal closer to Chicago? How much closer? Montreal; 32 miles

Jason and Sheila drove from Vancouver to New York in 6 days.

19. How many miles did they drive? 3070 miles
20. How many miles did they average each day? 512 miles
21. Jason and Sheila used 135 gallons of gasoline. To the nearest mile, how many miles per gallon did they average? 23 miles per gallon
22. If they paid an average of $ 0.95 per gallon, how much did Jason and Sheila spend for gas? $128.25

Objective 140

SUBTRACTING ON A NUMBER LINE

This interpretation of subtraction shows the source of the word *difference* as an answer to a subtraction problem.

Subtract, using a number line. $6 - 8$

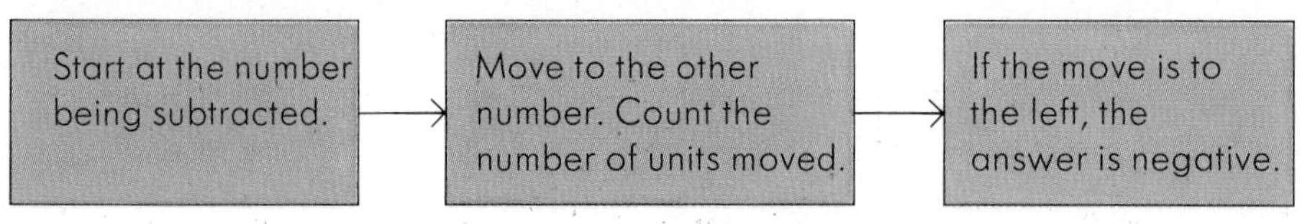

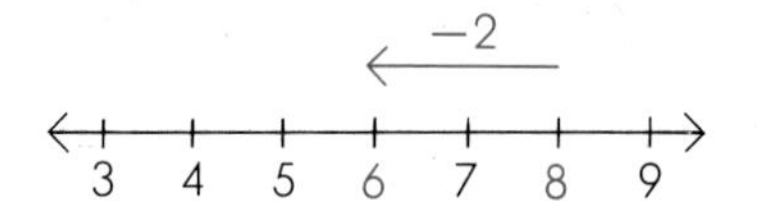

$6 - 8 = -2$

EXAMPLE 1. Subtract, using a number line. $-1 - 5$

$-1 - 5 = -6$

−6

−3 −2 −1 0 1 2 3 4 5

TRY THIS

Subtract, using a number line.

1. $4 - 7$ −3 **2.** $-2 - 6$ −8 **3.** $-6 - 3$ −9

Subtract, using a number line. $-3 - (-7)$

Start at the number being subtracted. → Move to the other number. Count the number of units moved. → If the move is to the right, the answer is positive.

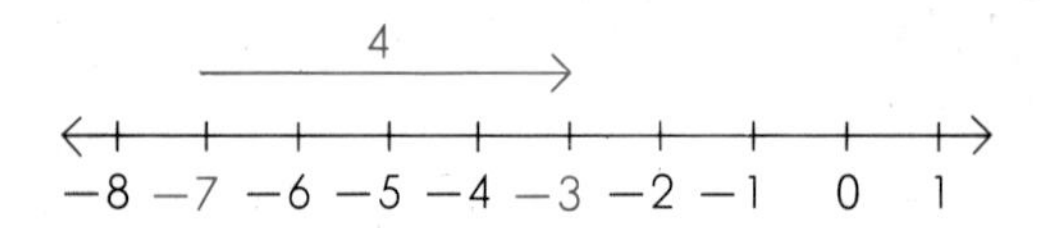

$-3 - (-7) = 4$

EXAMPLE 2. Subtract, using a number line. $4 - (-4)$

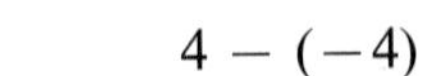

$4 - (-4) = 8$

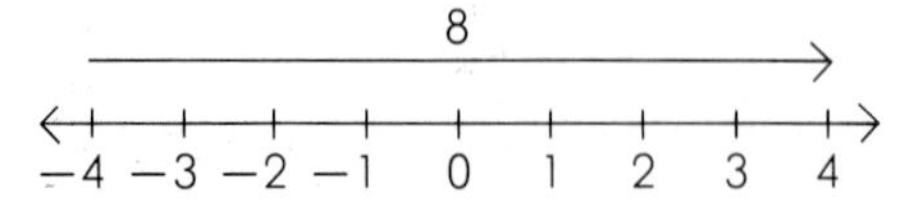

TRY THIS

Subtract, using a number line.

4. $-2 - (-3)$ 1 **5.** $7 - (-3)$ 10 **6.** $5 - (-6)$ 11

PRACTICE

Subtract, using a number line.

1. $2 - 5$ -3 **2.** $1 - 5$ -4 **3.** $3 - 5$ -2 **4.** $5 - 9$ -4

5. $-5 - 4$ -9 **6.** $-3 - 5$ -8 **7.** $-2 - 4$ -6 **8.** $-1 - 3$ -4

9. $-7 - (-2)$ -5 **10.** $-4 - (-9)$ 5 **11.** $-5 - (-3)$ -2 **12.** $-6 - (-4)$ -2

13. $-5 - 5$ -10 **14.** $-2 - 2$ -4 **15.** $-3 - 3$ -6 **16.** $-8 - 8$ -16

17. $8 - (-8)$ 16 **18.** $6 - (-6)$ 12 **19.** $10 - (-10)$ 20 **20.** $9 - (-9)$ 18

21. $8 - (-3)$ 11 **22.** $7 - (-2)$ 9 **23.** $6 - (-7)$ 13 **24.** $5 - (-4)$ 9

25. $0 - 5$ -5 **26.** $0 - 4$ -4 **27.** $0 - 10$ -10 **28.** $0 - 8$ -8

29. $8 - 12$ -4 **30.** $7 - 10$ -3 **31.** $6 - 14$ -8 **32.** $8 - 9$ -1

33. $-8 - 1$ -9 **34.** $-9 - 4$ -13 **35.** $-7 - 6$ -13 **36.** $-9 - 2$ -11

37. $-5 - (-5)$ 0 **38.** $-8 - (-8)$ 0 **39.** $-11 - (-11)$ 0 **40.** $-12 - (-12)$ 0

PROBLEM CORNER

3. The answer depends on the order in which the numbers are subtracted.

1. The temperature one day was $-7°F$ and it dropped 4°. What was the new temperature? $-11°F$

2. The temperature one day was 9°C and it dropped 12°. What was the new temperature? $-3°C$

3. The temperature in St. Louis was $-3°F$. In Minneapolis it was $-7°F$. What was the difference between the temperatures of the two cities?
4°F or $-4°F$

Objective 141

SUBTRACTING

Numbers like 4 and -4 are called opposites. Their sum is 0. Another word for *opposite* is *inverse.*

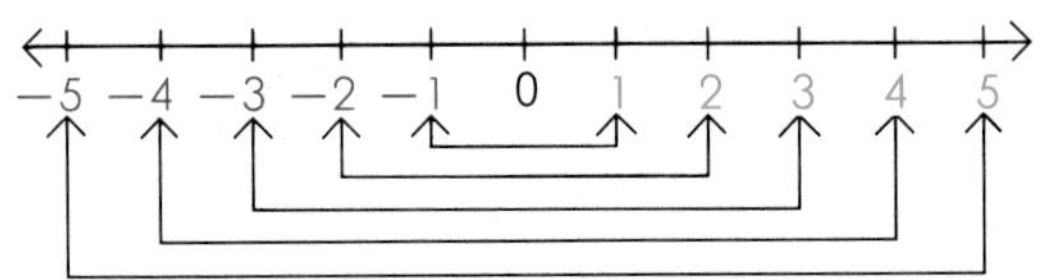

$2 + (-2) = 0$
$-1 + 1 = 0$
$5 + (-5) = 0$

EXAMPLES. Find the opposite.

1. -3 The opposite of -3 is 3.
2. 4 The opposite of 4 is -4
3. 0 The opposite of 0 is 0.

TRY THIS Find the opposite.

1. -5 5 **2.** -6 6 **3.** 6 -6 **4.** 12 -12 **5.** 0 0

Subtract. $3 - 8$

Find the opposite of the number being subtracted. → Then add.

The opposite of 8 is -8.

$$3 - 8 = 3 + (-8) = -5$$

EXAMPLES. Subtract.

4. $-5 - 4 = -5 + (-4) = -9$
5. $2 - 6 = 2 + (-6) = -4$
6. $5 - (-9) = 5 + 9 = 14$
7. $-7 - (-2) = -7 + 2 = -5$

TRY THIS Subtract.

6. $-6 - 3$ -9 **7.** $5 - 9$ -4
8. $-5 - (-9)$ 4 **9.** $-2 - (-2)$ 0

PRACTICE

For extra practice, use Making Practice Fun 83 with this lesson.

1. $3 - 4$ −1　**2.** $2 - 6$ −4　**3.** $7 - 9$ −2　**4.** $1 - 6$ −5

5. $-3 - 5$ −8　**6.** $-5 - 4$ −9　**7.** $-2 - 7$ −9　**8.** $-2 - 3$ −5

9. $-8 - (-3)$ −5　**10.** $-2 - (-7)$ 5　**11.** $-6 - (-4)$ −2　**12.** $-9 - (-1)$ −8

13. $-8 - 8$ −16　**14.** $-9 - 9$ −18　**15.** $-12 - 12$ −24　**16.** $-13 - 13$ −26

17. $9 - (-9)$ 18　**18.** $7 - (-7)$ 14　**19.** $10 - (-10)$ 20　**20.** $16 - (-16)$ 32

21. $6 - 9$ −3　**22.** $9 - 16$ −7　**23.** $7 - 10$ −3　**24.** $3 - 8$ −5

25. $0 - 12$ −12　**26.** $0 - 8$ −8　**27.** $0 - 14$ −14　**28.** $0 - 9$ −9

29. $7 - (-2)$ 9　**30.** $8 - (-3)$ 11　**31.** $5 - (-7)$ 12　**32.** $9 - (-4)$ 13

33. $-7 - 6$ −13　**34.** $-9 - 2$ −11　**35.** $-8 - 5$ −13　**36.** $-2 - 8$ −10

37. $-9 - (-9)$ 0　**38.** $-5 - (-5)$ 0　**39.** $-7 - (-7)$ 0

40. $28 - 53$ −25　**41.** $65 - 79$ −14　**42.** $15 - (-84)$ 99

PROBLEM CORNER

Use positive and negative numbers to solve each problem.

1. Rocco has $75 in his savings account and owes $88 on his credit card account. What is the difference between the balances of these accounts? −$13 or $13

2. Mae owes $34 on a layaway account. She also owes $17 on her gas credit card. What is the difference between the amounts Mae owes? $17 or −$17 The answer depends on the order in which the numbers are subtracted.

Objective 142

MULTIPLYING

To provide some motivation for multiplying positive and negative numbers, do the following. Multiply the integers 4 through −3 by 5. The integers decrease by 1 each time and the answers decrease by 5.

The product of two positive numbers is positive.

The product of two negative numbers is positive.

To see how to multiply a negative and a negative, multiply the integers 4 through −3 by −5. The integers decrease by 1 each time and the answers increase by 5.

Multiply. $-6 \times (-9)$

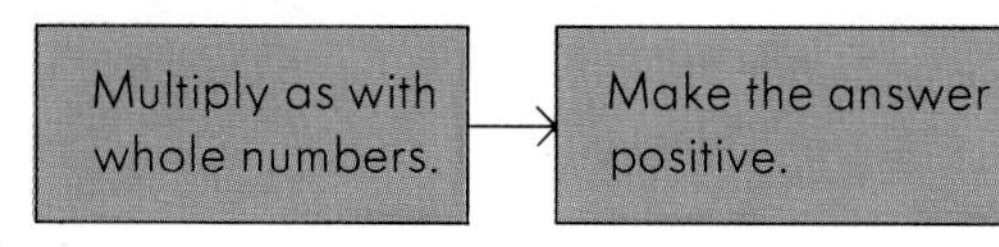

$6 \times 9 = 54$ $\qquad$ $-6 \times (-9) = 54$

EXAMPLES. Multiply.

1. $-4 \times (-5) = 20$ $\qquad$ **2.** $-7 \times (-11) = 77$
3. $-18 \times (-6) = 108$

TRY THIS Multiply.

1. $-3 \times (-8)$ 24 $\qquad$ **2.** $-9 \times (-14)$ 126 $\qquad$ **3.** $-12 \times (-4)$ 48

The product of a positive and a negative number is negative.

Multiply. -4×5

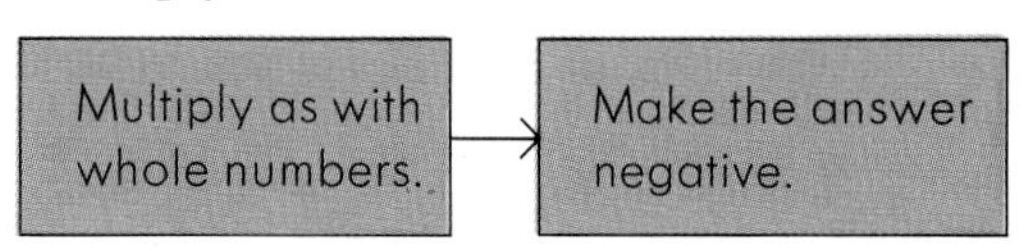

$4 \times 5 = 20$ $\qquad$ $-4 \times 5 = -20$

EXAMPLES. Multiply.

4. $-3 \times 5 = -15$ $\qquad$ **5.** $7 \times (-8) = -56$
6. $7 \times (-10) = -70$

TRY THIS Multiply.

4. -6×7 −42 $\qquad$ **5.** -3×6 −18 $\qquad$ **6.** $20 \times (-5)$ −100

PRACTICE

Multiply.

1. $-8 \times (-2)$ 16 **2.** $-2 \times (-5)$ 10 **3.** $-7 \times (-6)$ 42 **4.** $-9 \times (-2)$ 18

5. $-9 \times (-5)$ 45 **6.** $-8 \times (-3)$ 24 **7.** $-9 \times (-8)$ 72 **8.** $-10 \times (-3)$ 30

9. $-6 \times (-15)$ 90 **10.** $-8 \times (-22)$ 176 **11.** $-25 \times (-8)$ 200 **12.** $-35 \times (-4)$ 140

13. $8 \times (-3)$ −24 **14.** $9 \times (-5)$ −45 **15.** $8 \times (-9)$ −72 **16.** $3 \times (-10)$ −30

17. -2×5 −10 **18.** -8×2 −16 **19.** -7×6 −42 **20.** -9×2 −18

21. -8×5 −40 **22.** 9×-7 −63 **23.** -10×4 −40 **24.** $12 \times (-3)$ −36

25. $15 \times (-8)$ −120 **26.** -10×12 −120 **27.** $-25 \times (-40)$ 1000

28. -200×5 −1000 **29.** $125 \times (-8)$ −1000 **30.** $-420 \times (-9)$ 3780

31. -42×8 −336 **32.** $56 \times (-9)$ −504 **33.** -86×43 −3698

34. $-95 \times (-74)$ 7030 **35.** $-32 \times (-41)$ 1312 **36.** -63×27 −1701

37. $23 \times (-35)$ −805 **38.** $56 \times (-15)$ −840 **39.** -24×24 −576

40. $-34 \times (-89)$ 3026 **41.** $-32 \times (-49)$ 1568 **42.** $-58 \times (-125)$ 7250

PROBLEM CORNER

1. Rita spent \$38 each month for food during the first 6 months of the year. Find the total amount spent, using positive and negative numbers. −\$228 (She spent \$228.) −\$87 (He spent \$87.)

2. Victor spent \$29 each month for gas for 3 months. Use positive and negative numbers to find the total amount spent.

Objective 143

DIVIDING

The reasons behind these rules come from the definition of division. $a \div b$ is that number which when multiplied by b gives a. Thus, $-24 \div (-8) = 3$ because $-24 = 3 \times (-8)$. This also gives students a way to check their work.

The quotient of two positive numbers is positive.

The quotient of two negative numbers is positive.

Divide. $-24 \div (-8)$

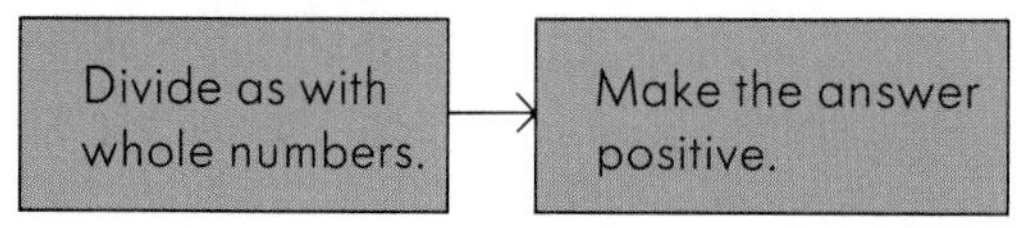

$24 \div 8 = 3$ $\qquad$ $-24 \div (-8) = 3$

EXAMPLES. Divide.

1. $-56 \div (-7) = 8$ $\qquad$ **2.** $-34 \div (-2) = 17$

3. $-42 \div (-6) = 7$

TRY THIS Divide.

1. $-21 \div (-3)$ 7 $\qquad$ **2.** $-28 \div (-2)$ 14 $\qquad$ **3.** $-36 \div (-9)$ 4

The quotient of a negative number and a positive number is negative.

Divide. $-48 \div 8$

Divide as with whole numbers. → Make the answer negative.

$48 \div 8 = 6$ $\qquad$ $-48 \div 8 = -6$

EXAMPLES. Divide.

4. $-56 \div 7 = -8$ $\qquad$ **5.** $28 \div (-2) = -14$

6. $-100 \div 25 = -4$

TRY THIS Divide.

4. $-24 \div 4$ −6 $\qquad$ **5.** $-48 \div 16$ −3 $\qquad$ **6.** $75 \div (-15)$ −5

PRACTICE

For extra practice, use Making Practice Fun 84 with this lesson.
Use Quiz 29 after this Practice.

Divide

1. $-8 \div (-2)$ 4

2. $-10 \div (-5)$ 2

3. $-15 \div (-3)$ 5

4. $-24 \div 6$ −4

5. $-30 \div 5$ −6

6. $-32 \div 4$ −8

7. $30 \div (-5)$ −6

8. $36 \div (-6)$ −6

9. $26 \div (-2)$ −13

10. $-16 \div (-8)$ 2

11. $-22 \div (-2)$ 11

12. $-63 \div (-9)$ 7

13. $-72 \div 9$ −8

14. $-84 \div 7$ −12

15. $-64 \div 2$ −32

16. $49 \div (-7)$ −7

17. $64 \div (-8)$ −8

18. $25 \div (-5)$ −5

19. $-100 \div (-50)$ 2

20. $-200 \div (-40)$ 5

21. $-400 \div (-80)$ 5

22. $-300 \div 75$ −4

23. $-300 \div 25$ −12

24. $200 \div (-8)$ −25

25. $90 \div (-18)$ −5

26. $-98 \div 14$ −7

27. $-84 \div 12$ −7

28. $-92 \div (-23)$ 4

29. $133 \div (-19)$ −7

30. $-128 \div (-4)$ 32

31. $55 \div (-11)$ −5

32. $-72 \div (-12)$ 6

33. $-65 \div (-13)$ 5

34. $432 \div (-18)$ −24

35. $-442 \div 17$ −26

36. $-756 \div (-28)$ 27

PROBLEM CORNER

−$2.20 (Each owes $2.20.)

1. The 10 members of the Bee-See Club sponsored a dance. They took in only $ 16, but their expenses were $ 38. If the club members decide to share the cost equally, how much does each one owe?

2. The *A Capella* Trio owes $ 57 for uniforms. If the 3 singers divide the cost evenly, how much does each one owe? −$19 (Each owes $19.)

Objective 192

APPLICATION

Maps are often available for classroom use from automobile associations, chambers of commerce, real estate offices, and map companies.

READING MAPS

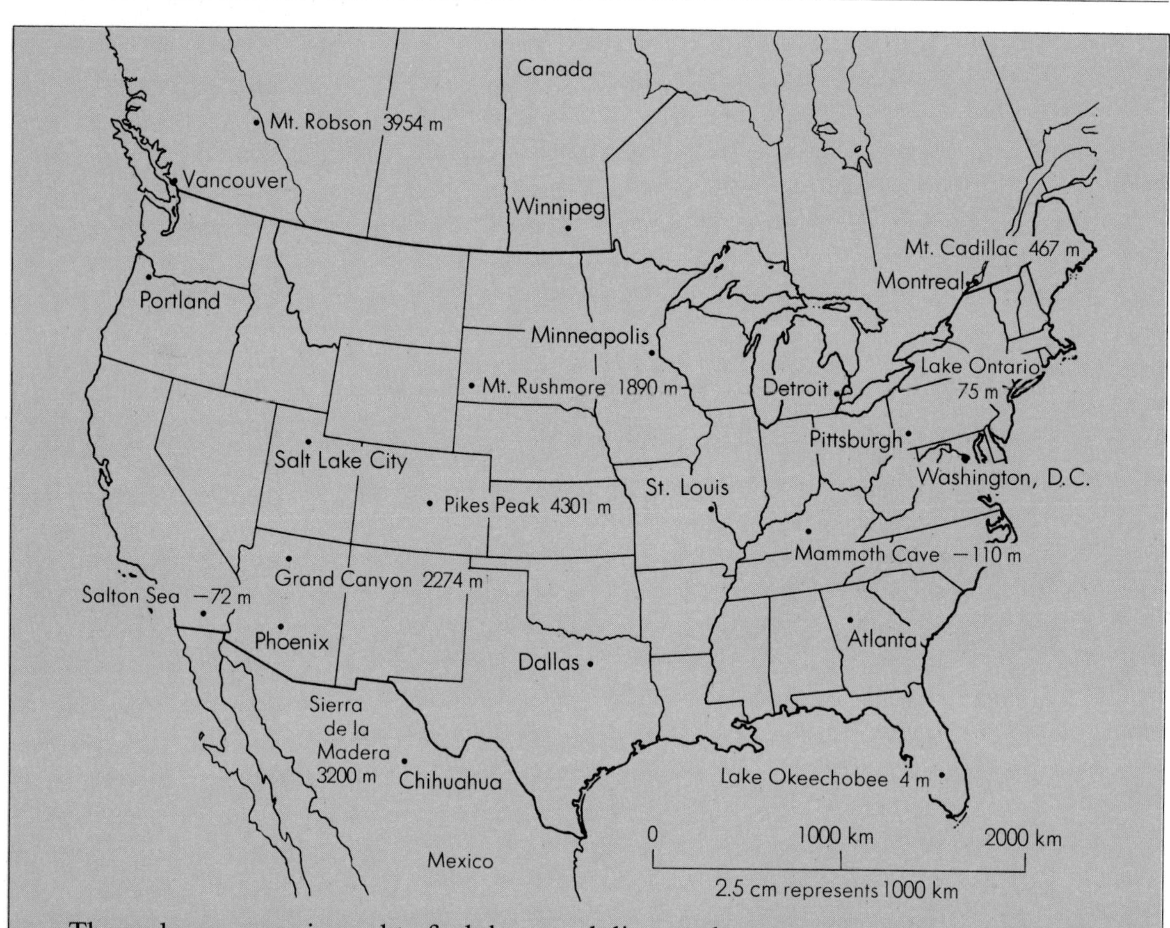

The scale on a map is used to find the actual distance between two places. First, measure the distance on the map. Then compare this measurement to the scale on the map. Actual distances are given. Student approximations will vary.

Approximate the actual distances between the following cities.

1. Detroit and Montreal 922 km
2. St. Louis and Washington, D.C. 1289 km
3. Vancouver and Minneapolis 2811 km
4. Chihuahua and Phoenix 1020 km
5. Pittsburgh and Atlanta 1121 km
6. Salt Lake City and Winnipeg 2209 km
7. Portland and Dallas 3287 km
8. Vancouver and Chihuahua 3411 km

A map may show the elevations of many landmarks and cities. What is the difference in elevation between these landmarks?

9. Lake Ontario, Mt. Cadillac 392 m
10. Mt. Robson, Mt. Rushmore 2064 m
11. Pikes Peak, Grand Canyon 2027m
12. Mammoth Cave, Lake Okeechobee 114 m
13. Sierra de la Madera, Salton Sea 3272 m
14. Mammoth Cave, Salton Sea 38 m
15. Both the highest and lowest points in the continental U.S. are in California. Mt. Whitney has an elevation of 4418 meters. Bad Water, California in Death Valley has an elevation of −86 meters. What is the difference in elevation between these two landmarks? 4504 m
16. The highest and lowest points on the earth's surface are in Asia. Mt. Everest in Nepal has an elevation of 8848 meters. The Dead Sea in Israel and Jordan has an elevation of −397 meters. What is the difference in elevation between these two places? 9245 m

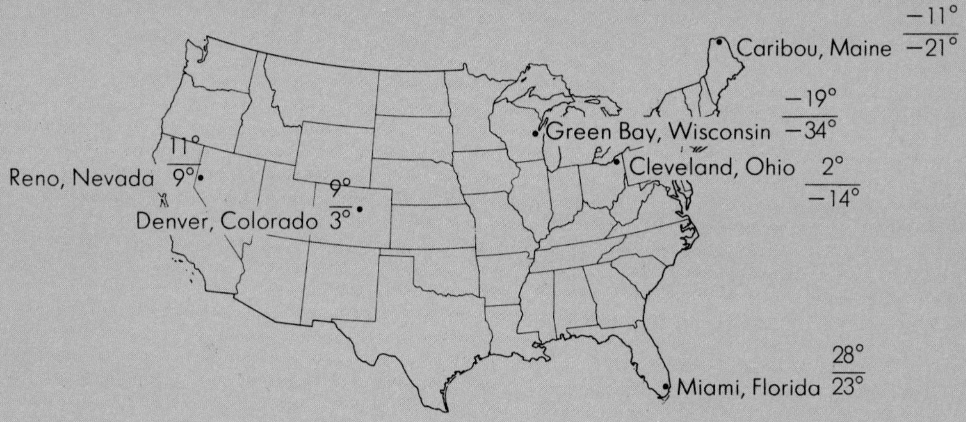

A weather map often gives high and low temperatures of an area. In this weather map each temperature is given in degrees Celcius.

29° ⟵ High temperature
16° ⟵ Low temperature

For each city, what is the difference between the high and low temperatures?

17. Green Bay, Wisconsin 15°
18. Reno, Nevada 2°
19. Cleveland, Ohio 16°
20. How much warmer was it in Reno, Nevada than in Denver, Colorado? 2°
21. Was it colder in Caribou, Maine or Green Bay, Wisconsin? How much colder? Green Bay, Wisconsin; 13°
22. How much colder was it in Cleveland, Ohio than in Miami, Florida? 37°
23. In which cities did the temperature not rise above 0° C? Green Bay, Wisconsin and Caribou, Maine

WELL DRILLER

Oil wells are drilled through land and ocean. Neal Morris works for World Petroleum Corporation. He builds oil-drilling derricks on land and from platforms on the ocean.

How far are these layers of material from the top of the oil rig?

1. 1st layer of rock 2180 ft
2. 2nd layer of rock 4180 ft
3. limestone and methane layer 3180 ft
4. petroleum layer 5180 ft

How far apart are these layers?

5. gravel and clay from petroleum 4000 ft
6. 1st rock layer from the 2nd 2000 ft
7. limestone and methane gas from the 2nd rock layer 1000 ft

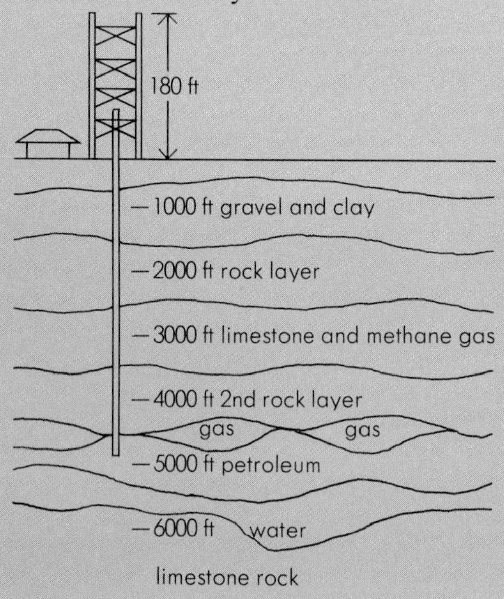

For a better understanding of depth, students could change the number of feet to miles and then name an object or place which is located that distance from the classroom.

Well depth is measured from ground level. It is common to have wells drilled in excess of 15,000 ft.

Neal is working on a well with a target depth of 24,087 ft. The well is in Tulsa, Oklahoma which is 668 ft above sea level.

8. How far below sea level is the target depth? 23,419 ft
9. After one month, the drill reaches 1784 ft. How far does Neal have to drill to get to the target depth? 22,303 ft
10. Neal expects the drilling to take 14 months. How many feet per month is this? 1720.5 ft

The drill bit needs to be cooled, since it heats up from friction and the heat in the ground. Neal cools the drill bit by pumping mud down the drill rod. At a depth of 12,500 ft the well temperature was 130° F. What is the difference between the well temperature and each of the following surface temperatures?

11. −12° F 142° 12. 32° F (Freezing) 98° 13. 50° F 80° 14. 78° F 52°

Neal travels to many states and countries checking on drilling sites. To what elevation was each well drilled?

Location	Well Depth	Elevation of the Field
15. Tierra del Fuego, Argentina 10,500 ft	17,000 ft	6500 ft
16. Prudhoe Bay, Alaska 14,350 ft	15,000 ft	650 ft
17. Galveston, Texas 19,700 ft	20,000 ft	300 ft
18. Ghawar, Saudi Arabia 6270 ft	6920 ft	650 ft
19. Gippsland Shelf, Australia 7800 ft	7200 ft	−600 ft
20. Swan Hills, Alberta, Canada 2100 ft	8100 ft	6000 ft

CHAPTER 14 REVIEW

For extra practice, use Making Practice Fun 85 with this lesson.

Give a positive or negative number for each situation. p. 394

1. 500 ft below sea level. -500 **2.** 5 hours from now 5 **3.** 9° below zero $-9°$

Compare. Use < or >. p. 394

4. -6 and -5 < **5.** 8 and -4 > **6.** -4 and 0 <

Add. p. 398

7. $-3 + (-4)$ -7 **8.** $6 + (-2)$ 4 **9.** $-7 + 1$ -6

10. $19 + (-14)$ 5 **11.** $-23 + 10$ -13 **12.** $25 + (-17)$ 8

Subtract. p. 404

13. $3 - 8$ -5 **14.** $-5 - (-3)$ -2 **15.** $-4 - 5$ -9

16. $5 - (-7)$ 12 **17.** $-4 - 8$ -12 **18.** $-10 - (-12)$ 2

Multiply. p. 406

19. $-6 \times (-7)$ 42 **20.** $8 \times (-3)$ -24 **21.** -5×4 -20

Divide. p. 408

22. $-28 \div (-7)$ 4 **23.** $36 \div (-9)$ -4 **24.** $-80 \div 10$ -8

Application p. 410

25. Was it colder in Juneau or Anchorage? Juneau

26. What was the difference between the high and low temperatures in each city? Anchorage: 39°; Juneau: 40°

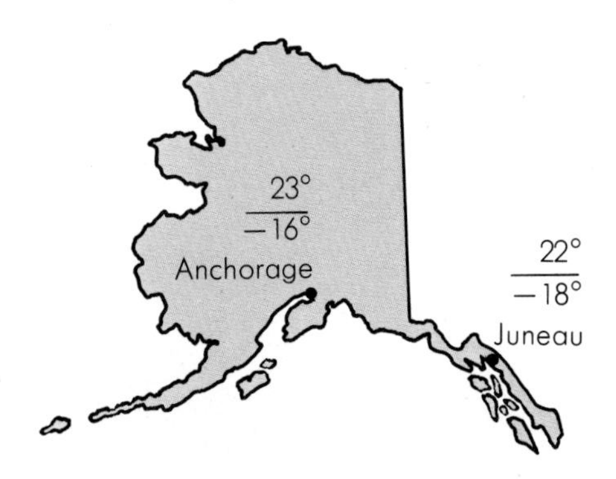

CHAPTER 14 TEST

Give a positive or negative number for each situation.

1. a gain of \$ 678 \$678 **2.** 135 ft below sea level −135 **3.** 7 hours ago −7

Compare. Use $<$ or $>$.

4. -9 and -10 $>$ **5.** -7 and 0 $<$ **6.** 10 and -3 $>$

Add.

7. $-2 + 5$ 3 **8.** $-4 + (-6)$ −10 **9.** $-8 + 3$ −5

10. $-18 + 23$ 5 **11.** $14 + (-18)$ −4 **12.** $-17 + (-21)$ −38

Subtract.

13. $-7 - 5$ −12 **14.** $5 - 6$ −1 **15.** $-2 - (-3)$ 1

16. $5 - (-8)$ 13 **17.** $-3 - 11$ −14 **18.** $-12 - (-8)$ −4

Multiply.

19. $7 \times (-9)$ −63 **20.** $-5 \times (-10)$ 50 **21.** -8×7 −56

Divide.

22. $72 \div (-9)$ −8 **23.** $-64 \div 8$ −8 **24.** $-56 \div (-7)$ 8

Application

25. Was it colder in Syracuse or Buffalo? Buffalo

26. What was the difference between the high and low temperatures in each city? Syracuse: 33°; Buffalo: 39°

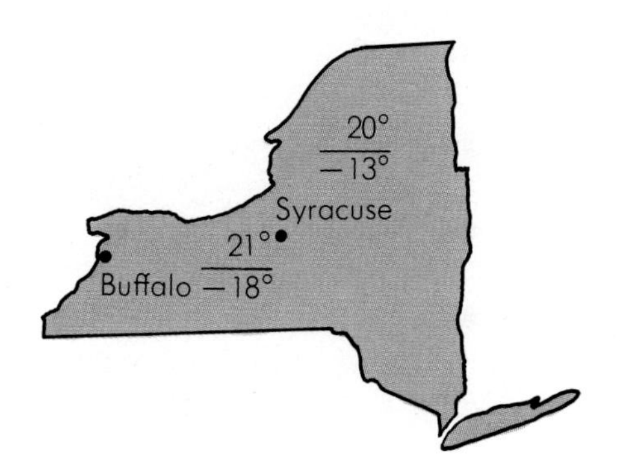

MORE PRACTICE

Give a positive or negative number for each situation. Use after page 395.

1. a loss of \$45 −\$45

2. a gain of \$68 \$68

3. 5 hours ago −5

4. 6 hours from now 6

5. 180 ft below sea level −180

6. 10,500 ft above sea level 10,500

7. 45° above zero 45°

8. 16° below zero −16°

9. 8 seconds before liftoff −8

10. 10 seconds after liftoff 10

11. lose 3 pounds −3

12. gain 7 yards 7

Compare. Use $<$ or $>$. Use after page 395.

13. -3 and 7 $<$

14. 7 and -3 $>$

15. -8 and -10 $>$

16. 0 and -3 $>$

17. 0 and 6 $<$

18. -7 and 0 $<$

19. -4 and 4 $<$

20. 5 and -5 $>$

21. -4 and -6 $>$

22. -10 and -9 $<$

23. -12 and -1 $<$

24. 8 and -3 $>$

25. -35 and -33 $<$

26. 14 and -1 $>$

27. -6 and 16 $<$

28. 0 and -1 $>$

29. -20 and -22 $>$

30. -7 and 7 $<$

Add. Use after page 399.

31. $-8 + 6$ -2

32. $-2 + (-4)$ -6

33. $-2 + 10$ 8

34. $8 + (-8)$ 0

35. $-9 + 2$ -7

36. $-3 + 7$ 4

37. $-6 + 7$ 1

38. $-11 + 9$ -2

39. $-3 + (-5)$ -8

40. $-14 + (-8)$ -22

41. $7 + (-5)$ 2

42. $-20 + (-15)$ -35

43. $100 + (-53)$ 47

44. $-80 + (-64)$ -144

45. $-97 + 34$ -63

46. $-58 + (-22)$ -80

47. $-72 + (-109)$ -181

48. $-47 + (-207)$ -254

Subtract. Use after page 405.

1. $-8 - 6$ −14 **2.** $-2 - (-4)$ 2 **3.** $-2 - 10$ −12

4. $-5 - 9$ −14 **5.** $-3 - 7$ −10 **6.** $-3 - (-9)$ 6

7. $-12 - (-5)$ −7 **8.** $-13 - 7$ −20 **9.** $-6 - (-8)$ 2

10. $7 - (-5)$ 12 **11.** $-20 - (-15)$ −5 **12.** $17 - 35$ −18

13. $1 - 100$ −99 **14.** $10 - (-20)$ 30 **15.** $-1 - 100$ −101

16. $-17 - 19$ −36 **17.** $14 - 75$ −61 **18.** $-23 - (-19)$ −4

Multiply. Use after page 407.

19. $-5 \times (-8)$ 40 **20.** -13×5 −65 **21.** $-6 \times (-8)$ 48

22. $-6 \times (-6)$ 36 **23.** $-5 \times (-5)$ 25 **24.** $6 \times (-22)$ −132

25. -10×14 −140 **26.** $9 \times (-9)$ −81 **27.** $8 \times (-8)$ −64

28. $-120 \times (-14)$ 1680 **29.** $-35 \times (-45)$ 1575 **30.** -44×85 −3740

31. -75×19 −1425 **32.** $46 \times (-18)$ −828 **33.** -25×25 −625

34. -24×6 −144 **35.** $-15 \times (-21)$ 315 **36.** $12 \times (-8)$ −96

37. $-140 \times (-7)$ 980 **38.** -13×195 −2535 **39.** $-39 \times (-56)$ 2184

Divide. Use after page 409.

40. $-48 \div 12$ −4 **41.** $-72 \div (-6)$ 12 **42.** $-54 \div 6$ −9

43. $48 \div (-16)$ −3 **44.** $49 \div (-7)$ −7 **45.** $-64 \div 8$ −8

46. $-98 \div 49$ −2 **47.** $-104 \div 4$ −26 **48.** $-125 \div (-25)$ 5

49. $400 \div (-10)$ −40 **50.** $-300 \div 75$ −4 **51.** $-58 \div (-29)$ 2

52. $-456 \div 12$ −38 **53.** $-232 \div 4$ −58 **54.** $306 \div (-17)$ −18

55. $255 \div (-17)$ −15 **56.** $-196 \div (-7)$ 28 **57.** $-308 \div 28$ −11

CALCULATOR CORNER

MYSTERY NUMBERS

Mystery Number A

If the available calculators have a constant addition mode, then it would be possible to obtain some numbers by pressing (=) over and over rather than using the particular addend over and over.

Clue 1 If you enter 14 on your calculator and press + 10 = over and over, Mystery Number A will appear.

Clue 2 If you start with 0 on your calculator and press + 4 = over and over, Mystery Number A will appear.

List some numbers that Mystery Number A could be.

Clue 3 Mystery Number A is the largest such number less than 1000.

What is Mystery Number A? 984

Mystery Number B

Clue 1 If you start with 0 on your calculator and press + 4 = over and over, Mystery Number B will appear.

Clue 2 If you enter 13 on your calculator and press + 5 = over and over, Mystery Number B will appear.

List some numbers that Mystery Number B could be.

Clue 3 Mystery Number B is the largest such number less than 30,000.

What is Mystery Number B? 29,988

Use your calculator to make up several mystery number problems.

CHAPTER 15 PRE-TEST

Evaluate each expression when $x = -2$ and $y = 3$. p. 422

1. $2x + y$ −1

2. $x - 3y$ −11

3. $-4xy$ 24

Solve. p. 426

4. $x + 8 = 5$ −3

5. $t - 7 = 15$ 22

6. $m + \frac{2}{3} = \frac{3}{4}$ $\frac{1}{12}$

Solve. p. 428

7. $8x = -48$ −6

8. $-7x = -63$ 9

9. $\frac{t}{8} = -9$ −72

Solve. p. 430

10. $5x + 4 = 39$ 7

11. $-2x + 40 = 28$ 6

12. $1.3x - 7 = 0.28$ 5.6

Solve. p. 432

13. Eight plus what number is −48? −56

14. The length of a rectangle is 18 meters and the area is 252 square meters. What is the width? 14 meters

15. A bicycle is on sale for $ 60. This is 80% of the original price. What was the original price? $75

16. An antique clock was sold for $ 450. This was 5 times the original price. What was the original price? $90

Graph the equation. p. 436

17. $y = 2 + x$ See answer section.

Application pp. 424, 434

18. Ben keeps $ 1000 in a savings account for 4 months at 6%. Find the simple interest. $20

19. Lisa still owes $ 800 on her credit card account. Interest is $1\frac{1}{2}$% per month. How much interest will she owe next month? How much will she owe in all? $12; $812

15 EXPRESSIONS AND EQUATIONS

GETTING STARTED

EXPRESS IT

WHAT'S NEEDED 31 squares of paper or cardboard. Write the numbers -15 through 15 on the squares, one number on each square. Put them in a box and mix well.

THE RULES

1. On a sheet of paper each player makes a scorecard. The players each draw a number from the box.
2. The leader writes an expression from the list on the blackboard.
3. Each player evaluates the expression, using his or her number. The players record their answers on their scorecards.
4. Each round the leader writes a new expression on the blackboard.

Play three rounds and total the scores. The player with the highest total wins.

SAMPLE Les has the number -3.

Round 1	Round 2	Round 3
$-2x - 3$	$5 - \frac{x}{3}$	$0.4x + 1$
$(-2)(-3) - 3$	$5 - \frac{-3}{3}$	$0.4(-3) + 1$
$6 - 3 = 3$	$5 - (-1)$	$-1.2 + 1 = -0.2$
	$5 + 1 = 6$	

Scorecard	
1.	3
2.	6
3.	-0.2
Total	8.8

EXPRESSIONS

$-2x - 3$

$5 + 4x$

$3 - 8x$

$7x - 5$

$-6x + 9$

$\frac{x}{4} - 3$

$5 - \frac{x}{3}$

$9x - 1$

$0.3x - 1.5$

$-4x + 10$

$7 - 0.6x$

$4 - \frac{x}{6}$

$5x - 9$

$8 - \frac{x}{5}$

$8x + 3$

$0.4x + 1$

Objective 144

EVALUATING EXPRESSIONS

Evaluate the expression, $a + b - 8$, when $a = 2$ and $b = -3$.

Replace each letter with its value. → Simplify.

$a + b - 8$
$2 + (-3) - 8$

$2 + (-3) - 8$
$-1 - 8 = -9$

EXAMPLE 1. Evaluate $2x - y - 54$ when $x = 4$ and $y = -3$.

$$\begin{aligned} 2x - y - 54 &= 2 \times 4 - (-3) - 54 \\ &= 8 - (-3) - 54 \\ &= 8 + 3 - 54 \\ &= 11 - 54 \\ &= -43 \end{aligned}$$

Replace x with 4 and y with −3. Simplify.

TRY THIS Evaluate each expression when $a = -4$ and $b = 9$.

1. $a - b + 29$ 16 **2.** $3b - a - 21$ 10

We write *ab* as an abbreviation for $a \times b$ or $a \cdot b$.

EXAMPLE 2. Evaluate $-9ab$ when $a = 3$ and $b = -6$.

$$\begin{aligned} -9ab &= -9 \cdot 3 \cdot (-6) \\ &= (-27) \cdot (-6) \\ &= 162 \end{aligned}$$

Replace a with 3 and b with −6. Multiply.

EXAMPLE 3. Evaluate $-4ab + 9$ when $a = -8$ and $b = -7$.

$$\begin{aligned} -4ab + 9 &= (-4) \cdot (-8) \cdot (-7) + 9 \\ &= 32 \cdot (-7) + 9 \\ &= -224 + 9 = -215 \end{aligned}$$

Replace a with −8 and b with −7. Multiply first. Then add.

TRY THIS Evaluate each expression when $x = -6$ and $y = 10$.

3. $-5xy$ 300 **4.** $7xy - 4$ −424

PRACTICE

For extra practice, use Making Practice Fun 86 with this lesson.

Evaluate each expression when $x = 4$.

1. $x + 8$ 12 **2.** $5x$ 20 **3.** $9 - x$ 5 **4.** $16 \div x$ 4

Evaluate each expression when $y = -3$.

5. $15 \div y$ −5 **6.** $19 - y$ 22 **7.** $-8y$ 24 **8.** $2y + 10$ 4

Evaluate each expression when $a = 5$ and $b = -6$.

9. ab −30 **10.** $5ab$ −150 **11.** $\frac{1}{2}ab$ −15 **12.** $-3ab$ 90

13. $a + b$ −1 **14.** $a - b$ 11 **15.** $b - a$ −11 **16.** $b + a$ −1

17. $a - b - 27$ −16 **18.** $a + b + 45$ 44 **19.** $89 - a + b$ 78 **20.** $100 + a - b$ 111

Evaluate each expression when $r = 8$ and $d = 16$.

21. $3.14d$ 50.24 **22.** $6.28r$ 50.24 **23.** $3.14 \times r \times r$ 200.96 **24.** $0.785 \times d \times d$ 200.96

Evaluate each expression when $b = 14$ and $h = 15$.

25. bh 210 **26.** $\frac{1}{2}bh$ 105 **27.** $\frac{1}{2} \times (b + h)$ $14\frac{1}{2}$ **28.** $\frac{1}{2}b + h$ 22

Evaluate each expression when $x = -3$ and $y = -4$.

29. $2xy - 16$ 8 **30.** $51 + x + y$ 44 **31.** $-2 - 3xy$ −38

32. $-5xy - 3y$ −48 **33.** $2x + y - x$ −7 **34.** $5y - y + 4y$ −32

35. $0.3x + 0.4y$ −2.5 **36.** $1.2x - 3.2y$ 9.2 **37.** $5.01x + y$ −19.03

PROBLEM CORNER

1. Regina traveled for 6 hours at 88 km per hour. How far did she travel? (Hint: $d = r \cdot t$) 528 km

2. Mario traveled for 7 hours at 78 km per hour. How far did he travel? 546 km

Objectives 193 and 194

APPLICATION

Many banks, savings and loan associations, and credit unions have literature available concerning interest.

INTEREST

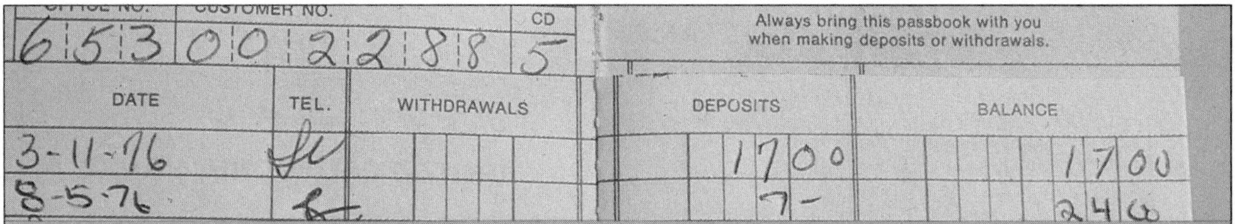

Interest is paid to the lender for the use of money. When you borrow money, you must pay interest for the use of the money. If you put money in a savings account, the bank pays you interest for the use of the money.

$I = Prt$ is the formula for simple interest.
I = interest
P = the principal, the amount loaned or borrowed
r = interest rate
t = length of time, in years

Julio put \$ 200 in a savings account for 6 months at 8% interest. Find the simple interest.

$$I = Prt$$
$$= \$200 \times 8\% \times \frac{6}{12} \qquad 6 \text{ months} = \frac{6}{12}\text{yr} = \frac{1}{2}\text{yr}$$
$$= \$200 \times 0.08 \times \frac{1}{2}$$
$$= \$200 \times 0.04$$
$$= \$80$$

Find the simple interest.

1. Sara kept \$ 100 in a savings account at $7\frac{1}{2}\%$ for 1 year. \$7.50
2. Bernie bought a \$ 500 savings certificate for 6% that matures in 6 months. \$15.00
3. Martha borrowed \$ 350 for 4 months at 8%. \$9.33
4. Larry borrowed \$ 500 for 2 months at 9%. \$7.50
5. Vera deposited \$ 1000 in a savings account for 1 year at 5.5%. \$55
6. Fred put \$ 150 in a savings account for 2 years at $6\frac{1}{4}\%$. \$18.75

Compound interest is money paid on the principal and the interest already earned. Interest compounded semiannually means that there are 2 payment periods per year. Interest compounded quarterly means that interest is paid 4 times per year.

Sumi has $ 100 in a savings account at 6% interest compounded semiannually. Find how much is in her account after 1 year. Find the compound interest.

1st payment period $I = Prt$

$= \$100 \times 6\% \times \frac{1}{2}$

$= \$100 \times 0.06 \times \frac{1}{2}$

$= \$100 \times 0.03 = \$3$

2nd payment period $P = \$100 + \$3 = \$103$

$I = Prt$

$= \$103 \times 6\% \times \frac{1}{2}$

$= \$103 \times 0.06 \times \frac{1}{2}$

$= \$103 \times 0.03 = \$3.09$

After 1 year $P = \$103 + \$3.09 = \$106.09$

The amount in the account is $ 106.09

Compound Interest = Compound Amount − Original Principal
= $ 106.09 − $ 100 = $ 6.09

The compound interest is $ 6.09

For the given amount and interest rate, compounded semiannually, find how much is in each account after 1 year. Find the compound interest.

7. $ 400; 8% **$467.95; $67.95**
8. $ 1000; 7% **$1071.23; $71.23**
9. $ 800; 6% **$848.72; $48.72**
10. $ 1500; 5% **$1575.94; $75.94**

Find the compound amount and compound interest. The interest is compounded annually.

11. $ 500 at 7% for 3 years **$612.52; $112.52**
12. $ 700 at 8% for 3 years **$881.80; $181.80**
13. $ 200 at 6.5% for 2 years **$226.85; $26.85**
14. $ 1200 at $7\frac{1}{2}$% for 2 years **$1386.75; $186.75**
15. Suppose in #11–14 the interest was not compounded. Find the simple interest for each problem. How much more interest was earned by compounding the interest?
$105; $7.52 12. $168; $13.80 13. $26: $0.85 14. $180; $6.75

Objective 145

THE ADDITION PRINCIPLE

The number 2 is a solution of the equation $x + 5 = 7$. When we replace x with 2, we get a true equation: $2 + 5 = 7$

If the replacement number makes the equation true, then it is called a solution.

We use the following principle to solve equations.

THE ADDITION PRINCIPLE

If an equation $a = b$ is true, then $a + c = b + c$ is true for any number c.

EXAMPLE 1. Solve. $x + 9 = -2$

$$x + 9 = -2$$
$$x + 9 + (-9) = -2 + (-9)$$ *Add -9 on both sides*
$$x + 0 = -2 + (-9)$$ *to get rid of the 9.*
$$x = -11$$ *The solution is -11.*

Check: $x + 9 = -2$

$-11 + 9$	-2
-2	

Replace x with the solution, -11.

TRY THIS Solve.

1. $x + 8 = -4$ -12 **2.** $x + 14 = 25$ 11

EXAMPLE 2. Solve. $y - 7 = 15$

$$y - 7 = 15$$
$$y - 7 + 7 = 15 + 7$$ *Add 7 on both sides to get rid of the -7.*
$$y + 0 = 22$$
$$y = 22$$ *The solution is 22.*

Check: $y - 7 = 15$

$22 - 7$	15
15	

Replace y with the solution, 22.

TRY THIS Solve.

3. $y - 8 = 27$ 35 **4.** $m - 5 = -4$ 1

PRACTICE

Solve.

1. $x + 2 = 6$ 4
2. $x + 3 = 9$ 6
3. $x + 4 = 10$ 6
4. $y + 9 = 4$ −5
5. $m + 8 = 2$ −6
6. $t + 10 = 5$ −5
7. $t + 12 = -7$ −19
8. $y + 11 = -4$ −15
9. $x + 14 = -8$ −22
10. $x + 39 = 58$ 19
11. $y + 59 = 86$ 27
12. $m + 47 = 83$ 36
13. $x - 8 = 3$ 11
14. $x - 9 = 15$ 24
15. $x - 7 = 14$ 21
16. $p - 8 = -3$ 5
17. $q - 9 = -15$ −6
18. $t - 7 = -14$ −7
19. $x - 2.3 = 7.4$ 9.7
20. $x - 3.7 = 8.4$ 12.1
21. $x - 2.6 = 8.3$ 10.9
22. $19 + x = 11$ −8
23. $17 + y = 26$ 9
24. $22 + p = 14$ −8
25. $234 + t = 709$ 475
26. $567 + y = 1123$ 556
27. $t - 899 = 1045$ 1944
28. $x - \frac{5}{6} = \frac{7}{8}$ $1\frac{17}{24}$
29. $x - \frac{1}{3} = \frac{1}{7}$ $\frac{10}{21}$
30. $r + \frac{2}{3} = 1$ $\frac{1}{3}$

PROBLEM CORNER

1. Julio can finish his experiment by solving the equation $x - 16 = 9$, where x represents the number of milliliters of water. How much water does he need? 25 milliliters
2. To find the amount of salt solution, Rosa uses the formula $27 + y = 33$. In the equation y represents the number of liters of solution. How much salt solution is needed? 6 liters

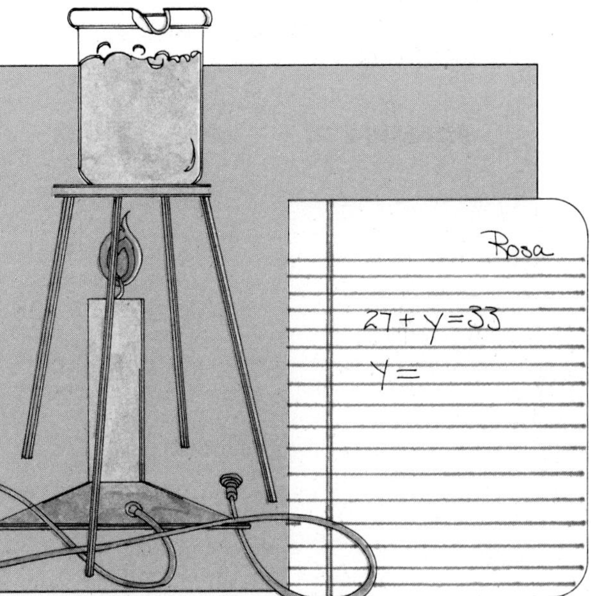

More Practice, page 442

Objective 146

THE MULTIPLICATION PRINCIPLE

Here is another principle for solving equations.

THE MULTIPLICATION PRINCIPLE

If an equation $a = b$ is true,
then $a \cdot c = b \cdot c$ is true for any number c.

EXAMPLE 1. Solve. $-8x = 32$

$$-8x = 32$$

$$-\frac{1}{8} \cdot (-8x) = -\frac{1}{8} \cdot 32$$ *Multiply by $-\frac{1}{8}$ on both sides to get rid of the -8.*

$$1x = -4$$

$$x = -4$$ *The solution is -4.*

Check: $-8x = 32$

$(-8) \cdot (-4)$	32
32	

Replace x with the solution, -4.

TRY THIS

Solve.

1. $-6x = 48$ -8 **2.** $12x = -72$ -6

EXAMPLE 2. Solve. $\frac{m}{7} = 38$

$$\frac{m}{7} = 38$$

$$\frac{1}{7}m = 38$$ $\frac{m}{7} = \frac{1}{7}m$

$$7 \cdot \frac{1}{7}m = 7 \cdot 38$$ *Multiply by 7 on both sides to get rid of the $\frac{1}{7}$.*

$$1m = 266$$

$$m = 266$$ *The solution is 266.*

Check: $\frac{m}{7} = 38$

$\frac{266}{7}$	38
38	

Replace m with the solution, 266.

TRY THIS

Solve.

3. $\frac{m}{9} = 45$ 405 **4.** $\frac{t}{8} = -13$ −104

PRACTICE

Solve.

1. $6x = 36$ 6 **2.** $3x = 39$ 13 **3.** $4x = 28$ 7

4. $9x = -36$ −4 **5.** $7x = -49$ −7 **6.** $9x = -63$ −7

7. $-8m = 64$ −8 **8.** $-9y = 81$ −9 **9.** $-7t = 70$ −10

10. $-12x = -72$ 6 **11.** $-15x = -105$ 7 **12.** $-16x = -80$ 5

13. $\frac{t}{2} = 15$ 30 **14.** $\frac{y}{4} = 9$ 36 **15.** $\frac{n}{7} = 8$ 56

16. $\frac{x}{14} = 12$ 168 **17.** $\frac{m}{13} = 15$ 195 **18.** $\frac{y}{18} = 15$ 270

19. $\frac{x}{9} = 8$ 72 **20.** $\frac{y}{3} = 8$ 24 **21.** $\frac{t}{6} = 9$ 54

22. $\frac{x}{10} = -18$ −180 **23.** $\frac{t}{12} = -16$ −192 **24.** $\frac{m}{14} = -15$ −210

25. $3.3y = 66$ 20 **26.** $2.8t = 56$ 20 **27.** $7.4p = 74$ 10

28. $6.3y = 44.1$ 7 **29.** $38.7x = 309.6$ 8 **30.** $29.4m = 235.2$ 8

PROBLEM CORNER

Beto finds the total amount he owes for his stereo by solving the equation $\frac{m}{12} = 15$. He lets m represent the total amount of money. How much does he owe for the stereo? $180

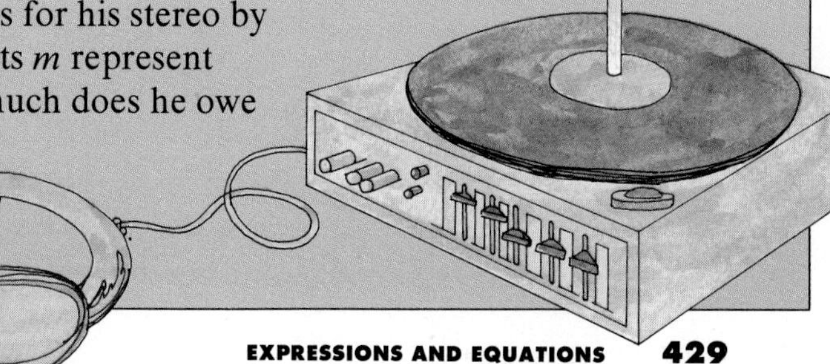

Objective 147

USING THE PRINCIPLES TOGETHER

When we need to use both the Addition and Multiplication Principles, we add first and then multiply.

EXAMPLE 1. Solve. $3x + 4 = 13$

$$3x + 4 = 13$$
$$3x + 4 + (-4) = 13 + (-4)$$ *Add -4 on both sides to get $3x$ alone.*
$$3x + 0 = 9$$
$$3x = 9$$
$$\frac{1}{3} \cdot 3x = \frac{1}{3} \cdot 9$$ *Multiply by $\frac{1}{3}$ on both sides.*
$$x = 3$$ *The solution is 3.*

Check: $3x + 4 = 13$

$3 \cdot 3 + 4$	13
$9 + 4$	
13	

Replace x with the solution, 3.

TRY THIS

Solve.

1. $9x + 6 = 51$ 5 **2.** $8y + 7 = 79$ 9

EXAMPLE 2. Solve. $-5x - 6 = 19$

$$-5x - 6 = 19$$
$$-5x - 6 + 6 = 19 + 6$$ *Add 6 on both sides to get $-5x$ alone.*
$$-5x + 0 = 25$$
$$-5x = 25$$
$$-\frac{1}{5} \cdot (-5x) = -\frac{1}{5} \cdot 25$$ *Multiply by $-\frac{1}{5}$ on both sides.*
$$x = -5$$ *The solution is -5.*

Check: $-5x - 6 = 19$

$-5 \cdot (-5) - 6$	19
$25 - 6$	
19	

Replace x with the solution, -5.

TRY THIS

Solve.

3. $-8x - 4 = 28$ -4 **4.** $3x - 4 = 8$ 4

PRACTICE

For extra practice, use Making Practice Fun 87 with this lesson.

Solve.

1. $5x + 6 = 31$ 5
2. $3x + 6 = 30$ 8
3. $4x + 7 = 27$ 5
4. $4x - 6 = 34$ 10
5. $6x - 3 = 15$ 3
6. $8x - 7 = 41$ 6
7. $-6t + 5 = 59$ −9
8. $-2m + 5 = 55$ −25
9. $-3n + 9 = 72$ −21
10. $7k + 2 = -54$ −8
11. $5y + 4 = -41$ −9
12. $9t + 8 = -64$ −8
13. $-10x - 2 = 78$ −8
14. $-9x - 5 = 58$ −7
15. $-7x - 8 = 76$ −12
16. $12y - 6 = 30$ 3
17. $15t - 9 = 66$ 5
18. $14m - 4 = 66$ 5
19. $3.2x - 5 = 7$ 3.75
20. $1.6y + 7 = 11$ 2.5
21. $6.4t + 5 = 13$ 1.25
22. $\frac{x}{7} - 3 = 9$ 84
23. $\frac{t}{4} + 8 = 3$ −20
24. $\frac{m}{9} - 2 = 10$ 108
25. $0.5x + 0.8 = 1.6$ 1.6
26. $0.4x - 0.9 = 0.3$ 3
27. $0.2x - 0.6 = 1.4$ 10
28. $\frac{y}{6} - 19 = 2.3$ 127.8
29. $\frac{x}{4} - 42 = 5.6$ 190.4
30. $\frac{t}{5} + 39 = -42$ −405

PROBLEM CORNER

What you could wish for if you had one wish: 10 . . .

To answer the riddle, solve each equation.
Then cross out the false answers in the chart. wishes

$2x - 12 = 8$
$2y + 4 = 18$
$6w - 9 = 9$
$5t + 6 = 26$
$-3m + 10 = -5$
$-3a + 9 = 6$

$x = 10$	W
$x = 2$	A
$y = 7$	I
$y = 1$	P
$w = 3$	S
$w = 5$	T
$t = 4$	H
$t = 10$	B
$m = 5$	E
$a = 4$	T
$a = 1$	S

Objective 148

SOLVING PROBLEMS WITH EQUATIONS

To solve problems we can translate to an equation and solve.

EXAMPLE 1. 4 times what number plus 7 is 27?

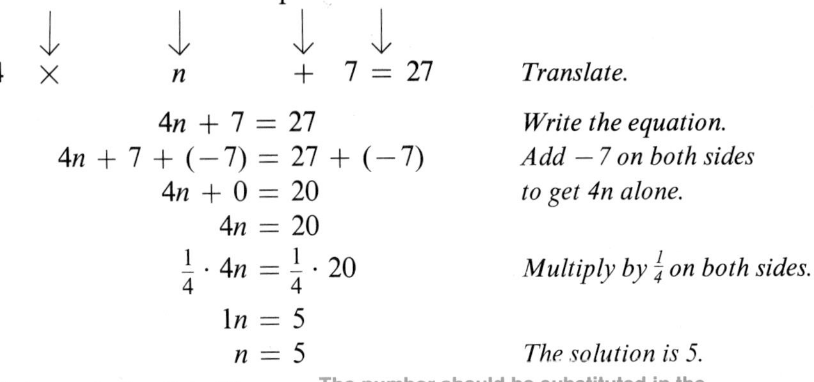

$4 \times n + 7 = 27$ *Translate.*

$$
\begin{aligned}
4n + 7 &= 27 && \textit{Write the equation.}\\
4n + 7 + (-7) &= 27 + (-7) && \textit{Add } -7 \textit{ on both sides}\\
4n + 0 &= 20 && \textit{to get 4n alone.}\\
4n &= 20\\
\frac{1}{4} \cdot 4n &= \frac{1}{4} \cdot 20 && \textit{Multiply by } \tfrac{1}{4} \textit{ on both sides.}\\
1n &= 5\\
n &= 5 && \textit{The solution is 5.}
\end{aligned}
$$

4 times 5 plus 7 is 27.

The number should be substituted in the conditions of the problem to check if it is the solution.

EXAMPLE 2. Marcel received the same number of points on 6 tests. On another test he received 86 points. His total number of points was 518. How many points did Marcel receive on each of the 6 tests?

6 times number of points per test plus 86 is total points

$6 \times n + 86 = 518$

$$
\begin{aligned}
6n + 86 &= 518 && \textit{Write the equation.}\\
6n + 86 + (-86) &= 518 + (-86) && \textit{Add } -86 \textit{ on both sides}\\
6n &= 432 && \textit{to get 6n alone.}\\
\frac{1}{6} \cdot 6n &= \frac{1}{6} \cdot 432 && \textit{Multiply by } \tfrac{1}{6} \textit{ on both sides.}\\
1n &= 72\\
n &= 72 && \textit{The solution is 72.}
\end{aligned}
$$

Marcel received 72 points on each test.

TRY THIS

Solve.

1. -6 times what number minus 8 is 46? -9
2. Cici earned the same amount of money for 4 days. Another day she earned \$ 18. She earned \$ 114 in all. How much did she earn on each of the 4 days? \$24

PRACTICE

Solve.

1. What number plus 9 is 13? 4
2. What number plus 7 is 22? 15
3. What number plus 13 is 8? −5
4. What number plus 19 is 11? −8
5. What number minus 7 is 11? 18
6. What number minus 8 is 19? 27
7. 5 times what number is 80? 16
8. 8 times what number is 96? 12
9. 9 times what number is −54? −6
10. 7 times what number is −63? −9
11. 9 times what number plus 6 is 51? 5
12. 8 times what number plus 7 is 79? 9
13. −3 times what number minus 9 is −72? 21
14. −7 times what number minus 8 is 76? −12
15. When 18 is subtracted from six times a certain number, the result is 96. What is the number? 19
16. When 28 is added to five times a certain number, the result is 232. What is the number? 52
17. The width of a rectangle is 7 cm and the area is 84 square centimeters. What is the length? 12 cm
18. The length of a rectangle is 27 ft and the area is 108 square ft. What is the width? 4 ft
19. A shirt is on sale for $ 9.60. This is 60% of the original price. What was the original price? $16.00
20. A blouse is on sale for $ 9.24. This is 66% of the original price. What was the original price? $14.00
21. Debbie earned the same amount of money for 3 days. Another day she earned $ 24. She earned $ 84 in all. How much did she earn on each of the 3 days? $20
22. Marcos earned the same amount of money for 5 days. Another day he earned $ 32. He earned $ 122 in all. How much did he earn on each of the 5 days? $18
23. Kirk sold his painting for $ 42. This was 6 times what he paid for the materials. What did he pay for the materials? $7
24. Jamie sold her ivy plant for $ 18. This was 3 times what she paid for it. What did she pay for the plant? $6

More Practice, page 443

Objective 195

Try to have some examples of the credit plans available from local banks, savings and loan associations, credit card companies, and department stores.

CREDIT COSTS

Over 2000 banks and credit card companies offer credit cards. In addition thousands of individual stores have their own credit cards. Credit card purchases are paid for by monthly payments. The interest rate is often $1\frac{1}{2}\%$ per month, which sounds small, but amounts to 18% per year.

After one month Trent has a balance of $400 in his credit card account. How much interest does he owe? How much does he owe in all?

$I = Prt$ Refer to pages 424 and 425 for lessons on the interest formula.

$I = \$400 \times 18\% \times \frac{1}{12}$

$I = \$400 \times 1\frac{1}{2}\%$ $\qquad P = \$400 + \$6$

$I = \$400 \times 0.015 = \$6$ $\qquad P = \$406$

Trent owes the credit card company $ 406.

For each balance, compute the interest and the amount owed after one month.

1. $ 800 $12 $812
2. $ 1000 $15 $1015
3. $ 400 $6 $406
4. $ 700 $10.50 $710.50

A color TV sells for $ 400 cash. It can be purchased on credit for $ 90 down and $ 40 per month for 9 months.

5. What is the cost of buying the TV in installments? The cost is the total amount paid. $450
6. How much interest is paid? $50
7. What percent of the $ 400 is the interest? $12\frac{1}{2}\%$

Sal bought a CB radio for $ 150. The sales tax was 3.5%. Sal made a down payment of $ 25 and charged the rest on his charge card. The bill each month was $ 30 plus $1\frac{1}{2}\%$ interest on the unpaid balance.

8. What was the purchase price including tax? $155.25
9. How much was charged to the credit card? $130.25
10. What was the first monthly payment? $31.96
11. What was the second monthly payment? $31.48
12. What was the third monthly payment? $31.01

How much was the interest for each of the following months?

13. the first month $1.96
14. the second month $1.48
15. the third month $1.01
16. Estimate how many months it took Sal to pay for the radio. 4 months

Rita bought new auto tires for $ 150 and charged them on her charge card. She paid $ 14.75 per month for 12 months.

17. How much did Rita pay in all for the tires? $177
18. How much did she pay in interest? $27

Some credit card companies also make money by charging the merchant. Nadine takes a group of people to The Spaghetti Shop for dinner. The bill is $ 57.85. Nadine charges it on her charge card. The Spaghetti Shop has to pay the charge card company 6% of the bill.

19. How much does The Spaghetti Shop have to pay the charge card company? $3.48
20. How much does The Spaghetti Shop actually get from the dinner bill? $54.37

Objective 149

GRAPHING EQUATIONS

Discuss with your students the process of choosing values for x. They should realize that two points determine a line. At least one other point should be used as a check point. 0 is usually an easy number to use as one of the values.

The graphs of many equations are straight lines.

EXAMPLE 1. Graph $y = 2x - 1$.

1. Make a table. Choose a value for x. In the equation replace x with that value. Solve for y. Repeat for other values of x.

$y = 2x - 1$
$y = 2 \cdot (-2) - 1$
$y = -4 - 1$
$y = -5$

$y = 2x - 1$
$y = 2 \cdot (-1) - 1$
$y = -2 - 1$
$y = -3$

$y = 2x - 1$
$y = 2 \cdot 0 - 1$
$y = 0 - 1$
$y = -1$

$y = 2x - 1$
$y = 2 \cdot 1 - 1$
$y = 2 - 1$
$y = 1$

$y = 2x - 1$
$y = 2 \cdot 2 - 1$
$y = 4 - 1$
$y = 3$

x	y
−2	−5
−1	−3
0	−1
1	1
2	3

2. Graph and label the ordered pairs in the table.

To graph an ordered pair like $(-2, -5)$, start at $(0,0)$. Go to the left 2 units, then down 5 units.

3. Connect the points.

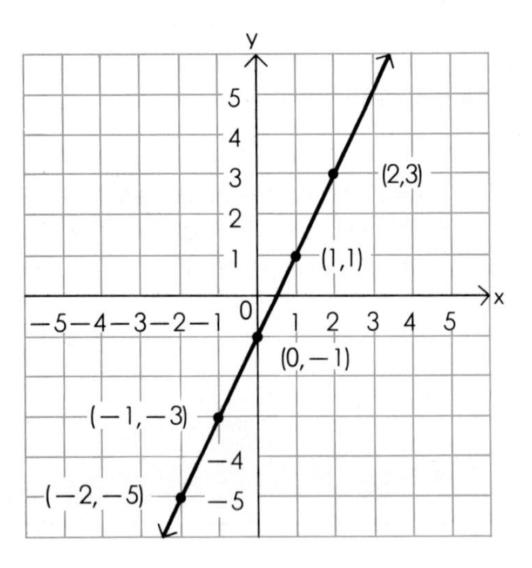

TRY THIS

Graph each equation. See answer section.

1. $y = x + 1$ **2.** $y = 1 - 2x$

PRACTICE

Use Quiz 30 after this Practice.

Graph each equation. See answer section.

1. $y = x + 4$ **2.** $y = x + 2$ **3.** $y = x + 3$

4. $y = x - 2$ **5.** $y = x - 4$ **6.** $y = x - 5$

7. $y = 2x$ **8.** $y = x$ **9.** $y = -1 \cdot x$

10. $y = -3x$ **11.** $y = -2x$ **12.** $y = -4x$

13. $y = 2x + 1$ **14.** $y = 3x + 1$ **15.** $y = 2x + 2$

16. $y = 4 - x$ **17.** $y = 3 - x$ **18.** $y = 5 - x$

19. $y = 10 - 3x$ **20.** $y = 9 - 2x$ **21.** $y = 8 - 3x$

22. $y = 3x - 5$ **23.** $y = 4x - 3$ **24.** $y = 5x - 4$

25. $y = -2x + 1$ **26.** $y = -2x + 2$ **27.** $y = -2x - 1$

PROBLEM CORNER

1. Luis is using the equation $y = x^2$ in his experiment. He chose these values for x: $-3, -2, -1, 0, 1, 2, 3$. Make a table and find the values for y. See answer section.

2. Graph $y = x^2$. See answer section.

More Practice, page 443

CAREER CAREER CAREER CAREER CAREER

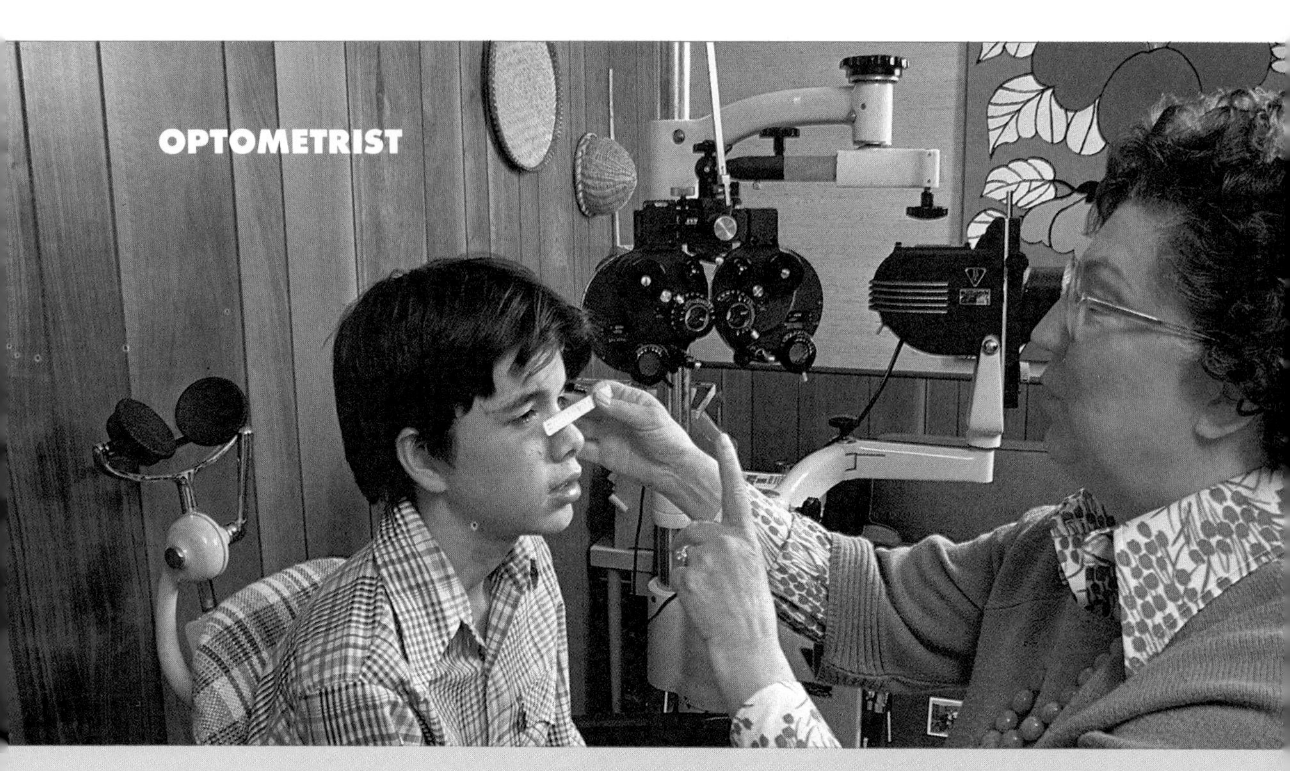

OPTOMETRIST

Dr. Mary Chaney is an optometrist. She relies on precise measurements and calculations to determine eyeglass prescriptions.

One important measurement is the pupillary distance, PD. This is the distance, in millimeters, between the pupils of the eyes. This distance is widest when a person looks into the distance. It is narrowest when a person looks at objects close by.

Patient	Far PD	Near PD	Difference	Average
Elaine Tyler	65	58	7 mm	61.5 mm
Doug Wong	70	65	5 mm	67.5 mm
Angela Perez	61	53	8 mm	57 mm
Mike Barnes	59	51	8 mm	55 mm
Sammy Cole	69	64	5 mm	66.5 mm
Julia Spinola	63	56	7 mm	59.5 mm

1. For each patient find the difference of the measures of the far and near PDs.
2. For each patient find the average of the near and far PDs.
3. For the six patients what is the largest PD? 70 mm

4. What is the smallest PD? 51 mm
5. What is the difference between the largest and smallest PD? 19 mm
6. What is the average near PD? 57.8 mm
7. What is the average far PD? 64.5 mm

Objects are not always the same size as they appear to be. Dr. Chaney uses this fact in fitting lenses to make objects appear larger or smaller.

The angular size of an object is the size that it appears to be. It is the ratio of the height of the object, *h* to its distance, *d*, from the eye.

Angular size $= \frac{h}{d}$ NOTE: *h* and *d* must be measured in the same unit.

Find the angular size of each of the following.

8. A $\frac{3}{4}$-inch penny held 24 inches from the eye 0.03
9. A 9-inch plate held 24 feet from the eye 0.03
10. A full moon, 2400 miles in diameter, 240,000 miles from the eye 0.01
11. How do the angular sizes of the penny and the plate compare? They are the same.
12. The angular size of the penny is about how many times as large as that of the moon? 3
13. Which of these has the larger angular size?
    a 6-ft baseball player seen from the bleachers 95 yards away or a 2-inch image of the player seen on a TV screen, 6 ft from the eye? The player seen on the TV screen

We can use the angular size to find other lengths.

14. A pen, 13 centimeters long, has an angular size of 0.5. How far is it from the eye? 26 centimeters
15. A pencil is held 76 centimeters from the eye and has an angular size of 0.25. How long is the pencil? 19 centimeters
16. You are standing 100 feet from a tree. The tree has an angular size of 0.12. How tall is the tree? 12 feet

CHAPTER 15 REVIEW

For extra practice, use Making Practice Fun 88 with this lesson.

Evaluate each expression when $x = -2$ and $y = -4$. p. 422

1. $18 + x - y$ 20 **2.** $5x + 4y$ −26 **3.** $8xy$ 64

Solve. p. 426

4. $x + 9 = -4$ −13 **5.** $y - 3.8 = 9.4$ 13.2 **6.** $t - \frac{5}{6} = \frac{2}{3}$ $1\frac{1}{2}$

Solve. p. 428

7. $\frac{t}{-3} = 8$ −24 **8.** $1.4x = 56$ 40 **9.** $-9x = -81$ 9

Solve. p. 430

10. $5x - 6 = 34$ 8 **11.** $-8n + 9 = 81$ −9 **12.** $\frac{m}{8} - 3 = 10$ 104

Solve. p. 432

13. −4 times what number plus 7 is 61? −13.5

14. The sales tax on a record is \$ 0.40. This is 5% of the purchase price. What is the purchase price? \$8.00

15. The length of a rectangle is 19 meters and the area is 323 square meters. What is the width? 17 meters

16. A painting was sold for \$ 140. This was 7 times the cost of the materials. What was the cost of the materials? \$20

Graph the equation. p. 436

17. $y = -2x$ See answer section.

Application pp. 424, 434

18. Ellen keeps \$ 900 in a savings account for 8 months at 5%. Find the simple interest. \$30

19. Tom still owes \$ 500 on his credit card account. Interest is $1\frac{1}{2}$ % per month. How much interest will he owe next month? How much will he owe in all? \$7.50; \$507.50

CHAPTER 15 TEST

Evaluate each expression when $x = 4$ and $y = -5$.

1. $-3xy$ 60

2. $x - 3y$ 19

3. $5x + 2y - 40$ −30

Solve.

4. $t - 5 = -12$ −7

5. $x + \frac{1}{3} = \frac{5}{6}$ $\frac{1}{2}$

6. $y - 8.4 = 2.9$ 11.3

Solve.

7. $3.2x = 64$ 20

8. $-8t = 96$ −12

9. $\frac{n}{4} = -13$ −42

Solve.

10. $3x + 7 = 40$ 11

11. $-4m - 9 = -48$ 9.75

12. $-2.5y - 7 = 23$ −12

Solve.

13. −8 times what number plus 9 is −47? 7

14. The width of a rectangle is 4.8 feet and the area is 72 square feet. What is the length? 15 feet

15. A stereo is on sale for $ 240. This is 75% of the original price. What was the original price? $320

16. A horse was sold for $ 1200. This was 3 times the price of the horse as a two-year-old. What was the price of the horse as a two-year-old? $400

Graph the equation.

17. $y = 3 - x$ See answer section.

Application

18. Darren keeps $ 700 in a savings account for 6 months at 8%. Find the simple interest. $28

19. Samantha still owes $ 600 on her credit card account. Interest is $1\frac{1}{2}$ % per month. How much interest will she owe next month? How much will she owe in all? $9; $609

MORE PRACTICE

Evaluate each expression when $x = -6$. Use after page 423.

1. $32 \div x$ $-5\frac{1}{3}$
2. $4x$ -24
3. $-3x + 10$ 28
4. $20 - x$ 26

Evaluate each expression when $a = -2$ and $b = -6$. Use after page 423.

5. $a - b + 40$ 44
6. $-7ab$ -84
7. $8ab$ 96
8. $2a + b$ -10
9. $5a + 7b$ -52
10. $5a - 4b$ 14
11. $5 \times a \times a$ 20
12. $-2a - 10b$ 64

Solve. Use after page 427.

13. $x + 6 = 11$ 5
14. $z + 4 = 1$ -3
15. $x + 5 = -2$ -7
16. $h + 27 = 63$ 36
17. $m + 28 = 19$ -9
18. $t + 17 = -10$ -27
19. $x - 9 = 4$ 13
20. $r - 11 = -7$ 4
21. $g - 3.4 = 9.7$ 13.1
22. $t - 7 = -23$ -16
23. $v - 23 = -48$ -25
24. $24 + y = 56$ 32
25. $x + 2348 = 5667$ 3319
26. $y + 103 = 5467$ 5364
27. $x + \frac{1}{4} = \frac{2}{3}$ $\frac{5}{12}$
28. $x + \frac{7}{8} = 1$ $\frac{1}{8}$
29. $t - \frac{3}{4} = \frac{5}{6}$ $1\frac{7}{12}$
30. $m - 1 = \frac{2}{5}$ $1\frac{2}{5}$

Solve. Use after page 429.

31. $4x = 20$ 5
32. $9x = 72$ 8
33. $-6x = 54$ -9
34. $2y = -18$ -9
35. $6z = -18$ -3
36. $-9t = -45$ 5
37. $\frac{t}{3} = 17$ 51
38. $\frac{x}{7} = -9$ -63
39. $\frac{y}{-5} = 13$ -65
40. $\frac{y}{10} = -13$ -130
41. $\frac{z}{-45} = -81$ 3645
42. $10x = 7$ $\frac{7}{10}$
43. $1.8y = 90$ 50
44. $4.4z = 88$ 20
45. $5.3d = 121.9$ 23
46. $-25t = 75$ -3
47. $-35y = 2100$ -60
48. $14h = 35$ $2\frac{1}{2}$
49. $2.8a = -126$ -45
50. $5.6b = -151.2$ -27
51. $3.4x = 255$ 75

Solve. Use after page 431.

1. $2x + 5 = 11$ 3
2. $3x + 8 = 20$ 4
3. $5t + 6 = -34$ −8

4. $4y - 7 = 25$ 8
5. $-3m - 25 = 7$ $-10\frac{2}{3}$
6. $8t - 10 = -82$ −9

7. $-5t - 11 = 69$ −16
8. $1.8y - 12 = 24$ 20
9. $6.4m + 2 = 10$ 1.25

10. $\frac{x}{-3} + 8 = 29$ −63
11. $\frac{x}{6} - 5 = 11$ 96
12. $\frac{y}{4} - 4 = 13$ 68

13. $3.2t - 4 = 8$ 3.75
14. $0.2x - 0.7 = 1.3$ 10
15. $0.8x + 0.4 = 2.9$ 3.125

16. $0.5y + 0.9 = 1.7$ 1.6
17. $45t + 13 = 1067$ 23.42
18. $12x - 24 = 672$ 58

Solve. Use after page 433.

19. What number plus 9 is 23? 14

20. Six times what number is −36? −6

21. −5 times what number minus 12 is 68? −16

22. Nine times what number plus 9 is 81? 8

23. When 37 is subtracted from five times a certain number, the result is 243. What is the number? 56

24. When 16 is added to 3 times a certain number, the result is 64. What is the number? 16

25. The length of a rectangle is 14 cm and the area is 378 cm^2. What is the width? 27 cm

26. The width of a rectangle is 19 m and the area is 494 m^2. What is the length? 26

27. A book is on sale for $ 11.96. This is 80% of the original price. What was the original price? $14.95

28. The sales tax on a plant is $ 0.39. This is 6% of the purchase price. What is the purchase price? $6.50

29. Leslie sold her bookcase for $ 21. This was 3 times what she paid for the wood. What did she pay for the wood? $7

30. Trevor sold his coin collection for $ 340. This was 4 times what he paid for the coins. What did he pay for the coins? $85

Graph each equation. Use after page 437. **See answer section.**

31. $y = x - 1$
32. $y = x + 1$
33. $y = 3x$
34. $y = -2x$

35. $y = 2x - 1$
36. $y = -2x + 2$
37. $y = 6 - 3x$
38. $y = -3x - 4$

SKILLS REVIEW

Make these changes. Refer to pages 286-299

1. 9 km = 9000 m	**2.** 7 L = 7000 mL	**3.** 8268 m = 8.268 km
4. 5 kg = 5000 g	**5.** 748 g = 0.748 kg	**6.** 4678 mL = 4.678 L
7. 6 t = 6000 kg	**8.** 48 kg = 0.048 t	**9.** 98 mm = 9.8 cm
10. 7 cm = 70 mm	**11.** 3 g = 3000 mg	**12.** 4286 mg = 4.286 g
13. 4 ft = 48 in.	**14.** 18 in. = $1\frac{1}{2}$ ft	**15.** 20 yd = 60 ft
16. 10 ft = $3\frac{1}{3}$ yd	**17.** 4 gal = 16 qt	**18.** 24 oz = $1\frac{1}{2}$ lb
19. 3 mi = 5280 yd	**20.** 6 qt = $1\frac{1}{2}$ gal	**21.** 5 ton = 10,000 lb
22. 2 mi = 10,560 ft	**23.** 5 lb = 80 oz	**24.** 2500 lb = $1\frac{1}{4}$ ton

Find the perimeter. Refer to page 312.

25 m
25.

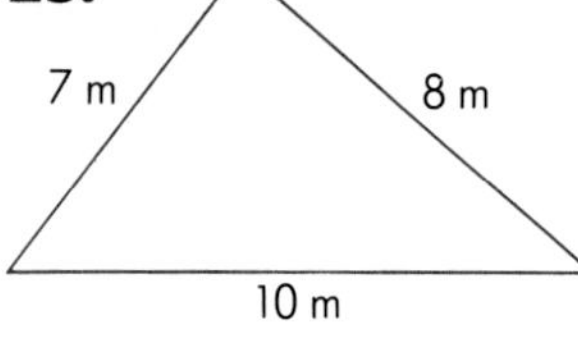

15 in.
26.
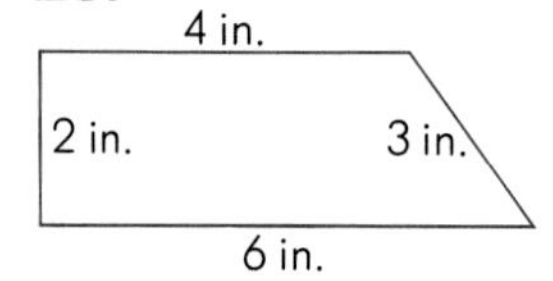

31 cm
27.
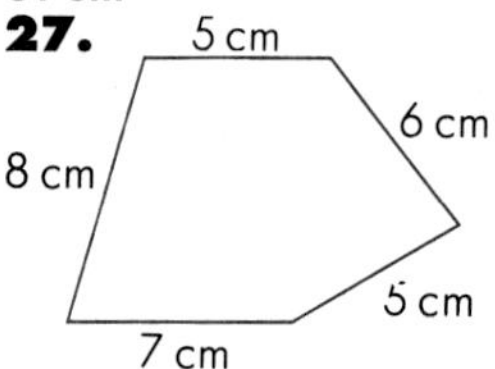

28. Rectangle
$l = 3$ ft 10 ft
$w = 2$ ft

29. Rectangle
$l = 23.7$ cm 72.6 cm
$w = 12.6$ cm

30. Square
$s = 3.4$ m 13.6 m

Find the circumference. Refer to page 314.

37.68 in.
31.
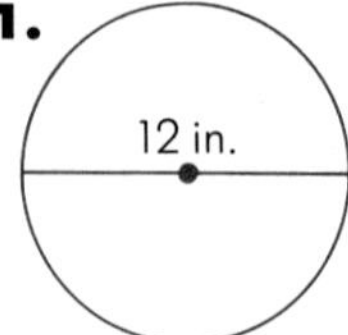

25.12 cm
32.
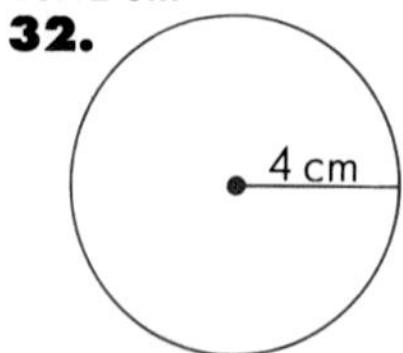

125.6 m
33.
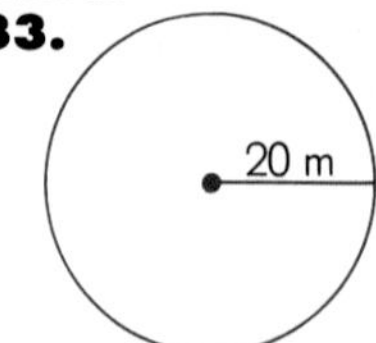

34. Circle
$r = 7$ ft 43.96 ft

35. Circle
$d = 25$ mm 78.5 mm

36. Circle
$r = 3$ yd 18.84 yd

Find the area. Refer to pages 318–327.

24 sq ft
1.

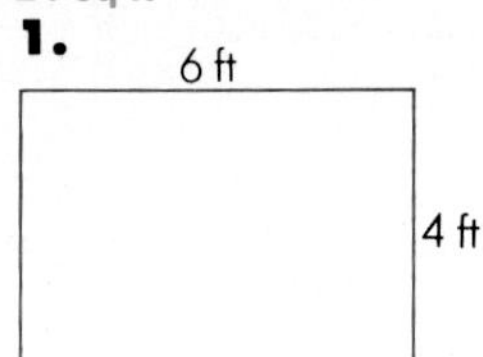

15.21 m²
2.

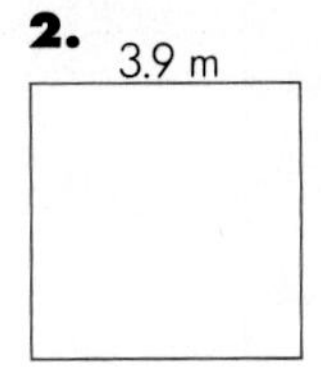

91.8 cm²
3.

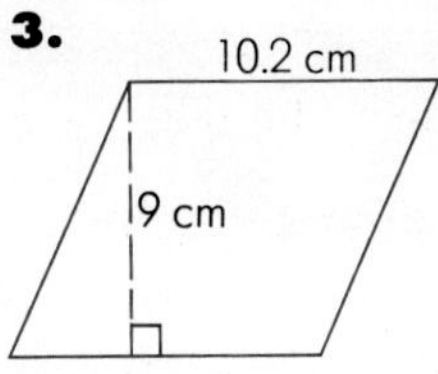

100 mm²
4.

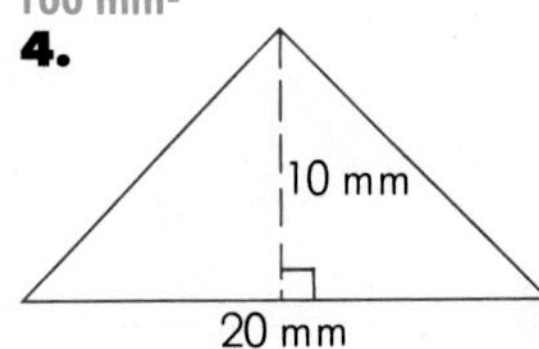

64 sq in.
5.

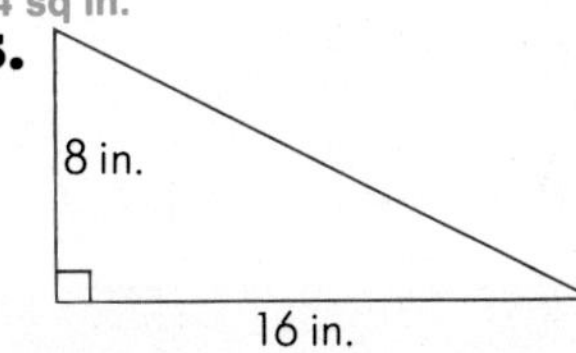

36 sq ft
6.

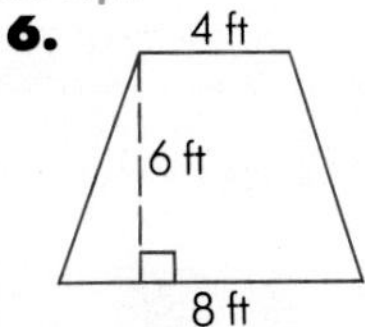

7. Trapezoid
$b_1 = 7$ m 34.4 m²
$b_2 = 9$ m
$h = 4.3$ m

8. Circle
$r = 18$ cm 1017.36 cm²

9. Circle
$d = 10$ yd 78.5 sq yd

Find the volume. Refer to pages 340–349.

280 cm³
10.

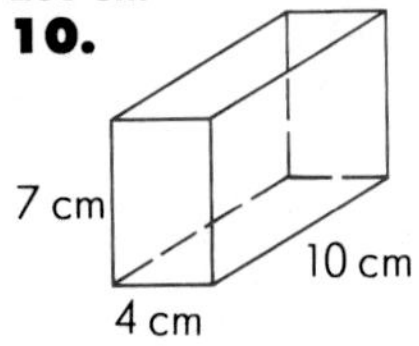

216 m³
11.

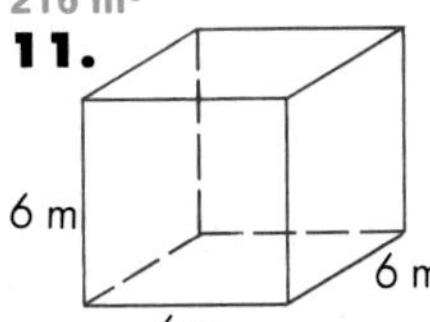

1607.68 cu in.
12.

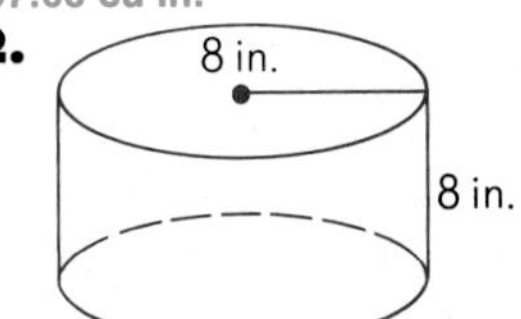

200 m³
13.

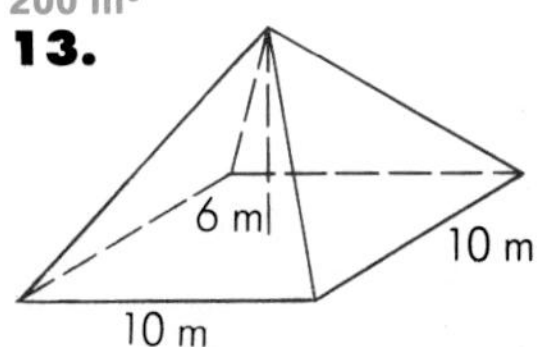

837.33 cu in.
14.

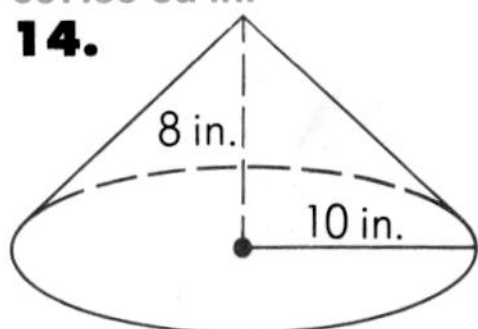

2143.57 cu in.
15.

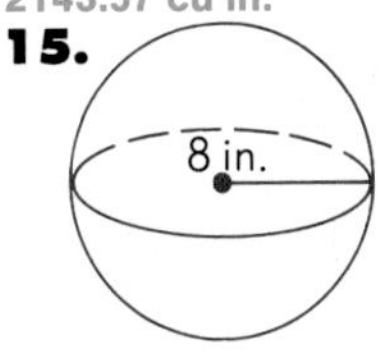

16. Rectangular prism
$l = 4$ m
$w = 1$ m
$h = 8$ m 32 m³

17. Cube
$s = 12$ in. 1728 cu in.

18. Cylinder
$d = 12$ cm 791.28 cm³
$h = 7$ cm

19. Pyramid
$l = 12$ yd
$w = 7$ yd
$h = 9$ yd 252 cu yd

20. Cone
$r = 12$ cm
$h = 12$ cm 1808.64 cm³

21. Sphere
$d = 18$ cm 3052.08 cm³

1. Make a line graph for the data in the table. Refer to page 370. See answer section.

National Longevity Records			
Country	Years	Country	Years
United States	113	U.S.S.R.	110
Canada	113	France	109
Japan	112	Germany	108
South Africa	111	Finland	107
Australia	110	Spain	106

2. Make a bar graph for the data in the table Refer to page 372. See answer section.

Moons of the Sun's Planets			
Planet	Number of Moons	Planet	Number of Moons
Mercury	0	Saturn	9
Venus	0	Uranus	4
Earth	1	Neptune	1
Mars	2	Pluto	0
Jupiter	9		

3. Find the angle for each percent in the table and make a circle graph.
Refer to page 374. See answer section.

Time Spent in One Week	
Activity	Percent
Sleeping	40%
School	25%
Recreation	25%
Eating	4%
Other	6%

Find the mean. Refer to page 378.

4. 2, 4, 6, 0 3

5. 9, 9, 12, 15, 19 12.8

6. 37, 48, 65, 27 44.25

Find the median. Refer to page 380.

7. 15, 7, 9, 10, 12 10

8. 24, 498, 76, 98, 376, 105, 381 105

9. 75, 86, 94, 88, 87 87

10. 6, 6, 8, 8, 7, 7, 9 7

11. 3, 7, 9, 10 8

12. 12, 4, 6, 11, 10, 9 9.5

Find the mode. Refer to page 380.

1. 4, 7, 4, 8, 4, 9 4

2. 7, 6, 5, 6, 4, 3, 2 6

3. 21, 48, 74, 21, 76, 90 21

4. 4, 3, 2, 4, 3, 2, 4 4

5. 3, 9, 4, 9, 3, 7 3 and 9

6. 6, 6, 7, 8, 12, 7, 11 6 and 7

Give a positive or negative number for each situation. Refer to page 394.

7. a loss of $76 −$76

8. a gain of $10 $10

9. 5 hours ago −5

10. 17° below 0 −17°

11. 25° above 0 25°

12. 90 ft above sea level 90

13. 130 ft below sea level −130

14. 12 sec after liftoff 12

15. 3 min before liftoff −3

Compare. Use $>$ or $<$. Refer to page 394.

16. 5 and -5 $>$

17. 0 and -4 $>$

18. -6 and 7 $<$

19. 4 and -5 $>$

20. -6 and -5 $<$

21. 9 and -12 $>$

22. -4 and -10 $>$

23. 0 and 7 $<$

24. 0 and -7 $>$

25. -7 and 7 $<$

26. -9 and -10 $>$

27. -4 and -1 $<$

Add. Refer to page 398.

28. $-6 + 7$ 1

29. $-11 + 7$ -4

30. $-6 + 8$ 2

31. $-8 + 6$ -2

32. $-3 + (-7)$ -10

33. $-8 + (-6)$ -14

34. $-3 + (-5)$ -8

35. $-4 + (-6)$ -10

36. $16 + (-3)$ 13

37. $9 + (-9)$ 0

38. $7 + (-12)$ -5

39. $6 + (-8)$ -2

Subtract. Refer to page 404.

40. $4 - 5$ -1

41. $-4 - 5$ -9

42. $-9 - (-4)$ -5

43. $-7 - 7$ -14

44. $8 - (-8)$ 16

45. $7 - 10$ -3

46. $0 - 7$ -7

47. $8 - (-3)$ 11

48. $-8 - 15$ -23

49. $-9 - (-9)$ 0

50. $37 - 43$ -6

51. $-20 - 10$ -30

Multiply. Refer to page 406.

52. 9×6 54

53. $9 \times (-4)$ -36

54. -3×6 -18

55. $-8 \times (-3)$ 24

56. $-8 \times (-6)$ 48

57. $-7 \times (-15)$ 105

58. $12 \times (-7)$ -84

59. -100×7 -700

Divide. Refer to page 408.

1. $-8 \div 4$ −2
2. $-8 \div (-4)$ 2
3. $-24 \div 3$ −8
4. $-16 \div (-2)$ 8
5. $72 \div (-9)$ −8
6. $42 \div (-6)$ −7
7. $-30 \div 6$ −5
8. $35 \div (-5)$ −7
9. $-9 \div (-3)$ 3
10. $28 \div (-4)$ −7
11. $-60 \div 12$ −5
12. $-72 \div (-6)$ 12
13. $-300 \div 50$ −6
14. $-480 \div (-60)$ 8
15. $400 \div (-25)$ −16

Evaluate each expression when $a = 5$ and $b = -4$. Refer to page 422.

16. $a + b$ 1
17. $a - b$ 9
18. $b - a$ −9
19. ba −20
20. $a \div b$ −1.25
21. $b \div a$ −0.8
22. $2a + b$ 6
23. $2(a + b)$ 2
24. $2a - b$ 14
25. $\frac{1}{2}b + a$ 3
26. $\frac{1}{2}(b + a)$ $\frac{1}{2}$
27. $a + b + 17$ 18
28. $3a + 3b$ 3
29. $3a - 3b$ 27
30. $2a + 3b + 10$ 8

Solve. Refer to page 426.

31. $x + 3 = 6$ 3
32. $y + 8 = 4$ −4
33. $t + 13 = -7$ −20
34. $x - 9 = 4$ 13
35. $p - 9 = -3$ 6
36. $x + 3 = 8$ 5
37. $y + 12 = -4$ −16
38. $x + 4 = 11$ 7
39. $t + 10 = 11$ 1
40. $p - 9 = -4$ 5
41. $x - 3.1 = 4.4$ 7.5
42. $y + 4.5 = 9$ 4.5

Solve. Refer to page 428.

43. $7x = 35$ 5
44. $6x = -36$ −6
45. $-8x = 24$ −3
46. $7a = -35$ −5
47. $-8y = 64$ −8
48. $7x = 28$ 4
49. $-6t = 60$ −10
50. $6x = 66$ 11
51. $7x = -56$ −8
52. $\frac{t}{3} = 15$ 45
53. $\frac{a}{4} = 8$ 32
54. $\frac{b}{10} = -2$ −20

Solve. Refer to page 430.

1. $3x + 3 = 12$ 3

2. $4x + 7 = 31$ 6

3. $9x - 6 = 30$ 4

4. $-4t + 7 = -1$ 2

5. $-6a + 9 = 45$ −6

6. $8b + 3 = -5$ −1

7. $-4x - 3 = 17$ −5

8. $-5a + 7 = 42$ −7

9. $-6x - 5 = 25$ −5

10. $1.2x - 3 = 1.8$ 4

11. $3.5b + 1 = -2.5$ −1

12. $7.6a + 3 = 25.8$ 3

13. $\frac{a}{3} - 1 = 8$ 27

14. $\frac{b}{2} + 3 = -6$ −18

15. $\frac{x}{5} + 5 = 6$ 5

Solve. Refer to page 432.

16. What number plus 8 is 14? 6

17. What number plus 10 is −1? −11

18. What number plus 12 is 7? −5

19. What number plus 6 is 0? −6

20. What number minus 7 is 13? 20

21. What number minus 10 is 10? 20

22. What number minus 8 is −1? 7

23. What number minus −4 is 11? 7

24. Six times what number is 30? 5

25. Seven times what number is 56? 8

26. Five times what number is −15? −3

27. −4 times what number is −8? 2

28. −6 times what number is 42? −7

29. What number divided by 5 is 5? 25

30. What number divided by 7 is −1? −7

31. What number divided by −3 is 8? −24

32. Six times what number plus 7 is 49? 7

33. −8 times what number minus 7 is 57? −8

34. When 15 is subtracted from 7 times a certain number, the result is 27. What is the number? 6

35. The length of a rectangle is 9 feet. The area is 72 square feet. What is the width of the rectangle? 8 feet

Graph each equation. Refer to page 436. See answer section.

36. $y = x + 5$

37. $y = x + 10$

38. $y = x - 7$

39. $y = x - 11$

40. $y = 3x$

41. $y = -2x$

42. $y = 4x + 1$

43. $y = 4x - 1$

44. $y = 7 - x$

APPENDIX

MEASUREMENT TABLES AND FORMULAS

METRIC MEASURES

Length

1 kilometer (km) = 1000 meters (m)
1 meter (m) = 100 centimeters (cm)
1 centimeter (cm) = 10 millimeters (mm)

Capacity

1 liter (L) = 1000 milliliters (mL)

Mass

1 metric ton (t) = 1000 kilograms (kg)
1 kilogram (kg) = 1000 grams (g)
1 gram (g) = 1000 milligrams (mg)

Temperature

0°C = freezing point of water
37°C = normal body temperature
100°C = boiling point of water

CUSTOMARY MEASURES

Length

12 inches (in.) = 1 foot (ft)
3 feet = 1 yard (yd)
36 inches = 1 yard
5280 feet = 1 mile (mi)
1760 yards = 1 mile

Capacity

8 fluid ounces (fl oz) = 1 cup (c)
2 cups = 1 pint (pt)
2 pints = 1 quart (qt)
4 quarts = 1 gallon (gal)

Weight

16 ounces (oz) = 1 pound (lb)
2000 pounds = 1 ton

Temperature

32°F = freezing point of water
98.6°F = normal body temperature
212°F = boiling point of water

FORMULAS

Perimeter of a rectangle	$P = 2l + 2w$
Perimeter of a square	$P = 4s$
Circumference of a circle	$C = \pi d$ or $C = 2\pi r$
Area of a rectangle	$A = lw$
Area of a square	$A = s \times s$
Area of a parallelogram	$A = bh$
Area of a triangle	$A = \frac{1}{2}bh$
Area of a trapezoid	$A = \frac{1}{2}h(b_1 + b_2)$
Area of a circle	$A = \pi r^2$
Volume of a rectangular prism	$V = lwh$
Volume of a cube	$V = s \times s \times s$
Volume of a cylinder	$V = \pi r^2 h$
Volume of a pyramid	$V = \frac{1}{3}lwh$
Volume of a cone	$V = \frac{1}{3}\pi r^2 h$
Volume of a sphere	$V = \frac{4}{3}\pi r^3$
Interest	$I = Prt$

SKILLS TEST

Add. (pp. 14, 16, 22)

1. $58 + 94$ 152

2. $17.95 + 6.49 + 8.75$ \$33.19

3. $430{,}532 + 769{,}298 + 97{,}305$ 1,297,135

4. $9.78 + 4.056$ 13.836

Subtract. (pp. 40, 42, 44, 50)

5. $90 - 28$ 62

6. $37{,}329 - 19{,}293$ 18,036

7. \$90.50 − 38.76 \$51.74

8. $19.3 - 8.567$ 10.733

Multiply. (pp. 70, 72, 82, 84)

9. \$9.58 × 74 \$708.92

10. 897×535 479,895

11. 96.2×0.55 52.91

12. 1000×5.13 5130

Divide. (pp. 104, 108, 110, 116)

13. $7365 \div 3$ 2455

14. $9.8 \div 7$ 1.4

15. $6.9 \div 100$ 0.069

16. $0.16\overline{)9.6}$ 60

Estimate. (pp. 28, 54, 76, 112)

17. $7341 + 8956$ 16,000

18. $5.298 - 2.99$ 2

19. 71×39 2800

20. $8.37\overline{)31.796}$ 4

What part is shaded? (p. 136)

21.  $\frac{3}{4}$

22. 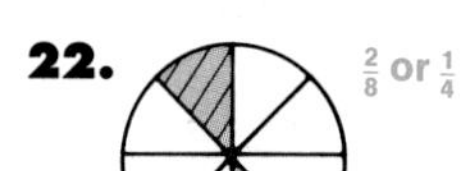 $\frac{2}{8}$ or $\frac{1}{4}$

23. 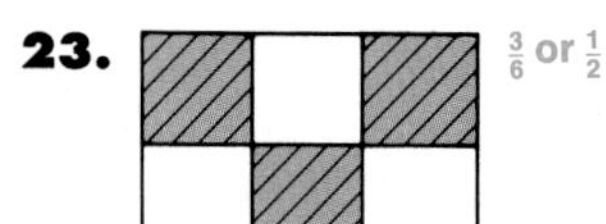 $\frac{3}{6}$ or $\frac{1}{2}$

Compare. Use $<$ or $>$. (p. 142)

24. $\frac{5}{8}$ and $\frac{3}{5}$ $>$

25. $\frac{5}{6}$ and $\frac{8}{9}$ $<$

Convert to a fraction. (p. 150)

26. $1\frac{2}{3}$ $\frac{5}{3}$

27. $4\frac{3}{5}$ $\frac{23}{5}$

Convert to a mixed number. (p. 152)

28. $\frac{19}{5}$ $3\frac{4}{5}$

29. $\frac{49}{10}$ $4\frac{9}{10}$

Convert to a decimal. (p. 154)

30. $9\frac{7}{100}$ 9.07

31. $18\frac{4}{5}$ 18.8

Multiply and simplify. (pp. 174, 176)

32. $\frac{3}{4} \cdot \frac{2}{5}$ $\frac{3}{10}$

33. $\frac{5}{8} \cdot \frac{7}{10}$ $\frac{7}{16}$

34. $10 \times 3\frac{2}{5}$ 34

35. $2\frac{2}{3} \times 2\frac{1}{4}$ 6

Divide and simplify. (pp. 180, 182)

36. $18 \div \frac{1}{3}$ 54 **37.** $\frac{6}{5} \div \frac{2}{3}$ $1\frac{4}{5}$ **38.** $4\frac{1}{5} \div \frac{3}{10}$ 14 **39.** $4\frac{1}{4} \div 1\frac{1}{8}$ $3\frac{7}{9}$

Add and simplify. (pp. 196, 200, 202)

40. $1\frac{3}{8} + \frac{7}{8}$ $2\frac{1}{4}$ **41.** $\frac{5}{6} + \frac{3}{4}$ $1\frac{7}{12}$ **42.** $6\frac{3}{5} + 8\frac{9}{10}$ $15\frac{1}{2}$ **43.** $\frac{3}{4} + \frac{2}{5} + \frac{3}{10}$ $1\frac{9}{20}$

Subtract and simplify. (pp. 206, 208, 210)

44. $\frac{7}{8} - \frac{3}{8}$ $\frac{1}{2}$ **45.** $14\frac{5}{6} - 9\frac{5}{8}$ $5\frac{5}{24}$ **46.** $9 - 6\frac{3}{5}$ $2\frac{2}{5}$ **47.** $8\frac{1}{3} - 1\frac{5}{6}$ $6\frac{1}{2}$

Write the ratio. Simplify, if possible. (p. 224)

48. 240 kilometers in 3 hours $\frac{80}{1}$

49. 20 sandwiches for 12 people $\frac{5}{3}$

Find the unit rate. (p. 226)

50. $ 18.40 is received for 5 hours of work. What is the amount per hour? $3.68

51. If 3 tennis balls cost $3.27, how much does 1 tennis ball cost? $1.09

Solve. (pp. 230, 232)

52. $\frac{2}{3} = \frac{4}{x}$ 6 **53.** $\frac{x}{7} = \frac{9}{8}$ 7.875 **54.** $\frac{3}{5} = \frac{x}{16}$ $9\frac{3}{5}$

55. A car travels 800 kilometers in 3 days. How far would it travel in 15 days? 4000 km

56. If 7 tickets cost $ 45.50, how much do 17 tickets cost? $110.50

Convert to a decimal. (p. 248)

57. 89% 0.89 **58.** 6.7% 0.067

Convert to a fraction. (p. 250)

59. 84% $\frac{21}{25}$ **60.** $33\frac{1}{3}\%$ $\frac{1}{3}$

Convert to a percent. (pp. 248, 252)

61. 0.75 75% **62.** 1.08 108% **63.** $\frac{3}{8}$ 37.5% **64.** $\frac{19}{20}$ 95%

Solve. (pp. 256, 258, 262)

65. What is 11% of 39? 4.29

66. What percent of 50 is 16? 32%

67. 45 is 20% of what number? 225

68. 180 is what percent of 400? 45%

Find the percent of increase or decrease. (p. 264)

69. From 4 to 5 25% **70.** From 15 to 12 20% **71.** From 30 to 10 $66\frac{2}{3}\%$

Make these changes. (pp. 286, 290, 292, 298)

72. 8 km = ___8000___ m

73. 9.3 mm = ___0.93___ cm

74. 105 cm = ___1050___ mm

75. 5307 m = ___5.307___ km

76. 1.32 L ___1320___ mL

77. 72.16 mL = ___0.07216___ L

78. 45.7 g = ___0.0457___ kg

79. 931.2 kg = ___931,200___ g

80. 8 t = ___8000___ kg

81. 998.4 kg = ___0.9984___ t

82. 4 ft = ___48___ in.

83. 60 in. = ___5___ ft

84. 7 ft = ___$2\frac{1}{3}$___ yd

85. $8\frac{1}{2}$ qt = ___17___ pt

86. 6 mi = ___31,680___ ft

87. 80 oz = ___5___ lb

Find the perimeter. (p. 312)

88. 20.5 m

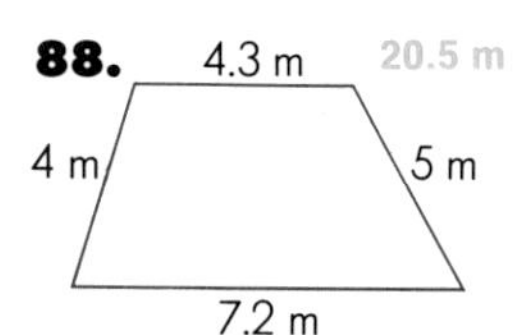

89. 160 yd

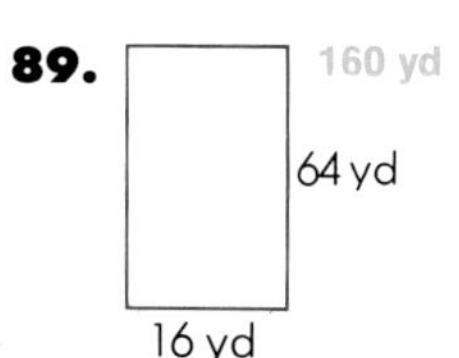

Find the circumference. (p. 314)

90. 43.96 mm

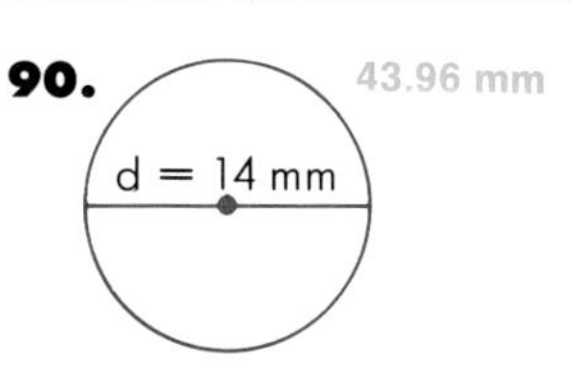

Find the area. (pp. 318, 320, 322, 326)

91. Rectangle
$l = 8$ km
$w = 3.4$ km 27.2 km²

92. Square
$s = 2\frac{1}{2}$ in.
$6\frac{1}{4}$ sq. in.

93. Parallelogram
$b = 8.5$ mm
$h = 3.4$ mm 28.9 mm²

94. Triangle
$b = 12$ ft
$h = 8$ ft
48 sq ft

95. Trapezoid
$h = 40$ m
$b_1 = 14$ m
$b_2 = 18$ m
640 m²

96. Circle
$r = 10$ cm
314 cm²

Find the volume. (pp. 340, 344)

97. 768 cm³

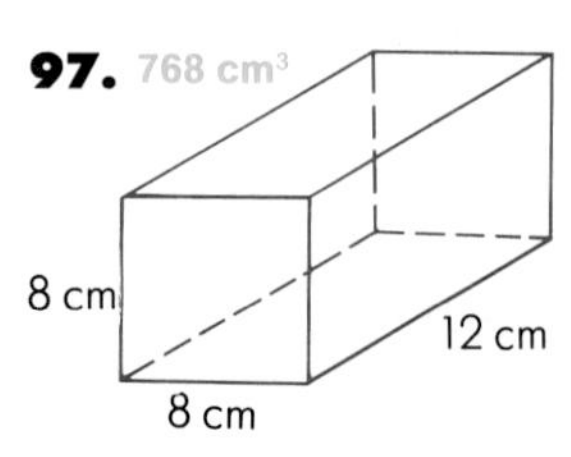

98. 703.36 cu ft

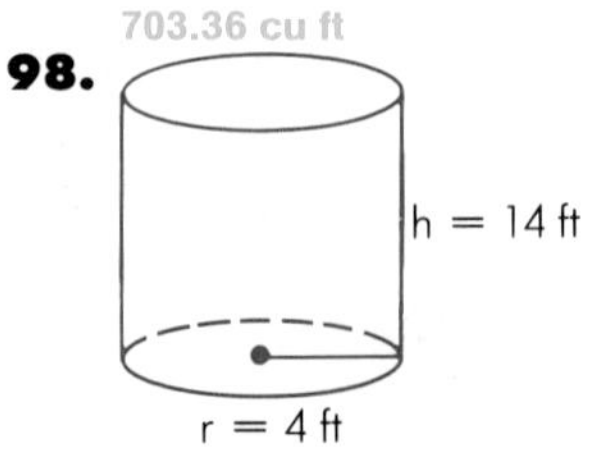

99. 1331 m³

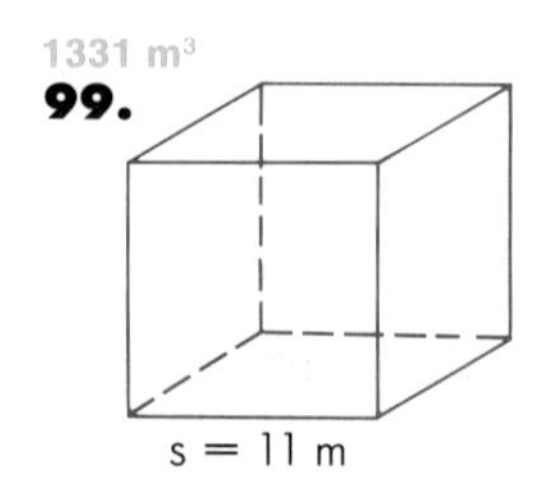

See answer section.

100. Make a bar graph. (p. 372)

Animal	Longest Known Lifetime (years)
Cat	34
Dog	27
Human	114
Horse	50
Chicken	30
Rattlesnake	20

101. Make a line graph. (p. 370)

Continent	Highest Point of Elevation (feet)
Asia	29,028
South America	22,834
North America	20,320
Africa	19,340
Europe	18,510
Antarctica	16,860
Australia	7,310

Find the mean. (p. 378)

102. 17, 20, 15, 22, 16, 14 $17\frac{1}{3}$ **103.** 92, 69, 80, 75 79

Find the median. (p. 380)

104. 17, 20, 15, 22, 16, 14 16.5

Find the mode. (p. 380)

105. 77, 52, 46, 52, 90, 83 52

Add. (p. 398)

106. $-12 + 8$ -4 **107.** $-14 + (-18)$ -32 **108.** $23 + (-9)$ 14

Subtract. (p. 404)

109. $3 - 8$ -5 **110.** $9 - (-5)$ 14 **111.** $-4 - (-7)$ 3

Multiply. (p. 406)

112. $-8 \times (-7)$ 56 **113.** $9 \times (-6)$ -54 **114.** -14×8 -112

Divide. (p. 408)

115. $-56 \div (-14)$ 4 **116.** $-64 \div 8$ -8 **117.** $81 \div (-9)$ -9

Evaluate each expression when $a = -3$ and $b = -5$. (p. 422)

118. $-6ab$ -90 **119.** $2(a - b)$ 4 **120.** $7b + a$ -38 **121.** $-3a + 8b$ -31

Solve. (pp. 426, 428, 430, 432)

122. $x + 9 = -1$ -10 **123.** $-7p = 42$ -6 **124.** $\frac{t}{8} = -11$ -88

125. $-8x - 83 = 13$ -12 **126.** -3 times what number minus 4 is -34? 10

Graph each equation. (p. 436)

127. $y = x - 2$ See answer section. **128.** $y = -1 \cdot x$

APPLICATIONS SKILLS TEST

1. Write a check from Juanita Perez to City Gas and Electric Company in the amount of $98.76. Date it August 9, 1981. (p. 10)

2. Write a check from Joe Wong to Silver Art Supply in the amount of $19.57. Date it March 27, 1982. (p. 10)

See answer section.

3. Find the net deposit. (p. 18)

		Dollars	Cents
Cash	Currency	186	00
	Coin	48	75
Checks	12-29	38	76
	11-104	448	46
Total		721	97
Less Cash Received			-0-
Net Deposit		721	97

4. Find the balance carried forward. (p. 46)

Date Aug. 6, 1982 $98.76 325
To City Gas and Electric
For Gas Bill

	Dollars	Cents
Balance Forward	268	74
Amount of Deposit	198	70
Total	467	44
Amount of Check	98	76
Balance Carried Forward	368	68

5. Complete the list to give change.
Charge: $12.72 Paid with: $20
Machine computes $7.28 change. (p. 80)

Clerk gives	Clerk says
1 $5 bill	$5.00
1 dollar	$6.00
1 dollar	$7.00
1 quarter	$7.25
1 penny	$7.26
1 penny	$7.27
1 penny	$7.28

6. Is the change counted correctly? No
Charge: $6.39 Paid with: $10 (p. 80)

Clerk gives	Clerk says
Item	$ 6.39
1 penny	6.50
1 quarter	6.75
1 quarter	7.00
1 dollar	8.00
1 dollar	9.00
1 dollar	10.00

Period Ending	Hours	Rate	Earnings: Amount Earned Regular Rate	Earnings: Overtime and Other	Gross Earnings	Deductions: FICA	Deductions: Federal Withholding Tax	Deductions: State Withholding Tax	Deductions: Others	Net Pay
7-8-81	30	4.50	135.00	38.00	173.00	10.80	24.50	3.90	48.80	

Statement of Earnings and Deductions for Employee's Record—Detach Before Cashing Check

7. Find the gross earnings. (p. 106)
$173.00

8. Find the net pay. (p. 106)
$85.00

9. Wilma spent 2 hr, 40 min studying and 3 hr, 35 min playing the piano. How much time did she spend in all? (p. 149)
6 hr, 15 min

10. Fred ran a marathon in 4 hr, 3 min. Larry ran the marathon in 3 hr, 25 min. How much longer did Fred run than Larry? (p. 149)
38 min

11. When it is 1 AM in Chicago, Illinois, what time is it in Los Angeles, California? (p. 156) **11 PM**

12. When it is 2 PM in Albuquerque, New Mexico, what time is it in Baltimore, Maryland? (p. 156) **4 PM**

13. Rhonda cuts $3\frac{2}{5}$ feet off of a rope which is $8\frac{1}{4}$ feet long. How much rope is left? (p. 212) **$4\frac{17}{20}$ ft**

14. Mike can save $\frac{5}{8}$ of his gasoline budget of $ 1120 by taking the bus to work. How much can he save? (p. 178) **$700**

15. Five pounds of potatoes cost $ 1.48. Ten pounds cost $ 2.90. Which is the more economical buy? (p. 228) **Ten pounds for $2.90**

16. Three shirts cost $ 41.85. What is the unit price? (p. 236) **$13.95**

17. Tom earns a 10% commission on his monthly sales. How much commission does he earn on sales of $ 7987? (p. 260) **$798.70**

18. Stacey and Steven go out for lunch. The bill is $ 9.20. If they tip 15%, how much should they leave for the tip? (p. 260) **$1.38**

19. A camera which regularly costs $ 48.00 is on sale for 10% off. What is the sale price? (p. 266) **$43.20**

20. Blouses which regularly cost $ 20 are on sale for $ 15. Find the percent of decrease in the price. (p. 266) **25%**

21. Sales tax is 6%. What is the total price of a record album which costs $ 7.98? (p. 294) **$8.46**

22. Marlene budgets 23% of her income for housing. Her net income is $ 780 per month. How much does Marlene budget for housing? (p. 316) **$179.40**

23. What was the difference between the high and low temperatures in Minneapolis? **21°** in Duluth? **16°** (p. 410)

24. Was it colder in Minneapolis or Duluth? How much colder? (p. 410) **Duluth; 6°**

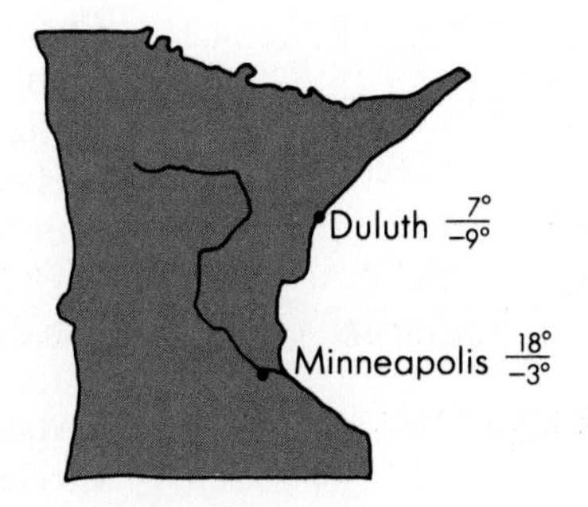

25. Louis still owes $ 550 on his credit card account. Interest is $1\frac{1}{2}$% per month. How much interest will he owe next month? **$8.25** How much will he owe in all? (p. 434) **$558.25**

26. Bella keeps $ 2000 in a savings account for 8 months at 7% interest. Find the simple interest. (p. 424) **$93.33**

INDEX

TRY THIS ANSWERS

CHAPTER 1

Pages 4–5

1. 10-21-36, 36-10-21, 21-10-36 **2.** 12-25-81, 81-12-25, 25-12-81 **3.** Social Security Number **4.** Zip Code **5.** Telephone Number **6.** 28:09 **7.** 7:16 **8.** 11:15 **9.** 9:09:09

Pages 6–7

1. tens, 40 **2.** thousands, 5000 **3.** ones, 5 **4.** ninety-eight **5.** three hundred seventy-six **6.** five thousand, two hundred eighty-seven **7.** six thousand, six. **8.** 3651 **9.** 5088 **10.** 3003

Pages 8–9

1. one cent **2.** seventeen dollars and forty-six cents **3.** two hundred thirty-six dollars **4.** 30¢ or $0.30 **5.** $30 or $30.00 **6.** $1.07 **7.** nine and $\frac{89}{100}$ dollars **8.** fifty-six and $\frac{04}{100}$ dollars **9.** one hundred seventy-four and $\frac{00}{100}$ dollars

Pages 12–13

1. ten-thousands, 70,000 **2.** hundred-millions, 700,000,000 **3.** hundred-billions, 100,000,000,000 **4.** 496 million, 287 thousand, 104 **5.** 380 billion, 87 million, 6 thousand, 100 **6.** 986,280,186 **7.** 4,000,000,000 **8.** 20,000,020

Page 14

1. 489 **2.** 1135 **3.** 1605 **4.** 371 **5.** $5.87 **6.** $1.72 **7.** $11.56 **8.** $16.61

Page 16

1. 22,376 **2.** 1,222,115 **3.** $1039.83 **4.** 1,399,339 **5.** $1510.24 **6.** 1,892,592

Pages 20–21

1. tenths, 0.4 **2.** hundredths, 0.09 **3.** thousandths, 0.006 **4.** 7 and 3 tenths **5.** 27 and 386 thousandths **6.** 7 and 492 ten-thousandths **7.** 0.06 **8.** 47.058

Page 22

1. 38.66 **2.** 10.681 **3.** 202.045 **4.** 1.655 **5.** 26.596 **6.** 10.656 **7.** 25.636 **8.** 2.134

Page 25

1. $168

Pages 26–27

1. 500 **2.** 6700 **3.** 14,400 **4.** 7.37 **5.** 9.49 **6.** 9 **7.** 342 **8.** $4 **9.** $41

Page 28

1. 140 **2.** 5600 **3.** 53 **4.** $49

CHAPTER 2

Page 40

1. 54 **2.** 831 **3.** 39 **4.** 288 **5.** $6.36 **6.** $0.21 **7.** $5.75

Page 42

1. 2,494 **2.** 18,899 **3.** 92,457 **4.** $20.86 **5.** $1128.19 **6.** $6,436.86

Page 44

1. 177 **2.** 7629 **3.** $20.56 **4.** 20,726 **5.** 50,759 **6.** $307.29

Page 48

1. 0.20 **2.** 0.70 **3.** 0.90 **4.** $<$ **5.** $>$ **6.** $=$

Pages 50–51

1. 1.55 **2.** 7.687 **3.** 4.46 **4.** 4.48 **5.** 5.3 **6.** 7.79

Page 54

1. 50 **2.** 3700 **3.** 590 **4.** 5 **5.** $9 **6.** 5

CHAPTER 3

Page 68

1. 1286 **2.** 156 **3.** 72,135 **4.** \$5.68 **5.** \$30.32 **6.** \$1646.33

Pages 70–71

1. 62,440 **2.** 2646 **3.** 57,168 **4.** \$8.76 **5.** \$212.50 **6.** \$3180.24

Pages 72-73

1. 87,400 **2.** 80,730 **3.** \$1524.75 **4.** 272,035 **5.** 291,150 **6.** \$5944.86

Page 76

1. 63,000 **2.** 700 **3.** 400,000 **4.** 400 **5.** 56,000 **6.** 2,000,000

Pages 78–79

1. \$24 **2.** \$480 **3.** \$150 **4.** \$3200 **5.** 350 **6.** 20 **7.** 2000

Page 82

1. 22.8 **2.** 6.966 **3.** 89.76 **4.** 0.024 **5.** 0.049 **6.** 0.00009

Pages 84–85

1. 630 **2.** 9678 **3.** 86.4 **4.** 0.043 **5.** 0.087 **6.** 0.00734

Page 86

1. \$30.35

CHAPTER 4

Page 99

1. 6 r 1 **2.** 6 **3.** 6 r 6 **4.** 46 r 1 **5.** 38 r 3 **6.** 80 r 8

Page 100–101

1. 157 r 5 **2.** 920 **3.** 4030 r 3 **4.** \$43.73 **5.** \$93.59 r 1¢ **6.** \$6.04

Page 103

1. 3 r 12 **2.** 2 r 26 **3.** 38 **4.** 20 r 67 **5.** 872 **6.** \$1.34

Page 104

1. 4 r 11 **2.** 4 **3.** 8 r 640 **4.** 11 r 382 **5.** 86 r 148 **6.** 30

Pages 108–109

1. 4.3 **2.** 2.3 **3.** 3.34 **4.** 0.5 **5.** 0.02 **6.** 0.002 **7.** 2.4 **8.** 0.65 **9.** 0.375

Page 110

1. 6 **2.** 30 **3.** 5.2 **4.** 6.874 **5.** 0.06 **6.** 0.0007

Pages 112–113

1. 30 **2.** 600 **3.** \$7 **4.** 4 **5.** 50 **6.** 5

Pages 116–117

1. 24.8 **2.** 41 **3.** 4321 **4.** 80 **5.** 600 **6.** 500

Page 119

1. 0.8 **2.** 23.5 **3.** 6.67 **4.** 50.71 **5.** 1.143 **6.** 13.286

Page 120

1. \$76,000

CHAPTER 5

Page 136

1. $\frac{3}{7}$ **2.** $\frac{2}{7}$ **3.** $\frac{5}{6}$ **4.** $\frac{1}{4}$

Pages 138–139

1. 1 **2.** 1 **3.** 0 **4.** 0 **5.** 9 **6.** 36 **7.** 0 **8.** Not possible **9.** 0 **10.** Not possible

Page 140

1. yes **2.** no **3.** yes

Page 143

1. $>$ **2.** $<$ **3.** $<$

Page 144
1. $\frac{679}{1000}$ 2. $\frac{37}{10}$ 3. $\frac{1795}{100}$ 4. $\frac{47,332}{10,000}$ 5. $\frac{32}{100}$ 6. $\frac{105}{100}$

Pages 146–147
1. 0.375 2. 0.063 3. 0.583 4. 4.917 5. 0.57 6. 1.7 7. 0.0987

Pages 150–151
1. $8\frac{3}{4}$ 2. $12\frac{2}{3}$ 3. $\frac{22}{5}$ 4. $\frac{61}{6}$ 5. $\frac{93}{10}$

Page 152
1. $2\frac{1}{3}$ 2. $1\frac{1}{10}$ 3. $18\frac{2}{6}$ or $18\frac{1}{3}$

Page 154
1. $4\frac{17}{100}$ 2. $13\frac{18}{1000}$ 3. $1\frac{9}{10}$ 4. 5.6 5. 22.8 6. 2.033

CHAPTER 6

Page 168
1. $\frac{8}{15}$ 2. $\frac{4}{6}$ 3. $\frac{1}{10}$ 4. $5\frac{5}{6}$ 5. $6\frac{2}{3}$ 6. $6\frac{2}{5}$ 7. $\frac{5}{8}$

Pages 170–171
1. $\frac{4}{6}, \frac{6}{9}, \frac{10}{15}$ 2. $\frac{14}{16}, \frac{21}{24}, \frac{28}{32}$ 3. $\frac{6}{9}$ 4. $\frac{18}{20}$ 5. $\frac{8}{32}$

Pages 172–173
1. $\frac{3}{5}$ 2. $\frac{3}{8}$ 3. $\frac{5}{6}$ 4. $\frac{8}{5}$ 5. 9

Page 174
1. $\frac{7}{12}$ 2. $\frac{1}{3}$ 3. $2\frac{2}{3}$ 4. 8 5. $10\frac{1}{2}$

Page 176
1. 39 2. $12\frac{4}{5}$ 3. $2\frac{3}{4}$ 4. $1\frac{5}{6}$ 5. $8\frac{1}{3}$ 6. $10\frac{5}{6}$

Page 180
1. $\frac{5}{2}$ 2. $\frac{2}{5}$ 3. 5 4. $\frac{1}{24}$ 5. $1\frac{1}{7}$ 6. $2\frac{2}{3}$ 7. $\frac{1}{10}$

Page 182
1. $3\frac{1}{9}$ 2. $\frac{9}{34}$ 3. $\frac{1}{10}$ 4. $\frac{12}{25}$

CHAPTER 7

Page 197
1. 1 2. $\frac{7}{12}$ 3. $\frac{7}{8}$

Pages 198–199
1. 18 2. 12 3. 24 4. 20

Page 200
1. $1\frac{5}{24}$ 2. $\frac{5}{9}$ 3. $1\frac{1}{2}$

Pages 202–203
1. $11\frac{4}{5}$ 2. $23\frac{2}{3}$ 3. $21\frac{7}{8}$ 4. $13\frac{7}{12}$ 5. $16\frac{1}{20}$

Pages 206–207
1. $\frac{1}{4}$ 2. $\frac{1}{2}$ 3. $\frac{1}{2}$ 4. $\frac{13}{18}$

Page 209
1. $5\frac{1}{4}$ 2. $12\frac{7}{8}$ 3. $3\frac{1}{6}$

Pages 210–211
1. $6\frac{5}{6}$ 2. $6\frac{3}{8}$ 3. $2\frac{23}{24}$ 4. $9\frac{1}{3}$

CHAPTER 8

Pages 224–225
1. $\frac{2}{3}$ 2. $\frac{3}{2}$ 3. $\frac{3}{5}$ 4. $\frac{12}{5}$ 5. $\frac{72}{60} = \frac{6}{5}$ 6. $\frac{5}{1}$ 7. $\frac{1}{4}$

Page 227
1. 19.6 miles/gal 2. 12¢ 3. $5.35 per hour

Pages 230–231
1. No 2. Yes 3. 20 4. $2\frac{1}{2}$ 5. $4\frac{2}{3}$

Page 233
1. 5 oranges 2. 424 miles

Page 234
1. 15 yds 2. 10.5 m

CHAPTER 9

Pages 248–249

1. 0.25 **2.** 1.00 **3.** 0.04 **4.** 0.045 **5.** 72% **6.** 700% **7.** 720% **8.** $66\frac{2}{3}$%

Pages 250–251

1. $\frac{31}{100}$ **2.** $\frac{3}{4}$ **3.** 4 **4.** $1\frac{1}{2}$ **5.** $\frac{1}{200}$ **6.** $\frac{1}{6}$ **7.** $\frac{1}{8}$ **8.** $\frac{103}{1000}$ **9.** $\frac{1}{250}$ **10.** $\frac{1}{2500}$

Page 252

1. 25% **2.** 30% **3.** 140% **4.** 230% **5.** $66\frac{2}{3}$% **6.** 62.5% **7.** $183\frac{1}{3}$% **8.** 287.5%

Pages 254–255

1. $x = 60\% \times 70$ **2.** $76 = 19\% \times n$ **3.** $18 = b \times 36$ **4.** $x = 0.5 \times 48$ **5.** $\frac{5}{8} \times a = 32$

Pages 256–257

1. 12.72 **2.** 0.185 **3.** $8\frac{1}{3}$ **4.** 18

Page 259

1. 50% **2.** $166\frac{2}{3}$% **3.** $77\frac{7}{9}$%

Page 263

1. 200 **2.** 18

Page 265

1. $66\frac{2}{3}$% **2.** 5% **3.** 20% **4.** 40%

CHAPTER 10

Pages 284–285

1. km **2.** mm **3.** cm **4.** m **5.** 30 cm **6.** 40 m **7.** 53 mm, 5.3 cm

Page 286

1. 0.187 **2.** 962 **3.** 0.296 **4.** 870

Page 290

1. 10 mL **2.** 400 L **3.** 3200 **4.** 500 **5.** 0.25

Page 292

1. 50 g **2.** 70 kg **3.** 2 t **4.** 250 mg **5.** 3.786 **6.** 7860 **7.** 1.286 **8.** 0.73

Page 296

1. F **2.** C **3.** 100°F **4.** 40°C

Page 298

1. 15 **2.** 2250 **3.** 10 **4.** $1\frac{1}{9}$

Page 300

1. right **2.** larger **3.** smaller **4.** 31° **5.** 145°

CHAPTER 11

Page 312

1. 21 cm **2.** 28 cm **3.** 32 ft

Page 314

1. 28.26 m **2.** 35.482 cm **3.** 18.84 cm **4.** 38.308 mm

Page 318

1. 18 km^2 **2.** 5 sq ft **3.** 6.25 cm^2 **4.** 12.25 sq yd

Pages 320–321

1. 72 cm^2 **2.** 32.1 mm^2 **3.** $7\frac{1}{2}$ sq ft **4.** 21.5 m^2 **5.** 7 km^2

Page 322

1. 24.4 m^2 **2.** 8 km^2

Page 326

1. 50.24 m^2 **2.** 452.16 km^2 **3.** 28.26 m^2 **4.** 78.5 sq ft

Page 328

1. 2

CHAPTER 12

Page 340

1. 24 cm³ **2.** 1000 m³

Page 344

1. 62.8 m³ **2.** 138.16 mm³

Page 346

1. 8 m³ **2.** 847.8 mm³

Page 348

1. 267.95 cm³ **2.** 113.04 m³

CHAPTER 13

Pages 368–369

1.–4.

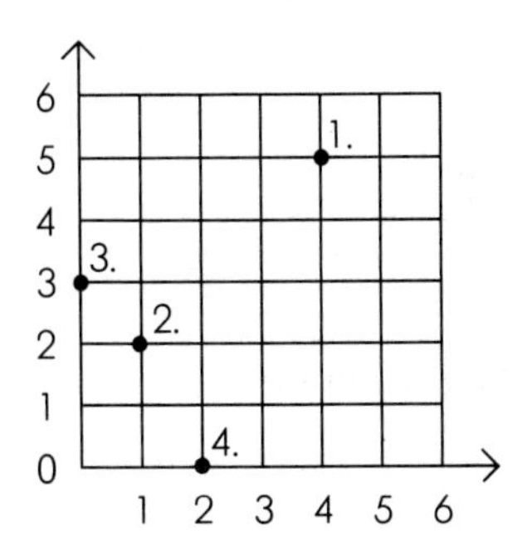

5. (5,9) **6.** (10,6) **7.** (0,7) **8.** (9,0)

Page 370

1.

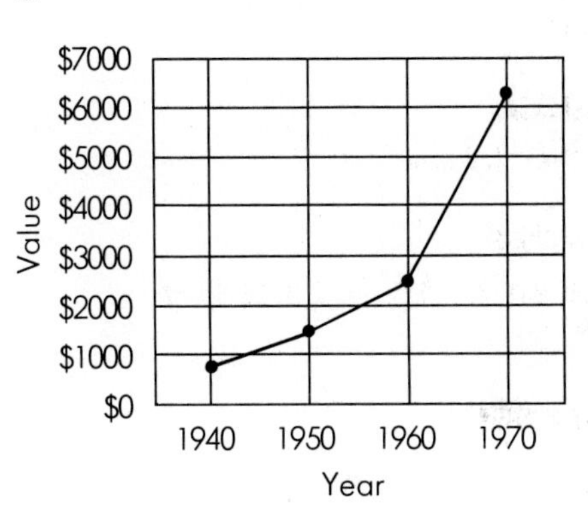

Page 372

1.

Ice Cream Flavor	Calories in a One-Scoop Cone
Chocolate Fudge	230
French Vanilla	220
Strawberry	170
Peach	170
Butter Pecan	200
Chocolate Mint	190

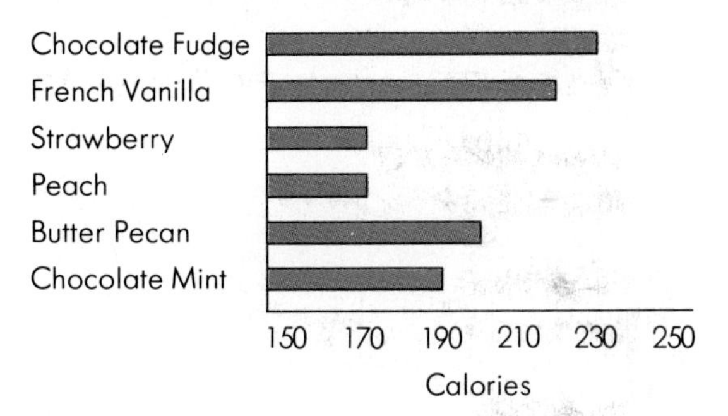

Page 374

1. 20%: 72°; 14%: 50°; 10%: 36°; 26%: 94°; 30%: 108°

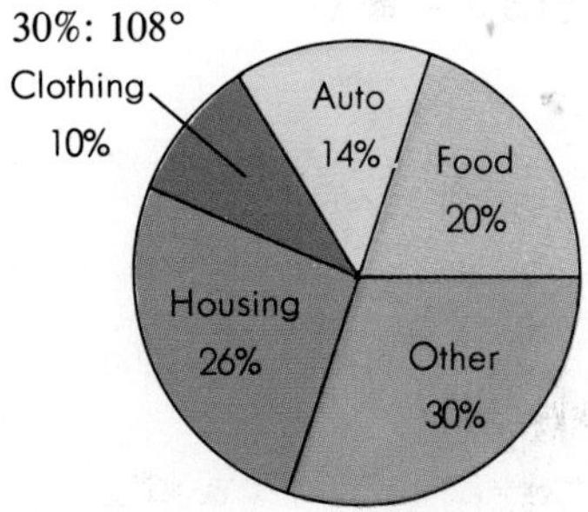

Page 378

1. 80.25 **2.** 152.82 **3.** 29 mi per gal

Page 380–381

1. 17 **2.** 37 **3.** 17 **4.** 8 **5.** 78 **6.** 101, 103

CHAPTER 14

Page 394–395

1. −8 **2.** −6 **3.** 6 **4.** −10 **5.** < **6.** > **7.** < **8.** < **9.** > **10.** <

Pages 396–397

1. −2 **2.** −4 **3.** 0 **4.** 0 **5.** 2 **6.** −4

Page 398

1. −11 **2.** −29 **3.** 3 **4.** −5 **5.** 7

Pages 402–403

1. −3 **2.** −8 **3.** −9 **4.** 1 **5.** 10 **6.** 11

Pages 404–405

1. 5 **2.** 6 **3.** −6 **4.** −12 **5.** 0 **6.** −9 **7.** −4 **8.** 4 **9.** 0

Page 406

1. 24 **2.** 126 **3.** 48 **4.** −42 **5.** −18 **6.** −100

Page 408

1. 7. **2.** 14 **3.** 4 **4.** −6 **5.** −3 **6.** −5

CHAPTER 15

Page 422

1. 16 **2.** 10 **3.** 300 **4.** −424

Page 426

1. −12 **2.** 11 **3.** 35 **4.** 1

Pages 428–429

1. −8 **2.** −6 **3.** 405 **4.** −104

Page 430

1. 5 **2.** 9 **3.** −4 **4.** 4

Page 432

1. −9 **2.** $24

Page 437

1–2.

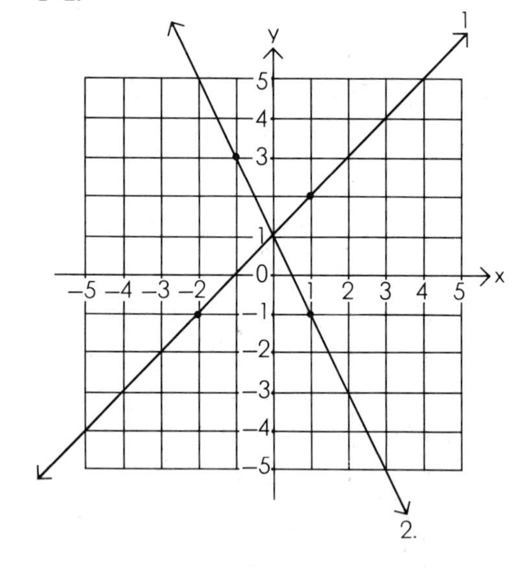